COUPLED ATMOSPHERE-OCEAN DYNAMICS: FROM EL NIÑO TO CLIMATE CHANGE

SHANG-PING XIE, Ph.D.

Scripps Institution of Oceanography,
University of California San Diego,
La Jolla, CA, USA.

Coupled Atmosphere-Ocean Dynamics: from El Niño to Climate Change
Elsevier
Radarweg 29, PO Box 211, 1000 AE Amsterdam, Netherlands
The Boulevard, Langford Lane, Kidlington, Oxford OX5 1GB, United Kingdom
50 Hampshire Street, 5th Floor, Cambridge, MA 02139, United States

ISBN: 978-0-323-95490-7

For information on all Elsevier publications visit our website at
https://www.elsevier.com/books-and-journals

Publisher: Candice G. Janco
Acquisitions Editor: Maria Elekiduo
Editorial Project Manager: Ali Afzal-Khan
Publishing Services Manager: Shereen Jameel
Project Manager: Vishnu T. Jiji
Senior Designer: Vicky Pearson Esser

Printed in the United States of America
Last digit is the print number: 9 8 7 6 5 4 3

CONTENTS

Acknowledgments

I am indebted to many mentors and colleagues for inspiration, encouragement, and support. Many students, postdocs, and collaborators—too numerous to be named here—contributed to the research presented in this book and in the process expanded my research horizons. Many helped with drawing figures with updated data, especially Dillon Amaya, Will Chapman, Yufan Geng, Yu Kosaka, Yun Liang, Jingwu Liu, Shang-Min Long, Qihua Peng, Ingo Richter, Ray Shi, Zihan Song, Chuan-Yang Wang, Liu Yang, and Zhen-Qiang Zhou, as acknowledged in figure captions. Louisa Munro, Vishnu T. Jiji, Ali Afzal-Khan, and Maria Elekidou of Elsevier skillfully guided the book project. Students at Hokkaido University, University of Hawaii, UCSD, and elsewhere who attended my lectures helped improve the clarity of the book by asking questions and participating in discussion.

My parents taught me work ethic and inspired me to go far. I would like to thank my wife Chang-Tu and daughter Cordelia for love, understanding, and support and for the perspective that the world is much bigger than my narrow science.

I am grateful for the long-term funding support from the National Science Foundation, National Aeronautics and Space Association, National Oceanic and Atmospheric Administration, and the Roger Revelle Chair endowment.

Preface

Climate affects the natural environment and human society. Climate variability drives, and climate change exacerbates, extreme events such as heatwaves, droughts, and flooding. As society becomes increasingly sophisticated and sensitive to climate variations, evaluating and understanding climate influence in a wide range of data is important and essential.

This book presents core coupled ocean-atmosphere dynamics that give rise to recurrent spatial patterns, preferred timescales, and predictability of our changing climate. It is based on a course I have taught for more than 25 years, mostly at graduate level but twice at upper division undergraduate level. I have lectured the course, in part or full, at summer/winter schools and training workshops around the world.

i. Ocean-atmosphere coupling

El Niño refers to the anomalous warming of the equatorial Pacific Ocean. From the oceanographic point of view, the Pacific warms because the atmospheric Southern Oscillation relaxes the prevailing easterly winds on the equator. From the meteorologic point of view, on the other hand, the easterly winds weaken because of the equatorial Pacific warming (El Niño). This circular argument implies that El Niño is not merely an oceanic phenomenon, nor the Southern Oscillation an atmospheric one; they are the two sides of the same phenomenon arising from their mutual interaction. This revolutionary idea led to the coinage of the term El Niño and the Southern Oscillation (ENSO) to emphasize the coupled nature of the phenomenon. Thus a new field of study, coupled atmosphere-ocean dynamics, was born.

ENSO research led to the integration of meteorology—the atmospheric science of weather—and physical oceanography. Meteorology and physical oceanography share a common set of concepts and principals of geophysical fluid dynamics (GFD), for which the effect of Earth rotation known as the Coriolis force is important. The dynamic approach led to numerical weather prediction and explained why major currents are found near the western boundaries of ocean basins. Excellent texts treating atmospheric and ocean dynamics from the unified view of GFD include Pedlosky (1982), Gill (1982), and Vallis (2017).

Coupled ocean-atmosphere dynamics culminated into the successful prediction of El Niño by Cane et al. (1986). I entered graduate school that year and spent the first year reading Gill's (1982) classic text, where "fundamental physical ideas and dynamical theory are skillfully interwoven with observed phenomena" (Batchelor and Hide 1988). Gill's (1982) text was published just before the dawn of coupled dynamics. The term

ENSO was yet been coined, but the field was at the threshold of a scientific revolution that views ENSO as a spontaneous, coupled ocean-atmosphere oscillation.

ii. Aims of the book

This book starts where Gill's (1982) text left off, with the title emphasizing our focus on exciting advances in the coupled approach to climate variability and change. There are excellent monographs on ENSO (Philander 1990; Clarke 2008; Sarachik and Cane 2010; McPhaden et al. 2020), but the field has expanded to encompass climate variability in other ocean basins and anthropogenic climate change. This book aims to synthesize coupled dynamics of both natural variability and anthropogenic change, in and out of the tropics, including but beyond ENSO. It builds the physical foundation for broad climate impact studies.

For students and researchers of ocean, atmospheric, and climate sciences, the book retains the rigor by including key equations with references to classic texts (Gill 1982; Holton 2004) or the original literature where the full mathematical treatment can be found. Equations are introduced and explained in plain, descriptive language so readers with limited prior exposure to fluid mechanics can follow the discussion of essential physical processes and mechanisms at work. The large number of illustrations and the narratives further aid the conceptual understanding.

iii. Organization

The book consists of three parts that are closely interrelated. Part I (Chapters 2—6) examines tropical and subtropical climates mostly from an atmospheric perspective. Chapter 2 starts from the planetary energy balance, illustrates the characteristics and mechanisms of ocean-atmospheric energy transports, and highlights the role of the meridional overturning circulation. Chapter 3 examines atmospheric deep convection, the heat source that drives the global atmospheric circulation. Equatorial waves are introduced, and their role in atmospheric adjustments to convective heating is illustrated. Chapter 4 studies the 30- to 60-day Madden-Julian Oscillation (MJO), a planetary-scale mode spontaneously arising from interactions of atmospheric deep convection and circulation. Chapter 5 introduces monsoons, a product of land-atmosphere-ocean interactions that affects more than half of the world population. We discuss atmospheric dynamics that makes each regional monsoon unique and at the same time connects them across the great span from Africa to Asia. Chapter 6 shifts the focus to subtropical regions of subsidence, where a temperature inversion caps the atmospheric boundary layer. Low-level clouds and the underlying ocean interact, giving rise to positive feedback important for a range of phenomena.

Part II (Chapters 7—11) is the core of the book and investigates ocean-atmosphere interactions in tropical climatology and interannual variability. Chapter 7 introduces a reduced-gravity model of the upper ocean and discusses the ocean adjustments to changing winds. Chapter 8 revisits the tropical rainfall distribution first outlined in Chapter 3 but from the perspective of coupled ocean-atmosphere interaction. Questions to be addressed include: Why is the tropical rain band—known as the intertropical convergence zone (ITCZ)—displaced north of the equator, and what causes the marked annual oscillation in temperature and rainfall on the equator over the eastern Pacific and Atlantic? Chapter 9 opens with a description of ENSO as a coupled ocean-atmospheric phenomenon and then illustrates the positive Bjerknes feedback that arises from the coupling. It further discusses how slow ocean adjustment processes, not in equilibrium with the wind variations, cause the tropical Pacific to oscillate between El Niño and La Niña. The last point enables prediction at seasonal lead times of ENSO and its global effects. Chapters 10 and 11 discuss regional climate and interannual variability over the tropical Atlantic and Indian oceans, respectively. The mean climate is distinct in important ways between these tropical oceans, but they share a common equatorial mode of Bjerknes feedback. A unique coupled mode across the Indo-western Pacific explains a mysterious post-ENSO effect on summer monsoons from India to China.

Chapter 12 shows that climate variability is dynamically distinct in and outside the tropics. In the extratropics, atmospheric internal variability is strong and organized into coherent spatial patterns because of the zonal variations in the mean westerly jet. It drives oceanic variability, while ocean feedbacks are nonlocal and weak. While tropical teleconnections to the extratropics are well established, we highlight recent ideas that extratropical variability has important effects the other way around on tropical climate. Anthropogenic emissions of greenhouse gases and aerosols perturb the planetary energy balance, causing Earth's climate to change at unprecedented rates. While global-mean surface temperature increase has traditionally been emphasized, we focus on regional climate change unleashed by global warming. After a brief introduction to climate feedback on global mean surface temperature, Chapter 13 discusses robust changes expected in a warmer climate that do not fundamentally depend on the ocean-atmospheric circulation change (so-called thermodynamic effect). Chapter 14 focuses on regional climate change, including a comparison of forcing by well-mixed greenhouse gases vs. highly distributed aerosols. Spatial patterns of ocean heat uptake and warming are important factors driving regional variations in tropical rainfall change.

iv. Pedagogical features

This book aims at a comprehensive and systematic treatment of large-scale ocean-atmosphere interactions. This allows a wide range of comparative views: climate modes among and across different tropical ocean basins (Chapters 9—11), the effects of the basin size and degree of continental/monsoon influence, zonal vs. meridional mode that peaks/vanishes on the equator, ocean feedback on the atmosphere in vs. out of the tropics

(Chapter 9 vs. 12), spontaneous internal oscillations vs. externally forced climate change (Chapter 13), and greenhouse vs. aerosol forcing (Chapter 14). Such comparative views, some discussed in Epilogue, offer unique insights into the mechanisms for climate variability and predictability.

Side boxes are used to introduce related topics and provide historical accounts of major breakthroughs in climate dynamics that have changed our view of the climate system. I know many of the pioneers who led the breakthroughs and asked them about what inspired them to probe what seemed unimportant, to connect what had been apart, to predict beyond the chaos of weather. These historical accounts offer real-life examples that knowledge is not static but in constant expansion, driven by curiosity, daring imagination, and clever innovation.

The book can be used as a stand-alone or in combination with other material for a range of courses. My graduate course at UCSD typically covers the first 12 chapters with some simplifications. Some of the chapters (3, 5, 9, and 12) may require two lectures to cover fully. It is also possible to skip Chapters 4 and 5 and cover Chapters 13 and 14. The 10-week course consists of 16 80-minute lectures and a 160-minute colloquium in which each student is required to make a presentation on a paper(s) related to a class topic. A list of suggested papers is available in online resources (http://sxie.ucsd.edu/book2023.html). Climate is local, and regional climate inspires interest in coupled atmosphere-ocean dynamics. Instructors may wish to supplement material to meet interests in local climate. At UCSD, for example, subtropical climate (Chapter 6) gets the full coverage while skipping the sections on East Asian and African monsoons (Chapter 5) and treating lightly chapters on the tropical Atlantic and Indian oceans. In Asia, on the other hand, my lectures give Asian monsoons and the Indian Ocean the full treatment while omitting chapters on subtropical climate and the Atlantic Ocean.

Review questions are provided at the end of chapters to reinforce key concepts and applications. Online resources include figures in the book, PowerPoint slides for my UCSD class, homework problems, and errata.

Cover. El Niño and the global influence are represented by anomalies of sea surface height (color shading) in December 1997 and geopotential height (line contours) in the upper troposphere (~ 10 km above the sea level) during December 1997-February 1998. Rising global surface temperature in time series from 1880 to 2021. Image credit: Q. Peng.

CHAPTER 1

Introduction

Contents

Mark Twain famously said, "Climate is what you expect, and weather is what you get." Formally, climatology is defined as the 30-year averages of temperature, precipitation, wind, and other ocean/atmospheric variables. When planning for long-distance travel, we check the climatology of the destinations to pack our suitcase and then check the local weather forecast to decide what to wear in the morning. Koppen (1884) developed a system to classify climate types based on temperature and precipitation. For example, San Diego, California, on the west coast of North America, is classified as Mediterranean climate with a winter rainy season; Shanghai, China, on the east coast of Asia, is a subtropical humid climate with a summer rainy season.

The climatology is supposed to be stable and unchanging—"lasts all the time" as Twain said—but in reality, climate varies markedly from one year to another even within the 30-year period during which a climatology is defined. Over the eastern equatorial Pacific Ocean, for example, sea surface temperature (SST) deviates as much as 4°C from the climatology, causing widespread anomalous conditions worldwide. In 1998, following such a major event of equatorial Pacific warming (known as El Niño), a great lake emerged in the desert on the Pacific coast of Peru, and the Yangtze River of China suffered one of the worst floods in recent history.

The World Meteorological Organization calls for decadal updates of 30-year mean climate normals. The current climatology is about 1°C warmer in global mean surface temperature (GMST) than a century ago (Fig. 1.1). The planetary warming coincides

Coupled Atmosphere-Ocean Dynamics: from El Niño to Climate Change
ISBN 978-0-323-95490-7, https://doi.org/10.1016/B978-0-323-95490-7.00001-1

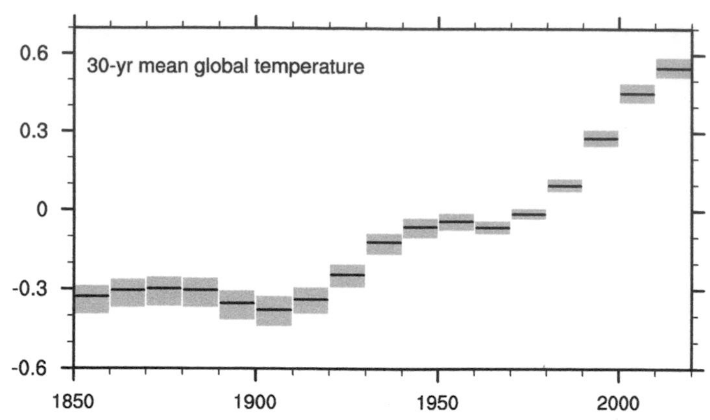

Fig. 1.1 Evolving 30-year averages of global mean surface temperature anomaly (°C), calculated each decade from the HadCRUT4 dataset. The black lines indicate the medium, and grey shading the 95% confidence interval of bias uncertainty computed from the 100-member ensemble. *(Courtesy ZQ Zhou.)*

with and is caused mostly by the increased greenhouse gas (GHG) concentrations in the atmosphere due to combustion of fossil fuels. The anthropogenic warming is raising global sea level, melting mountain glaciers and Arctic sea ice, and elevating the frequency and intensity of heatwaves on land and in the ocean.

Our desire to live in coastal metropolitan areas with sunny weather intensifies the exposure to risks of climate change. Coastal communities are vulnerable to inundations due to rising seas and hurricanes. Urbanization reduces surface evaporation and elevates the risks of heatwaves. Megacities in dry regions (e.g., Beijing and Los Angeles) face the risks of droughts and wildfires and often resort to megaengineering solutions such as water transfers over long distances that result in further environmental degradation.

The recognition that climate is changing and the concern that climate change exacerbates society's vulnerability to climate perturbations in ways humanity has never experienced have heightened the interest in climate science in the scientific community as well as among the general public, stakeholders, and policymakers. This book provides the scientific foundation for those who are interested in fundamental questions such as the following:

1. Why does climate vary from one year to another?
2. What gives rise to recurrent patterns of climate variability?
3. How will climate change in the face of increasing GHG in the atmosphere?
4. How predictable is climate, and how are reliable climate predictions made?

We focus on climate physics based on conservation laws that govern variations of fundamental variables such as temperature and velocity.

1.1 Role of the Ocean in Climate

Surface waves are among the most visible phenomena of ocean-atmosphere interaction. Ocean water moves slowly (current velocity <1 m/s) compared to air, and the velocity difference between the water and air (wind shear) makes the air-sea interface unstable, generating wind waves. As they propagate away from the storms that generate them, the waves become smooth swells of long wavelengths and periods. Of short horizontal (100 m) and time (10 s) scales, wind waves and swells are not the focus of this book, although they are important for turbulent exchanges of momentum and heat between the ocean and atmosphere.

My hometown is 200 km from the coast. Every summer a typhoon or two came with heavy rains and high winds enough to uproot trees. What I did not know is that these typhoons are the product of ocean-atmosphere interaction, developing and intensifying over the warm ocean (Fig. 1.2). They travel thousands of km over the ocean, grow on heat from the ocean, and cause severe weather conditions and storm surges upon landfall. Through evaporation, the ocean supplies water vapor that fuels typhoons with latent heat of condensation in towering cumulonimbus clouds on the eyewall and in spiral rain bands.

For us who dwell on land, the perception of climate is atmospheric; air temperature, wind, and precipitation are important environmental variables that affect us. The ocean influence is obvious in coastal cities such as San Diego. The daily weather forecast in San Diego is made separately for four distinct geographical zones: coastal, inland, mountain, and desert (east of the coastal mountain range on the California-Arizona border). The following facts imply the ocean's importance for climate.

- The entire mass of the atmosphere (1036 hectopascals [hPa]) is equivalent to 10 m of sea water.

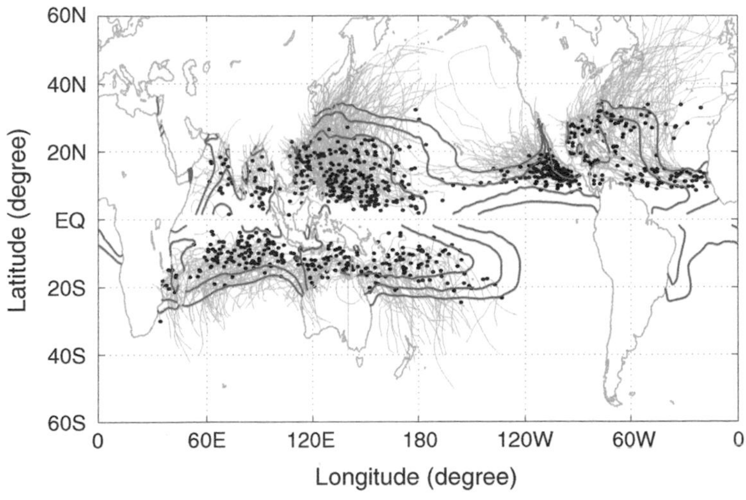

Fig. 1.2 Genesis *(black dots)* and tracks of tropical cyclones during 1996-2005, along with sea surface temperature in local summer (red contours for 27°C, 28°C, and 29°C). *(Courtesy W Mei.)*

- The heat capacity of the entire atmospheric column is equivalent to that of a 4-m water column.
- The ocean is on average 4000 m deep.
- The part of the ocean that is involved in the seasonal variation is 100 m deep.

Here the heat capacity (C) refers to the heat needed to raise the temperature (T) of an object by 1°C, $C = \rho c_p H$, with ρ the density, c_p the specific heat at constant pressure, and H the depth of the fluid column.

Ocean temperature is stably stratified, with warm water floating on cold water below (Fig. 1.3). The thin layer that separates the warm surface and cold deep water is called the thermocline. In winter, the reduced solar radiation cools the surface water. Vertical convection induced by the surface cooling erodes the temperature stratification, and a surface mixed layer that is typically 50 to 100 m deep develops above the thermocline. Over land, by contrast, the seasonal variation in temperature decreases rapidly with depth and is confined to the top few meters of the soil (of a heat capacity equivalent to 1–2 m of water); deep caves in the desert remain cool even on hot summer days. Because of the large difference in heat capacity between the water and soil layers involved, the seasonal cycle in surface air temperature is generally much smaller over ocean than land. For example, daily mean temperature in Nanjing (inland China) swings from freezing 3°C in winter to sweltering 28°C in summer, while it stays within the comfortable range from 14°C to 22°C in San Diego (Pacific coast) (Fig. 1.4), even though the two cities are at the same latitude (32°N).

The empirical orthogonal function (EOF) method identifies recurrent spatial patterns, each associated with a time series called the principal component (PC) that describes the phase variation of the spatial pattern (see later). Fig. 1.5 shows the results of an EOF

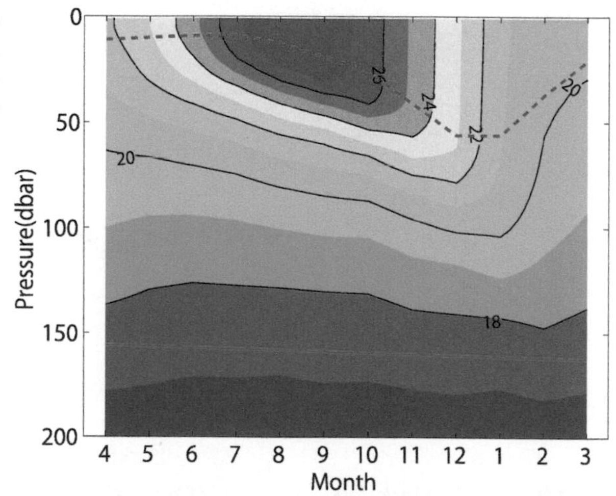

Fig. 1.3 Time-depth section of ocean temperature variations (°C) at 25–30°N, 165–185°E. The gray dashed line indicates the base of the mixed layer. *(From Hosoda et al. [2015].)*

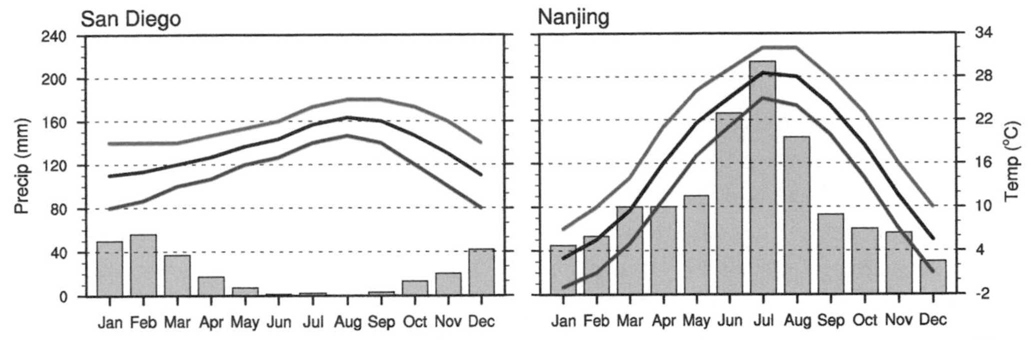

Fig. 1.4 Long-term climatology of surface air temperature (daily max in *red*, mean *black*, and min *blue*, °C) and precipitation (bars, mm/month) at San Diego, California (United States) and Nanjing, China. *(Courtesy ZQ Zhou.)*

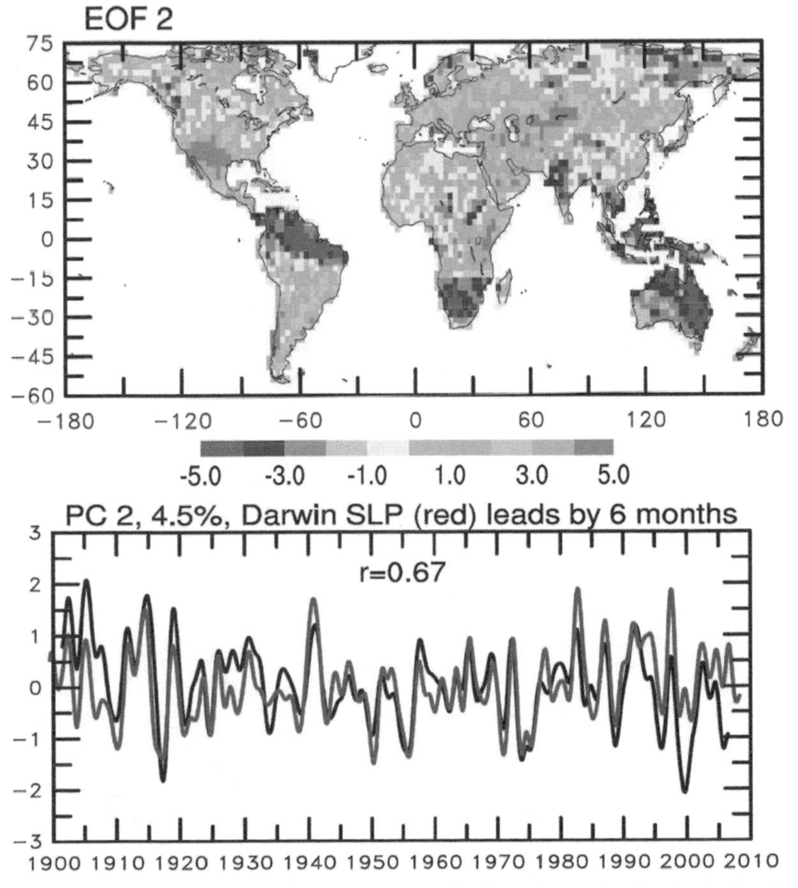

Fig. 1.5 Second empirical orthogonal function *(EOF)* of Palmer Drought Sevrity Index from 1900 to 2008. Red (blue) areas are dry (wet) for a positive value of the principal component *(PC, bottom)*. *SLP,* Sea level pressure. *(From Dai [2011].)*

analysis performed in the global domain on the Palmer Drought Severity Index, which measures the soil dryness and is calculated from temperature and precipitation measurements. The second mode is highly correlated with atmospheric sea level pressure (SLP) at Darwin in northern Australia ($r = 0.63$ for >100 years), an index that tracks El Niño and the Southern Oscillation (ENSO). It is quite remarkable that global soil dryness covaries closely with ENSO, which is a spontaneous oscillation arising from tropical Pacific Ocean—atmosphere interactions (see Chapter 9). Specifically, when the equatorial Pacific is in warm phase (El Niño) and Darwin SLP is anomalously high, the maritime continent (referred to tropical islands between the Indian and Pacific oceans) is drier while southwest North America is wetter than normal (see Fig. 1.5b). To the extent that the global atmospheric circulation is driven by latent heat release in deep convection, it does not come as a surprise that changes in tropical ocean condition lead to atmospheric anomalies.

Ocean affects the atmosphere through SST and sea ice. Global atmospheric general circulation models (GCMs; see later), originally developed for weather forecast, can be used to evaluate the SST effect on climate. We compare a pair of runs, one forced with the observed evolution of SST/sea ice, and one with the repeating monthly SST/sea ice climatology ($SST' = 0$, with the prime denoting the deviation from the climatology). Fig. 1.6 compares the standard deviation for year-to-year variability in June-July-August seasonal mean precipitation and zonal wind between atmospheric GCM simulations with and without interannual SST variability. Over the tropical oceans, SST variability increases precipitation variability, while the SST effect is not clear in the extratropics (see Fig. 1.6a—b). The SST effect on zonal wind variability is apparent, especially over the equatorial Pacific and Indo-western Pacific north of the equator (see Fig. 1.6c—d).

Atmospheric GCMs were originally developed to predict weather initialized with observed current atmospheric conditions. The skill of such initialized numerical weather

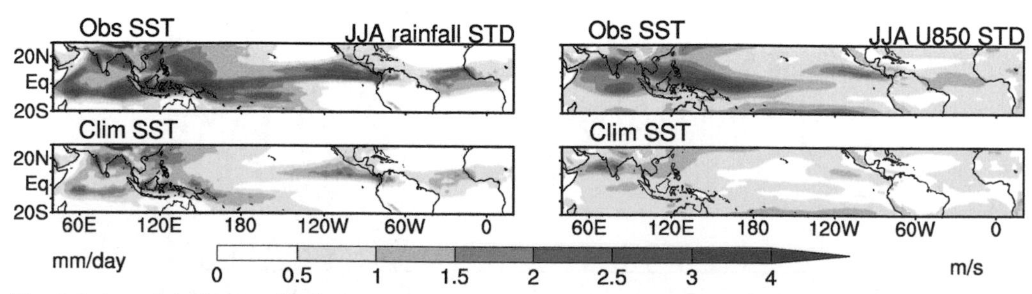

Fig. 1.6 June-July-August *(JJA)* seasonal mean standard deviation *(STD)* of (*left*) rainfall (mm/day) and (*right*) 850 hPa zonal wind velocity (m/s) in atmospheric general circulation model simulations *(top)* with and *(bottom)* without year-to-year sea surface temperature *(SST)* variability. *(Courtesy ZQ Zhou.)*

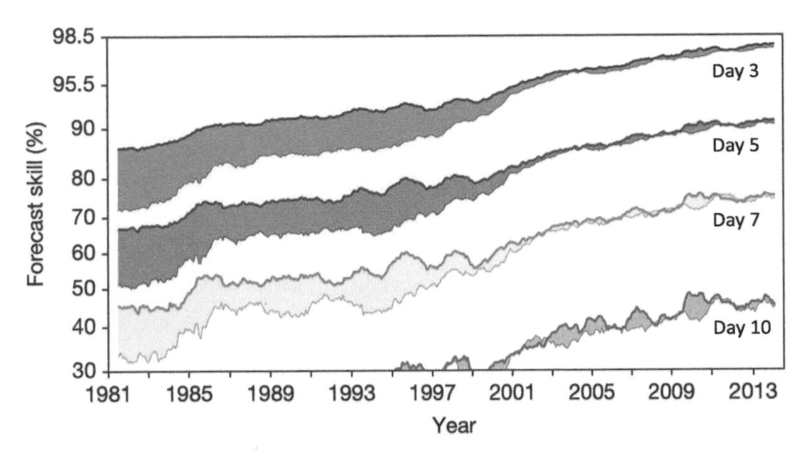

Fig. 1.7 Skill of European Centre for Medium-Range Weather Forecasts operational forecast at lead times of 3, 5, 7, and 10 days in the Northern *(thick curves)* and Southern Hemispheres. Forecast skill is the correlation between the forecasts and the verifying analysis of the geopotential height of the 500-hPa level, expressed as the anomaly from the climatology. The convergence of the curves for the Northern and Southern Hemispheres after 1999 indicates the breakthrough in exploiting satellite data. Today's day 5 forecast is as good as day 3 forecast 30 years ago. *(From Bauer et al. [2015].)*

prediction is apparent and increases steadily (Fig. 1.7). Weather forecast is a scientific triumph enabled by advances in understanding atmospheric dynamics (e.g., quasi-geostrophic theory), real-time observations (for both better understanding and initial condition), and numerical modeling (including the mathematic formulation, data assimilation, computation methods, and computers). The atmosphere is a chaotic system, limiting the skill of deterministic weather forecast to less than 2 weeks. While weather conditions (e.g., temperature and precipitation on a given day and hour) are not predictable beyond 2 weeks, the above-mentioned atmospheric GCM results suggest that statistics of atmospheric variability (e.g., monthly mean precipitation anomalies) might be predictable as long as SST anomalies are known. The ocean and atmosphere are coupled, and their mutual interactions determine the joint variability. This book is about coupled dynamics that give rise to the variability and predictability.

1.2 Climate in the News

Each year anomalous climate events take place around the world, resulting in large socioeconomic and environmental impacts. Following are highlights of some recent events. The annual supplements of the *Bulletin of the American Meteorological Society,* entitled "State of the Climate" and "Explaining Extreme Events," provide more comprehensive accounts.

- GMST in 2020 was on par with that of 2016, the warmest year on the instrumental record boosted by a major antecedent El Niño. By now, all seven highest annual mean GMST happened in the recent 7 years. The Intergovernmental Panel on Climate Change (IPCC) Sixth Assessment Report (2021) attributed most of GMST rise for the past century to anthropogenic emissions of carbon dioxide and other GHGs.
- Hurricane Dorian struck the Bahamas in September 2019 and is the most intense North Atlantic storm at landfall with 185 mph winds. Other recent record-making tropical cyclones include Haiyan (the most intense storm at landfall, with 190 mph winds, in November 2013), Patricia (the most intense storm of all time at 215 mph, in October 2015), and Fantala (the most intense tropical cyclone of the Indian Ocean, in April 2016). The maximum intensity of tropical cyclones increases with increasing temperature (see Section 10.5).
- On July 25, 2019, Paris recorded its all-time high temperature of 42.6°C. The heat forces authorities to impose speed limits as rails buckled and highways softened. On June 28, 2021, in the Pacific Northwest, new temperature records were set: 47°C at Portland and 42°C at Seattle-Tacoma Airport (the previous June record was 36°C in 2017 for comparison). On the following day, Lytton, British Columbia, set the new all-time Canadian record at 49.6°C. Heatwaves that last for days are associated with subsidence in high-pressure systems aloft, called the blocking highs. The background greenhouse warming increases the frequency and severity of baking conditions in heatwaves. Hot and dry conditions are conducive to wildfires. In 2020, fires in the western United States burned a record 10.2 million acres, equivalent in area to the state of Rhode Island. On September 9, 2020, smoke from wildfires turned the sky in the bay area of San Francisco, California, apocalyptically rust-colored as if it were on Mars.
- A severe coldwave swept across the United States in February 2021 as the polar vortex migrated south. On February 16, temperatures dropped to −11°C in Houston, more than 20°C below normal and the coldest since 1989. Millions of homes and businesses lost power for days, resulting in $200 billion in damage.
- Positive SST anomalies exceeding 3°C (3 standard deviations) over the midlatitude North Pacific, dubbed the Big Blob, developed in winter 2013–2014 (Amaya et al. 2016). While these extratropical SST anomalies are comparable in magnitude with those of El Niño on the equator, they had limited impact downstream on North America. Unlike El Niño that can be predicted months in advance, the predictability of extratropical SST anomalies is low (see Chapter 12).
- From January to March 2017, torrential rains devastated the usually dry Pacific coast of northern Peru, causing extreme flooding and widespread landslides. At Piura, accumulated rainfall reached 631 mm during the 3 months, as compared to the annual

climatology of 75 mm. Extremely high SSTs above 28.5°C were observed off Peru, promoting atmospheric deep convection. Similar events of extreme rainfall and coastal warming took place in 1983 and 1998, but they were each associated with a strong basin-wide El Niño. The strong coastal El Niño of 2017 was highly unusual in that it was accompanied by a weak basin scale La Niña (Peng et al. 2019).

- Widespread bushfires ravaged Australia from June 2019 to January 2020, driving koalas and kangaroos away from their habitats. Smoke from the fires darkened the skies in New Zealand. An anomalous SST pattern called the Indian Ocean dipole (see Section 11.2) contributed to dry and hot conditions over Australia during local summer and spring, while increased rainfall on the other side of the Indian Ocean was instrumental in the outbreak of desert locusts across eastern Africa.
- From June to July 2020, rainfall in the 1600-km-long Yangtze River basin from Chongqing to Shanghai was 70% above normal, and the severe flooding displaced millions of residents. The record-high summer rain was fueled by water vapor transport from the tropics by the anomalous southwesterly winds, a recurrent pattern to be discussed in Section 11.5.
- In July 2021, widespread severe floods devastated Europe. Cologne, Germany, observed 154 mm of rain in 24 hours. About the same time on July 20 on the other side of the continent, Zhengzhou, China, logged a record hourly rainfall of 202 mm—equivalent to 30% of the annual amount—causing hundreds of deaths. The enormous downpour was a result of an atmospheric river impinging on steep Taihang Mountains immediate to the west of the metropolis of 6.5 million. The extreme rainfall was forecast days in advance as Typhoon In-fa east of Taiwan streamed a river of water vapor toward Zhengzhou.

This book provides the dynamic foundation to identify the physical causes and evaluate predictability of extreme climate events. For example, initialized dynamic climate models predicted the wet summer of 2020 along the Yangtze River a few months in advance, while such seasonal forecast was not possible for European floods of July 2021. Some of the extreme events are due to naturally occurring variability, and some receive a boost from the growing greenhouse warming. In the literature, anthropogenic global warming effect is being increasingly detected from the type of extreme events related to temperature increase, such as terrestrial and marine heatwaves, droughts, and wildfires. With more water vapor in a warming atmosphere, precipitation rate is expected to increase, elevating the risks of flooding.

1.3 Fundamentals

This section briefly reviews some basics of ocean and atmospheric dynamics, which are further developed in Chapters 2, 3, and 7. This is followed by a brief introduction to

GCMs (see later) and some basic statistical methods (see later). Readers familiar with the material can skip these sections.

1.3.1 Geophysical Fluid Dynamics

Large-scale motions of the atmosphere and ocean are subject to strong influence of Earth rotation. The Coriolis force $\boldsymbol{C_o}$ is an apparent force introduced for noninertial coordinates fixed on the Earth that rotate around the North Pole. It is proportional to the horizontal velocity relative to the Earth and directs to the right (left) of the velocity in the Northern (Southern) Hemisphere

$$\boldsymbol{C_o} = f(\mathrm{v}, \ -\mathrm{u})$$

where (u, v) is the velocity vector with the eastward and northward components, $f = 2\Omega\sin\varphi$ is the Coriolis parameter (positive in the Northern Hemisphere and negative in the Southern Hemisphere), Ω is the angular velocity of Earth rotation, and φ the latitude. We adopt Cartesian coordinates (x, y, z), where (x, y) are the zonal/eastward and meridional/northward coordinates, and z is the vertical height from the sea level. In oceanography, velocity refers to the direction the current flows toward, while in meteorology it refers to the direction the wind comes from. A flow with u greater than 0 is an eastward current but a westerly wind.

In the interior ocean below a thin surface mixed layer (50–100 m deep) and the free atmosphere above the planetary boundary layer (~1 km deep), turbulent mixing is small. Large-scale motions are in geostrophic balance between the Coriolis and pressure (p) gradient forces

$$-fv = -\frac{1}{\rho}\frac{\partial p}{\partial x} \tag{1.1}$$

$$fu = -\frac{1}{\rho}\frac{\partial p}{\partial y} \tag{1.2}$$

In the Northern Hemisphere, facing down the geostrophic velocity, higher pressure is on the right side. Consider a circular storm with low pressure at the center. By itself, the pressure gradient would drive a fluid parcel toward the center; in reality, however, the storm is associated with a counterclockwise flow, which generates the Coriolis force to balance the pressure gradient force (Fig. 1.8).

The geostrophic balance is a good approximation of the horizontal momentum equations, but the diagnostic relation cannot predict the evolution of the balanced flow. For such a prediction we need to consider deviations from the balance and develop quasi-geostrophic theory of the slowly varying Rossby wave (Section 7.7 of Holton 2004; Chapter 5 of Vallis 2017).

Fig. 1.8 Geostrophic balance between the pressure gradient *(P)* and Coriolis *(C_o)* forces around a low-pressure *(L)* center in the Northern Hemisphere. ***u*** is the wind velocity. *(Image: NASA)*

1.3.2 Ocean

Upper ocean circulation is predominantly driven by wind stress $\boldsymbol{\tau} = (\tau_x, \tau_y)$. The momentum equations vertically integrated over a thin surface Ekman layer can be approximated as the balance between the wind stress and Coriolis force

$$0 = fv_E + \frac{\tau_x}{\rho_o H_E} \tag{1.3}$$

$$0 = -fu_E + \frac{\tau_y}{\rho_o H_E} \tag{1.4}$$

where ρ_o is the water density and H_E is the layer thickness (nominally ~50 m). Here we have assumed that vertical mixing, due partly to the strong vertical shear, is limited in the Ekman layer. The pressure gradient is neglected because the horizontal scale of the atmospheric wind field is much larger than that of the ocean current (1000 km vs 100 km). The Ekman theory was developed in the early 20th century to explain the peculiar observation that Arctic icebergs moved not down but to the right of the wind. In the Northern Hemisphere, the surface Ekman flow is directed to the right of the wind stress vector so the resultant Coriolis force balances the wind stress. The Ekman flow is southward in the midlatitude westerly winds while northward in the tropical easterly trade winds. The Ekman flow converges in the subtropics between the prevailing trades and westerlies. Indeed, that is where the great Pacific garbage patch is located due to the accumulation of floating microplastics.

From the continuity equation for mass conservation of incompressible fluid,

$$\frac{\partial u}{\partial x}+\frac{\partial v}{\partial y}+\frac{\partial w}{\partial z}=0 \tag{1.5}$$

the divergence of the surface Ekman flow drives vertical motion at the bottom of the Ekman layer (called the Ekman pumping),

$$w_E=\frac{1}{\rho_o}curl\left(\frac{\tau}{f}\right) \tag{1.6}$$

A cyclonic wind stress field (e.g., around a low-pressure atmospheric storm) drives a positive (upward) Ekman pumping. Much of the upper ocean (the top 100s m) is not directly forced by the wind stress but indirectly through the Ekman pumping (see Chapter 7).

Off the US west coast in summer, the northerly winds prevail, driving the surface water offshore (the Ekman flow) (Fig. 1.9). The continuity requires an upwelling of cold water from beneath, keeping the coastal water cool even in the summer; SST at Santa Barbara, California (34°N) is only 18°C, about 10°C lower than on the western side of the Pacific south of Tokyo. The upwelling shoals the thermocline on the coast and the e-folding cross-shore scale is the Rossby radius of deformation,

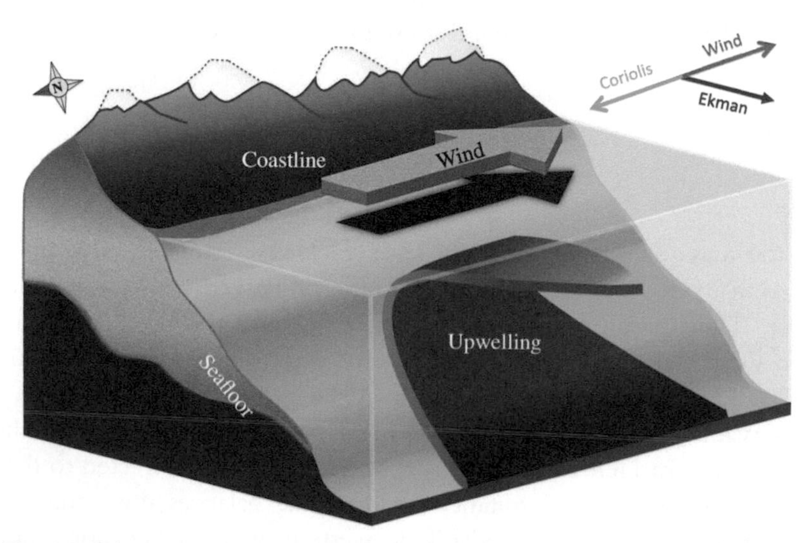

Fig. 1.9 Schematic of coastal upwelling off California. *(Adapted from Wikipedia.)*

$$R_f = \frac{c_o}{f} \tag{1.7}$$

where $c_o = \sqrt{\frac{\Delta\rho}{\rho_o} gH}$ is the phase speed of internal long gravity waves (see Chapter 7). Here $\Delta\rho$ is the density difference across the thermocline of the mean depth H. Similar coastal upwelling is also observed on the west coasts of South America and Africa, the Somali coast, and many other places.

The equator is the singularity for Ekman pumping defined in Eq. 1.6. Consider the uniform easterly winds. From Eq. 1.3, the Ekman flow is poleward on either side of the equator, causing a divergence and upwelling at the equator. The upwelling of cold water from beneath keeps the surface equatorial Pacific and Atlantic cool. Chapters 7 and 8 provide detailed discussions of the equatorial upwelling and its climatic effect.

1.3.3 Atmosphere

Sea water is nearly incompressible, with temperature and density nearly conservative following a moving parcel. The atmosphere is compressible, with density varying with temperature and pressure following the ideal gas law,

$$\rho = \frac{p}{RT} \tag{1.8}$$

where R is the gas constant, ρ and p are the density and pressure of air. In the vertical, the atmosphere is in hydrostatic balance between the pressure gradient and gravitational force,

$$\frac{\partial p}{\partial z} = -\rho g = -\frac{p}{RT}g \tag{1.9}$$

For an isothermal atmosphere,

$$p = p_s \exp\left(-\frac{z}{H_s}\right) \tag{1.10}$$

where $H_s = RT/g \sim 7.5$ km is called the scale height. Atmospheric pressure decreases exponentially with height.

Isobaric Coordinates

In meteorology, pressure is used as the vertical coordinate instead of vertical height (see Chapter 3, Holton 2004). In isobaric coordinates, the geostrophic balance is expressed as

$$-fv = -\frac{\partial \Phi}{\partial x} \tag{1.11}$$

$$fu = -\frac{\partial \Phi}{\partial y} \tag{1.12}$$

where $\Phi \equiv gz$ is the geopotential and z the geopotential height. Here the horizontal gradient of geopotential is taken on an isobaric surface (p = const.). The advantage of isobaric coordinates is obvious compared to the expression in height coordinates where the coefficient $1/\rho$ increases exponentially with height in Eqs. 1.1 and 1.2. This is why meteorologists use weather charts on isobaric surfaces.

The continuity equation on isobaric coordinates is also simple:

$$\frac{\partial u}{\partial x}+\frac{\partial v}{\partial y}+\frac{\partial \omega}{\partial p} = 0 \tag{1.13}$$

where $\omega \equiv \frac{dp}{dt}$ is the pressure velocity (positive downward). Here, d/dt is the material derivative following the moving air parcel. The hydrostatic equation equivalent to Eq. 1.9 is

$$\frac{\partial \Phi}{\partial p} = -\frac{RT}{p} \tag{1.14}$$

It states that the thickness between two isobaric surfaces is proportional to the temperature of the layer.

Thermodynamic Variables

A rising air parcel without heat exchange (adiabatic) with the environment expands in volume as pressure drops. The work it does to the outside lowers its internal energy, and hence temperature, according to the first law of thermodynamics. Although temperature of an adiabatic air parcel varies with ambient pressure, the dry static energy defined as

$$s \equiv c_p T + gz \tag{1.15}$$

is conservative, where c_p is the specific heat at constant pressure. This is equivalent to the conservation of potential temperature,

$$\theta \equiv T\left(\frac{p_0}{p}\right)^{\kappa} \tag{1.16}$$

defined as the temperature of the air parcel when brought adiabatically to a reference pressure level (usually p_0 = 1000 hPa). Here, $\kappa = R/c_p = 2/7$. The dry static energy and potential temperature are related as $ds = c_p T\, d(\ln\theta)$.

The rate of temperature decreases with height is called the lapse rate, $\Gamma \equiv -dT/dz$. For an air parcel in adiabatic ascent ($ds = 0$ in Eq. 1.15), the lapse rate is $\Gamma_d = g/c_p = 9.8$ K/km. This dry adiabatic lapse rate is observed in the planetary boundary layer (the first km from the surface) in the afternoon when the heated ground surface forces convection.

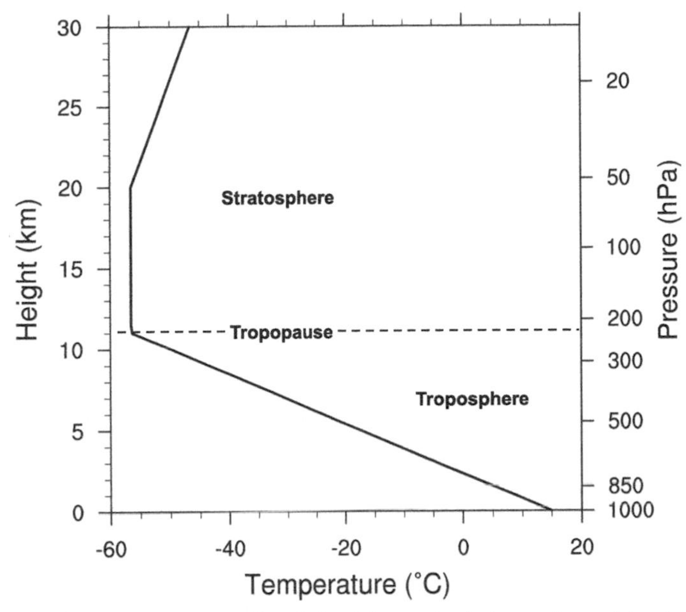

Fig. 1.10 Vertical temperature profile of the US standard atmosphere. *(Courtesy ZQ Zhou.)*

Further above, the lapse rate is on average 6.5 K/km (Fig. 1.10), much smaller than the dry rate because of the release of latent heat of condensation in rainy clouds. Air temperature reaches a vertical minimum at the tropopause, the height deep moist convection reaches as visualized by the flat top of anvil clouds. In the stratosphere above the tropopause, air temperature starts to increase with height because of ozone absorption of solar radiation in the stratosphere. This book concerns the troposphere between the ground surface and tropopause, in which weather variations take place.

Because potential temperature in a stably stratified atmosphere increases with height ($d\theta/dz > 0$), subsidence under high-pressure systems at upper levels often causes high surface temperatures in summer when the horizontal advection is weak. The atmospheric surface heatwave is analogous to surface cooling in ocean upwelling zones, both due to the vertical advection of background temperature stratification. On land, the subsidence-induced surface heatwave is exacerbated by reduced cloud cover that increases the downward solar radiation and soil drying that limits evapotranspiration.

1.3.4 Air-Sea Exchange

The atmosphere and ocean are in contact exchanging momentum and heat. Wind stress at the ocean surface is often expressed as

$$\boldsymbol{\tau} = \rho_a C_d W \boldsymbol{u} \tag{1.17}$$

where ρ_a is the air density, C_d the nondimensional aerodynamic drag coefficient, W the time average of scalar wind speed, which is generally greater than the scalar speed of the time mean vector wind velocity $\boldsymbol{u}$. The ocean and atmosphere are in turbulent exchange of sensible heat and water vapor. The sensible heat flux is formulated as

$$Q_H = \rho_a C_H W(T - T_a) \tag{1.18}$$

and the latent heat flux through surface evaporation as

$$Q_E = \rho_a C_E WL(q_s - q_a) \tag{1.19}$$

where C_H and C_E are the exchange coefficients, the subscript a denotes the atmospheric variable nominally measured at 10 m above the surface, and q_s is the saturation mixing ratio at the sea surface. The exchange coefficients (C_d, C_H, and C_E) are about 1.3×10^{-3} but functions of wind speed (due to interaction with wind waves) and the air-sea temperature difference (a measure of static stability).

1.4 General Circulation Models

A GCM refers to a set of governing equations for momentum, energy, water, and other conservative quantities for the atmosphere, ocean, and their coupled system, usually under some approximations. The effects of unresolved processes are represented with so-called parameterizations based on empirical and theoretical relationships. In a coupled GCM the atmospheric model passes the wind stress, heat, and freshwater fluxes to the ocean model, while the ocean model passes the SST and sea ice fields to the atmospheric model (Fig. 1.11). The land surface model is often treated as part of the atmospheric model, exchanging heat and water with each other. GCMs are so complex that only numeric solutions are possible on a global grid system of finite spacing. GCMs represent the state of the art of our understanding of the climate system both at process level and in terms of macro behavior and characteristics. Indeed, GCMs show skills in simulating and predicting weather (see Fig. 1.7) and climate variability (see Chapter 9) as well as radiatively induced climate change (see Chapter 13).

For climate, we examine monthly or seasonal averages. Generally, atmospheric variability $P(x,y,p;\, t)$ is comprised of its internal variability (the so-called butterfly effect) and the response to external forcing, $P = P_I + P_F$. For the atmosphere, SST and sea ice distributions are an important forcing at the bottom boundary. The SST/sea ice variability could be due to the interaction with the ocean (e.g., El Niño) and is implicitly encoded with the effect of other external forcing such as insolation, GHG concentrations, and aerosols.

A perturbed initial condition ensemble (PICE) refers to a set of atmospheric GCM runs that share the same observed evolution of SST and sea ice but with different initial

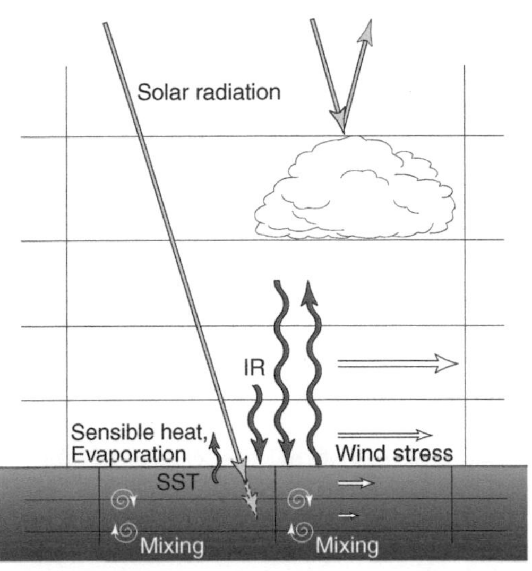

Fig. 1.11 Schematic of a coupled ocean-atmosphere model through energy fluxes and wind stress. *IR*, infrared radiation; *SST*, sea surface temperature. *(From Neelin [2011].)*

conditions. The difference among the member runs samples the internal variability P_I while the ensemble mean represents the forced variability P_F,

$$P_F = \frac{1}{M}\sum_{m=1}^{M} P_m \tag{1.20}$$

$$P_I^m = P^m - P_F, \; m = 1, \; 2, \; \ldots, \; M \tag{1.21}$$

where m denotes the mth member run, and M is the ensemble size. Alternatively, one can evaluate the internal variability by forcing the atmospheric GCM with climatologic SST and sea ice. (See Fig. 1.6 for such an example.) The comparison with a parallel atmospheric GCM run with "observed" SST evolution indicates that seasonal mean atmospheric variability in the deep tropics is largely forced by SST variability, especially over the ocean.

With ocean GCMs, the same method can be applied to isolate SST variability that is forced by the atmosphere, T_F. At 1° horizontal resolution the ocean internal variability T_I is small and atmospherically forced SST variability dominates (see Fig. 9.8), while at ocean eddy-permitting resolution (grid spacing $\leq 0.25°$) ocean internal variability becomes important.

For the coupled ocean-atmosphere system, SST is an internal variability, whereas insolation, GHGs, and aerosols are external forcing. The PICE ensemble with a coupled GCM can be used to evaluate the radiatively forced response T_F and internal variability T_I.

1.5 Statistical Methods

A time series of N elements long can be decomposed into a mean known as the climatology and deviations denoted with a prime,

$$T_i = \overline{T} + T_i', \; i = 1, \; 2, \; \ldots, \; N \tag{1.22}$$

where $\overline{T} \equiv \frac{1}{N}\sum_{i=1}^{N} T$, and T′ is also known as anomaly. The variance is

$$\sigma_T^2 \equiv \frac{1}{N}\sum_{i=1}^{N} T_i'^2 \tag{1.23}$$

where σ is called the standard deviation, a measure of the amplitude of temporal variability.

1.5.1 Correlation

Covariance of two time series x_i and y_i is defined as

$$R_{xy} \equiv \frac{1}{N}\sum_{i=1}^{N} x_i y_i \tag{1.24}$$

The cross-correlation,

$$r_{xy} = \frac{R_{xy}}{\sigma_x \sigma_y} = \frac{1}{N}\sum_{i=1}^{N} \frac{x_i}{\sigma_x}\frac{y_i}{\sigma_y} \tag{1.25}$$

is a measure of how closely x and y are related to each other. Obviously, $|r_{xy}| \le 1$. The correlation often needs to be tested for statistical significance. We say that a correlation is significant at the 95% level when the null hypothesis that the time series are unrelated can be rejected 95% of times. For a given confidence level, the critical correlation coefficient can be determined based on Student's t-test if the degree of freedom is known.

The degree of freedom is calculated as $N_E - 2$. Here the effective degree of freedom is generally not the same as the sample size. Consider a sine oscillation. The effective degree of freedom is only two within one period regardless of how many times the oscillation is sampled during the time period. An effective sample size considers the lag-1 autocorrelation r_x,

$$N_E = N\frac{1 - r_x}{1 + r_x} \tag{1.26}$$

Bretherton et al. (1999) give the effective sample size for evaluating a cross-correlation,

$$N_E = N\frac{1 - r_x r_y}{1 + r_x r_y} \tag{1.27}$$

1.5.2 Empirical Orthogonal Function

The EOF analysis decomposes a time-varying field $T(x,y,t)$ into a set of mutually orthogonal spatial base functions $R_n(x,y)$ and time series $P_n(t)$

$$T(x, y, t) = \sum_{n=1}^{N} P_n(t) R_n(x, y) \tag{1.28}$$

Here $R_n(x,y)$ is called the nth EOF mode and $P_n(t)$ the PC. In climate data, the first few EOF modes usually explain the majority of total variance in the domain of analysis. In practice, the spatial field $T(x,y)$ is represented by a spatial vector T of $I \times J$ elements long on a regular grid of $i = 1, 2, \ldots I$, and $j = 1, 2, \ldots J$.

The EOF analysis can be performed on two or more field variables by forming a joint vector of $M \times I \times J$ long, where M is the number of variables. The EOF analysis seeks a set of joint modes, say between SST and precipitation, each sharing the same PC. Specifically, one may choose the joint variables to be the monthly sequence of the same variable, say SST, to identify the seasonal evolution of an interannual mode. This is called the month-reliant or cyclostationary EOF analysis (Kim et al. 2015).

Instead of sharing the same PC, the singular value decomposition (SVD) method identifies a pair of spatial patterns for two related field variables, say between SST and precipitation, by maximizing the temporal covariance of the PCs (Wallace et al. 1992). It is also called the maximum covariance analysis (MCA).

CHAPTER 2

Energy balance and transport

Contents

2.1 Planetary energy balance and greenhouse effect

Earth receives radiative energy from the Sun. The spectrum of blackbody radiation follows Plank's law, with the wavelength of peak radiation $\lambda_m = b/T$, where T is the temperature in Kelvin and b is a constant. With surface temperature at 5800 K, the Sun radiates energy that peaks in the visible range (0.4—0.7 μm in wavelength) while at −18°C, Earth radiates energy at much longer wavelengths in the infrared range of 5 to 50 μm. Thus the solar radiation is often called the shortwave radiation and the terrestrial radiation the longwave radiation (Fig. 2.1). The clear atmosphere is nearly

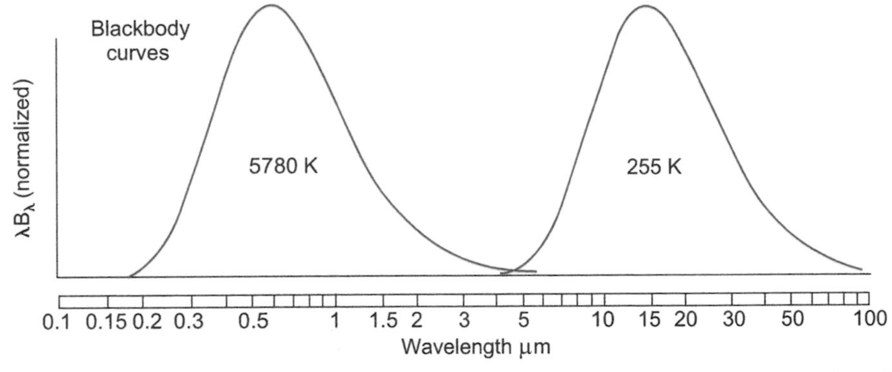

Fig. 2.1 Blackbody spectra representative of the Sun *(left)* and the Earth *(right)*. The wavelength scale is logarithmic, and the ordinate has been multiplied by wavelength to retain the proportionality between areas under the curve and intensity. In addition, the intensity scales for the two curves have been scaled to make the areas under the two curves the same. *(From Wallace & Hobbs [2006].)*

Coupled Atmosphere-Ocean Dynamics: from El Niño to Climate Change
ISBN 978-0-323-95490-7, https://doi.org/10.1016/B978-0-323-95490-7.00002-3

transparent to the shortwave solar radiation, which is why we can see through the air during the day, but strongly absorptive of the longwave terrestrial radiation.

Consider a simple case where the atmosphere is modeled as a glass sphere (Fig. 2.2). Like a greenhouse, the glass case is transparent to solar radiation but acts as a blackbody for terrestrial infrared radiation. Let Q_s be the solar radiation absorbed on the ground surface,

$$Q_s = (1 - a)I_0/4 \tag{2.1}$$

where $I_0 = 1361\ \mathrm{W/m^2}$ is called the solar constant and $a = 0.3$ the planetary albedo that measures the reflectivity. At the top of the atmosphere (TOA), the net incoming solar radiation equals the outgoing longwave radiation (OLR). This gives the glass sheet temperature in Kelvin,

$$\sigma T_1^4 = Q_S \tag{2.2}$$

where σ is the Stefan-Boltzmann constant and T_1 is the temperature the ground surface would be if not for the atmosphere. At the ground surface under the glass sheet, the energy balance is

$$\sigma T_S^4 = \sigma T_1^4 + Q_S = 2Q_S \tag{2.3}$$

In addition to solar radiation, the ground surface receives additional longwave radiation from the atmosphere. This is called the atmospheric greenhouse effect, which raises the ground temperature from freezing T_1 (−22C) to comfortable $\sqrt[4]{2}T_1$ (15°C). It can be shown that adding more glass layers increases the ground temperature further according to $\sqrt[4]{n+1}T_1 = \sqrt[4]{(n+1)Q_S/\sigma}$, where n is the number of glass sheets.

Fig. 2.3 shows the energy budget of the global atmosphere. Of the incoming solar radiation at the TOA, 31% is reflected back to space, 49% is absorbed on the ground, while the atmosphere absorbs only 20%. In addition to solar radiation, the ground surface

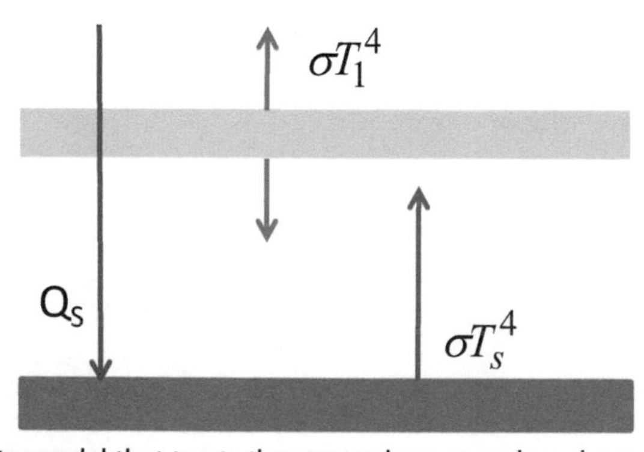

Fig. 2.2 A simple climate model that treats the atmosphere as a glass sheet, transparent to solar (Q_s) but black for infrared radiation (*blue and brown arrows*).

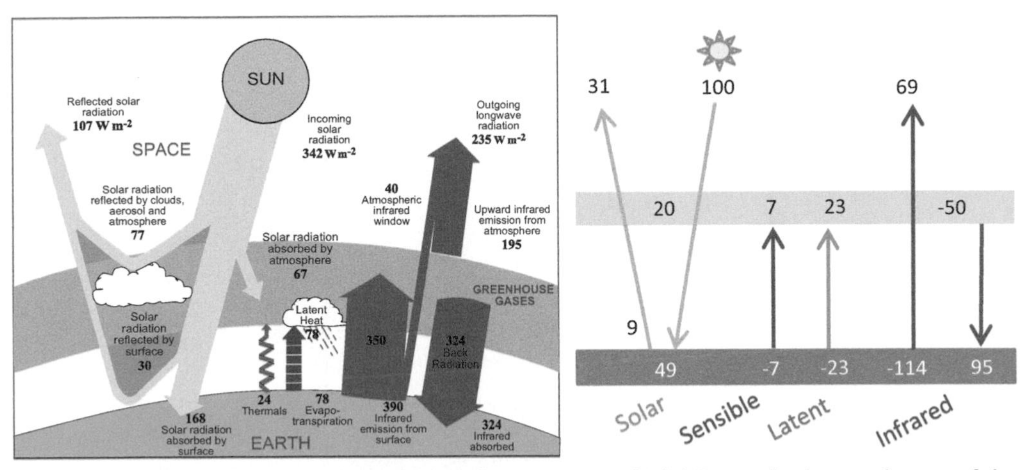

Fig. 2.3 *(left)* Energy flow though the global climate system. *(right)* Energy budget at the top of the atmosphere (TOA), for the atmosphere (*cyan shading*) and at the ground surface (*brown shading*). The fluxes are normalized by the downward solar radiation at TOA. Most of solar radiation absorption takes place at the Earth surface, and the atmosphere is mostly heated from beneath. *(From* (left) *Hartmann [2016].)*

receives 95 units of infrared radiation from the atmosphere. This greenhouse effect increases the ground surface temperature.

Consider a static atmospheric column in radiative equilibrium that does not allow vertical convection. Because the atmosphere is transparent to sunlight, it is heated from beneath at the ground surface by the absorbed solar energy. Water vapor is a strong greenhouse gas, and its concentration increases with temperature. Because temperature decreases with altitude from the solar-heated ground surface, so does water vapor content. This is equivalent to adding infrared absorbing layers near the surface. As a result, the temperature profile in radiative equilibrium increases sharply toward the surface (Fig. 2.4, *red curve*), so much so that the atmospheric column becomes gravitationally unstable and vertical convection takes place.

We consider a simple model of the atmosphere that is completely transparent to solar radiation but opaque to terrestrial infrared radiation. In this so-called two-stream gray atmosphere, the radiative equilibrium temperature profile follows

$$\sigma T^4 = \frac{F_0}{2}\left(\frac{5}{3}\tau + 1\right) \tag{2.4}$$

where $F_0 = 235\ \mathrm{W/m^2}$ is the OLR that equals the net downward solar radiation at the TOA. The infrared optical depth

$$\tau = \tau_s e^{-z/h} \tag{2.5}$$

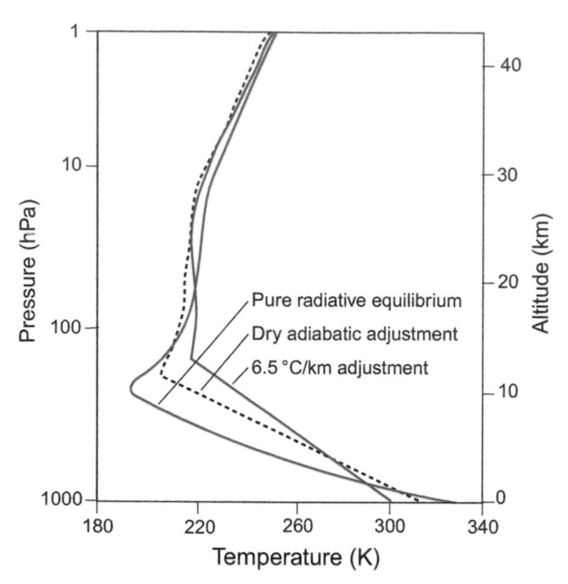

Fig. 2.4 Pure radiative *(red line)* and radiative-convective equilibria. For the latter, the dry adiabatic *(dashed line)* and moist adiabatic *(blue line)* lapse rates are used to represent the convective effect. *(From Wallace & Hobbs (2006).)*

increases exponentially toward the surface due mostly to infrared-absorbing water vapor. Here $h = 2.5$ km can be viewed as the water-vapor scale height, and typically $\tau_s = 3$. Thus air temperature in radiative equilibrium increases rapidly toward the surface. With these parameters typical of the Earth atmosphere, the surface air temperature reaches 334 K (61°C), much higher than observed.

The radiative equilibrium is gravitationally unstable with Γ greater than Γ_d. The resultant convection adjusts the lapse rate to the moist adiabat in the troposphere (see Fig. 2.4, *blue line*). Further above, the stratosphere remains in radiative equilibrium and is stably stratified. The ozone absorption of solar radiation causes temperature to increase with altitude in the lower stratosphere, creating a temperature minimum at the tropopause that separates the convective troposphere below and the stratosphere above. Convection mixes both potential temperature and water vapor in the troposphere. Of 49 units of solar radiation received, the ground surface returns the energy to the atmosphere through infrared radiation and turbulent mixing at a ratio of 19:30 (see Fig. 2.3). The turbulent mixing causes surface fluxes of heat and water vapor due to air-sea surface differences in temperature and specific humidity, respectively (see Chapter 1). As latent heat of condensation is released when the water vapor condenses in the atmosphere, evaporation from the surface represents a flux of latent heat. The combined sensible and latent heat fluxes are also called the turbulent flux.

Carbon dioxide (CO_2) is a strong greenhouse gas. Massive combustion of fossil fuels since the industrial revolution has increased atmospheric CO_2 concentration from 280 to over 420 parts per million. The resultant increase in atmospheric optical depth (τ_s) causes surface temperature to rise (Eqs. 2.4 and 2.5). Chapters 13 and 14 discuss the problem of global warming induced by anthropogenic increase in atmospheric greenhouse gas concentrations.

2.2 Radiative imbalance and energy transport

Assume that preindustrial climate is nearly in equilibrium. Globally integrated, the net downward solar radiation equals the OLR at the TOA. At a given location, however, such a TOA radiative balance is not achieved because of the divergence of lateral energy transport by ocean and atmospheric motions. In annual mean, the solar radiation absorbed by the ocean-atmosphere system peaks at the equator at 300 W/m^2 and decreases poleward to below 50 W/m^2 at the poles. The OLR distribution, however, is flatter, with a broad maximum of 250 W/m^2 in the tropics and polar minima of 100 W/m^2 (Fig. 2.5). As a result, the tropics suffers a deficit of energy flux, and the polar regions enjoy a net energy gain at the TOA. This TOA energy imbalance results from poleward energy transport by atmospheric winds and ocean currents.

Consider two atmosphere-ocean columns in the tropical and polar regions, respectively (Fig. 2.6). The energy balance for each column follows

$$C\frac{\partial T_1}{\partial t} = \lambda(T_1 - T_{1E}) - d(T_1 - T_2) \tag{2.6}$$

$$C\frac{\partial T_2}{\partial t} = \lambda(T_2 - T_{2E}) - d(T_2 - T_1) \tag{2.7}$$

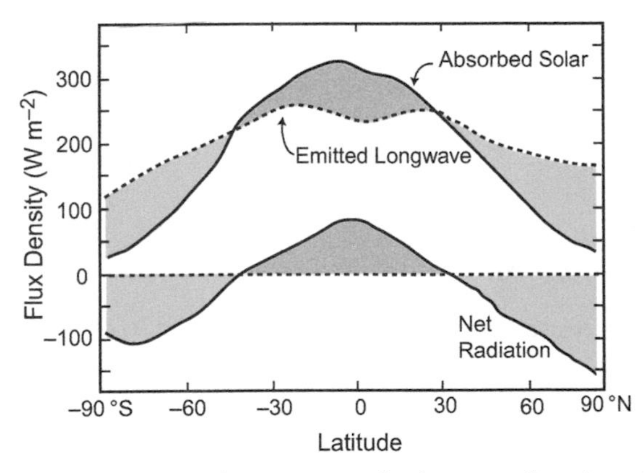

Fig. 2.5 The annual and zonal mean radiative energy budget as a function of latitude. In the zonal mean, downward solar radiation exceeds outgoing longwave radiation in the tropics, but the budget reverses in high latitudes implying energy transport by atmospheric and ocean motions. *(From Wallace & Hobbs (2006).)*

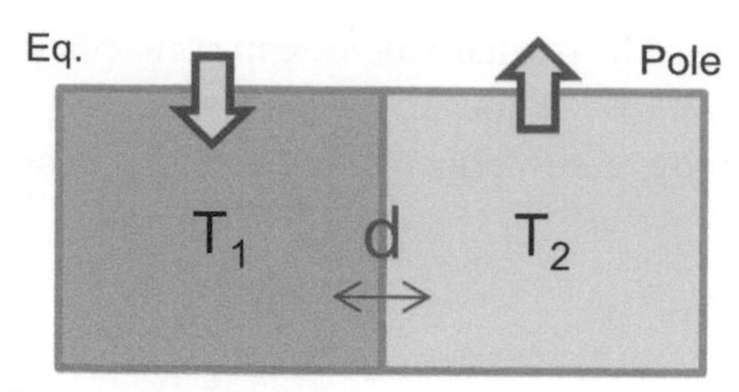

Fig. 2.6 Schematic of a two-box model that represents tropical and extratropical columns with a constant coefficient for lateral energy exchange.

where T_1 and T_2 denote sea surface temperature (SST) in the tropical and polar columns, respectively, the subscript E denotes the value in local radiative-convective equilibrium, C is the ocean heat capacity (which is much larger than that of the atmosphere and other components of the climate system), λ represents the radiative feedback at the TOA ($\lambda \sim -1.2$ Wm^{-2}K^{-1}; see Chapter 13), and the ocean-atmospheric transport effect is modeled as a newtonian damping with a coefficient d. The steady-state solution gives the tropical-to-pole temperature difference,

$$\widehat{T} \equiv T_1 - T_2 = \frac{\lambda}{\lambda - d}\widehat{T}_E \tag{2.8}$$

Thus the meridional energy transport ($d > 0$) reduces the tropical-to-pole temperature gradient ($\widehat{T} < \widehat{T}_E$), resulting in TOA radiative imbalance of $\lambda(T - T_E) = \pm d\widehat{T}$ (downward for the tropical and upward for the polar column).

Generally in the steady state, the ocean-atmospheric energy transport $F(y)$ divergence is balanced by the radiative energy imbalance at the TOA,

$$\frac{\partial F}{\partial y} = R \tag{2.9}$$

The energy transport can be further decomposed into the atmospheric and ocean components, $F = F_a + F_o$. In the following, we discuss the distribution and dominant mechanisms for each component.

2.3 Ocean heat transport

Sea water is nearly incompressible. The energy conservation is expressed as

$$\frac{\partial T}{\partial t} + \frac{\partial}{\partial x}(uT) + \frac{\partial}{\partial y}(vT) + \frac{\partial}{\partial z}(wT) = \frac{\partial}{\partial z}\left(\kappa \frac{\partial T}{\partial z}\right) \tag{2.10}$$

where the material derivative on the left-hand side (LHS) is cast in the flux form, and κ is the turbulent diffusivity in the vertical. In steady state, integrating through the entire water column along a latitudinal circle yields

$$\rho c_p \frac{\partial}{\partial y}[vT]_o = \int_0^{2\pi} Q_{net} \cos\varphi d\lambda \tag{2.11}$$

where $[\bullet]_o \equiv \int_0^{2\pi} \cos\varphi d\lambda \int_{-H}^{0} [\bullet] dz$ is the zonal and vertical integration, and $\rho c_p \kappa \frac{\partial T}{\partial z}\big|_{z=0} = Q_{net}$ is the net surface heat flux (downward positive). The heat transport can be decomposed into that in individual ocean basins. In an ocean basin spanning from the western (x_W) to eastern (x_E) boundary, the heat transport can be further decomposed into

$$[vT]_{x_W}^{x_E} = (x_E - x_W) \int_{-H}^{0} \overline{v}\overline{T} dz + \left[v^* T^*\right]_{x_W}^{x_E} \tag{2.12}$$

where the overbar denotes the zonal mean across the basin associated with the meridional overturning circulation (MOC) and the asterisk is the deviation from the zonal mean that represents the horizontal gyre circulation. In general, the heat transport by the zonal-mean MOC is much larger than the gyre transport in climatology.

Consider a two-layer MOC (Fig. 2.7). The upper (lower) layer is h_1 (h_2) thick. The zonally integrated volume transport in the upper layer V is balanced by that in the lower layer,

$$V \equiv (x_E - x_W) \int_{-h1}^{0} \overline{v} dz = -(x_E - x_W) \int_{-(h1+h2)}^{-h1} \overline{v} dz$$

The basin-integrated heat transport is approximated as

$$[vT]_{x_W}^{x_E} = V\left(\overline{T}_1 - \overline{T}_2\right) \tag{2.13}$$

where $\overline{T}_1 \equiv \int_{-h1}^{0} \overline{v}\overline{T} dz \Big/ \int_{-h1}^{0} \overline{v} dz$ is the velocity-weighted vertical mean of temperature in the upper layer, and $\overline{T}_2$ in the lower layer. Thus the ocean heat

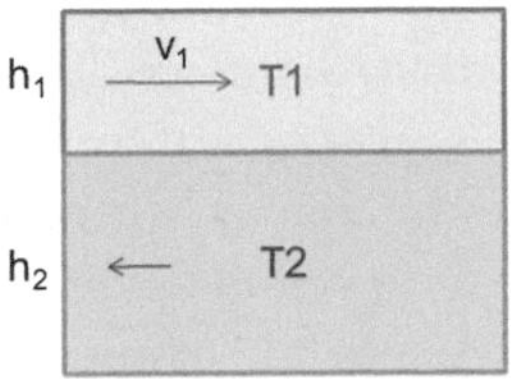

Fig. 2.7 Schematic of heat transport by an ocean overturning circulation.

transport is the MOC volume transport times the vertical temperature stratification over the MOC depth (upper minus lower layer temperature difference).

2.3.1 Ocean meridional overturning circulations

The ocean features two types of MOCs, one shallow and one deep. The shallow MOC is wind driven. In the surface Ekman layer ($H_E \sim 50$ m deep), the Coriolis force deflects the surface Ekman current to flow to the right (left) of the wind stress vector in the Northern (Southern) Hemisphere (see Eqs. 1.3 and 1.4). In the tropics, the easterly trade winds drive a poleward Ekman flow in the surface layer (Fig. 2.8A). In the subtropical gyre equatorward of the maximum westerly winds, a deep mixed layer forms in winter due to the surface cooling and convection. The equatorward geostrophic flow advects cold water from the winter mixed layer into the thermocline. While most of the water subducted into the thermocline recirculates in the anticyclonic subtropical gyre, part of it eventually upwells on the equator due to the Ekman divergence (see Chapters 1 and 7). Because of a large excess of evaporation over precipitation, surface salinity is high in the surface subtropical ocean (Box 2.1). The high-salinity tongue that extends equatorward along the thermocline (say, the 20°C isotherm) is indicative of the lower branch of the shallow MOC (see Fig. 2.8A). A large temperature difference exits between the warm poleward Ekman flow (25°C) and cold equatorward flow in the thermocline ($\sim$15°C), the shallow MOC transports heat poleward.

At the equator, the abyssal water is near the freezing point. Such cold water cannot form locally but is formed in the winter subpolar North Atlantic (Box 2.1) and spreads into the Pacific and Indian basins via the Southern Ocean. Much of the North Atlantic deep water (NADW) upwells in the Southern Ocean and is advected equatorward by the westerly wind-induced surface Ekman flow and subducts into the thermocline as implied by a tongue of low salinity in Fig. 2.8B. The NADW is a water mass characterized by its high salinity in the 1000- to 3000-m layer. The deep Atlantic MOC (AMOC) is an inter-hemispheric circulation, transporting heat across the equator from the Southern into the Northern Hemisphere. The interhemispheric AMOC spans a temperature range of 3°C to 10°C in the vertical. See Talley et al. (2011) for a detailed discussion of ocean water-masses and circulation.

2.3.2 Sea surface heat flux

The meridional divergence of zonally integrated heat transport by ocean currents needs to be balanced by a downward surface heat flux into the ocean (see Eq. 2.11). Indeed the equatorial oceans receive heat from the atmosphere as the cold thermocline water upwells and comes in contact with the atmosphere (Fig. 2.9; see also Fig. 2.8A). In the subtropics, the anticyclonic gyre circulation causes a large contrast in surface heat flux between the western and eastern basins, especially in the Northern

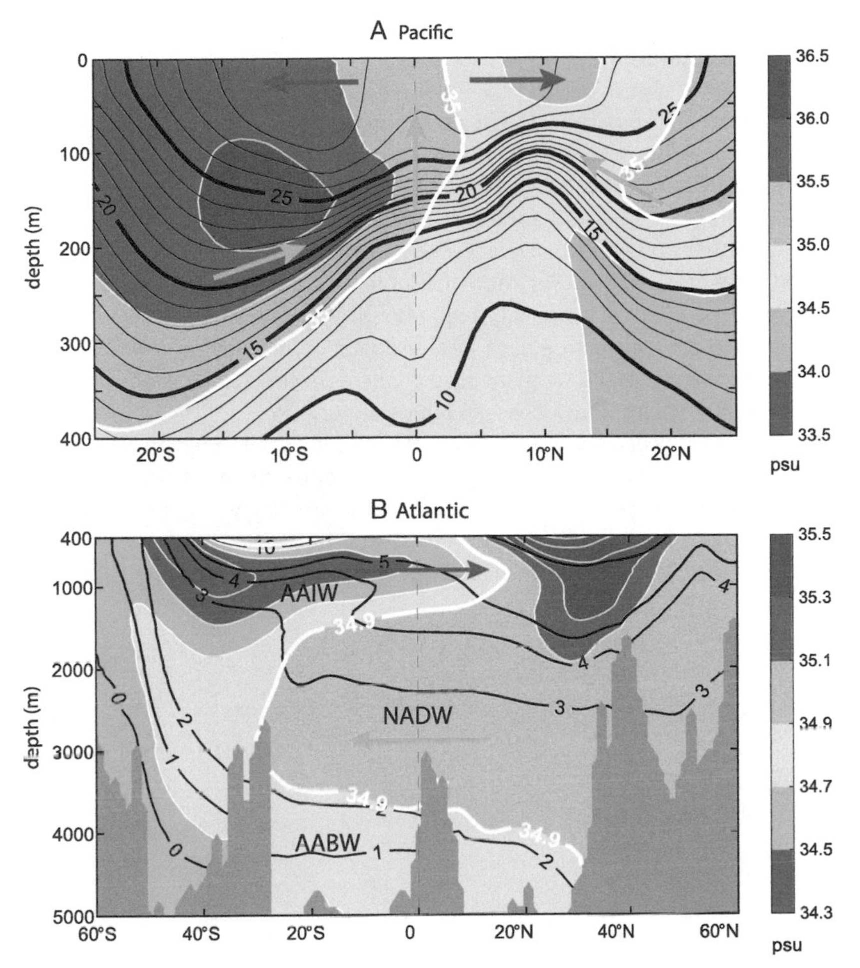

Fig. 2.8 (A) Temperature *(black contours)* and salinity *(color shading, white contours)* in a tropical Pacific transect at 160°W. (B) Same but for the Atlantic along 30°W. In the Pacific, MOC is shallow and transports heat poleward, with warm surface water flowing poleward and cold thermocline water toward the equator. In the Atlantic, deep water forms in subpolar northern latitudes and spreads southward and then into much of the world ocean. The deep southward flow is compensated by a northward upper flow, of which the Gulf Stream is a part. As a result, there is a net northward heat transport across the equator. The shallow and deep MOCs leave clear signatures in the salinity distribution. The NADW is of higher salinity than either the Antarctic intermedium water *(AAIW)* or Antarctic bottom water *(AABW)*. The World Ocean Atlas 2018 is used. *(Courtesy ZH Song.)*

Hemisphere and in winter. The ocean releases huge amounts of heat in poleward-flowing, warm western boundary currents (e.g., the Kuroshio and Gulf Stream) and the extensions while being heated in the equatorward-flowing, cold eastern boundary currents in the presence of coastal upwelling (see Fig. 2.9, *top*). Compared to the

BOX 2.1 Surface salinity

Surface evaporation is large in the tropics and decreases rapidly poleward with decreasing temperature (Box Fig. 1A). Evaporation (E) has a weak minimum on the equator because of low temperature due to ocean upwelling and low wind speeds. Compared to smooth variations in evaporation, precipitation (P) is highly variable in the meridional direction, featuring a sharp peak just north of the equator and a broad maximum in the storm track in each hemisphere. Globally, E-P vanishes on long-term average. The meridional distribution of sea surface salinity largely follows that of E-P (see Box Fig. 1B). In the subtropics, E-P reaches a broad maximum, and the excessive evaporation keeps salinity high. Along the northward-displaced ITCZ, on the other hand, heavy precipitation exceeds local evaporation, freshening surface water. Much lower salinity is observed in the subpolar oceans where precipitation in storm tracks exceeds evaporation that is kept small by low ocean temperatures.

Much of the world ocean below 1 km depth is filled with water of remarkably uniform temperatures ($T \sim 2°C$). For cold water ($T < 5°C$), the temperature effect on density is weak, and the salinity effect becomes important (see Box Fig. 1B). Surface salinity is 33 practical salinity unit (PSU) or less in the subpolar North Pacific, and 35 PSU in the North Atlantic. The difference in surface salinity and hence density explains why the deep water forms in the North Atlantic not in the North Pacific.

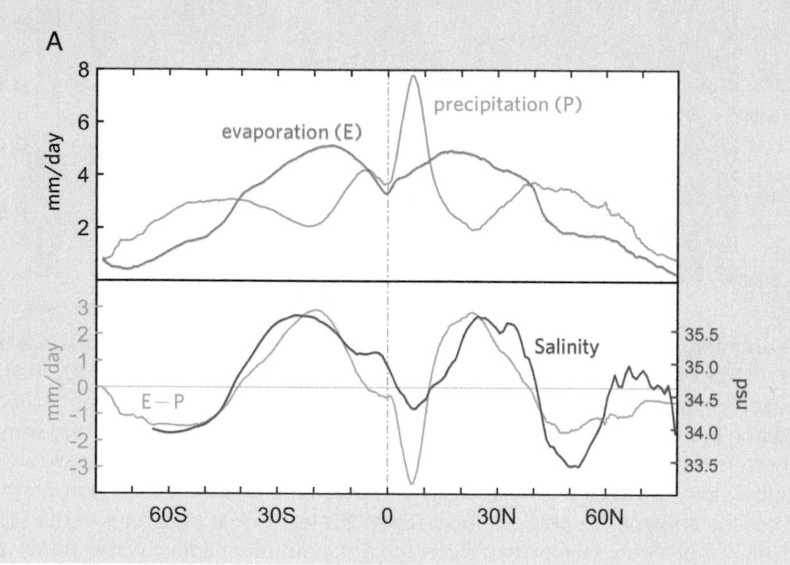

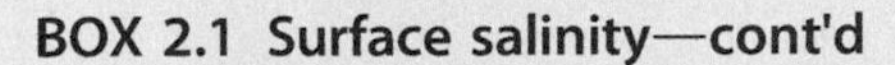

BOX 2.1 Surface salinity—cont'd

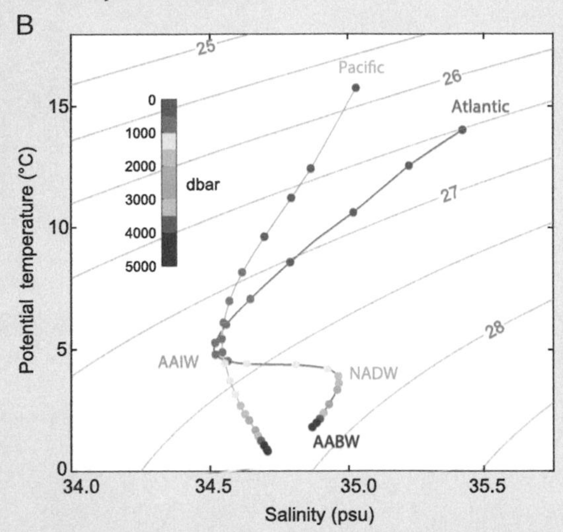

Box Fig. 1 *((A) Annual and zonal-mean distributions of precipitation, evaporation, and their difference (mm/day, ERA5 for 1979-2020), along with surface salinity (psu, Argo for 2004-2020). (B) Vertical soundings of salinity and temperature at the equator in the Atlantic (0°,* orange line*) and Pacific (180°,* blue line*), on the background of potential density (σ_0) contours. The colors of dots indicate the depth (starting from 200 m). The Antarctic intermedium water* (AAIW) *appears as a vertical minimum in salinity, and the* NADW *as a vertical maximum below 500 m in the Atlantic. The World Ocean Atlas 2018 is used. (A-B, Courtesy ZH Song.))*

North Pacific, the subpolar North Atlantic loses heat due to the northward heat transport by the AMOC.

Three-dimensional thermal structures of the ocean are poorly measured before the Argo era (which began nominally in 2002), especially considering the abundance of mesoscale (100s km) eddies in the ocean. The surface variables that affect surface heat fluxes (cloud cover, SST, air surface temperature and humidity, and wind speed) are much better sampled. In practice, ocean heat transport is inferred from surface heat flux observations by integrating Eq. 2.11 from the northern boundary. Without the deep MOC in the Pacific, ocean heat transport is nearly symmetric about the equator due to the wind-driven shallow MOC in either hemisphere. In the Atlantic basin, the interhemispheric deep MOC drives a strong northward heat transport across the equator.

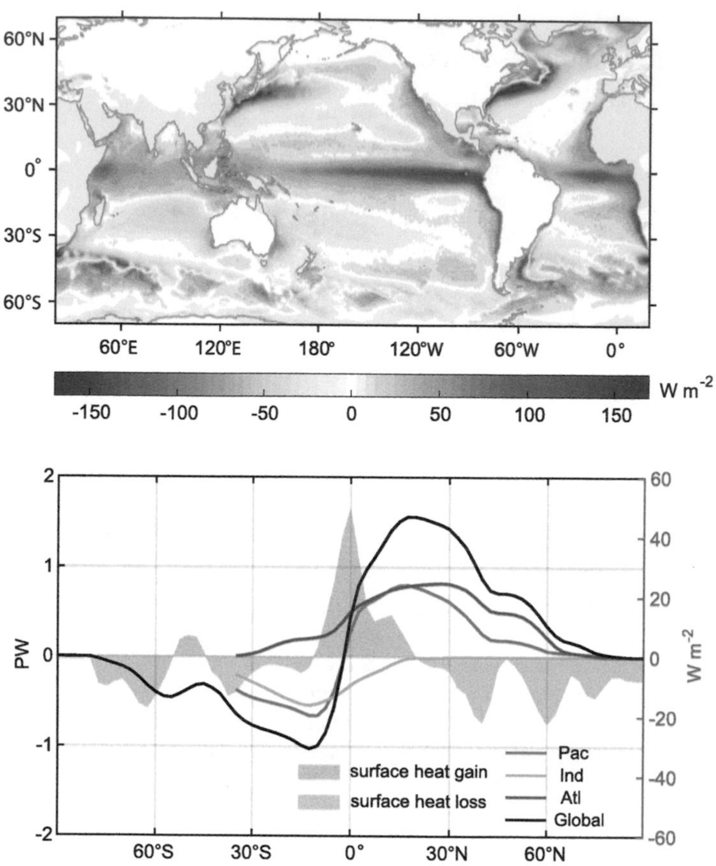

Fig. 2.9 *(top)* Annual-mean net surface heat flux (W/m^2) based on the ERA5 for 1979-2020. *(bottom)* Annual-mean and zonal-integrated ocean heat transport (*color lines*, 1 PW = 10^{15} W) and net surface heat flux (*shaded*, W/m^2) from the fully coupled CESM1 preindustrial run. The meridional ocean heat transport is calculated from integration of net surface heat flux over the ocean basin. *(Courtesy ZH Song.)*

The pole-to-pole integration of Eq. 2.11 yields a global constraint,

$$\int_{-\pi/2}^{\pi/2} d\varphi \int_{0}^{2\pi} Q_{net} \cos \varphi d\lambda = 0 \tag{2.14}$$

In a steady state, the globally integrated sea surface heat flux needs to vanish. This global constraint is applied to adjust surface heat flux calculations, which contain considerable uncertainties in formulation and input data. With the anthropogenic increase in TOA radiative forcing, the steady-state assumption is no longer valid. In fact, the globally integrated ocean heat content, $\int_{-\pi/2}^{\pi/2} d\varphi \int_0^{2\pi} \cos\varphi d\lambda \int_{-H}^{0} T dz$, is observed to increase (see Chapters 13 and 14), contributing to global sea level rise.

2.4 Atmospheric energy transport

Air is compressible, and water vapor, upon condensation, increases air temperature through latent heat release. The moist static energy (MSE) of an air parcel, $m \equiv c_p T + gz + Lq$, includes the latent energy of water vapor it contains. Here q is the specific humidity in kg/kg (Box 2.1). Much like the global ocean heat budget (see Eq. 2.11), integrating the MSE conservation equation zonally and vertically from the surface to TOA gives

$$\frac{\partial}{\partial y}[vm]_a = \int_0^{2\pi} (R_{TOA} - Q_{net}) \cos\varphi d\lambda \qquad (2.15)$$

where $[\cdot]_a \equiv \int_0^{2\pi} \cos\varphi d\lambda \int_0^{p_s} [\cdot] dp/g$ represents the zonal and vertical column integration. The atmospheric energy transport divergence is balanced by the net energy flux into the atmospheric column, from the TOA and surface

2.4.1 Tropics

In the tropics a vertical overturning circulation called the Hadley circulation dominates in the zonal mean, driven by latent heating in the rainy rising branch. The surface winds transport moisture equatorward to form a rain band in the intertropical convergence zone (ITCZ) (Fig. 2.10). In the annual mean, the Hadley circulation features one cell in either hemisphere, but the rising branch and the ITCZ are slightly displaced north of the equator. The zonal-mean ITCZ marking the rising branch of the Hadley circulation migrates back and forth across the equator, largely following the seasonal march of the solar radiation.

The MSE in the tropics features a vertical minimum in the midtroposphere (~600 hPa) (Fig. 2.11B), below which the downward increase is due to increasing water vapor. As a result of active deep convection, the moist energy at the tropopause is restored toward that in the atmospheric boundary layer near the surface (see Chapter 3).

Because the Hadley circulation transports high moist energy in both the upper and lower branches but in opposite directions, the net energy transport is small (see Fig. 2.11A), that is,

$$[vm]_a = V(m_u - m_l) \qquad (2.16)$$

BOX 2.2 Thermodynamic variables of the moist atmosphere

Water vapor is an important greenhouse gas with large variations in three dimensions and time. After a clear, long winter night, the ground surface often frosts as a result of strong radiative cooling. The latent heat of phase change between water vapor, liquid water, and ice drives the atmospheric circulation. The amount of water vapor is often measured by the mixing ratio or specific humidity in kg/kg, $q \equiv \rho_v/\rho = \varepsilon e/p$, where ρ is density with v denoting water vapor, e the partial pressure of water vapor, and $\varepsilon = R/R_v$ with $R = 287$ and $R_v = 461$ $\text{Jkg}^{-1}\text{K}^{-1}$ being the gas constants for dry air and water vapor, respectively. Relative humidity is defined as $R_H \equiv e/e_s$, where the saturation vapor pressure e_s follows the Clausius-Clapeyron equation and is an exponential function of temperature. This temperature dependency dictates that most water vapor is confined near the surface and in the tropics.

In an ascending parcel, the adiabatic cooling causes water vapor to condense. The potential temperature of a rising saturated air parcel is no longer conservative but increases due to latent heat of condensation. The MSE includes latent energy of water vapor,

$$m \equiv c_p T + gz + Lq = s + Lq$$

where L is the latent heat of vaporization. Equivalent potential temperature (θ_e) is the potential temperature of an air parcel after all its moisture has condensed,

$$\theta_e = \theta \exp\left(\frac{Lq}{c_p T_{LCL}}\right)$$

where LCL stands for the lifting condensation level (the cloud base). A rising air parcel from the surface becomes saturated at LCL. In a rising moist air parcel, θ_e is conserved. The MSE and equivalent potential temperature are related as $dm = c_p T\, d(\ln\theta_e)$.

Saturated equivalent potential temperature (θ_e^*) is similar to θ_e but with in situ humidity replaced by the value in saturation

$$\theta_e^* = \theta \exp\left(\frac{Lq_s}{c_p T}\right)$$

The static stability can be determined from the vertical sounding of temperature as follows:

1. $d\theta_e^*/dz > 0$, absolutely stable (near the tropopause)
2. $d\theta_e^*/dz < 0$, conditionally unstable for saturated parcels (e.g., in the lower troposphere)
3. $d\theta/dz < 0$, absolutely unstable (e.g., in the afternoon just above the ground surface)

The moist adiabatic lapse rate

$$\Gamma_s = \Gamma_d/(1 + L/c_p \cdot dq_s/dT)$$

is always smaller than the dry adiabatic rate $\Gamma_d = g/c_p = 9.8$ K/km because the condensational heat slows the cooling in a rising saturated parcel. ($\Gamma_s \sim 4$ K/km in warm humid air mass in the lower troposphere, ~6–7 K/km in the midtroposphere, and approaches the dry rate of 9.8 K/km in the upper troposphere.) The difference between the dry and moist adiabatic rates becomes very small in the upper troposphere because cold temperature there holds very little moisture. Because of active convection, the lapse rate in the troposphere above the boundary layer is close to the moist adiabatic.

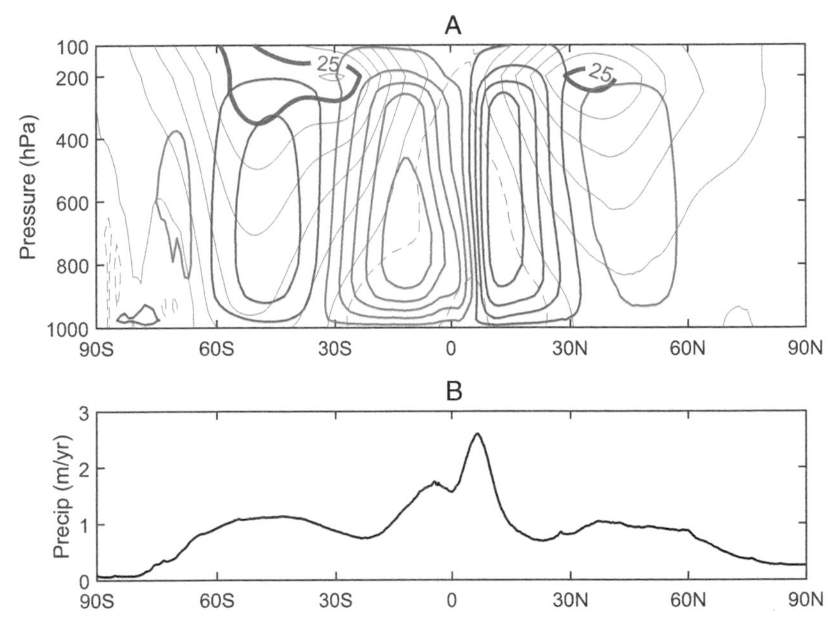

Fig. 2.10 (A) Annual and zonal-integrated mass transport function (positive/negative in *red/blue*, at intervals of 2×10^{-10} kg/s starting at 10^{-10} kg/s) and zonal wind velocity (*grey* contours at 5 m/s intervals; the 25 m/s contour *thickened* and -2.5 m/s *dashed*). (B) Annual and zonal-mean precipitation (m/yr). ERA5 for 1979-2020 is used. *(Courtesy ZH Song.)*

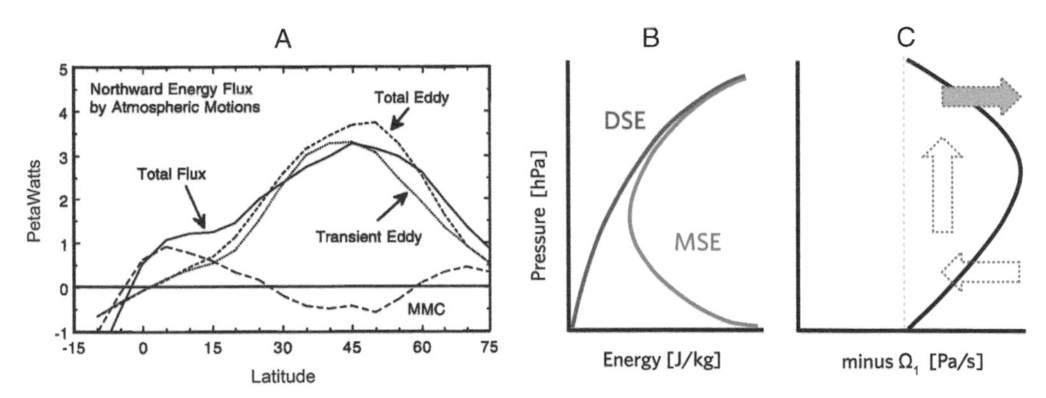

Fig. 2.11 (A) Atmospheric energy transport as a function of latitude, along with the contributions by the mean meridional circulation *(MMC)* and eddies. Vertical profiles of (B) dry *(DSE)* and moist static energy *(MSE)*; and (C) vertical velocity in the tropics. *(A, From Hartmann (1994); B-C, Inoue & Back (2015). © American Meteorological Society. Used with permission.)*

where $V = [v]_0^{pm}$ is the upper branch mass transport, $m_u = [vm]_0^{pm}/V$, and $m_l = -[vm]_{pm}^{ps}/V$. Here, pm ~600 hPa is a midtropospheric level of maximum vertical velocity or zero horizontal velocity of the overturning Hadley circulation. Because the vertical maximum of vertical velocity—or, equivalently, the interface of the upper and lower branches of the Hadley cells—is generally above the moist energy minimum (see Fig. 2.11), $m_u - m_l > 0$. In other words, the net moist energy transport by the deep overturning cell, although small, is in the direction of the transport by the upper limb, poleward for the Hadley circulation (see Fig. 2.11C).

The Hadley circulation transports dry energy poleward but latent energy of water vapor equatorward. The net transport of moist energy is a small fraction of the dry energy transport,

$$[vm]_a = \widehat{M}[vs]_a$$

where

$$\widehat{M} \equiv \frac{[vm]_a}{[vs]_a} = \frac{m_u - m_l}{s_u - s_l} \ll 1 \tag{2.17}$$

is called the gross moist stability (GMS). It represents the ratio of column-integrated moist to dry energy transport, or equivalently the upper minus lower troposphere difference in moist energy normalized by the dry static stability. For the deep overturning circulation, GMS is small but positive. As a result, in the tropics the atmospheric energy transport by the deep Hadley circulation is weak while the ocean transport dominates because of a strong shallow MOC and strong temperature difference between the surface and thermocline.

Here we note a similarity in energy transport by the atmospheric and ocean MOCs. Near the convective region where the static energy difference is small in the vertical, the MOC energy transport is small. For this reason, the ocean energy transport is weak in the high-latitude regions. In the atmosphere, GMS is kept low by deep convection and dynamical adjustment in the broad tropics within 15° (see Chapter 3) on either side of the equator.

The poleward flow in the upper branch of the Hadley circulation induces westerly winds. Neglecting friction that is weak in the upper troposphere, we obtain from the zonal momentum equation,

$$v\left(\frac{\partial u}{\partial y} - f\right) = 0 \tag{2.18}$$

The meridional profile of the westerly velocity is

$$u = \beta y^2/2 \tag{2.19}$$

The poleward-moving air parcel in the upper branch of the Hadley circulation gains a westerly velocity due to the Coriolis effect. Here the beta-plane approximation $f = \beta y$ is used, where $\beta = df/dy$ is taken as constant and of the value at the equator.

On a sphere, as it moves poleward and the radius shrinks, a zonal ring of air gains westerly velocity out of the need of conserving the angular momentum around the axis of Earth rotation. This is much like a spinning skater who pulls the arms inward. The conservation of angular momentum, $(\Omega a\cos\varphi + u)a\cos\varphi$, yields the zonal wind profile,

$$u(\varphi) = \Omega a \sin^2\varphi/\cos\varphi \qquad (2.20)$$

where a is the radius of Earth and φ is latitude. The westerly velocity of an angular momentum conserving ring of air reaches 133 m/s at 30° latitude, a velocity much larger than observed. Eq. 2.19 is the beta-plane approximation appropriate for low latitudes.

2.4.2 Extratropics

With the annual mean solar radiation decreasing poleward, one might imagine a thermally driven Hadley circulation that extends from the equator all the way to poles. The observed Hadley circulation, however, terminates at 30°N in either hemisphere because baroclinic instability breaks out as the poleward flow accelerates the westerlies in the upper troposphere (see Eq. 2.19). Large meanders of the westerly jet cause weather variations. The linear growth rate of the resultant zonal waves in the so-called Eady problem of baroclinic instability is

$$\alpha_{max} \propto \frac{f}{N}\frac{\partial \overline{u}}{\partial z} \qquad (2.21)$$

where N is the buoyancy frequency, $N^2 = \frac{g}{\theta_0}\frac{d\theta_0}{dz} = \frac{g}{T_0}(\Gamma_d - \Gamma_0)$, with the subscript 0 denoting the background temperature stratification and the overbar denoting the zonal mean. The vertical wind shear is in thermal wind balance with the meridional temperature gradient

$$f_0\frac{\partial \overline{u}}{\partial lnp} = R\frac{\partial \overline{T}}{\partial y} \qquad (2.22)$$

As the bottom boundary condition, surface wind is weak because of friction.

The westerly jet is in geostrophic balance and flows along the geopotential contours with high (low) values on the equatorward (poleward). A poleward (equatorward) meander of the geopotential contours is called a ridge (trough) in meteorology, corresponding to a positive (negative) geopotential perturbation (Fig. 2.12A).

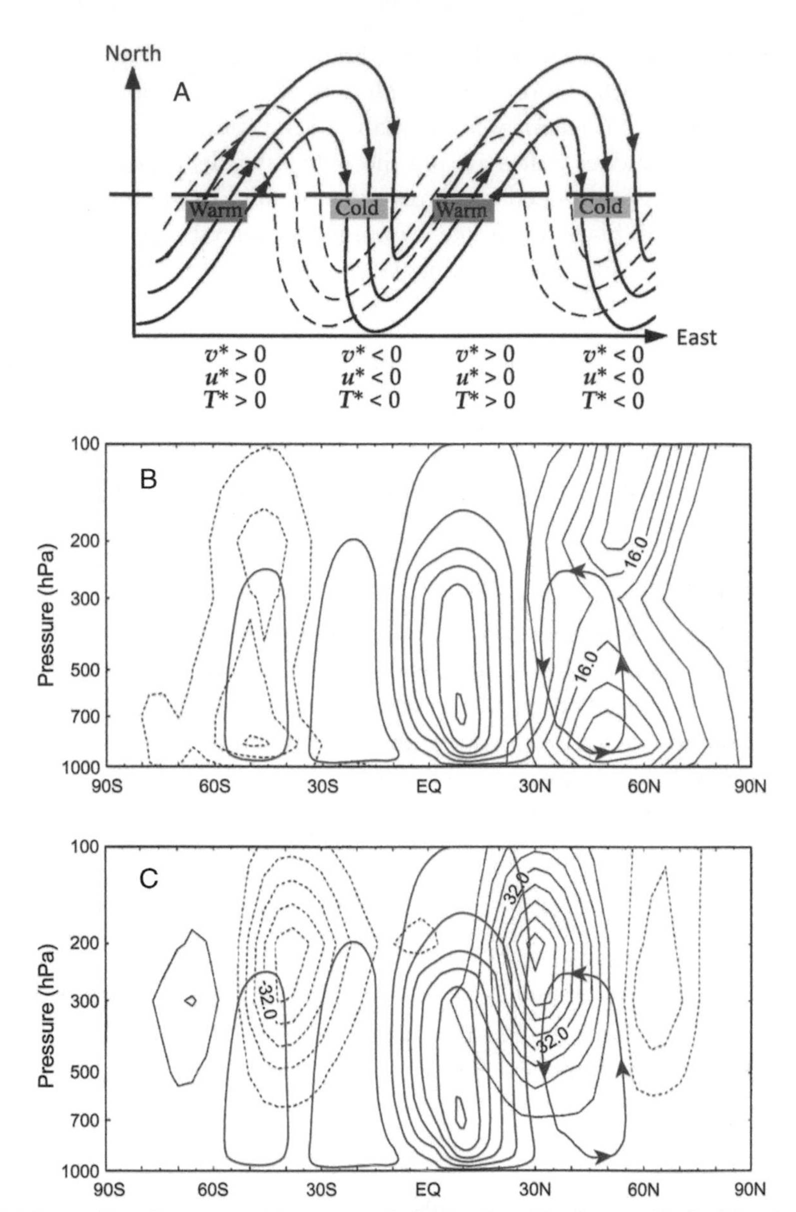

Fig. 2.12 (A) Streamlines/geopotential contours *(solid lines)* and isotherms *(dashed lines)* with midlatitude disturbances that transport momentum and heat northward. Zonal and December-February mean eddy transport of (B) heat (K m/s) and (C) westerly momentum (m^2/s) as a function of latitude and pressure (black contours, negative dashed, and zero omitted), superimposed on the mass transport function (color contours). *(A, From Hartmann (1994); B-C, Holton (2004).)*

Growing baroclinic eddies feature an up-westward phase tilt in geopotential perturbation; the midtropospheric trough is displaced west of the surface low, and the upward perturbation motion is in phase with and helps intensify the surface low. The streamlines and isotherms meander so that the poleward flow advects warm air and the equatorward flow advects cold air (see Fig. 2.12A) (i.e., $\overline{v'T'}>0$ in the storm tracks in 30°–60° latitude). Here the prime denotes temporal variations on the synoptic weather timescale (~1 week) and the overbar the zonal mean. Because temperature and water vapor perturbations are positively correlated in synoptic eddies, the eddy transport of sensible and latent heat reinforces each other and in fact is the dominant mechanism of poleward energy transport outside the tropics (see Fig. 2.11A). Through this energy transport, synoptic eddies act to reduce the meridional temperature gradient that causes baroclinic instability in the first place.

In the midlatitudes, a secondary overturning circulation called the Farrell cell forms to counter the meridional divergence of eddy energy flux. See Holton (2004) for a detailed discussion. Zonally integrating the thermodynamic equation at the midtroposphere yields

$$-S_p\overline{\omega} = -\frac{\partial\left(\overline{v'T'}\right)}{\partial y} \tag{2.23}$$

where $S_p = (\Gamma_d - \Gamma_0)/\rho_0 g$ is the stability parameter. Here the moist process is neglected for simplicity. Adiabatic warming in the descending motion (LHS), say at 30°N, balances the cooling due to eddy heat divergence on the equatorward flank of the storm track (the RHS) while adiabatic cooling in the ascending motion counters the eddy warming poleward of the storm track, say at 55°N (see Fig. 2.12B).

In upper-level weather charts, synoptic troughs and ridges are tilted southwestward (see Fig. 2.12A), resulting in a poleward transport of westerly momentum $\overline{u'v'}>0$. The eddy momentum flux is large in the upper troposphere on the southern flank of the storm track. The Coriolis deceleration associated with the equatorward flow in the upper branch of the Farrell cell, say at 45°N, balances the westerly acceleration due to the meridional convergence of eddy momentum flux (see Fig. 2.12C),

$$-(1 - R_o)f\overline{v} = -\frac{\partial\left(\overline{u'v'}\right)}{\partial y} \tag{2.24}$$

where $R_o = \frac{1}{f}\frac{\partial\overline{u}}{\partial y} \leq 1$ is the Rossby number. The eddy momentum flux divergence decelerates the westerlies induced by the poleward flow in the upper branch of the Hadley cell, say at 30°N. As a result, the peak zonal-mean velocity at the core of the subtropical westerly jet is only 40 m/s, much smaller than obtained from angular momentum conservation (133 m/s at 30°N/S). In contrast to the Hadley cell that is thermally driven by the equator-to-pole temperature gradient, the Farrell cell is driven by eddy fluxes of heat and momentum.

At the surface, friction is important, and the wind is weak compared to upper levels. We have

$$-f\overline{v} = \varepsilon\overline{u} \text{ at the surface.} \tag{2.25}$$

Thus the equatorward flow in the lower branch of the Hadley cell induces the easterly trade winds that peak around 15° latitude, while the poleward flow in the lower branch of the Farrell cell induces the westerly winds that peak at 40° (Fig. 2.13; see also Fig. 2.10). For the Earth rotation rate to remain constant, we have

$$\int_{-\pi/2}^{\pi/2} d\varphi \int_{0}^{2\pi} \tau_x cos\varphi d\lambda = 0 \tag{2.26}$$

at the surface by global integration of the zonal momentum equation. Here we have neglected the form drag due to the pressure difference between the west and east surfaces of mountain ranges. This is the global constraint on the zonal wind stress distribution. Specifically, the easterly wind stress in the tropics balances the westerly stress in the midlatitudes.

The above discussion shows that the synoptic storms not only bring about rich weather variations but are an essential element in maintaining the meridional structure

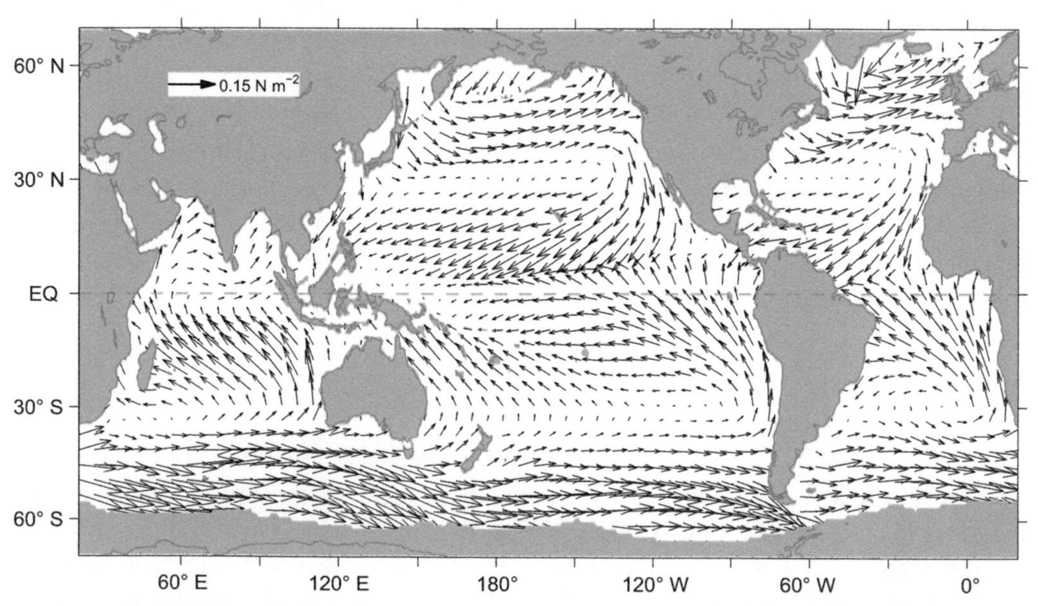

Fig. 2.13 Annual-mean wind stress over the ocean, based on QuikSCAT data. *(From Huang [2015].)*

of Earth climate through the transports of energy and momentum. The eddy effects are important for the meridional overturning (Hadley and Farrell) circulation, the subtropical westerly jet in the upper troposphere, and surface westerlies.

The tropopause marks the boundary between the troposphere where temperature decreases with height and the stratosphere where temperature increases upward. In the tropics, the tropopause height H_T is set by deep moist convection (i.e., $m|_{z=H_T} \approx m|_{z=0}$). In the midlatitudes, vertical convection from the surface is shallow, reaching at most 700 hPa over the warm ocean in winter cold surges off the east coasts of Asia and North America. The tropopause height is determined instead by baroclinic eddies. A discontinuity exists near the subtropical westerly jet, with the tropical tropopause ($H_T \sim 16$ km) dropping to the midlatitude tropopause ($H_T \sim 11$ km). Satellite observations of cloud top height capture this discontinuity in tropopause height (Fig. 2.14).

The tropopause is not a material surface, across which tropospheric (low ozone) and stratospheric (high ozone, strong stratification, high potential vorticity) air exchanges along isentropic (constant-θ) surfaces. For example, the $\theta = 350$ K isentropic surface is nearly flat in the troposphere in the tropics but in the stratosphere poleward of the subtropical jet. The 330 K isentropic surface, on the other hand, is in the stratosphere at high latitudes but slants downward toward the surface in the midlatitudes. Because of the exchanges with the stratosphere, ozone spikes of a stratospheric origin are often observed at the surface after the passage of the cold front of a synoptic low-pressure storm.

The combined energy transport by the ocean and atmosphere may be calculated from accurate measurements of TOA radiation by satellite. Atmospheric observations are dense

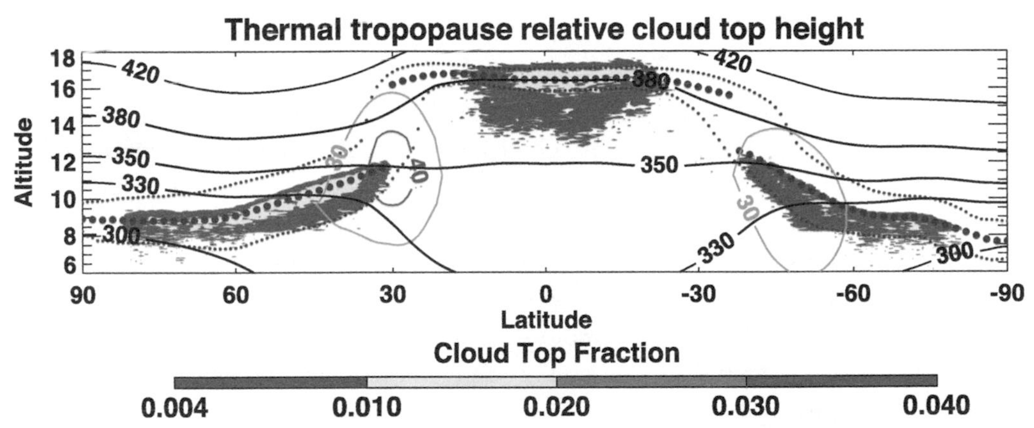

Fig. 2.14 Cloud top fractions from the Cloud-Aerosol Lidar and Infrared Pathfinder Satellite Observation (CALIPSO) for March 2007. Also shown are the zonal mean tropopause height (*large purple dots*, km) and its 10th and 90th percentiles (*small purple dots*), zonal wind speed (*cyan contours* for 30 and 40 m/s), and potential temperature (*black contours*, K). *(From Pan & Munchak (2011).)*

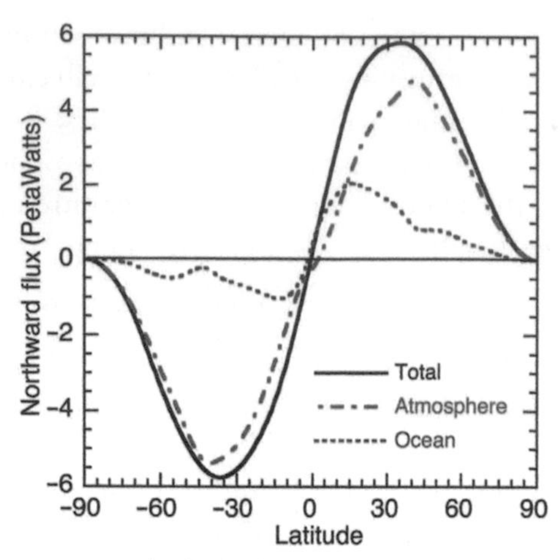

Fig. 2.15 The annual mean transport by latitude for the total *(black)*, atmosphere *(red)*, and ocean *(blue)*. *(From Hartmann [2016].)*

enough for direct calculations of MSE transport, while the ocean heat transport may be derived from surface heat flux calculations. Atmospheric and ocean transports are of similar magnitude in the deep tropics (Fig. 2.15). The ocean heat transport is proportional to the vertical temperature difference as illustrated by the two-layer model. Because of the poleward decrease in ocean temperature stratification, the ocean heat transport decreases from the subtropics poleward. The atmospheric transport dominates outside the tropics due mostly to stationary and transient eddies. This statement is valid for the climatology, but changes in deep ocean circulation and the heat transport are important driving global-scale climate change (see Chapters 12 and 14).

Review questions

1. Extend the glass-sheet radiative balance model discussed in the chapter by modeling the atmosphere as two sheets of blackbody glass. Calculate the ground temperature and the temperature of each glass sheet. Compare with one glass sheet model.
2. Extend the radiative balance model discussed in the chapter by considering atmospheric emissivity $\varepsilon < 1$. Instead of σT_1^4, the atmosphere emits infrared radiation at $\varepsilon\sigma T_1^4$. Note that the atmospheric absorption of ground infrared radiation is of the same efficiency ε. Calculate ground temperature as a function of solar radiation I and emissivity ε. Discuss how an increase in ε, say by increasing atmospheric CO_2 and water vapor, will change ground temperature.
3. Based on the global energy balance (solar absorption and surface heat flux), how can one make an argument for ocean's importance in climate?

4. Downward solar radiation at the top of the atmosphere (TOA) is nearly constant in 30°N to 90°N on summer solstice (June 21) (see figure). Discuss why the Siberia is still colder than Shanghai in summer.

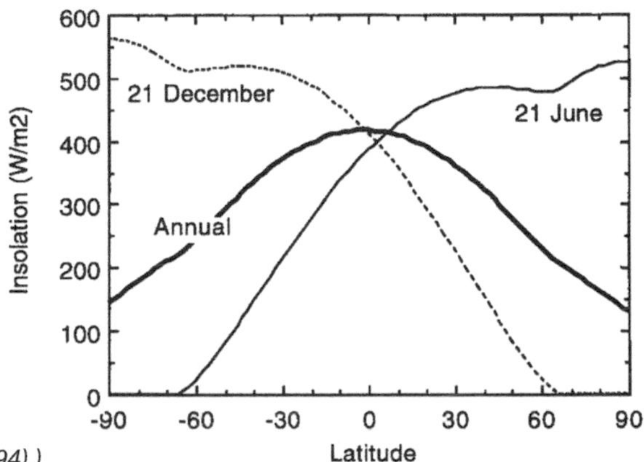

(From Hartmann (1994).)

5. Calculate TOA radiation in the two-box (tropical and polar) model (see Eqs. 2.6 and 2.7).
6. Discuss why the meridional heat transport in the ocean is small in the extratropics. *Hint:* The ocean heat transport is proportional to the temperature difference between the upper and lower layers.
7. Explain why the atmospheric meridional overturning circulation (MOC; e.g., Hadley) is inefficient in transporting moist static energy. Compare with the heat transport by ocean MOCs. Consider the annual mean for simplicity.
8. In which direction does the Hadley circulation transport energy?
9. Discuss how extratropical eddies transport energy. Are the eddy transports of dry (sensible) energy and water vapor (latent energy) in the same direction?
10. Discuss why the ocean heat transport peaks in the tropics and why the atmospheric energy transport peaks in the midlatitudes.
11. Discuss briefly why the Hadley circulation does not extend all the way to the poles.
12. Why are baroclinic instability waves weak at low latitudes? *Hint:* Consider the maximum growth rate of baroclinic instability.
13. At a fundamental level, consider why eddies are important in meridional transport of energy in the atmosphere but not in the ocean. *Hint:* What drives eddies and how do they feed back on the base state?
14. Consider which of the wind-driven shallow ocean MOC or thermohaline deep MOC is important for the following features of the net heat flux distribution at the ocean surface:

- **a.** a. Downward flux into the ocean in the deep tropics and heat release from the ocean in the midlatitudes in both hemispheres
- **b.** b. Within the Atlantic basin, overall downward flux in the South Atlantic and ocean heat release in the subpolar North Atlantic

CHAPTER 3

Tropical convection and planetary-scale circulation

Contents

In the rising branch of the Hadley circulation, air parcels expand and cool, causing water vapor to condense and water droplets to grow. Eventually raindrops grow so large that they fall as precipitation. Observed precipitation peaks near but not exactly on the equator. (We will explain the equatorial minimum and the northward-displaced maximum in Chapter 8.) There is another rainfall maximum in the midlatitudes associated with extratropical storm tracks in either hemisphere. The tropical rainfall maximum, especially the one north of the equator, is called the *intertropical convergence zone* (ITCZ) for reasons to be explained later. In the zonal mean, the ITCZ moves back and forth across the equator, following the seasonal march of the Sun (Fig. 3.1). It is interesting to note that while the ITCZ rainfall peaks in summer, midlatitude precipitation in storm tracks peaks in winter, indicating their differences in formation. Large amounts of latent heat of condensation are released in tropical rain bands. The heating in turn drives atmospheric (e.g., Hadley) circulation. This implies positive feedback between tropical convection and circulation, which generates spontaneous oscillations (see Chapter 4).

In the tropics, precipitation is mostly associated with deep convective systems, including cumulonimbus clouds that extend from the boundary layer to the tropopause. Such systems include narrow convective cores that often overshoot above the tropopause

Coupled Atmosphere-Ocean Dynamics: from El Niño to Climate Change
ISBN 978-0-323-95490-7, https://doi.org/10.1016/B978-0-323-95490-7.00003-5

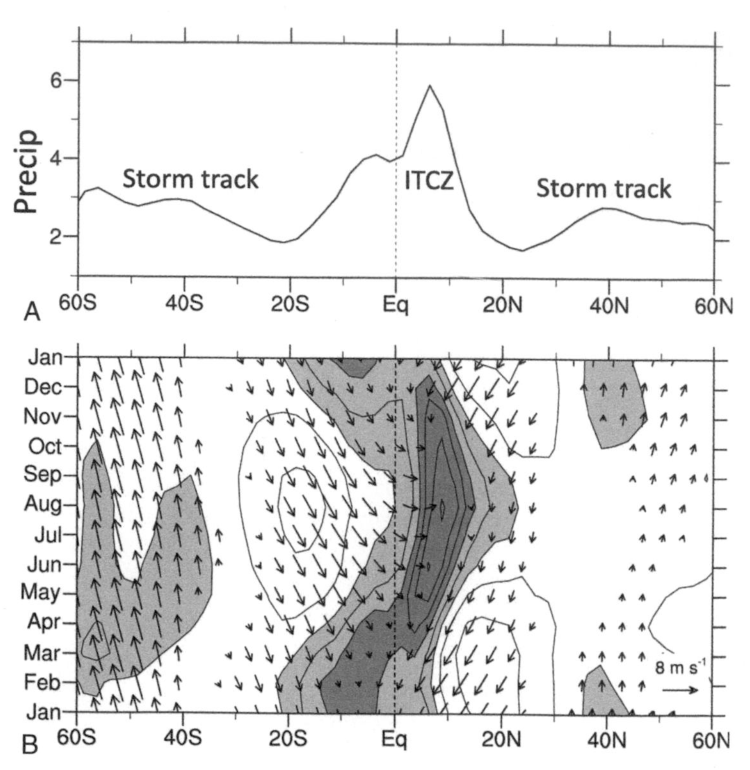

Fig. 3.1 Zonal-mean precipitation (mm/day) climatology. (A) Annual mean as a function of latitude. (B) Latitude-time diagram (light gray >3 mm/day, dark gray >5 mm/day; black contours at 1 mm/day interval) with zonal-mean surface wind velocity (v, u) (m/s). *(Courtesy ZQ Zhou.)*

and expansive anvil clouds with a flat cloud top capped by the tropopause (Fig. 3.2). From satellite infrared radiometers, deep convection is seen of cold cloud tops with very low outgoing longwave radiation (OLR) (Fig. 3.3B). In the monthly or longer mean, OLR = 250 W/m^2 marks the margins of tropical rain bands.

Precipitation averaged over 1 month or longer is organized into zonally oriented bands (see Fig. 3.3A). There is one ITCZ each in the tropical Indian, Pacific, and Atlantic oceans. In the Southern Hemisphere, the South Pacific convergence zone (SPCZ) extends from New Guinea toward the subtropical Southeast Pacific, and the South Atlantic convergence zone (SACZ) from the Amazon toward the midlatitude South Atlantic. The southeastward slanted rain bands are a hybrid of tropical convection and precipitation induced by the modulations of Rossby waves by a decelerating subtropical westerly jet (van der Wiel et al. 2015).

3.1 Water vapor budget

Tropical rain bands (e.g., the ITCZ) are associated with strong surface/low-level wind convergence as evident in ship-based and satellite observations of sea surface wind (see

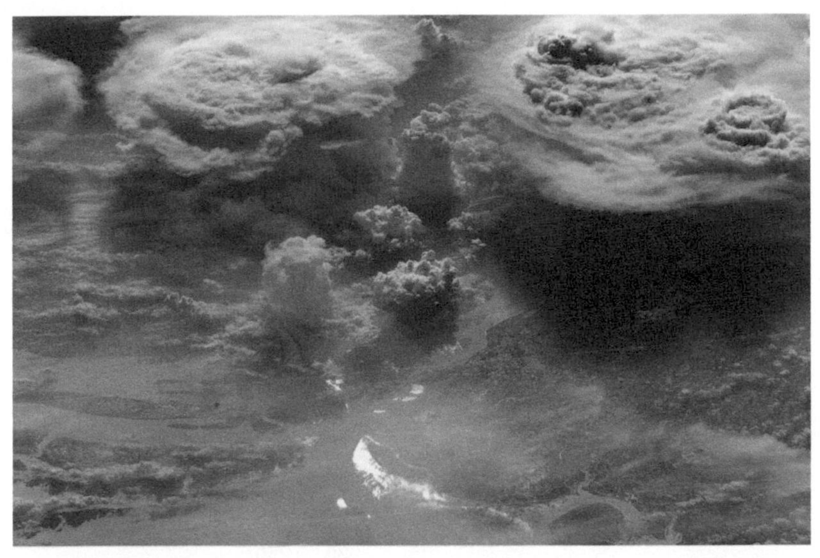

Fig. 3.2 Monsoon clouds over Hatia Island and Bangladesh on June 3, 2002 from the International Space Station. Flat anvil tops indicate the tropopause and emit much lower outgoing longwave radiation than the background. *(Image from NASA.)*

Fig. 3.3). For this reason, meteorologists often use the terms *tropical rain bands* and *convergence zones* interchangeably.

To explain this close association between tropical rainfall and surface wind convergence, we cast the moisture conservation equation in the flux form and integrate vertically through the atmospheric column

$$\partial\langle q\rangle/\partial t+\nabla\cdot\langle \boldsymbol{u}q\rangle = E-P \tag{3.1}$$

where the angular brackets denote the vertical integral $\langle \bullet\rangle\equiv\int_0^{ps}(\bullet)dp/g$ and average over a month or longer ($\partial\langle q\rangle/\partial t\approx 0$), $\boldsymbol{u}$ is the horizontal wind velocity vector, and q specific humidity in kg water vapor per kg air. In contrast to precipitation (P) that is confined in narrow rain bands, surface evaporation (E) is flat in space. As a result, the net water flux at the sea surface, $E-P$, largely follows the spatial patterns of precipitation. $P-E$ is strongly positive in the rain bands, indicating the importance of moisture convergence according to Eq. 3.1.

Let

$$\widetilde{u}\equiv\int_0^{ps} uqdp\Big/\int_0^{ps} qdp \tag{3.2}$$

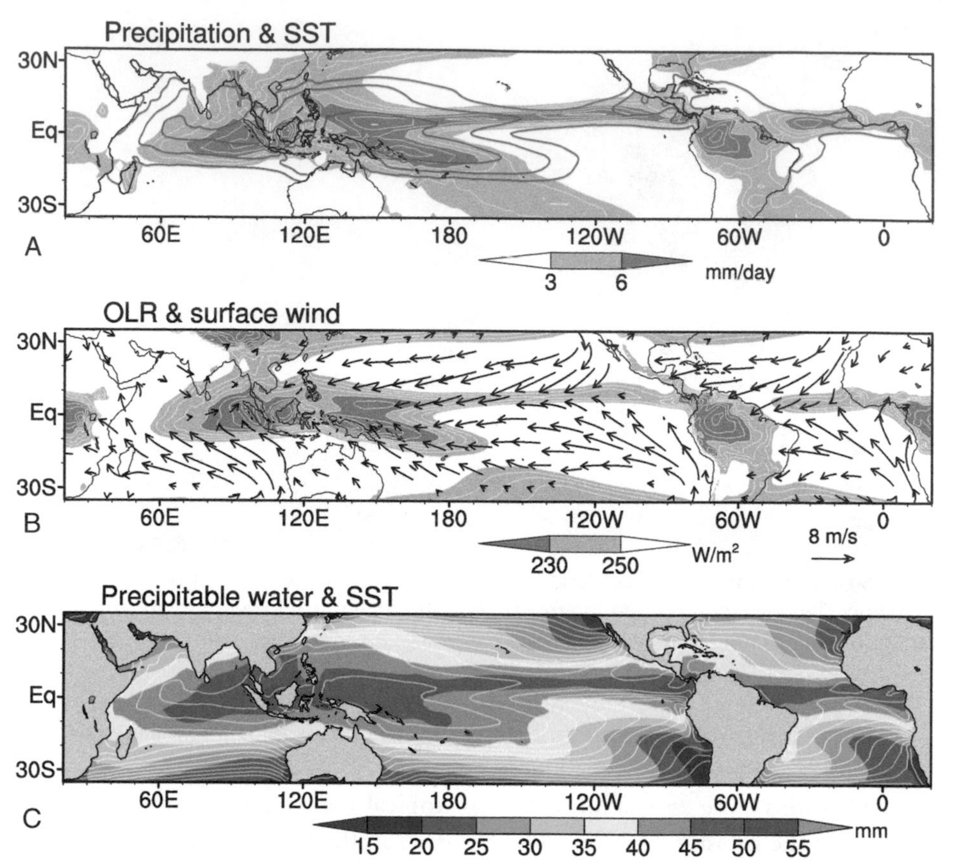

Fig. 3.3 Annual mean climatology. (A) Precipitation (grey shading, white contours at 1 mm/day interval) and sea surface temperature *(SST)* (red contours for 27°C, 28°C, and 29°C); (B) outgoing longwave radiation (*OLR*; grey shading, white contours at 5 W/m^2 interval) and surface wind velocity; (C) column-integrated precipitable water (color shading) and SST (white contours from 18°C to 29°C at 1°C interval). *(Courtesy ZQ Zhou.)*

and $\langle q \rangle \equiv \int_0^{ps} q/\rho_o dp/g$ is the column-integrated vapor path or total precipitable water often expressed in equivalent water height $w = \langle q \rangle / \rho_o$ in unit of mm, where ρ_o is the water density. The water balance for an atmospheric column is expressed as

$$\nabla \cdot [\tilde{\boldsymbol{u}} \langle q \rangle] = E - P \tag{3.3}$$

Note that $\tilde{\boldsymbol{u}}$ is the q-weighted average in the vertical. Since q is strongly trapped near the surface (Fig. 3.4), $\tilde{\boldsymbol{u}}$ represents the low-level wind and may be approximated by vertical-mean wind from the surface to 800 mb. This explains why low-level winds, not the winds at 500 hPa, matter for water vapor transport convergence and precipitation. The moisture convergence is further decomposed into terms of wind convergence and horizontal advection

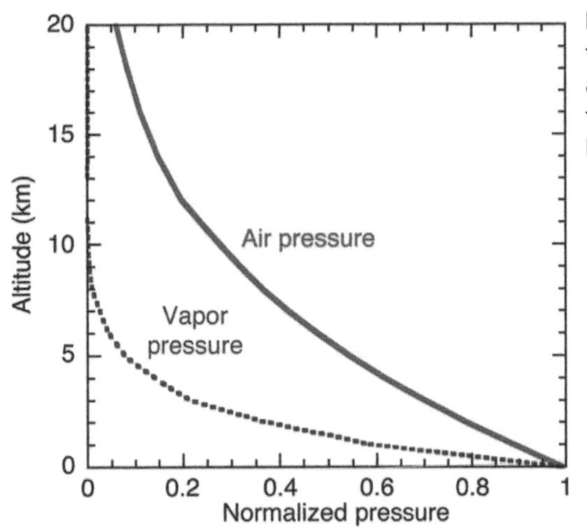

Fig. 3.4 Air pressure and partial pressure of water vapor as functions of altitude for globally and annually averaged conditions, normalized by the surface values of 1013.25 hPa and 17.5 hPa, respectively. *(From Hartmann [2016].)*

$$\nabla \cdot [\widetilde{\boldsymbol{u}}\langle q\rangle] = \langle q\rangle\nabla\cdot\widetilde{\boldsymbol{u}} + \widetilde{\boldsymbol{u}}\cdot\nabla\langle q\rangle \tag{3.4}$$

Let U be the velocity scale, L the horizontal scale, W the typical precipitable water, and δW the horizontal variations. The ratio of the second to first term on the right-hand side (RHS) is

$$\delta W \;\; U/L : W \;\; U/L \sim \delta W : W$$

Satellite observations of precipitable water show $\delta W/W \sim 0.2$ to 0.3 in the tropics (see Fig. 3.3C). Thus low-level wind convergence dominates the moisture convergence, with Eq. 3.3 approximated as

$$\langle q\rangle\nabla \cdot \widetilde{\boldsymbol{u}} \approx E - P \tag{3.5}$$

This explains the robust collocation of surface wind convergence and precipitation in the tropics.

The collocation of rain bands with low-level wind convergence seems to contradict our experience with thunderstorms in the late afternoon of a sunny summer day. In a thunderstorm, downpours are associated with divergent wind gusts in cold pools due to the melt of hailstones and evaporation of falling raindrops. On small scales of thunderstorms, the advection of moisture and hydrometeors is very important, and the approximation (see Eq. 3.5) is not valid. The advection is also important at the large scale in the extratropics because of increased horizontal gradients of water vapor (see Fig. 3.3C). The approximation Eq. 3.5 is valid in the tropics for temporal averages over 1 week or longer.

Globally averaged, precipitable water is about 20 mm. With an average rain rate of 1 m/year, the average residence time $W/P \sim 7$ days, which represents the average time for water vapor evaporated from the surface to fall back as precipitation.

3.2 Ocean temperature effect on convection

If one plots rainfall and sea surface temperature (SST) together, it is obvious that major tropical precipitation is confined to warm waters of SST greater than 27°C (see Fig. 3.3A). This is also consistent with the empirical observation that tropical cyclones (organized deep convection) require SST greater than 26.5°C to form (see Chapter 1). This strong SST dependency of deep convection indicates a tight coupling of the ocean and atmosphere in the tropics, which gives rise to El Niño and the Southern Oscillation (see Chapter 9).

Fig. 3.5 shows a typical atmospheric sounding over the warm ocean with frequent deep convection. The troposphere is stably stratified for dry convection (potential temperature increases upward, $d\theta/dz > 0$) but conditionally unstable for deep convection (saturation equivalent potential temperature decreases upward, $d\theta_e^*/dz < 0$) in the lower troposphere. Consider an adiabatic, rising air parcel without mixing dry ambient air. The temperature of the air parcel rising from the surface decreases at the dry lapse rate $\Gamma_d = 9.8$ K/km, eventually causing condensation and cloud formation at the lifting condensation level (LCL). At this point, the air parcel is still cooler than the surrounding. Lifted further up, the parcel is warmed by condensational heat. At the level of free convection (LFC), its temperature equals the ambient. Further above, the air parcel is warmer than the ambient because of the condensational heat, and the positive buoyancy allows free convection. The rising parcel loses buoyancy at the level of neutral buoyancy (LNB), where the convective plume detrains and forms anvil clouds with a flat top. Such deep convection is conditioned on an initial lift sufficiently large to raise the plume above the LFC. The initial lift could be

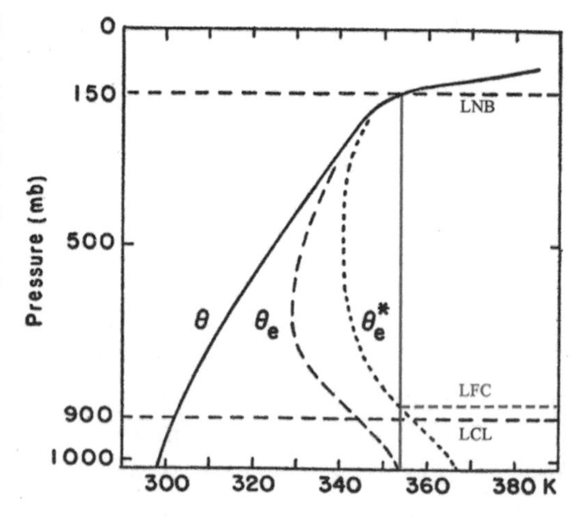

Fig. 3.5 Typical profiles over warm tropical oceans of potential temperature, equivalent potential temperature, and saturation equivalent potential temperature. For a rising air parcel (red vertical line) from the surface, the lifting condensation level *(LCL)*, level of free convection *(LFC)*, and level of neutral buoyancy *(LNB)* are marked. *(Adapted from Ooyama [1969]. © American Meteorological Society. Used with permission.)*

due to small-scale turbulence, organized cloud clusters, synoptic weather disturbances, and/or large-scale surface wind convergence such as the ITCZ. In addition to providing the initial lift, surface wind convergence helps the moist convection by increasing relative (and specific) humidity and boosting θ_e in the surface boundary layer. These are the additional reasons for why low-level convergence drives deep convection.

The temperature profile in the free troposphere is surprisingly uniform horizontally within the tropics (see discussion in Section 3.6), but θ_e at the surface and in the marine boundary layer is largely determined by SST as relative humidity (R_H) at the sea surface is around 80% when averaged over 1 month or longer. For a small change in surface temperature δT, we linearize the surface equivalent potential temperature $\theta_e = \theta\, exp\left(\frac{Lq}{c_p T_0}\right)$

$$\frac{\delta\theta_e}{\theta_e} = \frac{\delta m}{c_p T} = (1 + b_e)\frac{\delta T}{T} \tag{3.6}$$

where

$$b_e = \alpha\frac{Lq}{c_p}\sim 2.5 \tag{3.7}$$

is the inverse Bowen ratio (latent to sensible heat), $q = R_H q_s$, q_s the saturation mixing ratio following the Clausius-Clapeyron (CC) equation $\delta q_s/q_s = \alpha\delta T$ with the CC coefficient $\alpha = 0.06\ K^{-1}$, and T_0 the typical surface air temperature. Here we have used $R_H \sim 0.8$ and $q_s \sim 20.8$ g/kg at $T_0 = 26°C$. In other words, at the sea surface, a 1° rise in air temperature leads to an increase of 3.5 K in equivalent potential temperature. The amplification, of course, is due to the latent heat in water vapor that increases rapidly with temperature.

Moist convection neutralizes the conditional instability. In the midtroposphere, temperature stratification is indeed close to being moist adiabatic with $d\theta_e^*/dz \sim 0$, but θ_e^* is considerably smaller than θ_e at the surface or LCL due to entrainment of dry and low moist static energy (MSE) air from the ambient (Zhou et al. 2019). The convection advects water vapor from the moist boundary layer, rendering the convective plume of much higher humidity than the ambient. The puffy, popcorn-like edges of a cumulus cloud indicate violent mixing with the ambient generated by turbulence in the positively buoyant plume with strong upward motions.

With a reduced SST, an undiluted air parcel of smaller surface θ_e can only reach a lower LNB. For smaller surface θ_e than the midtropospheric θ_e* minimum, a rising air parcel never becomes positively buoyant and cannot convect freely. This explains the SST threshold for deep convection (Fig. 3.6A). Here we assume that the ambient temperature profile in the free troposphere does not depend on the local SST, which turns out to be true due to weak temperature gradient (Section 3.6). Considering entrainment

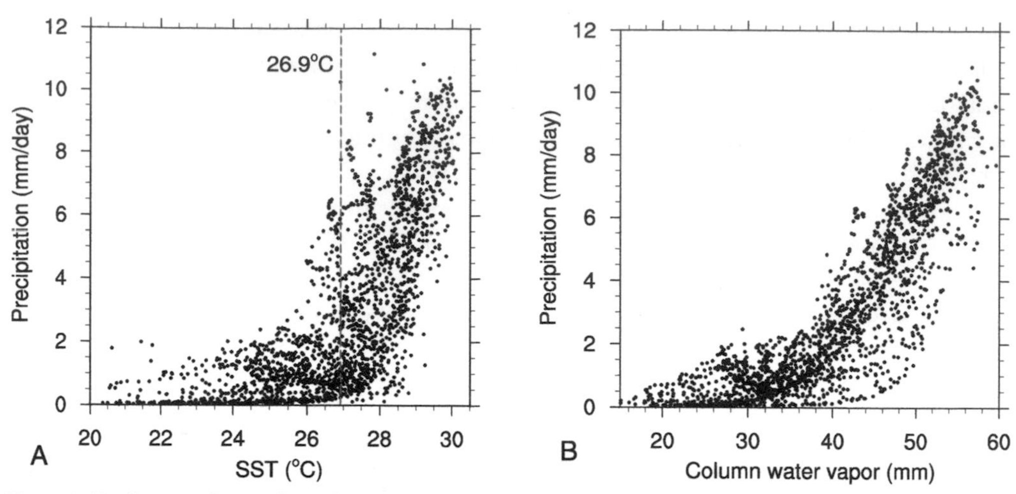

Fig. 3.6 Nonlinear relationship of monthly precipitation with (A) sea surface temperature *(SST)* and (B) column-integrated water vapor over tropical oceans (23.5°S–23.5°N) during December to February based on the 1997–2018 climatology on a 2.5° × 2.5° grid. The red dashed line marks the tropical mean SST (26.9°C). *(Courtesy Y Liang.)*

of dry ambient air requires even higher surface θ_e and hence SST for a convective plume to become positively buoyant and reach the tropopause. Likewise, a nonlinear relationship exists between the column-integrated water vapor and precipitation, with a threshold $w_c = 40$ mm (see Fig. 3.6B). By reducing the buoyancy loss from lateral entrainment, a moist atmospheric column is favorable for deep convection.

The SST effect on the height of convection is clear from the OLR distribution (see Fig. 3.3B): OLR $= 230\ \text{W/m}^2$ in the western Pacific and $250\ \text{W/m}^2$ in the ITCZ over the eastern Pacific and Atlantic where SSTs are lower and lateral mixing is stronger with water vapor decreasing sharply on either side.

3.3 Latent heat release in convection

The latent heating in deep convection can be diagnosed from observations. From the governing equations for dry static energy and water vapor, we define

$$Q_1 \equiv \frac{ds}{dt} = \frac{\partial s}{\partial t} + \nabla \cdot (s\boldsymbol{u}) + \frac{\partial}{\partial p}(s\omega) = Q_R + L(c-e) - \frac{\partial}{\partial p}\overline{s'\omega'} \tag{3.8}$$

$$Q_2 \equiv -L\frac{dq}{dt} = -L\left[\frac{\partial q}{\partial t} + \nabla \cdot (q\boldsymbol{u}) + \frac{\partial}{\partial p}(q\omega)\right] = L(c-e) + L\frac{\partial}{\partial p}\overline{q'\omega'} \tag{3.9}$$

Here Q_1 is called the apparent heat source, Q_2 the moisture sink, Q_R the radiative heating, c the condensation of water vapor, and e the evaporation. The prime denotes high-frequency turbulence unresolved by the soundings, and the overbar the time average. Q_1 and Q_2 are often diagnosed from an atmospheric sounding array and/or gridded reanalyses to infer how the atmospheric column is heated. The diagnosed Q_1 distribution can be imposed as diabatic heating to drive dry atmospheric models and study the circulation response (Section 3.5).

Fig. 3.7 shows the results over the convective region of the tropical Northwest Pacific based on a sounding network near Marshall Islands (165°E, 10°N). Radiation cools the tropospheric column at a rate of 1 to 2 K/day. Q_1 is much larger and positive, with a deep heating profile that peaks at 400 to 500 hPa. This is due to condensational heat in convection as indicated by large, positive Q_2 in the column. The Q_2 profile is consistently more bottom heavy than Q_1. From Eqs. 3.8 and 3.9, we have

$$Q_1 - Q_2 - Q_R = -\frac{\partial}{\partial p}\overline{m'\omega'} = g\frac{\partial F}{\partial p} \tag{3.10}$$

where $m = s + Lq$ is the moist static energy, and F is the upward turbulent transport of total energy

$$F(p) \equiv -\overline{m'\omega'}/g = \int_{p_T}^{p}(Q_1 - Q_2 - Q_R)dp/g \tag{3.11}$$

At the cloud top ($p = p_T$), the turbulent transport vanishes. Thus the upward displacement of Q_1 relative to Q_2 indicates that turbulent convection generates an upward flux of total energy at midlevels. At the surface,

$$F_0 = Q_H + LE \tag{3.12}$$

where Q_H is the sensible heat flux, and E the surface evaporation.

In the subsidence region over the southeast Pacific, radiative cooling is nearly vertically uniform from the top of the boundary layer to the tropopause (Fig. 3.8, *right*).

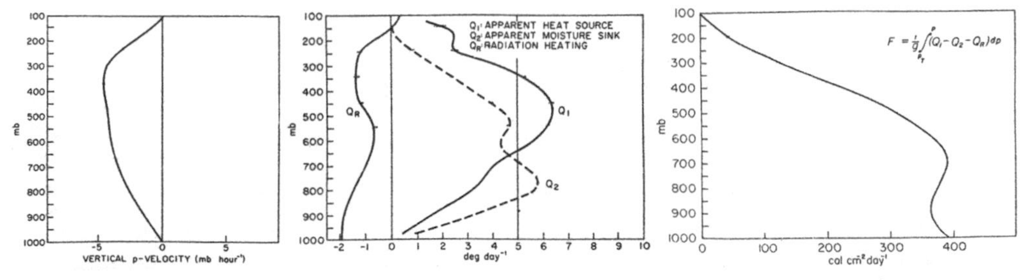

Fig. 3.7 Vertical profiles averaged from four-time daily soundings for April 15 to July 22, 1956, at Marshall Islands (165°E, 10°N): *(left)* pressure velocity; *(middle)* Q_1, Q_2 and radiative cooling; and *(right)* upward turbulent flux of MSE. *(From Yanai et al. [1973]. © American Meteorological Society. Used with permission.)*

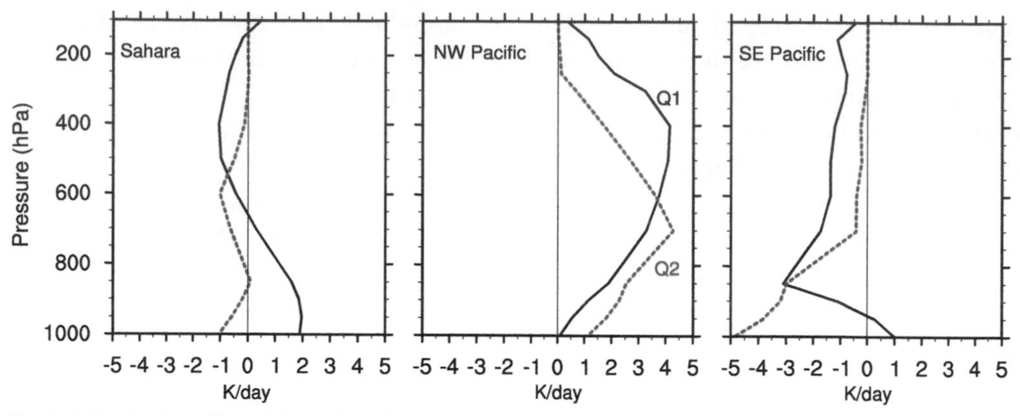

Fig. 3.8 Vertical profiles of Q_1/c_p (*solid black*, K/day) and Q_2/c_p (*dashed blue*, K/day) for June to August climatology over the Sahara Desert (18°E, 21°N), the Northwest Pacific warm pool (145°E, 5°N), and the Southeast Pacific (90°W, 5°S). *(Courtesy ZQ Zhou.)*

Positive Q_1 at the bottom indicates a sensible heating by the ocean, while negative Q_2 indicates surface evaporation, which moistens the lower troposphere above the boundary layer as indicated by the sharp vertical gradient in Q_2. Over the dry Sahara Desert, Q_2 is nearly 0 throughout the atmosphere. The lower atmosphere is heated by sensible heat (positive Q_1 trapped near the surface) while radiative cooling dominates above the boundary layer (see Fig. 3.8A).

Vertically integrating Eqs 3.8 and 3.9 yields

$$\langle Q_1 \rangle = \langle Q_R \rangle + LP + Q_H \quad (3.13)$$

is the column-integrated diabatic heating, and

$$\langle Q_2 \rangle = L(P - E) \quad (3.14)$$

is the column-integrated moisture sink. In the tropical convective region, latent heating due to precipitation dominates and as a result, $<Q_1>$ and $<Q_2>$ both resemble the precipitation distribution (see Fig. 3.3). $<Q_1>$ and $<Q_2>$ are not identical and their difference is

$$<Q_1 - Q_2> = <dm/dt> = R_{TOA}\downarrow - R_{sfc}\downarrow + Q_H + LE = R_{TOA}\downarrow - Q_{net} \quad (3.15)$$

where R is the radiative flux and Q_{net} is the net surface heat flux (downward positive). Vertical integrated $Q_1 - Q_2$ difference represents the net vertical energy flux into the atmospheric column (at the TOA and surface), which is balanced by the horizontal divergence of atmospheric energy transport $<dm/dt>$ (see also Eq. 2.15). In warm, convective regions (e.g., the tropical Northwest Pacific), $R_{TOA}\downarrow \sim 100\ \mathrm{W/m^2}$

(because of small OLRs from deep convective clouds), while the net surface heat flux is weakly downward (to balance the vertical mixing with the thermocline water), Q_{net} ~20 W/m^2. As a result, <$Q_1 - Q_2$> ~80 W/m^2.

The vertical profile of convective heating (Q_1) is observed to depend on SST since convection reaches higher altitudes over warmer SSTs. The convective heating rate peaks in the mid-upper troposphere (400–500 hPa) in the western Pacific warm water pool (SST ~29°C) while it peaks at the lower troposphere (600–700 hPa) in the marine ITCZ over the Northeast Pacific and Atlantic warm pools with cooler SSTs (~27°C) (Fig. 3.9). Since latent heating rate is nearly balanced by vertical motion (which is diagnostically linked to horizontal convergence by continuity), wind convergence is much deeper in the warm western than in the cooler eastern Pacific. In the tropical Northeast Pacific, winds converging onto the ITCZ are confined in an inversion-capped boundary layer about 1 km deep (shallow convergence) both from the north and south. SSTs are low outside the narrow, zonal warm band that anchors the ITCZ.

3.4 Equatorial waves

Atmospheric response in temperature, pressure, and wind to convective heating is shaped by large-scale waves under the strong influence of Earth rotation. Waves can propagate to faraway places. Swells surfers enjoy in summer off the south-facing beaches of Hawaii and California come all the way from the stormy Southern Ocean near Antarctica, generated

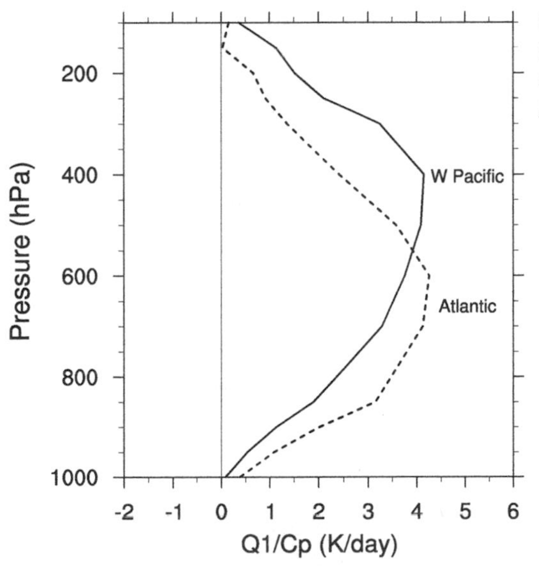

Fig. 3.9 Comparison of Q_1 profiles over the tropical western Pacific (145°E, 5°N) and Northeast Atlantic (40°W, 6°N, *dashed*). *(Courtesy ZQ Zhou.)*

by howling winds. This example illustrates that actions in one location can have major impacts on remote regions.

Throwing a rock in a deep pond generates circular wave fronts propagating outward. The circular wave form reflects that deep water is an isotropic media to gravity waves (wave properties, such as phase speed, do not change with direction). In a shallow pond, the bathymetry affects the wave speed and the shape of wave fronts.

What happens if one throws a big rock in the equatorial Pacific? Fig. 3.10 shows the dispersion of a thermocline depth depression (see Chapter 8 for ocean adjustment), 1000 km across centered on the equator. (This is also equivalent to the dispersion of a tropospheric temperature anomaly as will become clear soon.) The waves generated are clearly anisotropic. Perturbations propagate only in the east-west direction and are trapped in the north-south by the equator. The perturbations are further distinct in structure and phase speed between eastward and westward propagation. The equator serves as a waveguide for large-scale ocean-atmospheric disturbances because the Coriolis parameter $f = 2\Omega \sin\varphi$ vanishes there.

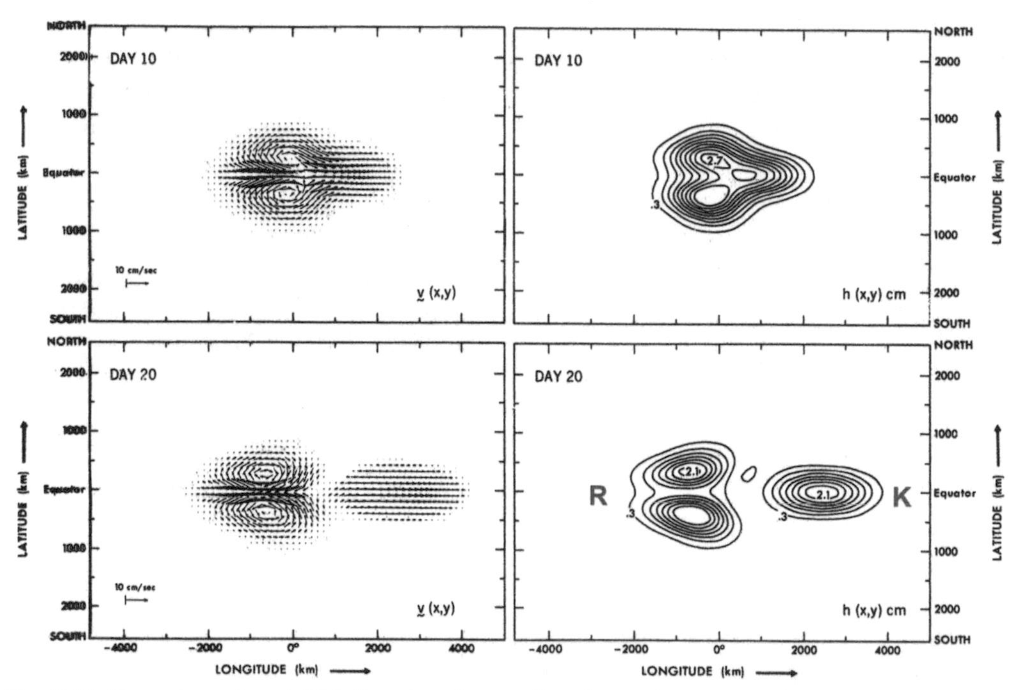

Fig. 3.10 Dispersion of an initial Gaussian thermocline depression centered at $x = 0$, equator, of e-folding scale of 500 km in both directions with no flow. *(left)* Current velocity and *(right)* thermocline depth anomalies. Geostrophy in the meridional direction prevents equatorial perturbations from propagating poleward. In the right bottom panel, K denotes the Kelvin wave, and K the Rossby waves. *(From Philander et al. [1984]. © American Meteorological Society. Used with permission.)*

3.4.1 Two-level model

The atmospheric flow in the tropics tends to have a baroclinic structure in the vertical (Fig. 3.11), with the horizontal velocity opposite in direction between the lower and upper troposphere while vertical velocity perturbations peak at the midtroposphere (e.g., the Hadley circulation). For such baroclinic disturbances, the wind convergence at the low level is associated with the wind divergence at the upper level, resulting in upward motion that peaks at the midlevel. The vertical velocity vanishes at the surface and tropopause (the rigid lid approximation).

A two-level model is often used to represent the baroclinic motion (Section 8.2 of Holton 2004; Chapter 11 of Gill 1982). The governing equations for horizontal velocity (u, v) and geopotential Φ at the low level $(p = p_3)$ are

$$\frac{\partial u}{\partial t} - fv = -\frac{\partial \Phi}{\partial x} \tag{3.16}$$

$$\frac{\partial v}{\partial t} + fu = -\frac{\partial \Phi}{\partial y} \tag{3.17}$$

$$\frac{\partial \Phi}{\partial t} + c^2 \nabla \cdot \boldsymbol{u} = -Q \tag{3.18}$$

$$c^2 = \left(\frac{p_2}{p_s}\right)^{\kappa} R \frac{\overline{\theta}_1 - \overline{\theta}_3}{2} = R\overline{T}_2 \frac{\overline{\theta}_1 - \overline{\theta}_3}{2\overline{\theta}_2} \tag{3.19}$$

where $c \sim 50$ m/s is the long gravity wave phase speed, and $Q = \kappa J$ is the diabatic heating rate at the midlevel (p_2). The Coriolis parameter may be approximated as $f = \beta y$ (the equatorial beta plane approximation), where $\beta = df/dy$ at the equator, and y is the northward distance from the equator. Here we assume a baroclinic structure (e.g., $\Phi = \Phi_3 = -\Phi_1$). From the hydrostatic approximation, the thickness is proportional to the tropospheric temperature,

$$\Phi_1 - \Phi_3 = RT_2 \tag{3.20}$$

Eqs. 3.16 through 3.18 are known as the shallow-water model of an equivalent depth

$$H_e = c^2/g \tag{3.21}$$

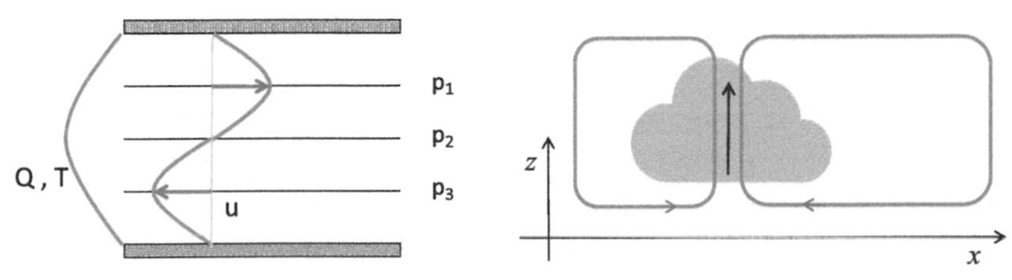

Fig. 3.11 *(left)* Two-level atmospheric model for *(right)* the baroclinic overturning circulation.

For internal waves of the baroclinic structure, the restoring force is the vertical difference in the mean potential temperature, $\overline{\theta}_1 - \overline{\theta}_3$. In Eq. 3.19, $R = 287$ J K^{-1} kg^{-1} is the gas constant for dry air, and $\kappa = R/c_p = 2/7$.

For convenience, we introduce nondimensional variables: $(x^*, y^*) = (x, y)/(c/\beta)^{1/2}$, $t^* = t(\beta c)^{1/2}$, $(u^*, v^*, \Phi^*) = (u/c, v/c, \Phi/c^2)$, where

$$R_E = (c/\beta)^{1/2} \sim 1500 \text{ km} \tag{3.22}$$

is the equatorial radius of deformation, a scale marking the meridional width of the equatorial waveguide. For planetary-scale (length scale >> R_E), low frequency (timescale >> $[\beta c]^{-1/2} \sim 0.3$ day) perturbations, the time derivative term in the meridional momentum equation can be dropped, known as the longwave approximation. The equations simplify in nondimensional form

$$\frac{\partial u}{\partial t} - yv = -\frac{\partial \Phi}{\partial x} \tag{3.23}$$

$$yu = -\frac{\partial \Phi}{\partial y} \tag{3.24}$$

$$\frac{\partial \Phi}{\partial t} + \nabla \cdot \boldsymbol{u} = -Q \tag{3.25}$$

Here we have omitted the asterisk for nondimensional variables for simplicity. Under the longwave approximation, the zonal flow and the meridional pressure gradient force are always in geostrophic balance, and no phase propagation is allowed in the meridional direction (see Fig. 3.10).

3.4.2 Kelvin wave

Consider the unforced problem first by letting $Q = 0$. There is a solution with $v = 0$,

$$(u, \Phi) = (u_0, \Phi_0)\exp[i(kx - \omega t)]\exp(-y^2/2) \tag{3.26}$$

where k is the zonal wavenumber and ω the angular frequency. In this so-called Kelvin wave solution, perturbation zonal velocity and pressure peak at the equator and decays poleward at an e-folding scale of $R_E/\sqrt{2}$ as in Eq. 3.26. The dispersion relationship $\omega(k)$ is

$$\omega = k \tag{3.27}$$

and the phase speed $c \equiv \omega/k = 1$. In dimensional space, it equals that of the long gravity wave ($c \sim 50$ m/s), with a circumglobal transit time of 9.3 days. The Kelvin wave is nondispersive (phase speed ω/k is not a function of k) and propagates eastward, with an equatorial trapped structure in the north-south direction.

In the zonal direction, the Kelvin wave ($v=0$) is a gravity wave ($\partial u/\partial t = -\partial \Phi/\partial x$) as if it did not feel the effect of Earth rotation. The zonal velocity and pressure are in phase; the easterly flow is associated with a low pressure (Fig. 3.12B). In the meridional direction, the Kelvin wave is in geostrophic balance ($fu = -\partial \Phi/\partial y$), and the poleward evanescent structure is due to Earth rotation. The Kelvin wave is sometimes called the semigeostrophic wave because it is geostrophic in the meridional but not in the zonal direction.

Mathematically, the westward propagating wave is also a solution, but it is unphysical because it entails a poleward growing structure.

A solid coast prohibits the cross-shore flow. In the ocean, coastal Kelvin waves with vanishing cross-shore velocity are commonly observed, propagating in the direction with the solid coast on the right (left) in the Northern (Southern) Hemisphere. During El Niño, anomalously high sea levels are observed to propagate poleward as coastal Kelvin waves along the west coast of North America from the equator all the way to Alaska. Even without a solid boundary, the Kelvin wave exists on the equator because it is a singularity for geostrophy with $f=0$. An inviscid wall along the equator would not affect the Kelvin wave, which propagates eastward with the equator on the right (left) in the Northern (Southern) Hemisphere.

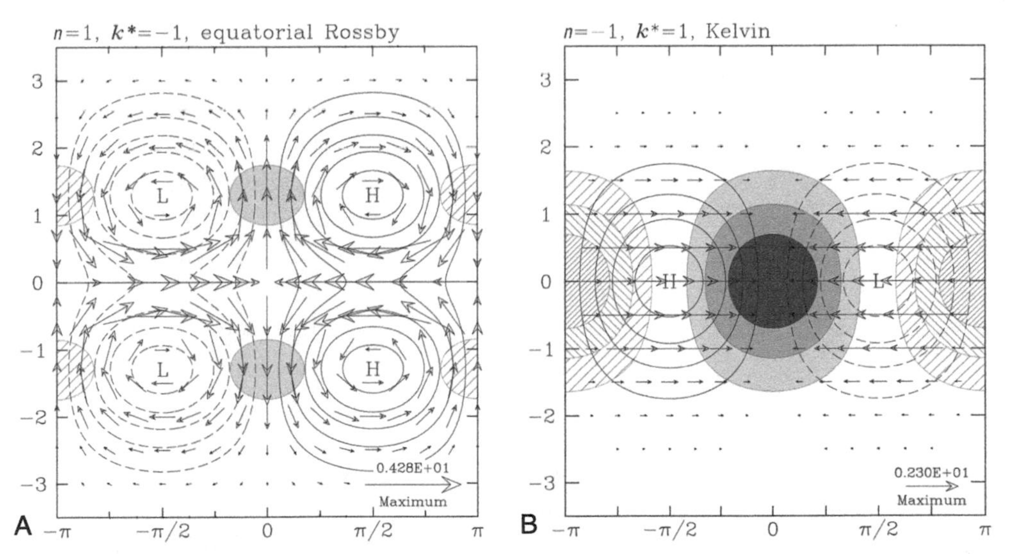

Fig. 3.12 Equatorially trapped (A) Rossby wave of $n=1$ and (B) Kelvin wave on nondimensionalized coordinates (equator $y=0$). Hatching is for divergence and shading for convergence. Line contours are for geopotential, with a contour interval of 0.5 unit. The largest wind vector is 2.3 units, as marked. The dimensional scales are as in Matsuno (1966). *(From Wheeler and Nguyen [2015].)*

3.4.3 Rossby waves

In general, $v \neq 0$. Eqs. 3.23 through 3.25 may be consolidated into one single equation for v

$$\frac{\partial}{\partial t}\left(y^2 v - \frac{\partial^2 v}{\partial y^2}\right) - \frac{\partial v}{\partial x} = 0 \tag{3.28}$$

The solutions are

$$v_n = 2^{-n/2} H_n(y) exp\left(-y^2 / 2\right) exp[i(kx - \omega_n t)] \tag{3.29}$$

$$\omega_n = -k/(2n+1), \; n = 1, 2, \ldots \tag{3.30}$$

The wave solutions are trapped by the equator (the Gaussian function in 3.29) and travel westward at phase speed $c_n = \omega_n/k = -1/(2n+1)$, much slower than the Kelvin wave. Here $H_n(y)$ are the Hermite polynomials,

$$H_0 = 1, \; H_1 = 2y, \; H_2 = 4y^2 - 2, \; H_3 = 8y^3 - 12y, \; \ldots \tag{3.31}$$

Rossby waves are symmetric in Φ and u about the equator for odd n, and antisymmetric for even n. Note that a symmetric wave features the meridional flow antisymmetric about and vanishing on the equator. The Rossby wave of the gravest meridional mode $n = 1$ is symmetric about the equator and travels westward at one-third of the Kelvin wave phase speed ($c_1 = -1/3$). The second Rossby wave mode is antisymmetric at $c_2 = -1/5$, and so on. The Kelvin wave may be viewed as a solution of mode $n = -1$, with $c_{-1} = 1$.

The Rossby waves have maximum pressure perturbations off the equator (see Fig. 3.12A). The wind and pressure perturbations are nearly in geostrophic balance; a cyclonic flow is associated with a low pressure, much like the midlatitude Rossby waves. For the first equatorial Rossby wave, the westerly wind is much larger on the equator than the easterly flow on the poleward flank of the off-equatorial low-pressure center because the Coriolis parameter decreases toward the equator. The meridional wind vanishes on the equator and peaks at $y = \pm 1$ (or $\pm R_E$ in dimensional space), in geostrophic balance with the off-equatorial pressure perturbations.

3.4.4 Wave dispersion

Now let us look back into the dispersion of an initial high-pressure perturbation in Fig. 3.10. In time, the isolated pressure perturbation symmetric about the equator splits into an eastward-propagating Kelvin wave with purely zonal flow and westward-propagating Rossby waves with anticyclonic circulation on either side of the equator. The Rossby waves propagate much more slowly westward than the Kelvin wave propagates eastward. In the meridional direction, wave propagation is not allowed because the zonal flow has fully adjusted in perfect geostrophic balance (see Eq. 3.24). This example illustrates that because of the meridional variations in the Coriolis parameter on a rotating sphere, not only the zonal and meridional directions are mutually distinct, but the east is also distinct from the west for planetary-scale perturbations, in phase propagation and structure.

The earlier discussion is based on the longwave approximation (see Eq. 3.24), which mathematically filters out shortwave and high-frequency perturbations. Matsuno (1966) obtained the full solutions to the shallow-water system (see Eqs. 3.16–3.18). Fig. 3.13 shows the full dispersion relationship in nondimensional zonal wavenumber-frequency space without making the longwave assumption. Here ω is always positive, but the zonal wavenumber k can be either positive (eastward propagating) or negative (westward propagating). The Kelvin wave solution remains intact, while the Rossby waves become dispersive with the westward phase speed decreasing with the increasing zonal wavenumber, a well-known property of midlatitude Rossby waves. On the high-frequency side $[\omega/(\beta c)^{1/2} > 1]$, there are inertial-gravity waves, or gravity waves under the influence of Earth rotation, of various meridional modes, $n = 1, 2, \ldots$. The dispersion curve that transverses the negative and positive wavenumber space represents the Yanai

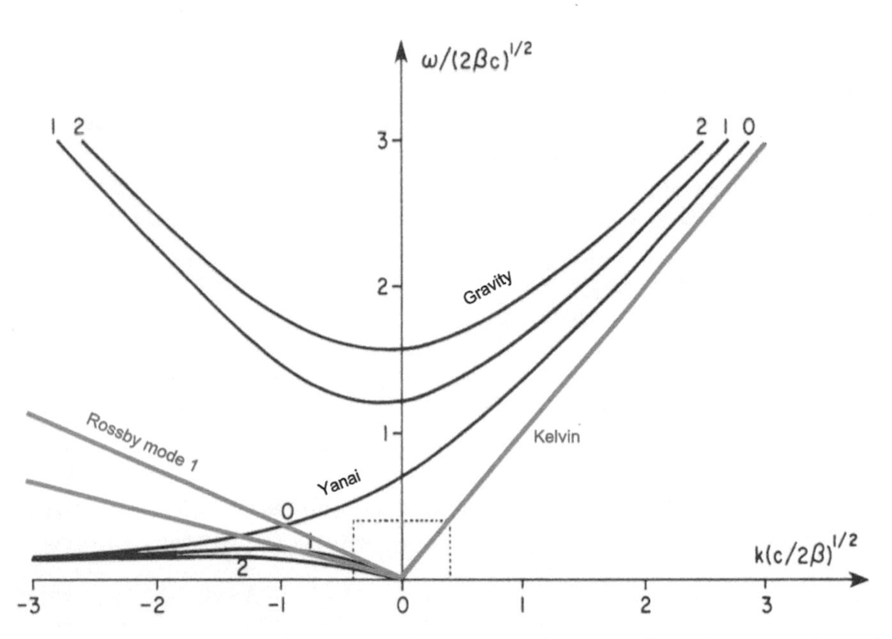

Fig. 3.13 Dispersion relationship for equatorial waves as a function of nondimensional zonal wavenumber and frequency. The dispersion curves under the longwave approximation in orange. The numerals on the dispersion curves denote meridional mode numbers. *(Adapted based on Gill [1982].)*

wave. Its wave structure is antisymmetric about the equator, with marked meridional flow but zero zonal flow on the equator. Its dispersion curve resembles gravity wave at the limit of large positive zonal wavenumber (far right side of Fig. 3.13) but resembles Rossby wave at the large negative wavenumber limit (far left side). For this reason, the Yanai wave is also called the mixed Rossby-gravity wave. It is unique to the equatorial waveguide and not found anywhere else.

Equatorial waves have been observed in the atmosphere (Box 3.1) and oceans (see Chapter 7). The equatorial wave theory is now an important foundation to understand and interpret ocean-atmospheric variability in the tropics. Strictly speaking, equatorial waves are dispersive ($c[k] \neq$ constant] except the Kelvin wave (see Fig. 3.13) but for large-scale, low-frequency variability that is the focus of this book, the longwave approximation (see Eq. 3.24) is valid and substantially simplifies the problem by filtering out inertial gravity and Yanai waves and making the Rossby waves nondispersive.

BOX 3.1 Discovery of equatorial waves

In November 1965, Taroh Matsuno submitted his paper on equatorial wave theory to the *Journal of the Meteorological Society of Japan* for publication.[12] It was part of his PhD thesis. He needed the degree to accept an offer of associate professorship at Kyushu University. Michio Yanai was the editor, and the Matsuno (1966) paper was swiftly published in February 1966. A few months later in October, Yanai and Maruyama (1966) published in the same journal an analysis of cross-equatorial wind variability now recognized as the Yanai wave. At the time, Matsuno and Yanai were on the faculty of the distinguished meteorology lab at University of Tokyo (Lewis 1993), respectively, as research associate and associate professor, but Yanai and Maruyama (1966) and Matsuno (1966) did not cross-reference each other. Soon afterward, Maruyama (1967), then a PhD student at the lab, identified the Yanai wave as a Matsuno solution.

On the other side of the Pacific at the University of Washington, Mike Wallace, a fledgling assistant professor just out of graduate school, and his student were analyzing the same atmospheric soundings over the tropical Northwest Pacific that Yanai and Maruyama used. Wallace and Kousky (1968) noted pronounced oscillations of 15- to 20-day periods in the lower-stratosphere zonal wind (Box Fig. 3A). James Holton, a professor at the same department, suggested that the oscillations might be equatorial Kelvin waves of Matsuno (1966) and asked about the meridional wind variability. In the power spectra on Kwajalein Island (see Box Fig. 3B) there is a sharp peak in the zonal wind but no power in the meridional wind! This marks the discovery of the equatorial Kelvin wave. Remarkably, Kwajalein is situated almost 1000 km away from the equator but dynamically still within the Rossby radius of deformation to detect the Kelvin wave.

BOX 3.1 Discovery of equatorial waves—cont'd

The equatorial wave theory opened a new chapter in dynamic meteorology, oceanography, and climatology, culminating in revolutionary studies of El Niño and the Southern Oscillation and other ocean-atmosphere coupled modes in the tropics. Remarkably, the discovery of equatorial waves originally had nothing to do with surface climate and was made in search of upward propagating waves believed to drive the quasi-biennial oscillation (QBO) in the stratosphere (Lindzen and Holton 1968), a phenomenon that was discovered just a few years earlier and whose amazing regularity fascinated meteorologists of the time.

I had the privilege of working with Matsuno in the mid-1990s at Hokkaido University, Japan, where he was the founding chair of the fledgling Department of Oceanic and Atmospheric Sciences. The department faculty lunched together regularly. Walking back from one of these lunches, I asked what motivated him to study equatorial waves. "I was just curious about what happens where *f* goes to zero," Matsuno reflected. Of course, he was well aware of the hypothesis that wave fluxes interact and drive the QBO at the equator. What a zoo of waves he discovered! On September 2010 in Seattle, I was in the audience for an incredible show with the dream cast: Yanai chaired the opening session and Matsuno gave the first talk at a symposium celebrating Wallace's 70th birthday.

Global observations of outgoing longwave radiation variability from space show a good fit with the dispersion relation of Matsuno's solutions, albeit at a slower phase speed due to latent heating in deep convection. Neglecting convective heating, Matsuno's theory misses an important mode of tropical variability known as the Madden-Julian oscillation (see Chapter 4).

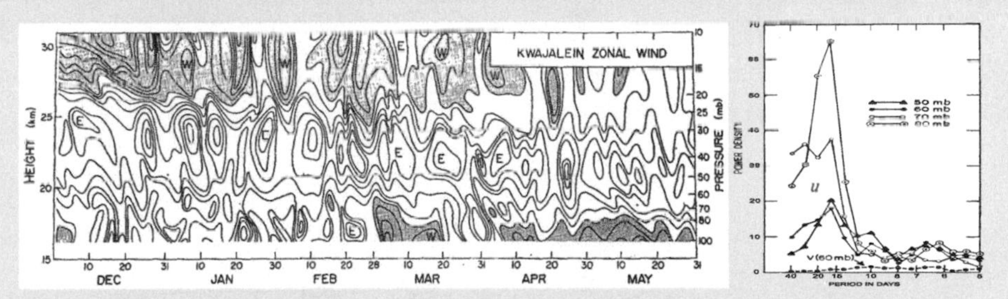

Box Fig. 3 *((A) Time-altitude section of zonal wind velocity (every 5 m/s, westerly shaded) in the lower stratosphere at Kwajalein Atoll (8°43′N, 167°44′E). (B) Power spectra of zonal* (solid line) *and meridional* (dashed line) *wind variability at Balbao (9°N, 80°W), Panama. From Wallace and Kousky [1968]. © American Meteorological Society. Used with permission.)*

Chapter 11 of Gill (1982) provides an elegant discussion of equatorial waves and their roles in the tropical circulation of the atmosphere and ocean. Chapter 11.4 of Holton (2004) also briefly discusses Matsuno wave theory.

3.5 Planetary-scale circulation

3.5.1 Response to an isolated heating

Convective heating peaks in the midtroposphere (see Fig. 3.7), warms the entire troposphere, and creates a high pressure aloft and a low-pressure perturbation at low levels. Sufficiently away from the equator, the warm core generated by the convective heating would be in thermal wind balance, with cyclonic flow at low levels and anticyclonic flow at the upper levels, as observed in a hurricane.

Gill (1980) considered the problem of how an isolated deep heat source drives the tropical circulation. For simplicity, we assume that the heating is stationary in time. The steady-state response to the heating is governed by equatorial wave equations under the longwave approximation on the beta-plane. From Eqs. 3.23 through 3.25, we obtain

$$\varepsilon u - yv = -\frac{\partial \Phi}{\partial x} \tag{3.32}$$

$$yu = -\frac{\partial \Phi}{\partial y} \tag{3.33}$$

$$\varepsilon \Phi + \nabla \cdot \boldsymbol{u} = -Q \tag{3.34}$$

where ε is the damping rate in the atmosphere (typically with a dimensional timescale of a few days). The damping on the left-hand side (LHS) of Eq. 3.34 may be viewed as due to radiation restoring tropospheric temperature toward the radiative-convective equilibrium. Assume that the heating is centered on the equator with a Gaussian profile in the meridional direction,

$$Q = Q_0(x)\exp\left(-y^2/2\right) \tag{3.35}$$

Let $q = \Phi + u$, and $r = \Phi - u$, and seek solutions

$$(q, r) = \sum [q_n(x), r_n(x)]\, D_n(y) \tag{3.36}$$

where parabolic cylinder functions $D_n(y) = 2^{-n/2} H_n(y)\exp\left(-y^2/2\right)$ are equatorially trapped and evanescent poleward. With a heating $Q_0(x)$ confined zonally to $-L < x < L$, the Kevin wave response

$$\varepsilon q_0 + \frac{\partial q_0}{\partial x} = -Q_0 \tag{3.37}$$

vanishes west of $x < -L$, grows eastward in the heating region, and decays east of $x = L$ with the zonal e-folding scale of $L_K = 1/\varepsilon$. On the other hand, the Rossby wave response

$$\varepsilon q_2 - \frac{1}{3}\frac{\partial q_2}{\partial x} = -\frac{1}{3}Q_0 \tag{3.38}$$

vanishes east of $x < L$, grows westward in the heating region, and decays west of $x = -L$ with the zonal e-folding scale of $L_R = 1/(3\varepsilon) = L_K/3$, which is three times shorter than that of the Kelvin wave east of the heating. The zonal velocity and pressure can be reconstructed from the Kelvin and Rossby responses (see Eqs. 3.37—3.38)

$$u = \left[q_0 + \left(2y^2 - 3\right)q_2\right]/2 \cdot \exp\left(-y^2/2\right) \tag{3.39}$$

$$\Phi = \left[q_0 + \left(2y^2 + 1\right)q_2\right]/2 \cdot \exp\left(-y^2/2\right) \tag{3.40}$$

The tropical atmospheric response on a rotating sphere is a hybrid of nonrotating and rotating response: It is meridionally trapped due to geostrophic balance (see Eq. 3.33) but extends far in the east-west direction along the equator (Fig. 3.14). On the equator, surface winds converge into the heat source from both the west and east, down the pressure gradient much like the nonrotating response, while off the equator to the west, surface winds are in geostrophic balance with pressure indicative of the strong influence of Earth rotation.

Because of the beta effect ($f = \beta y$), the response is highly asymmetric to the east and west (see Fig. 3.14). To the east, the Kelvin wave is excited with low pressure and easterly wind trapped on the equator at the low level. To the west, symmetric Rossby waves are excited, with a pair of cyclonic circulations astride the equator in geostrophic balance with low pressure centers off the equator. In the steady state (where the dissipation arrests the waves excited by the heating), the Kelvin wave to the east survives much farther away from the heat source, by a factor of 3 (the ratio of the Kevin:Rossby wave phase speed) or more.

An initial value problem helps appreciate the role of equatorial waves in the adjustment of an atmosphere initially at rest to the abrupt onset of a convective heating. Fig. 3.15 shows the longitude-time section of midtropospheric temperature response. To the east of the heating, the tropospheric warming is successively delayed, and the timing of the warming is consistent with the eastward propagation of the Kelvin wave front. Likewise, the successively delayed warming to the west of the heating is consistent with the slower westward propagation of the Rossby wave front. Such distance-time sections are called Hovmöller diagrams, named after the original creator, which are useful in tracking propagating signals. The Hovmöller diagram here shows that the Kelvin and Rossby wave fronts propagate out from the heating and are eventually frozen by dissipation. The frozen waves give rise to strong east-west asymmetry in temperature, pressure, and wind. The tropospheric warming due to the Kelvin wave to the east peaks on the equator while that due to the Rossby wave to the west peaks off the equator.

In the vertical transect on the equator, two circulation cells form (see Fig. 3.14C), due to the Kelvin and Rossby waves, respectively. The eastern cell has a much larger zonal extent than the western cell because of the larger phase speed of the Kelvin than Rossby wave. This two-cell vertical circulation is termed the Walker circulation. In the

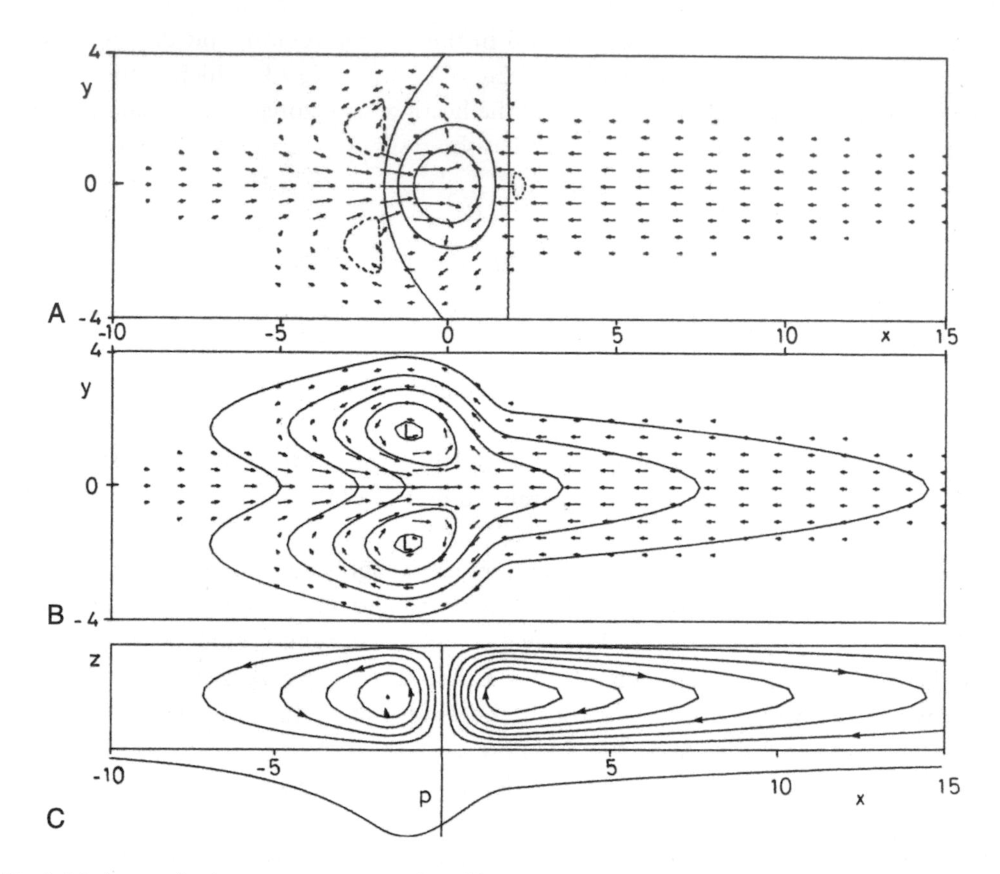

Fig. 3.14 Atmospheric response to an isolated heating centered on the equator. Low-level wind velocity along with (A) vertical velocity *(downward dashed)* and (B) low-level geopotential. (C) Overturning stream function and low-level geopotential on the equator. *(From Gill [1980].)*

convective region, latent heating is balanced by adiabatic cooling in the upward motion, while outside the heating region the adiabatic warming in the slow descent spreads the warm temperature in the form of equatorial waves.

The tropical atmospheric response to an isolated heat source is often called the Matsuno-Gill, or simply Gill pattern (see Fig. 3.14). The meridionally trapped and zonally asymmetric pattern is a good example illustrating the importance of equatorial waves. It also explains some important features of the observed tropical atmospheric circulation.

3.5.2 Observed tropical circulation

The annual-mean rainfall averaged within the Rossby radius of deformation (15°S−15 °N) shows pronounced zonal variations and is concentrated over the Indo-western Pacific where SSTs are high and above 27°C (see Fig. 3.3). Consistent with the zonally

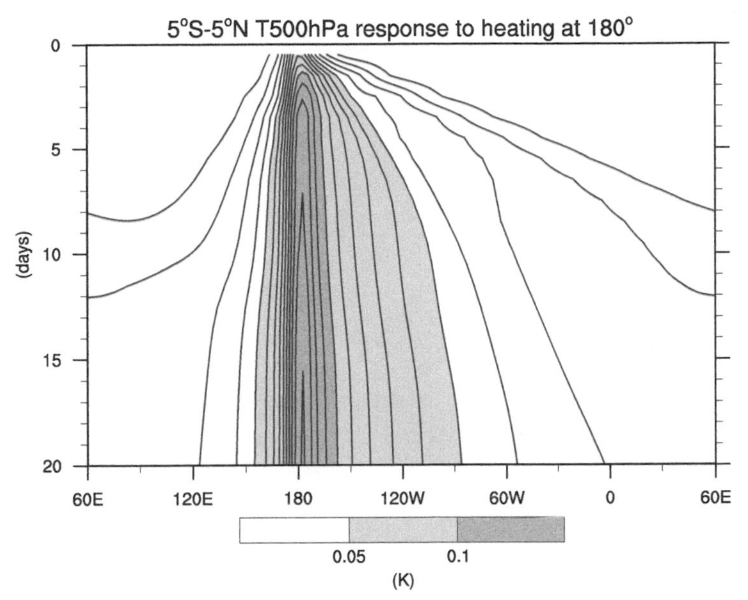

Fig. 3.15 Equatorial (5°S-5°N) 500-hPa temperature response to an abrupt onset of diabatic heating centered at 180° in a linear baroclinic model with the resting basic state. Contour interval: 0.01 K. *(Courtesy Y Kosaka.)*

enhanced convection over the Indo-western Pacific warm water pool, a two-cell Walker circulation is observed with the rising branch centered over the maritime continent (a region with large islands from Sumatra to New Guinea) (Fig. 3.16). As predicted by the Gill model, the Pacific cell is zonally more extensive than the Indian Ocean cell. The Walker circulation manifests strongly in the low-level wind field, which drives the upper ocean circulation and regulates SST as will be discussed in later chapters. To the east of the warm pool convection, the surface easterlies prevail over the tropical Pacific whereas weak westerlies prevail over the equatorial Indian Ocean (see Fig. 3.16). At low levels, the cyclonic Rossby gyre is visible over the tropical South Indian Ocean but not so much north of the equator because of the monsoonal winds that reverse direction between winter and summer (see Chapter 5). In the upper troposphere, a pair of the anticyclonic circulations is observed west of the Indo-western Pacific warm pool, on either side of the equator in the Indian Ocean sector. If the zonal-mean easterlies are removed, the westerly eddy flow in the upper branch of the Pacific Walker cell becomes clear. The upper-level mean westerlies over the eastern equatorial Pacific, called the westerly duct, allow low-frequency variability from the midlatitude storm tracks to propagate toward and across the equator (Chapter 7 of Webster 2020).

A similar Walker circulation, with reduced zonal extents, is observed in the Atlantic sector. At low levels, the easterlies prevail over the tropical Atlantic and converge onto

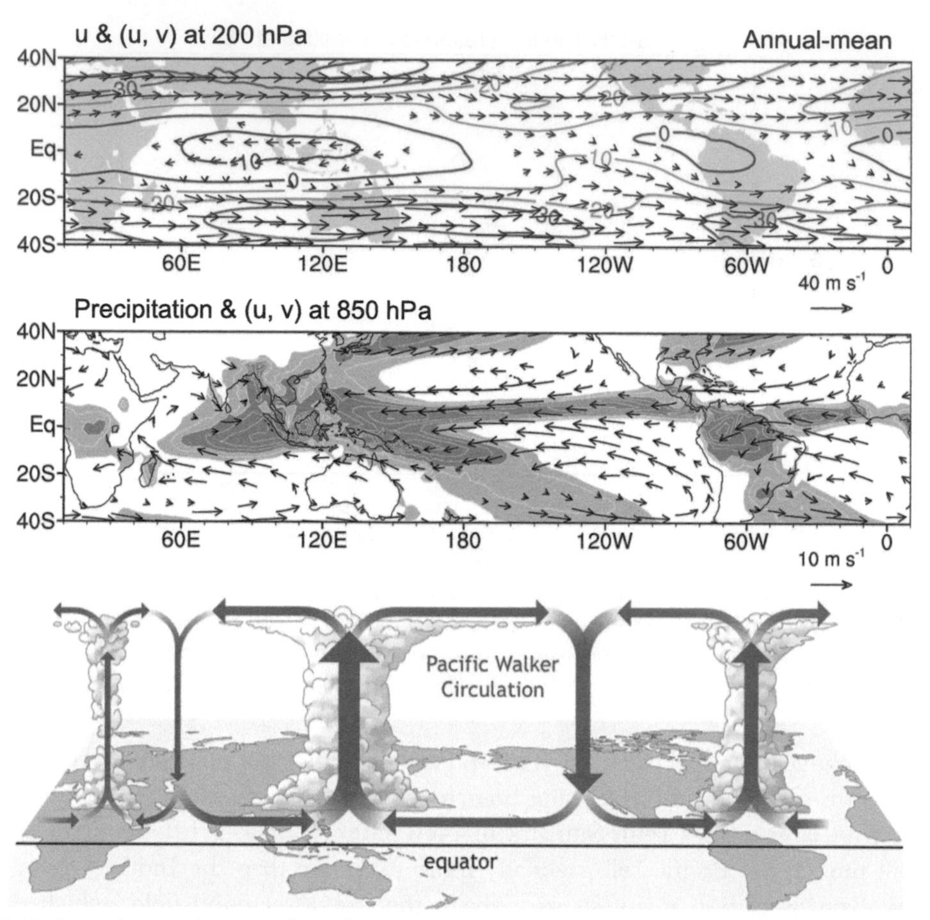

Fig. 3.16 Annual-mean horizontal wind velocity vectors (m/s) at *(top)* 200 hPa with zonal wind speed (contours), and *(middle)* at 850 hPa with precipitation (light gray >3 mm/day, dark gray >5 mm/day; white contours at interval of 1 mm/day). *(bottom)* Schematic overturning circulation on the equator. *(top/middle, Courtesy ZQ Zhou; bottom, NOAA.)*

convection over the Amazon. In the upper troposphere, the Amazon convection drives divergent flow that is easterly to the west and westerly to the east. As a result, there is another westerly duct in the tropical Atlantic.

It is quite remarkable that a simple linear model like Gill's explains several salient features of tropical circulation. The success of the model illustrates the importance of convective heating and Kelvin/Rossby wave adjustments for the tropical circulation. To a large extent, convective heating in the tropics is anchored by major warm water pools in the Indo-west Pacific and west Atlantic (and by the rainy Amazon). The prevalence of easterly winds in the equatorial Pacific and Atlantic can be viewed as a response to such zonal variations in convective heating and SST.

While small in the free tropical troposphere, horizontal temperature gradient can be large in the planetary boundary layer (typically 1–2 km deep), between land and sea, and maintained by SST variations. Air temperature gradients in the boundary layer, via hydrostatic pressure, are an important mechanism for low-level wind. Over the eastern equatorial Pacific and Atlantic especially, sharp SST gradients drive wind convergence in the boundary layer capped by a stable temperature inversion, fueling convection in the ITCZ (Back and Bretherton 2009).

3.5.3 Rotational and divergent flow

In general, any horizontal flow field may be decomposed into rotational flow $\boldsymbol{u}_\psi$ represented by the stream function (ψ) and divergent flow $\boldsymbol{u}_\chi$ by velocity potential (χ) (Fig. 3.17)

$$\boldsymbol{u} = \boldsymbol{u}_\psi + \boldsymbol{u}_\chi = \boldsymbol{k} \times \nabla\psi - \nabla\chi \tag{3.41}$$

ψ and χ can be obtained by solving the Poisson equations

$$D \equiv \nabla \cdot \boldsymbol{u} = -\nabla^2\chi \tag{3.42}$$

$$\zeta \equiv \boldsymbol{k} \cdot \nabla \times \boldsymbol{u} = \nabla^2\psi \tag{3.43}$$

where D is the divergence and $\zeta = \frac{\partial v}{\partial x} - \frac{\partial u}{\partial y}$ is the vorticity. By definition, the rotational flow is nondivergent, while the divergence is nonrotational. In the midlatitudes, geostrophy dominates, and Φ/f approximates the stream function, where Φ is geopotential. The thermally induced circulation in the tropics is divergent as the adiabatic cooling by vertical motion needs to balance diabatic heating. The divergent flow and velocity potential in the upper troposphere are often used to diagnose the thermally induced circulation.

The east-west Walker and north-south Hadley overturning circulations are best seen in velocity potential for the divergent flow. In the upper troposphere, large negative values of velocity potential are found over the Indo-western Pacific warm pool, South America, and Africa. The potential flow diverges from these regions, and converges strongly over the eastern Pacific and Atlantic (Fig. 3.18).

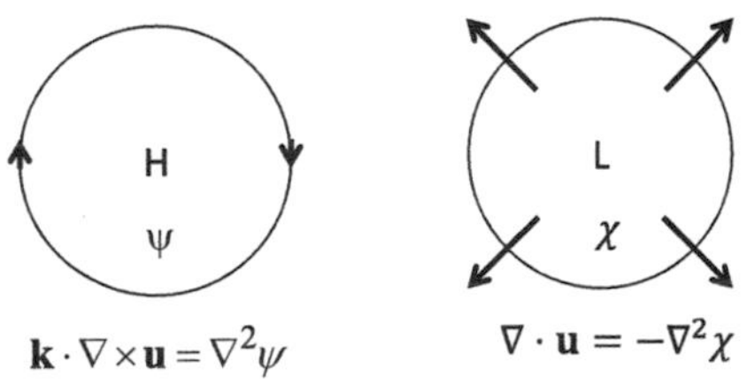

Fig. 3.17 Decomposition into *(left)* rotational flow represented by the stream function (ψ) and *(right)* potential flow by velocity potential (χ).

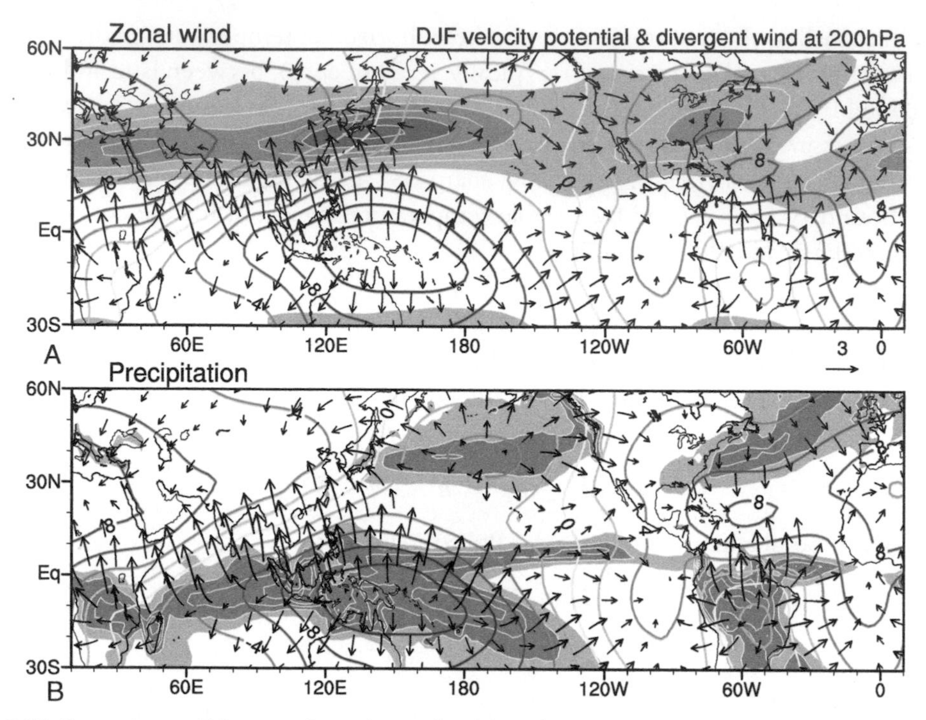

Fig. 3.18 December to February climatology of 200-hPa divergent wind (m/s) and velocity potential (color contours at interval of 2 from −10 to 10 × 10^6 m^2/s). (A) Zonal wind velocity (gray shading at 20, 40, and 60 m/s, white contours at 10 m/s interval) and (B) precipitation (light gray >3 mm/day; dark gray >5 mm/day, white contours at 2 mm/day interval). *(Courtesy ZQ Zhou.)*

Deep convection over the Indo-western Pacific warm pool drives divergent flow in the upper troposphere. The northward flow converges strongly onto East Asia, forming a vigorous local Hadley circulation. Not balanced by the pressure gradient, the northward divergent/ageostrophic flow accelerates the subtropical westerly flow over Japan (see Fig. 3.18) with the Coriolis effect, $du/dt - fv_\chi = 0$. Thus the local Hadley cell anchored by maritime continent convection is a major mechanism for the subtropical jet core over Japan, together with orographic forcing by the Tibetan Plateau and Rockies.

3.6 Weak temperature gradient and convective threshold

On the equator, the Coriolis parameter vanishes and the geostrophy collapses in the east-west direction, allowing fast zonal propagation of Kelvin and Rossby waves. Because of the fast wave propagation, the local time derivative is much larger than the horizontal advection; their ratio is $c/U > 1$ for the Kelvin and low-order Rossby waves, where $U \sim 10$ m/s is the scale of tropical wind. As a result, the nonlinear horizontal advection is small, and the linear wave dynamics as represented in Eqs. 3.16 through 3.18 dominate the tropics. This explains the success of the Matsuno wave theory. In the extratropics, by contrast, the advection by the westerly jet is of the first-order importance because of strong wind and horizontal gradient.

The fast-traveling equatorial waves flatten horizontal variations in free tropospheric temperature. Unlike the Gill solution, convective heating is not exactly localized but distributed with multiple centers, a zonal distribution that helps further smooth out temperature variations. (In the boundary layer, the strong SST forcing maintains large temperature gradients.) In March, 500 mb temperature is nearly uniform within 20°S and 10°N, with deviations of only less than 1 K (Fig. 3.19). Temperature is slightly higher near major convective zones such as the Indian and Pacific ITCZ, the SPCZ, and the Amazon, where strong condensational heat is released. These small temperature variations are all that is needed to drive rigorous tropical circulation because of small Coriolis force. Strong convection in the above regions restores the tropospheric temperature profile toward a moist adiabat, as most obvious in the midtroposphere where $\partial\theta_e^*/\partial p \approx 0$ (see Fig. 3.5).

In the meridional direction, consider an initial perturbation with a step-function jump in temperature or pressure (Fig. 3.20, *red dashed line*). Assume no zonal variation for simplicity. Without Earth rotation ($f = 0$), gravity waves would flatten the pressure gradient completely. On an f-plane, it is the classic problem of geostrophic adjustment; gravity waves propagate in the meridional direction, and the pressure gradient is smoothed but maintained across a scale of the Rossby radius in balance with zonal geostrophic flow. On a rotating sphere, the pressure gradient diminishes near the equator but is maintained around $y = \pm R_E$ in geostrophic balance with the zonal flow (see Fig. 3.20, *black curve*). This illustrates how gravity waves destroy meridional temperature gradient in the equatorial region. The shallow-water system (see Eqs. 3.16–3.18) without the longwave approximation features gravity waves (black dispersion curves with the nondimensional frequency >1 in Fig. 3.13).

Because of the weak temperature gradient, we can neglect the horizontal advection in the thermodynamic equation

$$\frac{\partial \theta}{\partial t} + \omega \frac{\partial \theta_0}{\partial p} = \left(\frac{p_s}{p}\right)^{\kappa} \frac{J}{c_p} \tag{3.44}$$

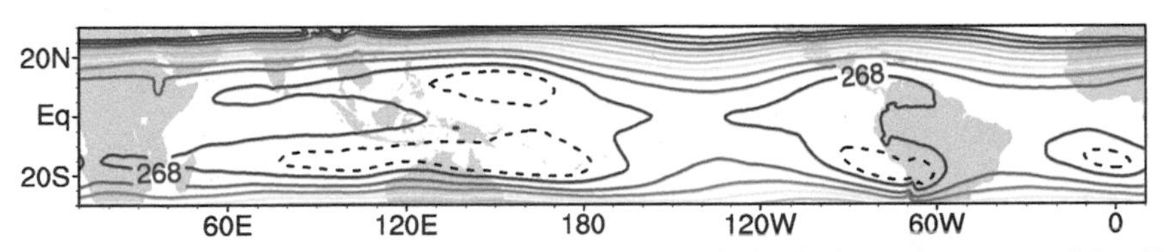

Fig. 3.19 March temperature (1 K intervals, 268.5 K contour dashed) climatology at 500 hPa. *(Courtesy ZQ Zhou.)*

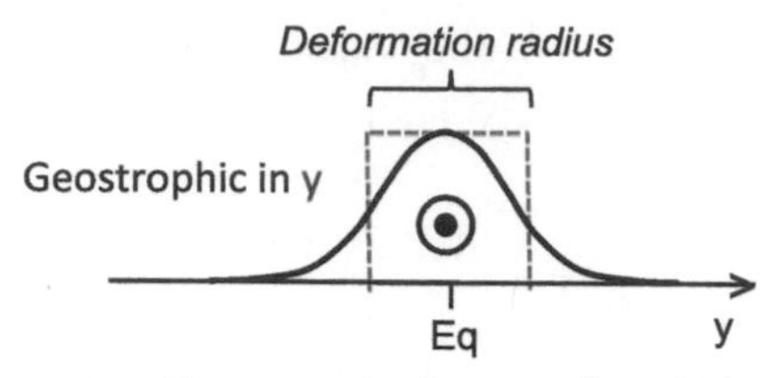

Fig. 3.20 Schematic of temperature adjustment in the meridional direction across the equator: the initial perturbation in red dashed and the adjusted profile in black. The Coriolis effect and temperature gradient are small within the radius of deformation from the equator.

where $J = Q_1$ is the diabatic heating rate. Here we have used the tropical mean temperature profile $\theta_0(p)$ for the vertical advection since perturbation vertical temperature gradient is much smaller. In quasi-steady state,

$$\omega \frac{\partial \theta_0}{\partial p} \approx \left(\frac{p_s}{p}\right)^{\kappa} \frac{J}{c_p} \tag{3.45}$$

In the tropics, the dominant thermodynamic balance is between the adiabatic cooling (LHS) and diabatic heating (RHS). In convective regions, radiative cooling is relatively small, and the convective heating is balanced by adiabatic cooling in the upward motion (see Fig. 3.7). In the nonconvective regions over cool waters, the radiative cooling is balanced by adiabatic warming associated with the slow subsidence (see Fig. 3.8, *right*).

In the tropical upper troposphere near the tropopause, water vapor content and horizontal gradients of temperature and geopotential are both small, so $\nabla m_T \approx \nabla s_T \approx 0$. Because entrainment is weak for deep convection, it is reasonable to assume that MSE at the tropopause equals surface MSE (see Fig. 3.5) averaged in the deep tropics (denoted by the overbar),

$$m_T = \overline{m}_s \tag{3.46}$$

At a given tropical location, the local convective instability may be measured by the MSE difference between the surface and tropopause

$$I = m_s - m_T = m_s - \overline{m}_s = c_p(1 + b_e)T^* \tag{3.47}$$

where

$$T^* \equiv T_s - \overline{T}_s \tag{3.48}$$

is called the relative SST referenced to the tropical mean. Here we have linearized q_s around the tropical mean SST $(\overline{T}_s)$ as in Eq. 3.6, with the inverse Bowen ratio $b_e = 2.5$. Thus the local convective instability is proportional to the local SST deviation from the tropical mean, positive where the local SST exceeds the tropical mean ($T^* > 0$). In other words, the tropical mean SST sets the threshold for convection (see

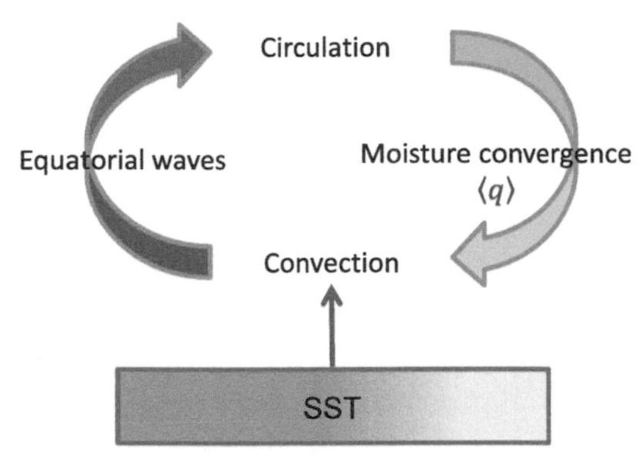

Fig. 3.21 Schematic of convection-circulation interactions along with the sea surface temperature *(SST)* effect.

Fig. 3.6). In current climate, the tropical-mean SST (20°S−20°N) is about 27°C, matching the observed SST threshold quite well (see Fig. 3.6A).

3.7 Outlook

Tropical convection and circulation are closely coupled. Convective heating drives the baroclinic overturning circulation as depicted in the Matsuno-Gill model, while the circulation sets up convection through the column water vapor and/or horizontal convergence (Fig. 3.21). Strong convective-circulation feedback generates spontaneous variability internal to the atmosphere, such as the 30- to 60-day Madden-Julian oscillation (see Chapter 4). On monthly and longer timescales, SST variations become important, modulating deep convection through a threshold behavior and pressure gradient in the atmospheric boundary layer (see Chapters 9−11). Land surface temperature responds strongly to the seasonal solar cycle, driving continental-scale monsoons (see Chapter 5). Summer monsoons show that MSE at the surface and integrated over the atmospheric column proves a better predictor for convective variability than surface temperature, especially on land where atmospheric relative humidity varies wildly in time and space.

Review questions

1. Why are major tropical rain bands called convergence zones?
2. Commercial aircrafts cruise at 250 hPa where air temperature is −50°C. What temperature will it reach when the outside air is compressed adiabatically to the cabin pressure of 800 hPa?
3. Explain that water vapor partial pressure decreases with height more sharply than air pressure (see Fig. 3.4). Consider an atmosphere of a constant relative humidity (R_H). Show the saturation water vapor partial pressure $e_s = e_0\ exp(-z/H_w)$, where

$H_w = (\alpha\Gamma)^{-1}$ is the scale height for water vapor pressure, $\Gamma \sim 6.5$ K/km is the lapse rate, and $\alpha = \frac{d}{dT} \ln e_s \approx 0.06$ K^{-1} is the CC coefficient.

a. Calculate H_w and compare with the scale height for atmospheric pressure $H_s = RT/g$ (see Eq. 1.10).

b. In the real atmosphere, $H_w \sim 2.5$ km. Compare with the calculation for (a). How does this imply in terms of relative humidity change with height?

4. What is the minimum surface θ_e^* for deep convection with the sounding in Fig. 3.5?

5. Make a brief argument for the SST effect on the height of moist convection.

6. Compare rainfall and OLR maps, then discuss their similarities and differences.

7. Answer the following questions about equivalent potential temperature $\theta_e = \theta \exp\left(Lq / c_p T_{LCL}\right)$.

a. Plot surface θ_e as a function of surface air temperature (SAT). Assume $T_{CLC} =$ SAT $-$ 5 K (equivalent to a cloud base of 500 m) and relative humidity of 80%. Use the Clausius-Clapeyron relationship $q_s = q_0 e^{\alpha(T-T_0)}$, where q_0 is the saturated specific humidity at reference temperature T_0.

b. What are the critical values of surface θ_e and SAT for deep convection? Refer to Fig. 3.5. Compare the critical SAT to the well-known threshold of 27°C and discuss what causes the difference.

8. Answer the following questions regarding midlatitude flow.

a. Show geostrophic flow on the f-plane ($f=$ const.) is nondivergent.

b. Why is it generally not feasible to compute vertical velocity from a network of atmospheric soundings using the continuity equation in the extratropics? *Hint:* The ratio of the geostrophic to ageostrophic flow is 10:1.

9. Calculate the wave speed of the baroclinic mode in Eq. 3.19 using the typical tropical sounding in Fig. 3.5. Calculate the equatorial radius of deformation.

10. A famous glaciologist cored glaciers on top of Mt. Kilimanjaro (5895 m) and the tropical Andes ($\geq$6000 m). He claims that temperature changes he measured in these ice cores were representative of tropical-wide conditions. Discuss the merits of his claim and your reasoning.

11. Based on Eq. 3.48, discuss how the SST threshold for convection may respond to global warming.

12. With increasing atmospheric CO_2, ocean temperatures are expected to increase. The tropical Pacific is twice as wide as either the Indian or Atlantic in the east-west direction. Suppose that in 2100, SST increases by 2.6°C in the tropical Pacific and by 2°C in both the tropical Indian and Atlantic oceans. How will the SST threshold for atmospheric deep convection change, and by how much?

13. Why is tropospheric temperature at a given pressure level nearly uniform within the tropics? What are the important consequences of this fact? Give two examples.

14. The moist adiabat refers to a vertical temperature profile with $d\theta_e^*/dp = 0$. Why are temperature soundings close to the moist adiabat in the free troposphere within the tropics, even over nonconvective regions of low SSTs?

15. In light of equatorial wave propagation and the weak temperature gradient, consider additional reasons for the fact that θ_e^* at the midtroposphere is smaller than surface θ_e^* in the warm convective region (i.e., the tropical western Pacific).

16. Apply a scaling analysis to show that the local time derivative is much greater than horizontal advection terms $\frac{\partial}{\partial t} >> \boldsymbol{u} \cdot \nabla$ for fast equatorial waves ($c/U >> 1$). This result indicates that the equatorial atmosphere is dominated by linear dynamics.

17. How is the tropical atmospheric response to a localized heating different between the east and west, and why?

18. Discuss one observed feature that Gill's solution explains successfully.

19. Now consider friction in the atmospheric boundary layer ($\sim$1 km deep). In the Kelvin response east of the imposed convective heating on the equator, discuss the frictional effect in the boundary layer flow. Is the meridional flow still zero in the boundary layer?

CHAPTER 4

Madden-Julian oscillation

Contents

Tropical convection is highly variable in space and time, organized into coherent structures (Fig. 4.1). Individual cumulonimbus clouds are ~10 km across in the horizontal and last for a few hours. Deep convective clouds aggregate into clusters of 100s km scale, and such mesoscale cloud clusters further aggregate into broad envelopes of several thousand km in zonal extent, also known as super cloud clusters. These convective envelops of enormous size propagate slowly eastward across the Indo-Pacific warm pool. This chapter focuses on spontaneous variability of tropical convection internal to the atmosphere; Chapter 5 discusses the summer monsoon forced by solar variations, and Chapters 9 through 11 year-to-year variability due to ocean-atmosphere coupling.

4.1 Convectively coupled waves

Outgoing longwave radiation (OLR) is a convenient proxy of convective activity, low above convective clouds and high over regions of suppressed convection. Satellite radiometer data are often converted into brightness temperature (T_b) using the Stefan-Boltzmann law. Low T_b values are indicative of precipitating deep convection. Generally, variability can be expressed in the Fourier form

$$T(x,t) = \int_0^\infty d\omega \int_{-\infty}^{\infty} \widetilde{T}(k,\omega) e^{i(kx-\omega t)} dk \tag{4.1}$$

Coupled Atmosphere-Ocean Dynamics: from El Niño to Climate Change
ISBN 978-0-323-95490-7, https://doi.org/10.1016/B978-0-323-95490-7.00004-7

Fig. 4.1 Hierarchy of tropical convective organizations captured by an infrared image from the Himawari geostationary satellite on January 18, 2022. Embedded in a convective envelop over the western Pacific are mesoscale (100s km) cloud clusters *(inset)*, each comprised of deep convective cumulonimbus clouds ($\sim$10 km). *(Image from Japan Meteorological Agency.)*

where $\widetilde{T}$ is the time-space spectrum, ω the angular frequency, and k the zonal wavenumber (positive for eastward and negative for westward propagation). Fig. 4.2 shows the time-space spectrum of equatorially symmetric T_b variability expressed as the ratio to a smooth background red spectrum (Wheeler and Kiladis 1999). Equatorial waves are important in organizing convective activity. For reference, Matsuno dispersion curves are shown for several different values of equivalent depth H_e, defined as $c^2 = gH_e$. For westward propagating perturbations ($k < 0$), enhanced power is found in the high-frequency (period <3 days) gravity wave regime *(upper*

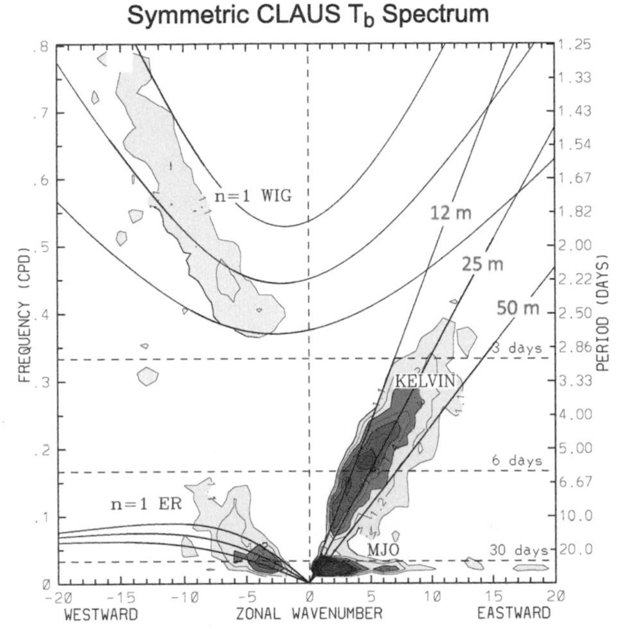

Fig. 4.2 Wavenumber–frequency power spectrum of the symmetric component of satellite brightness temperature (T_b) for July 1983 to June 2005, summed from 15°N to 15°S, plotted as the ratio between raw T_b power and the power in a smoothed red noise background spectrum (Wheeler and Kiladis 1999). Contour interval is 0.1, and contours and shading begin at 1.1, where the signal is significant at greater than the 95% level. Dispersion curves for the Kelvin, $n = 1$ equatorial Rossby *(ER)*, $n = 1$ westward inertio-gravity *(WIG)* waves are plotted for equivalent depths of 12, 25, and 50 m. *(From Wheeler and Nguyen [2015].)*

left), and low-frequency regime *(lower left)* that follows the longwave Rossby wave dispersion curve of H_e ~25 m or equivalently c = 16 m/s. Here the zonal wavenumber (k = 1) refers to a single sinusoidal wave along the equatorial belt. For eastward-propagating, positive wavenumbers, energetic perturbations are concentrated in two distinct regimes on the Wheeler-Kiladis diagram. One follows the nondispersive Kelvin wave dispersion of H_e ~20 m (k <5) and H_e ~25 m (k >5), and one is near the bottom (30- to 90-day period) slightly to the right from the center (k = 1~3). The 30- to 60-day variability is called intraseasonal variability, also known as the Madden-Julian oscillation (MJO) to be discussed in the next section.

Matsuno's theory does not consider the effect of moist convection. Remarkably, convective perturbations still follow the dispersion relation of the Kelvin wave, albeit at a much slower phase speed than the dry value of $c_d = 50$ m/s. In the following, we use $c_m = 15$ m/s ($H_e = 23$ m) as the nominal phase speed of convectively coupled (CC) Kelvin waves. Here we focus on the Kelvin wavelike variability and show that the interaction of Matsuno waves with convection slows the phase speed.

4.1.1 Phase speed slowdown

First consider water waves at the ocean surface. A rise in water surface causes pressure underneath to increase because water is much heavier than air above. The pressure perturbation forces currents to create the next crest in the direction of wave propagation. Thus the density difference between water and air under gravity is the restoring force for the water waves. Then consider waves at the interface between oil above and water below. Internal waves between oil and water propagate much like surface water waves except at a much slower phase speed because of a smaller density difference between the media. In general, the phase speed of a wave is in proportion to the square root of density difference across the interface, $c \propto \sqrt{\Delta \rho}$ (Section 7.1.3).

Now consider the two-level model to capture the tropical baroclinic circulation (see Chapter 3). The restoring force in a dry atmosphere is proportional to the dry static stability measured by the vertical gradient of potential temperature $\Delta\theta_0 = \theta_{400} - \theta_{800}$ ($\sim$30 K). If one brings an air parcel adiabatically from 800 to 400 hPa, its temperature is 30 K cooler than the surroundings, forcing the parcel back to the original altitude. When the air motion is spatially nonuniform, the temperature waves propagate due to this restoring force at $c_d \propto \sqrt{\Delta\theta_0}$ (see Eq. 3.19).

The column integrated water vapor budget (see Eq. 3.5) may be approximated as

$$P \approx -\langle q \rangle \nabla \cdot \boldsymbol{u} + E \tag{4.2}$$

where $\boldsymbol{u}$ is the low-level velocity. Neglecting surface evaporation perturbations ($E' = 0$) gives the perturbation latent heating at midlevel

$$Q = \kappa \left(\frac{p_2}{p_s}\right)^{\kappa} LP = -q^* c_d^2 \nabla \cdot \boldsymbol{u} \tag{4.3}$$

where q^* is the nondimensional column precipitable water that controls the strength of circulation-convection coupling

$$q^* \equiv L\langle \overline{q} \rangle / (c_p \Delta\theta_0) \tag{4.4}$$

where $\langle \overline{q} \rangle$ is the mean column-integrated water vapor path. The thermodynamic equation

$$\frac{\partial \Phi}{\partial t} + c_m^2 \nabla \cdot \boldsymbol{u} = 0 \tag{4.5}$$

is identical to that of the original shallow water system (see Eq. 3.18) except with the moist wave speed

$$c_m^2 = c_d^2 (1 - q^*) \tag{4.6}$$

When an adiabatic moist air parcel rises from the lower (p_3) to upper (p_1) troposphere, the latent heat of water vapor condensation warms the parcel, reducing the temperature difference from the ambient at p_1. This reduces the restoring force that tries to pull the parcel back to p_3, and the effective restoring force is proportional to $(1 - q^*)\Delta\theta_0$. The reduced restoring force due to latent heating slows the propagation of CC wave perturbations as in Eq. 4.6. The nondimensional precipitable water parameter $q^* = 0.91$ for $c_d = 50$ m/s and $c_m = 15$ m/s.

4.1.2 Kelvin wave

On the Hovmöller diagram of T_b over 63 days from April 1 through June 2, 1987 (Fig. 4.3), CC Kelvin waves are easily identifiable as super clusters of active convection propagating eastward at a constant phase speed (~15 m/s), often around much of the equatorial belt. Sometimes two such CC Kelvin waves coexist, with a typical zonal

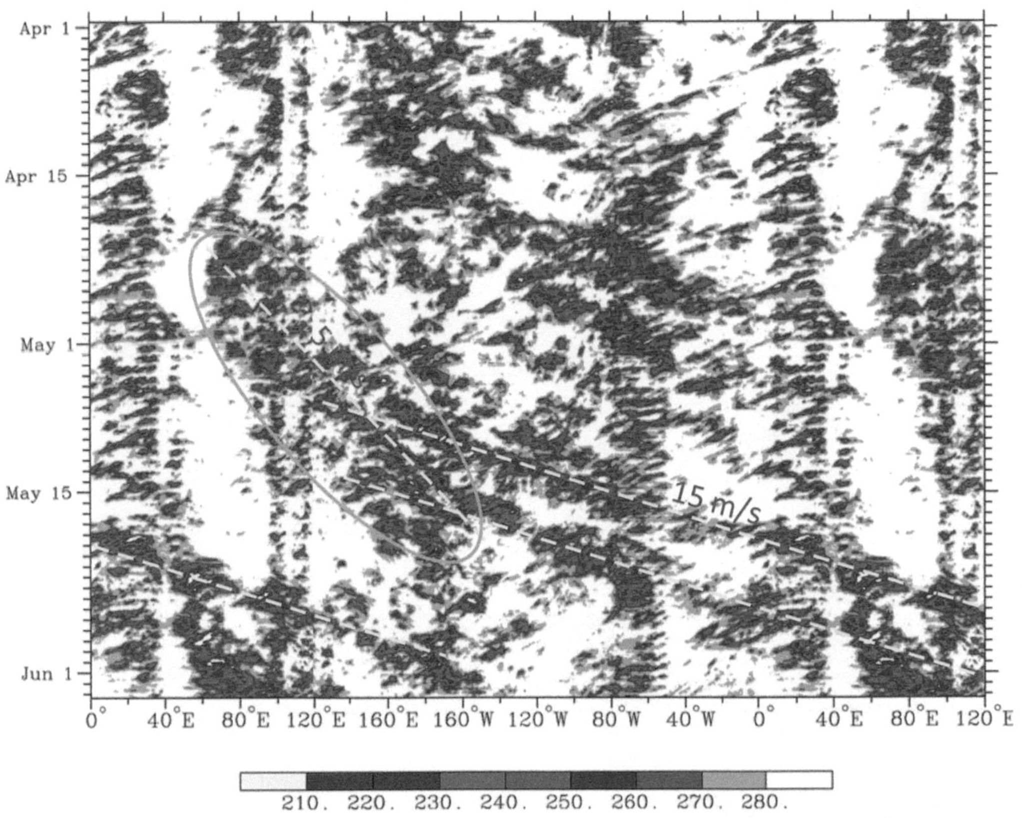

Fig. 4.3 Time-longitude section of brightness temperature (T_b, color bar in K), averaged from 2.5°S to 7.5°N from April 1 through June 2, 1987. The 5 m/s and 15 m/s phase velocities are highlighted. *(Adapted from Kiladis et al. [2009].)*

spacing of 120°, or $k = 3$. This, together with $c_m = 15$ m/s, explains that the variance of the CC Kelvin wave peaks around 6 days in period on the Wheeler-Kiladis diagram (see Fig. 4.2).

Deep convection is concentrated in the zonally elongated intertropical convergence zone (ITCZ). Convective anomalies of Kelvin waves are displaced north of the equator over the eastern Pacific and across the Atlantic during most of the year (Fig. 4.4). Despite the northward displacement of convective heating anomalies, the pressure and flow anomalies are surprisingly symmetric, centered on the equator, and dominated by the

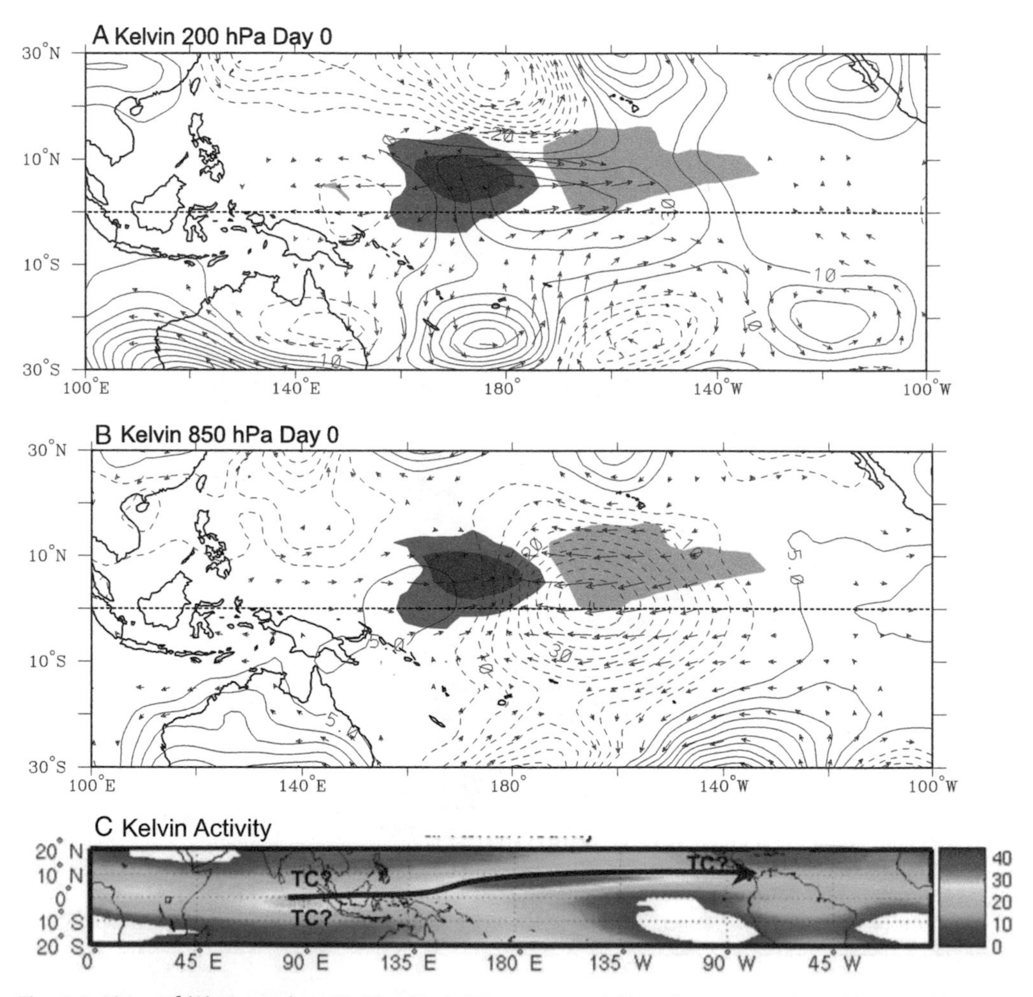

Fig. 4.4 Maps of (A) anomalous T_b (shading), (B) geopotential height (contours), and (C) wind (vectors) associated with a 20 K perturbation in Kelvin wave T_b at the base point 7.5°N, 172.5°E at 850 hPa (contour interval of 5 m, negative dashed) and 200 hPa (ci = 10 m). Dark (light) shading is for negative (positive) T_b perturbations of ±10 K and 3 K. In (C), Kelvin wave filtered T_b variance. *(From Kiladis et al. [2009].)*

zonal wind perturbations in phase with the pressure with little meridional wind perturbations, much as Matsuno's Kelvin wave solution.

The dominance of Kelvin wave can be explained by considering a Gill response of the dry atmosphere to a heat source moving eastward at the speed of $c_m = 15$ m/s. On the moving frame $x' = x - c_m t$, the Kelvin wave travels eastward at the reduced speed of $c'_K = c_d - c_m = 35$ m/s while the first Rossby mode travels westward at the increased speed of $c'_R = c_d/3 + c_m = 32$ m/s. The Kelvin and Rossby wave responses are in steady state, described by the Matsuno-Gill model (see Eqs. 3.37 and 3.38)

$$\varepsilon q_0 + r_K \frac{\partial q_0}{\partial x'} = -Q_0 \tag{4.7}$$

$$\varepsilon q_2 - \frac{r_R}{3}\frac{\partial q_2}{\partial x'} = -\frac{1}{3}Q_0 \tag{4.8}$$

where $r_K = c'_K/c_d = 1 - r_c$ and $r_R = 3c'_R/c_d = 1 + 3r_c$ are the ratios of the Doppler shifted to original wave speed, with $r_c = c_m/c_d$ the ratio of moist to dry wave speed. For simplicity, assume a zonally constant heating profile within the convective region $-L < x' < L$. For a narrow heating ($3\ L/L_K << 1$), the adiabatic cooling due to low-level convergence approximately balances the convective heating, and the radiative damping (first term on the left-hand side,.LHS) is negligible

$$q_0 \approx -Q_0(x' + L)/r_K \tag{4.9}$$

$$q_2 \approx Q_0(x' - L)/r_R \tag{4.10}$$

in the forcing region $-L < x' < L$. On the frame moving with the convective heating, the Kelvin wave slows in phase speed ($r_K = 0.7$), stays in the forcing region longer, and gains larger amplitude than with a stationary forcing. For the same reason, the Rossby wave speeds up ($r_R = 1.9$) with the strongly reduced amplitude. It is left as homework to show that the antisymmetric Rossby wave ($n = 2$) is even more strongly damped. This explains why the Kelvin wave response dominates even though the convective heating is markedly displaced north of the equator.

The circumglobal propagation of CC Kelvin waves suggests that zonal variations in the mean state (e.g., sea surface temperature [SST] and zonal wind) are secondary. Indeed, atmospheric model experiments on the so-called aquaplanet—an earth covered by water—with zonally uniform SST produce Kelvin wavelike super clusters that propagate eastward at a slow phase speed. Fig. 4.5 shows an example with $c_m = 23$ m/s or $r_c \sim 0.5$. As in observations, the Rossby wave response is strongly suppressed.

4.1.3 Evaporation-wind feedback

In such aqua-planet experiments, surface evaporation variability in Eq. 4.2 is not negligible and is important for the CC Kelvin wave. Surface evaporation can be cast as

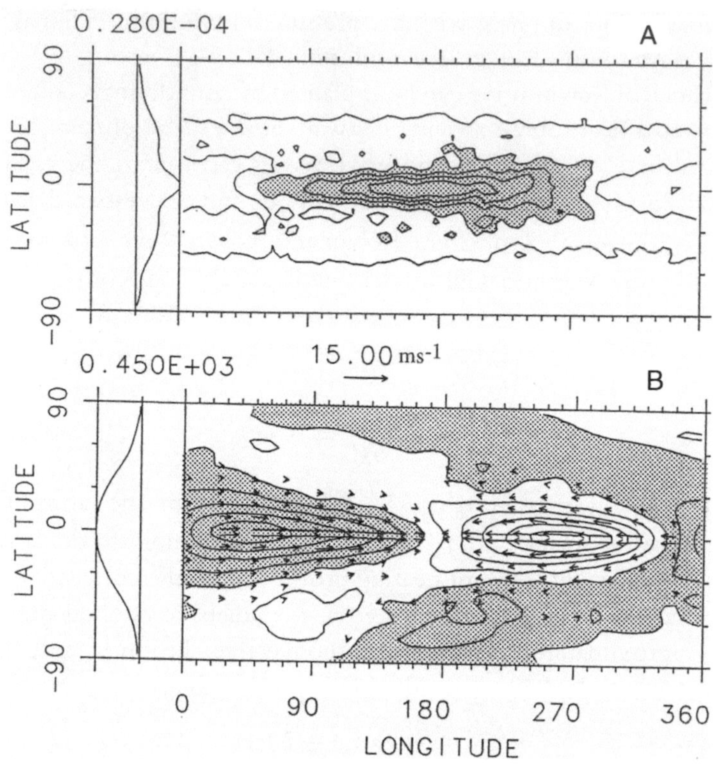

Fig. 4.5 Composite CC Kelvin wave on an eastward moving frame at 23 m/s in a two-level aquaplanet atmospheric model with zonally uniform SST: (a) convective heating (10^{-5} K/s), and (b) zonal variations in low-level wind (m/s) and geopotential height (contour interval: 10 m). The left panels show the zonal-mean: (a) heating rate and (b) geopotential height. *(From Xie et al. [1993].)*

$$E = \rho_a C_E |\boldsymbol{u}| (q_s - q_a) \tag{4.11}$$

where ρ_a is air density, C_E the aerodynamic coefficient for evaporation, q_s the saturated specific humidity at the sea surface, and q_a the surface air humidity. In the presence of the mean easterly wind ($\overline{u}$ <0) on the equator, surface evaporation perturbation may be linearized as

$$E' = \overline{E}u'/\overline{u} \tag{4.12}$$

where the overbar denotes the mean. Here we have neglected humidity variability. The perturbation convective heating (see Eq. 4.2) becomes

$$P = -\langle \overline{q} \rangle \nabla \cdot \boldsymbol{u} + \overline{E}u/\overline{u} \tag{4.13}$$

A generalized form of the thermodynamic equation (see Eq. 3.18) becomes

$$\frac{\partial \Phi}{\partial t} + c_m^2 \nabla \cdot \boldsymbol{u} = -\left(\frac{L\overline{E}}{\overline{u}}\right) u \tag{4.14}$$

Multiplying F on either side and zonal integration along the equator yields an energy equation

$$\frac{1}{2}\frac{\partial}{\partial t}\left(\overline{\Phi^2 + c_m^2 u^2}\right) = \left(\frac{L\overline{E}}{-\overline{u}}\right)\overline{u\Phi} \tag{4.15}$$

Note that $\overline{u\Phi} > 0$ since the perturbation zonal flow of a Kelvin wave is in phase with geopotential, where the overbar denotes the zonal average. As a result, the energy in the parentheses on the LHS must grow in time. The positive covariance between the surface wind speed ($-u_3$) and tropospheric temperature ($\propto -\Phi_3$), called the evaporation-wind feedback, destabilizes the CC Kelvin wave (Emanuel 1987; Neelin et al. 1987).

4.2 Madden-Julian oscillation

In the early 1970s, the increased computing power enabled calculating power spectra of long time series using the newly developed fast Fourier transform method. This allowed investigations into low-frequency atmospheric variability beyond the weather timescale (<10 days). Madden and Julian (1972) analyzed surface pressure observations at stations in the equatorial belt. They discovered a pronounced spectral peak at 30 to 60 days and envisioned a zonal wavenumber-one Walker circulation that is coupled with deep convection and propagates slowly eastward (Fig. 4.6). The MJO proves to be the dominant mode of atmospheric variability on the intraseasonal timescale. Satellite observations (e.g., OLR) and global atmospheric analysis brought out a long-lasting boom in MJO research in the mid-1980s (e.g., Lau and Chan 1985), revealing rich structures and global influences of this important phenomenon. The MJO remains a hot topic in atmospheric/climate research now, 50 years after its discovery.

In the longitude-time Hovmöller diagram (Fig. 4.3), the intraseasonal MJO is manifested as a broad convective envelope that propagates slowly eastward at a nominal phase speed $c_I = 5$ m/s. Typically an MJO envelope of active convection first appears in the equatorial Indian Ocean, grows, moves slowly across the maritime continent into the western Pacific, and dissipates near the International Dateline. The convective envelope and slow phase propagation of the MJO are limited to the Indo-western Pacific region (60°−180°E) where warm equatorial SSTs are in excess of the convective threshold. The MJO and CC Kelvin wave are distinct in several important ways. First, the eastward propagation of convective anomalies is limited to the Indo-western Pacific warm pool, while the Kelvin waves circumnavigate the equatorial belt. On the Wheeler-Kiladis diagram (see Fig. 4.2), the CC Kelvin wave follows a nearly constant phase speed with a

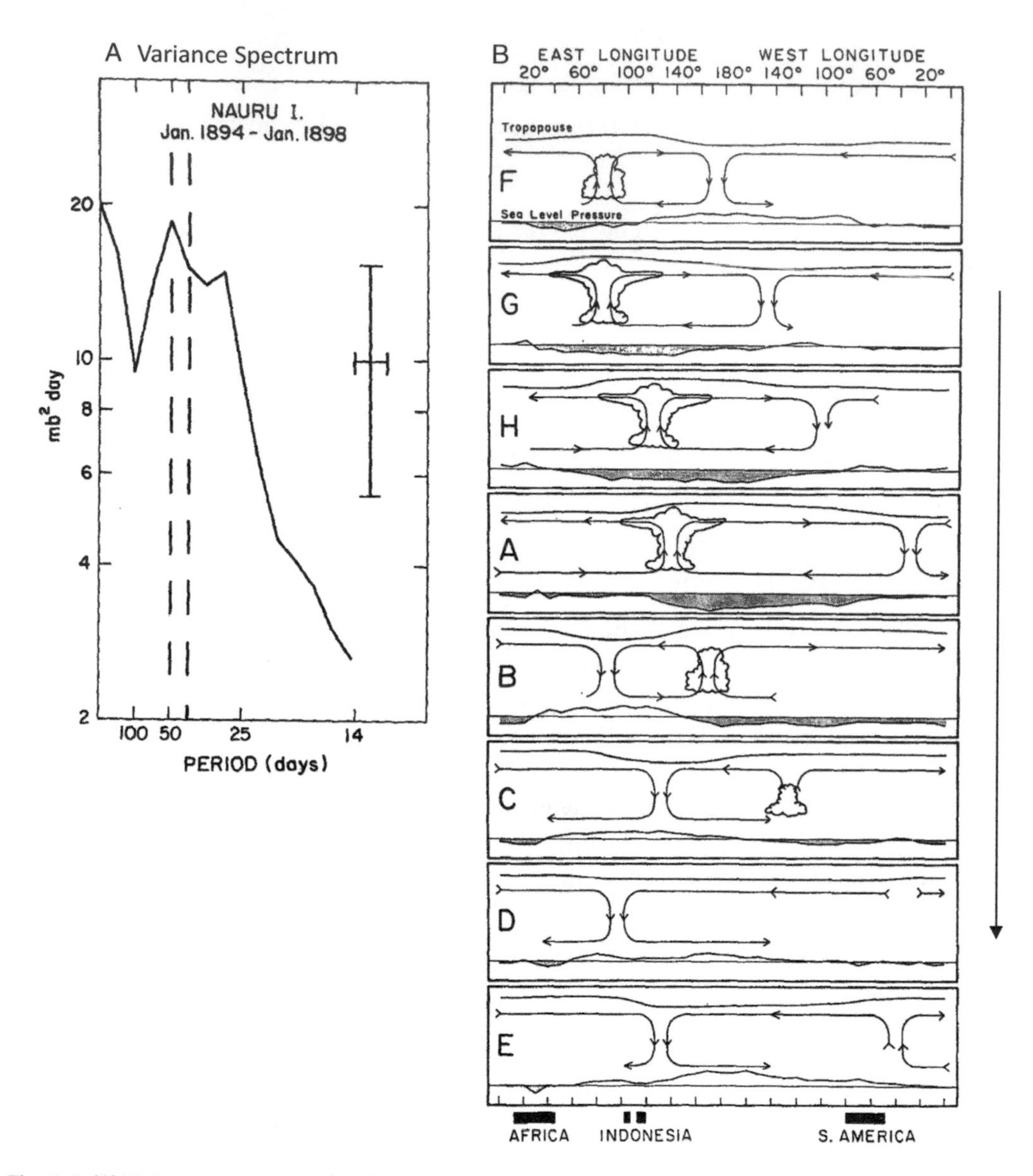

Fig. 4.6 (A) Variance spectrum of surface pressure at Nauru Island in the western equatorial Pacific. (B) Schematic of the 30- to 60-day oscillation. *(From Madden and Julian [1972]. © American Meteorological Society. Used with permission.)*

wide wavenumber-frequency range, while the MJO has a well-defined scale in space ($k = 1 \sim 2$) and time (30—60 days) but does not follow the Matsuno wave dynamics. Second, the phase speed is much smaller for the MJO ($c_I/c_d = 1/10$) than CC Kelvin wave ($c_m/c_d = 1/3$). The large difference in phase speed causes their distinct spatial structure, as illustrated later.

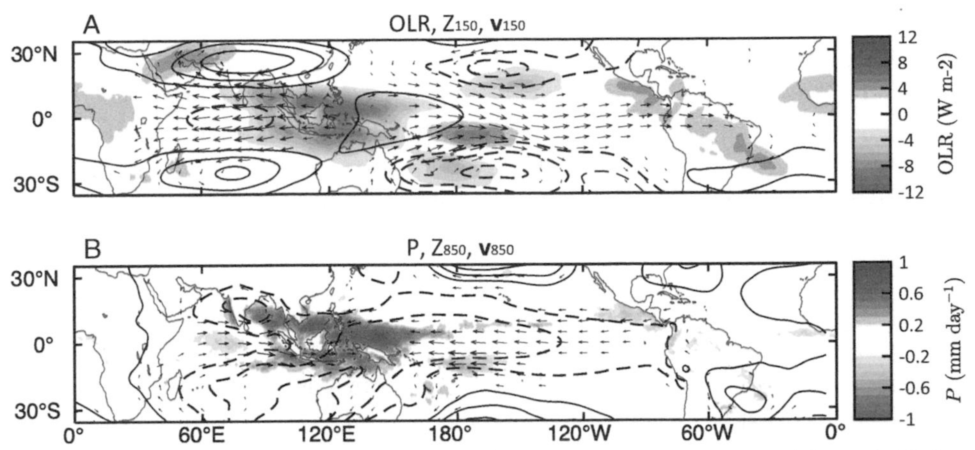

Fig. 4.7 MJO anomalies when convection is active over the Maritime Continent (phase 5). (A) Outgoing longwave radiation (*OLR*, shading), 150 hPa geopotential height (contours at 2-m interval) and horizontal winds *(arrows)*. (B) Precipitation (shaded) and 850 hPa geopotential height (contours at 1-m interval) and horizontal winds. *(From Adames and Maloney [2021].)*

4.2.1 Circulation structure

Fig. 4.7B shows the composite structure at 850 hPa when the MJO convective envelope is over the maritime continent. With the convective envelope of 60° across in longitude centered on the equator, the low-level flow field strongly resembles the Gill pattern. A pair of cyclonic Rossby gyres are found west of active convection while a Kelvin-like response of nearly pure zonal wind anomalies that extends far eastward along the equator. This Gill-type response with the Kelvin-Rossby couplet contrasts with the flow field of the CC Kelvin wave (see Fig. 4.4), where the Rossby response is strongly suppressed by the relatively fast propagation of the heating anomaly. Since the MJO phase speed is only 10% of the dry Kelvin wave, the Doppler effect is weak.

The upper-level circulation in the equatorial belt (see Fig. 4.7A) is similar to the low-level one but with the reversed sign. This is consistent with the baroclinic structure in the tropics. The Rossby gyres, however, show a zonal wavenumber-one structure, not confined to the west of the enhanced convective heating as observed at low levels. This may be due to weak dissipations at the upper level, and a wavenumber-one pattern in convection; the enhanced convection in the maritime continent/western Pacific is associated with a reduced activity over tropical Americas. Matsuno (1966) calculated the baroclinic response to wavenumber-one heating anomalies. With weak damping, the Rossby response is wavy, with the upper-level anticyclonic gyre displaced west of the positive heating anomaly. Remarkably, the observed upper-level MJO circulation resembles the wavy Matsuno pattern, while the low-level circulation is similar to the damped Gill pattern.

4.2.2 Zonal modulations

The MJO convective envelope is coupled with the Matsuno-Gill circulation pattern of a baroclinic vertical structure. Near the Dateline, the MJO convection gradually weakens because of cool SSTs underneath. Losing the convective anchor, the Kelvin-Rossby wave couplet disperses away in so-called free waves. CC Kelvin waves, in particular, radiate eastward from the dissipating MJO convective envelope. All this is captured in the Hovmöller diagram of OLR and upper-level zonal wind (Fig. 4.8). OLR and zonal wind anomalies show a slow (5 m/s) phase speed in the Indo-western Pacific sector, but the zonal wind perturbations propagate much faster at 15 m/s in the Western Hemisphere.

Although the MJO and CC Kelvin wave are distinct in horizontal structure and mechanism, their mixture in the MJO Hovmöller diagrams (e.g., upper-level velocity in Fig. 4.8) suggests that they are related in occurrence and phase. The radiation of fast-moving CC Kelvin waves from the slow-moving MJO envelope—especially when it approaches the Dateline—is evident in the Hovmöller diagram of equatorial T_b variability (see Fig. 4.3).

Embedded in the eastward moving MJO and CC Kelvin waves are westward moving cloud clusters of much smaller scales (~100s km). Such clusters last for one to a few days and are tracible over a few thousand km, each including multiple convective cores. Some of these clusters develop into tropical cyclones. Over the maritime, American and African continents, westward-moving diurnal convection is also visible. Fig. 4.9 is a closeup of the longitude-time section over the Indo-western Pacific sector, showing westward propagating cloud clusters embedded in a broad MJO envelope.

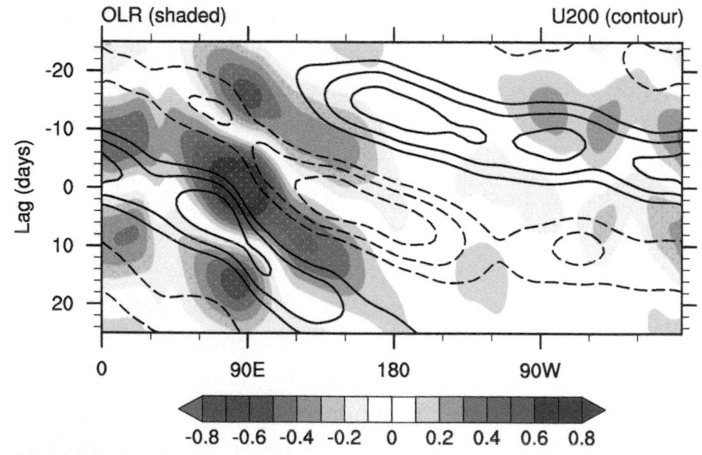

Fig. 4.8 Hovmöller (longitude-time lag) diagram of the correlation of outgoing longwave radiation *(OLR)* (shaded) and 200 hPa zonal wind (contours starting at ±0.2, interval = 0.1) anomalies against OLR variability averaged over the eastern Indian Ocean (75–100°E; 10°S–10°N) for 1997-2020. Fields have been averaged between 10°S and 10°N and 20–100-day band-pass filtered. *(Courtesy Y. Liang.)*

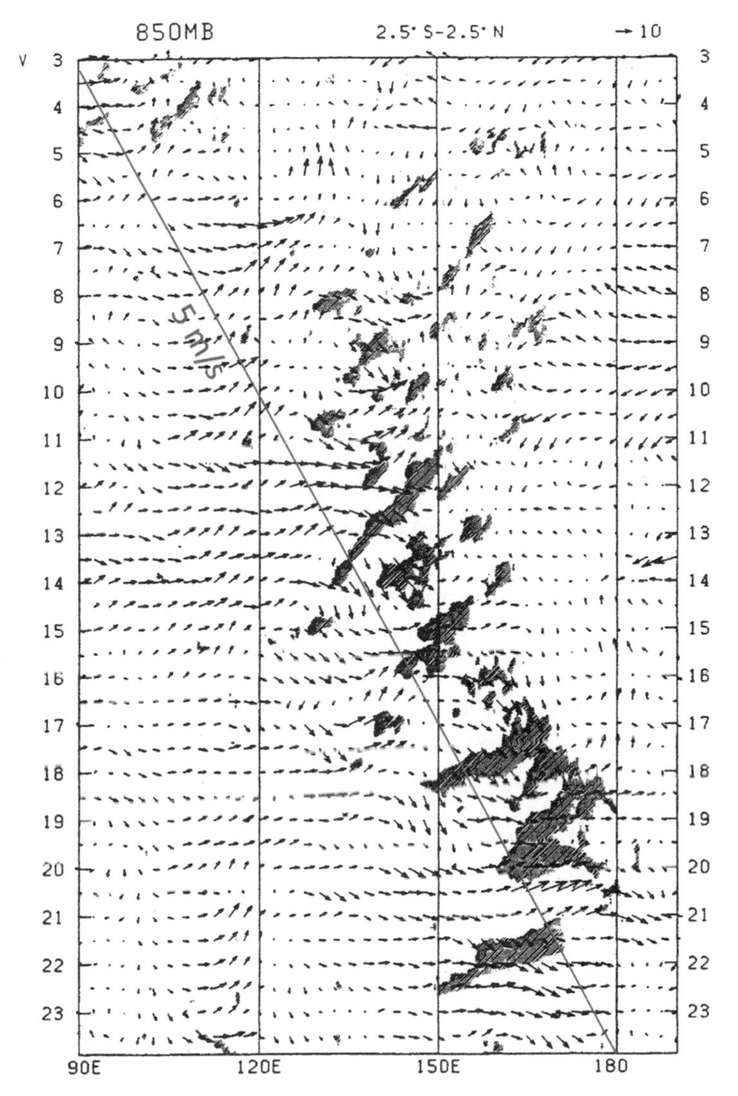

Fig. 4.9 Longitude-time section of infrared blackbody temperature (T_b <225 K shaded) and 850 hPa wind velocity (m/s). *(From Nakazawa [1988].)*

4.2.3 Index

The empirical orthogonal function (EOF) analysis separates spatiotemporal variability into mutually uncorrelated time expansion series $\widetilde{T}_i$ and mutually orthogonal spatial patterns P_i

$$T(x, y, t) = \sum \widetilde{T}_i(t) P_i(x, y)$$

The time expansion coefficients $\widetilde{T}_i$ are also called the principal components (PCs).

Joint EOF analysis of OLR, low (850 hPa) and upper (250 hPa) level zonal wind averaged in 15°S to 15°N is used to track the MJO. Convective activity peaks over the maritime continent in EOF1 while EOF2 is 90° shifted in zonal phase with an OLR dipole between the Indian and western Pacific Oceans. They each explain 12% to 13% of the total variance, and the PCs are well correlated at 10-day lead/lag (Fig. 4.10B). This implies that the pair of EOFs represents the MJO's eastward phase propagation. The pair of the leading PCs can be calculated in real time by projecting observations onto the EOFs, and is called the real-time multivariant MJO (RMM) indices.

Data are often displayed in a RMM1-RMM2 phase diagram (see Fig. 4.10A), where an eastward MJO follows a counterclockwise rotation, and the distance from the origin denotes the amplitude. At phases 2 and 3, RMM1 is small, a zonal dipole of convective anomalies appears over the Indo-western Pacific warm pool; convection is enhanced in the Indian Ocean sector but suppressed over the western Pacific (Fig. 4.11A). Enhanced convection has moved over the maritime continent at phase 5 and approaches the Dateline at phase 6. Then the MJO convective envelope weakens on the equator, while the convection strengthens along the South Pacific convergence zone at phases 7 and 8.

The velocity potential is the inverse Laplacian of horizontal divergence, rainfall, or OLR. The spatial integration twice makes velocity potential a smooth field easy to track. The zonal wavenumber-one dominates the MJO in velocity potential (see Fig. 4.11B). Velocity potential perturbations amplify with a robust eastward propagation at 5 m/s over the Indo-western Pacific warm pool but are weak in the Western Hemisphere because of loss of the convective anchor.

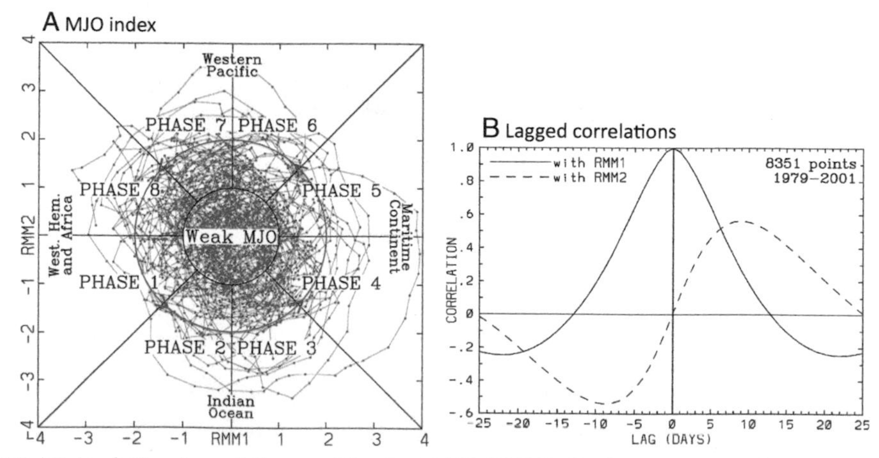

Fig. 4.10 (A) Evolution in real-time multivariant MJO (RMM1-2) phase space, with regions of active convection marked corresponding to eight phases. (B) Lagged correlations of RMM1 with itself *(solid line)* and RMM2 *(dashed line)* as a function of lag (days). *(From Wheeler and Hendon [2004]. © American Meteorological Society. Used with permission.)*

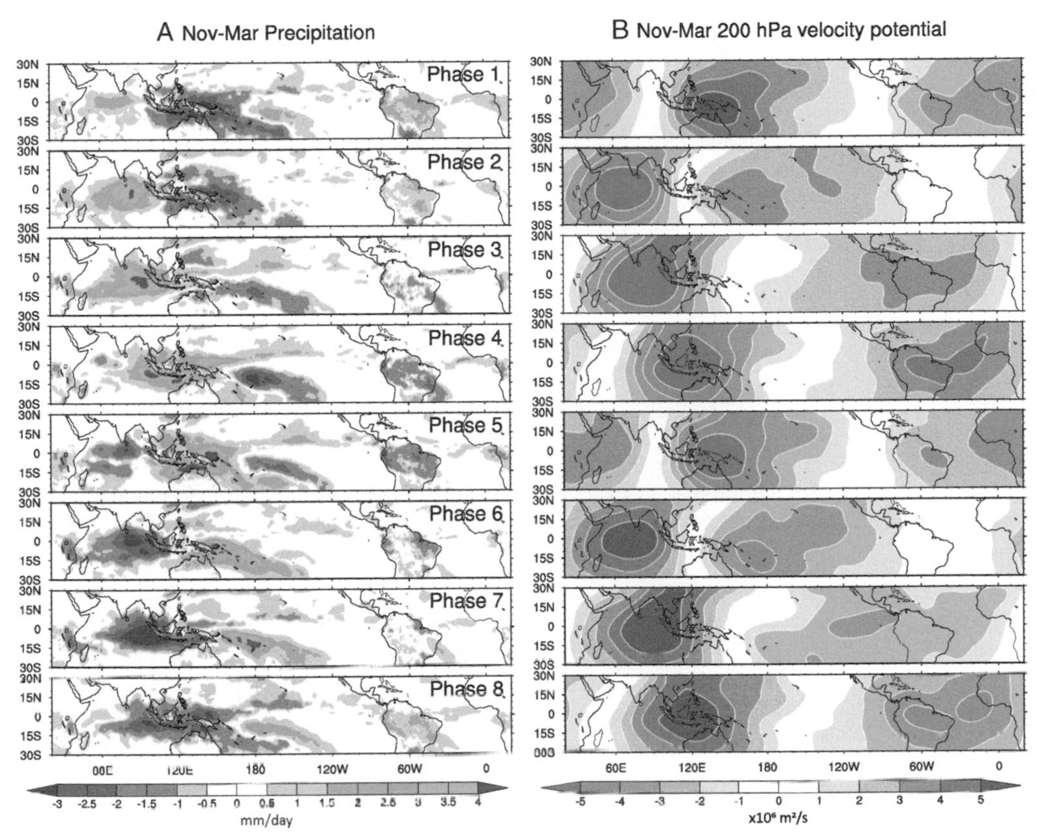

Fig. 4.11 MJO anomalies at eight phases for November to March 1997-2020: (A) precipitation (mm/day) and (B) 200 hPa velocity potential (10^6 m^2/s). *(Courtesy Y. Liang.)*

4.2.4 Seasonality

Convective variability of MJO is largely confined to the Indo-western Pacific warm pool encompassed by the SST = 28°C isotherm. In the meridional direction, the Indo-western Pacific warm pool is displaced into the summer hemisphere, and so is the band of high precipitation variance of MJO. The MJO is most pronounced during the boreal winter months (November—March) and the precipitation variance is larger than boreal summer months (May—September), especially over the western Pacific (Fig. 4.12). Our discussion of MJO so far applies largely to the Northern Hemisphere (NH) winter.

In the NH winter, the mean westerly flow in the extratropical upper troposphere serves as a strong waveguide for barotropic Rossby waves (see Chapter 9). When the convective center of MJO is located over the maritime continent (phase 4), a wave train emanates, with centers of action over the Aleutians (high pressure) and western Canada

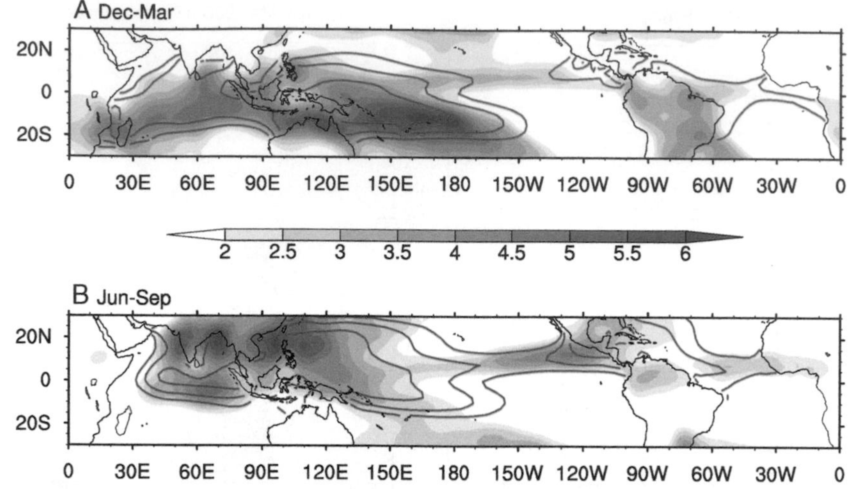

Fig. 4.12 MJO standard deviation of precipitation (mm/day, shaded) and mean sea surface temperature (27°, 28°, and 29°C contours) for (A) December to March and (B) June to September 1997-2020. *(Courtesy Y. Liang.)*

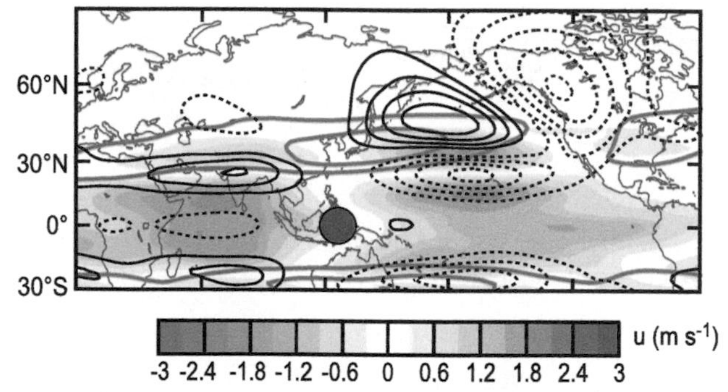

Fig. 4.13 MJO anomalies of zonal wind (shading) and Z (solid black contours at 5 m interval, the zero contour omitted) superimposed upon the annual-mean climatologic u (orange contours), all 100 to 300 hPa averaged. The orange contour interval is 10 m/s starting at 20 m/s. The red circle indicates velocity potential minimum at upper levels corresponding to MJO phase 4. *(From Adames and Wallace [2014]. © American Meteorological Society. Used with permission.)*

(low pressure) (Fig. 4.13). This wave train is barotropic (of the same phase through the troposphere) and called the Pacific-North American (PNA) teleconnection pattern. Here teleconnection refers to the connections among far-apart regions. The MJO-induced PNA pattern persists for 2 to 3 weeks, affecting North American weather pattern (e.g., the probability of a North Pacific blocking event is elevated during MJO phases 5–6). The MJO-excited teleconnection pattern (e.g., PNA) is pronounced in local winter when the mean westerly flow and the barotropic waveguide effect are strongest.

During May to September, MJO convective variability shifts into the Northern Hemisphere. At RMM phase 4, a northwestward-tilted rain band develops, stretching from India to the maritime continent and equatorial western Pacific (Fig. 4.14). This slanted band of active convection propagates eastward, causing an apparent northward phase propagation (Fig. 4.15). Yasunari (1979) and Sikka and Gadgil (1980) first discovered coherent northward propagating oscillations in the Indian Ocean sector from satellite cloud imageries.[7,8] The northward propagating MJO causes rainfall over India to oscillate between active and break phases at the intraseasonal timescale. For this reason, the boreal summer intraseasonal oscillation is also called the monsoon intraseasonal oscillation.

In NH summer, the intraseasonal convective variability is coherent across the Pacific north of the equator. During phases 4 to 6, convection is active across the Pacific ITCZ, while it is suppressed over the far eastern Pacific south of Mexico and the Caribbean (see Fig. 4.14). This apparent west-east seesaw of summer convective activity across the

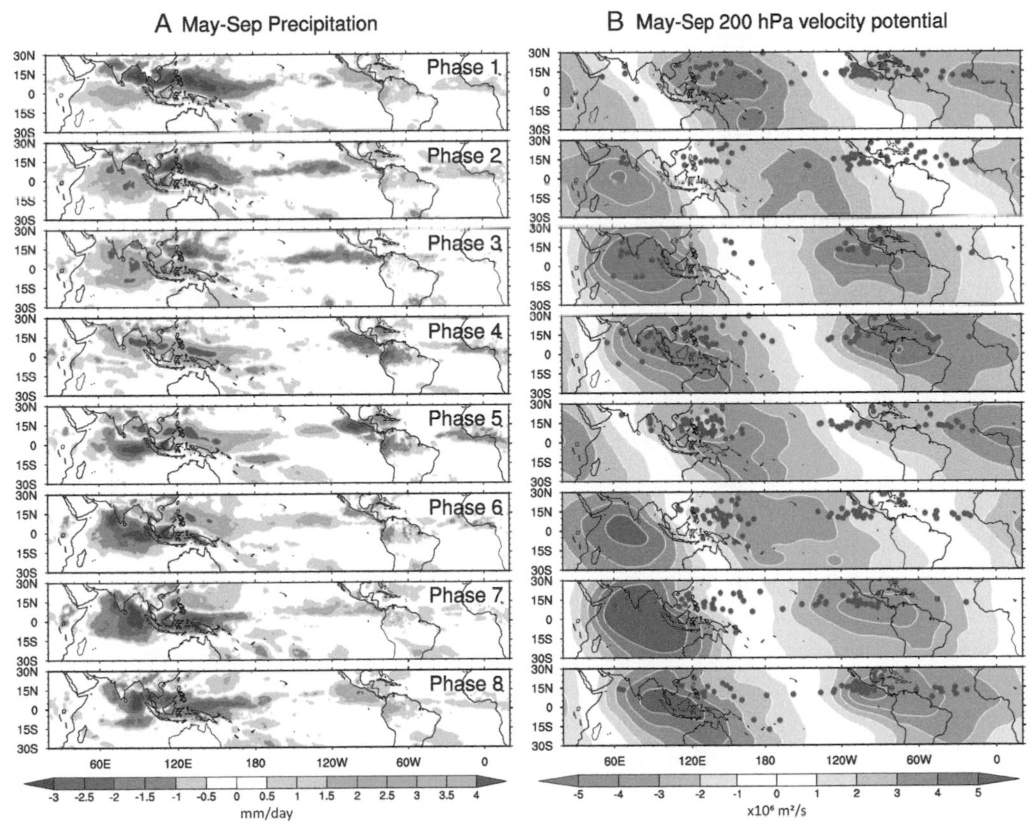

Fig. 4.14 MJO anomalies during May to September: (A) precipitation (mm/day) and (B) 200 hPa velocity potential (10^6 m^2/s) along with tropical cyclogenesis *(red dots)* for 1997-2020. *(Courtesy Y. Liang.)*

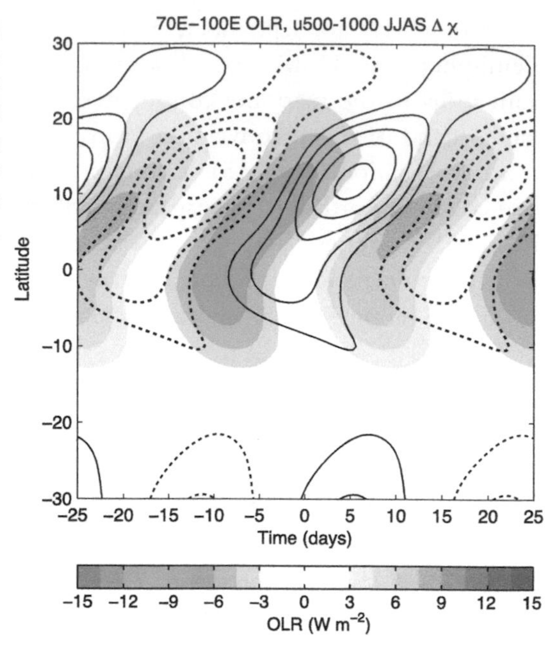

Fig. 4.15 Time—latitude section of outgoing longwave radiation *(OLR)* (shaded) and 500 to 1000 hPa averaged u (contours; interval is 0.2 m/s) for June-September, zonally averaged over 70° to 100°E. The diagram is based on linear combinations of PC1 and PC2 of velocity potential and a 40-day cycle. *(From Adames et al. [2016]. © American Meteorological Society. Used with permission.)*

tropical Pacific is mediated by equatorial Kelvin waves radiating from the MJO convection over the Indo-western Pacific.

Summer is the season of tropical cyclones, which are highly organized cloud clusters with strong cyclonic rotational flow in the lower troposphere. MJO modulates tropical cyclone statistics. Tropical cyclogenesis tends to occur when and where the MJO convection is active with upper-level divergence (see Fig. 4.14). Enhanced columnar water vapor path, active convection, and low-level cyclonic flow are large-scale environmental conditions in favor of tropical cyclogenesis (Section 10.5).

4.2.5 Ocean response

MJO is associated with pronounced variability in the zonal wind on the equator. MJO wind variability is large in the western half of the Pacific, with an amplitude of 1.5 m/s at the surface (Fig. 4.16A). The westerly wind anomalies in the western Pacific drive downwelling Kelvin waves (with an anomalously deepened thermocline; see Chapter 8) that propagate eastward at a phase speed of 2.2 m/s (see Fig. 4.16B). The typical amplitude of the oceanic Kelvin waves is 4 cm in sea surface height (or 13 m in the thermocline depth). The typical period of ocean waves is 60 days, on the longer side of the westerly wind pulses in the western Pacific. This apparent mismatch in frequency with the wind forcing is because the ocean Kelvin wave is more responsive to low-frequency wind forcing. The downwelling ocean Kelvin waves are associated with anomalous eastward

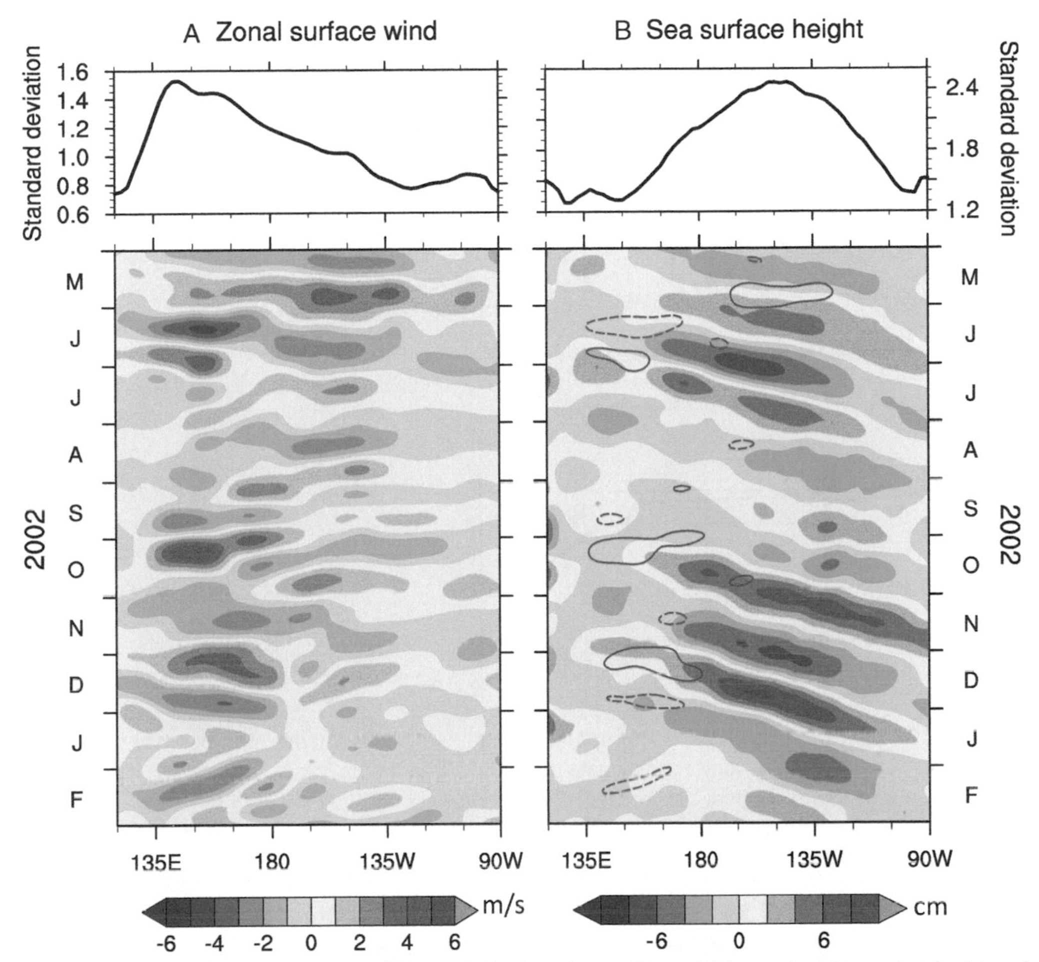

Fig. 4.16 Longitude-time sections of 20–120-day band-pass-filtered (A) zonal surface wind (m/s) and (B) sea surface height (cm) anomalies at the equator across the Pacific. The upper panels show the standard deviations for 1997-2020. In (B), the $u' = 2.5$ m/s contours are plotted in grey (negative dashed). *(Courtesy Y. Liang.)*

currents, which push the western Pacific warm pool to expand eastward through advection. The deepened thermocline further reduces the upwelling cooling, causing surface warming especially in the eastern Pacific where the mean thermocline is shallow and close to the surface.

Over the Indo-western Pacific warm pool, heavy rainfall often creates a shallow freshwater lens atop a deep isothermal layer. Fig. 4.17 shows an example in the summer Bay of Bengal. After a rainstorm in late July/early August 1999, a low-salinity layer forms within the 30-m deep isothermal layer, and the resultant density stratification shallows the surface mixed layer to only 10 to 20 m deep. The isothermal layer below the shallow

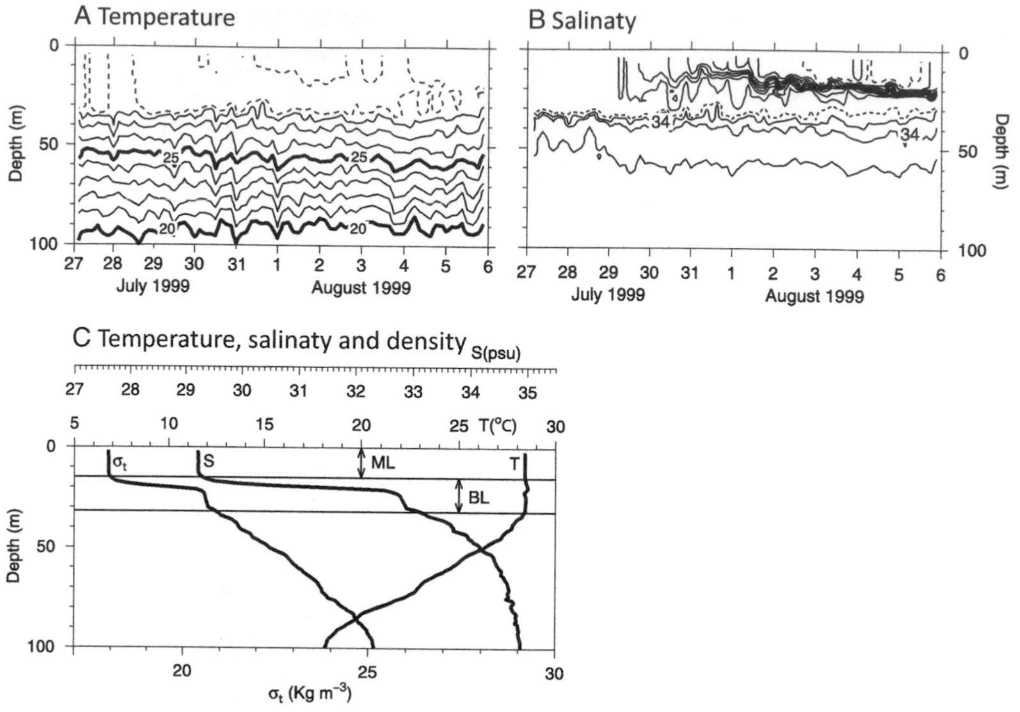

Fig. 4.17 Time-depth sections of (A) temperature and (B) salinity in the summer Bay of Bengal (17°30′N, 89°E), following a rainy event in late July 1999. (C) Schematic of a low-salinity mixed layer riding above a nearly isothermal barrier layer. *(From Vinayachandran et al. [2002].)*

halocline is called the barrier layer, which resists the mixing with the cold thermocline. Under calm sunny conditions after the rainstorm, the shallow halocline can cause large surface warming. As the next rainstorm approaches, surface cooling and wind mixing destroy the surface halocline. Only afterwards, the turbulent mixing may reach the isothermal barrier layer.

In the summer Bay of Bengal, the MJO-induced SST variability can exceed 1°C. Intraseasonal anomalies of zonal wind and convection are about 90° out of phase. The active convection is preceded by anomalous easterly wind (see Fig. 4.15), which reduces the background westerly wind. The reduced wind causes SST to rise at a 90° lag. Thus the SST and convection variability are nearly in phase, suggestive of positive feedback. Indeed in models, the ocean coupling seems to amplify the summer MJO over the Bay of Bengal.

In general, intraseasonal variability in SST and convection is not in phase, and the effect of ocean coupling is not that clear. Since ocean is a low-pass filter and does not respond strongly to intraseasonal atmospheric forcing, MJO-induced SST variability is generally modest in magnitude but can be enhanced due to ocean upwelling. Large intraseasonal SST variability is observed in the summer central-west South China Sea and winter tropical South Indian Ocean (~10°S). Both are regions of ocean upwelling.

4.2.6 Subseasonal prediction

The low-frequency and oscillatory nature of RMM indices (see Fig. 4.10) suggests that the MJO may be predictable at leads beyond 2 weeks, the latter considered the limit of deterministic weather forecast. Initialized dynamic models have been used to predict the MJO. Fig. 4.18 shows an example. The skill as measured by anomaly correlation coefficient between observed and predicted RMM indices decreases with the lead time but stays above 0.5 (considered as useful) at lead times up to 25 to 30 days. Of highest skill are the forecasts initialized at phase 3 (see Fig. 4.18) when the MJO convection peaks in the Indian Ocean (see Fig. 4.11). The skill is lowest for forecasts initialized at phase 1 when the convective anomalies are not yet fully developed over the Indian Ocean.

Weather forecast targets at the timing, location, and intensity of weather phenomena such as rainstorms and tropical cyclones. The chaotic nature of the atmosphere limits the useful weather forecast to lead times up to 2 weeks. Because of the MJO effect on tropical cyclogenesis and midlatitude stationary waves, MJO forecast offers hopes for extended weather forecast beyond 2 weeks. For example, week 4 forecast of winter surface air temperature over North America is skillful at MJO phase 7 (Johnson et al. 2014), following the MJO-forced PNA pattern (see Fig. 4.13).

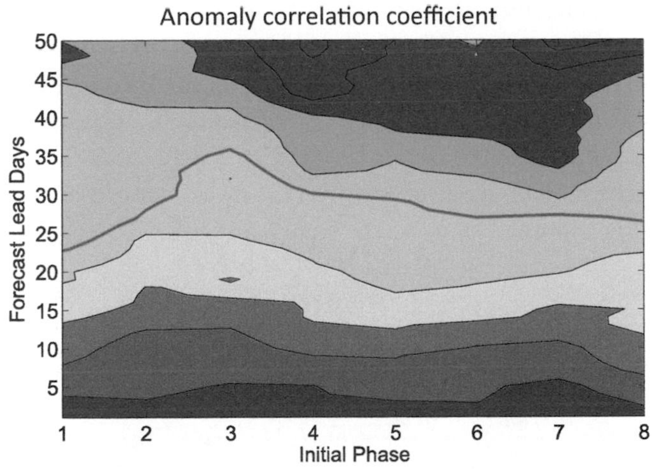

Fig. 4.18 Anomaly correlation coefficient (red solid contour = 0.5, ci = 0.1) for real-time multivariant MJO indices as a function of the MJO phase and forecast lead time (days). *(From Xiang et al. [2015]. © American Meteorological Society. Used with permission.)*

4.3 Moisture mode theory

From a longitude-time section of convective variability (see Fig. 4.3), it is unclear a priori why there are two physically distinct types of eastward-moving convective perturbations: fast, circumglobal Kelvin waves and slow MJO envelops confined to the Indo-western Pacific sector. The analysis of the horizontal structure reveals that the fast phase speed allows the perturbations to retain much of the Kelvin wave structure (see Fig. 4.4) and travel circumglobally across considerable variations in the underlying SST. The Matsuno theory is based on the budget of column-integrated dry static energy ($s = c_pT + gz$) (see Eq. 3.6)

$$\frac{\partial\langle s\rangle}{\partial t} - \widetilde{\omega}\widehat{M}_s = -\langle R\rangle + LP \tag{4.16}$$

where $\langle\bullet\rangle$ denotes the column-integration, $\widehat{M}_s = -\langle\Omega\frac{\partial\bar{s}}{\partial p}\rangle$ is the dry static stability, which provides the restoring force for equatorial waves. Here $\omega = \widetilde{\omega}(x, y, t)\bullet\Omega(p)$, where $\Omega(p)$ represents the vertical structure of the perturbation pressure velocity and peaks around the mid-troposphere. We have neglected the horizontal advection terms by invoking the weak temperature gradient approximation.

The CC Kelvin wave and MJO are dynamically distinct. While the prognostic temperature term in Eq. 4.16 is essential for Matsuno waves, it is negligible for the slow MJO

$$-\widetilde{\omega}\widehat{M}_s = -\langle R\rangle + LP \tag{4.17}$$

Indeed for MJO, the flow is in quasi-steady equilibrium with the convective heating and displays the characteristic Matsuno-Gill pattern of Kelvin-Rossby couplet (see Fig. 4.7). Column-integrated water vapor path $\langle q\rangle$ turns out to be a slow variable strongly coupled with deep convection; a moist atmospheric column is favorable for deep convection by reducing the dry entrainment. Indeed, precipitation increases exponentially when the column vapor path exceeds a threshold (see Fig. 3.6B).

The column-integrated budget of perturbation moist static energy (MSE) ($m = s + Lq$) (see Eq. 3.15) is expressed as

$$\left\langle\frac{dm}{dt}\right\rangle = R_{TOA}\downarrow - Q_{net} \tag{4.18}$$

where

$$\left\langle\frac{dm}{dt}\right\rangle = \frac{\partial\langle m\rangle}{\partial t} - \widetilde{\omega}\widehat{M}_s\widehat{M} + \langle\boldsymbol{u}\bullet\nabla m\rangle \tag{4.19}$$

$$\widehat{M} \equiv \frac{\widehat{M}_s - \widehat{M}_q}{\widehat{M}_s} \tag{4.20}$$

is the gross moist stability normalized by the gross dry stability, and $\widehat{M}_q = L\left\langle \Omega \frac{\partial \overline{q}}{\partial p} \right\rangle$ measures the moisture stratification. Since the effects of background $\overline{s}$ and $\overline{q}$ on moist stability oppose each other, $\widehat{M} \ll 1$ (Section 2.4.1). For MJO, perturbation MSE is dominated by latent energy of water vapor, $\langle m \rangle \approx L\langle q \rangle$. The MSE budget can be approximated as

$$L\frac{\partial \langle q \rangle}{\partial t} - \widetilde{\omega}\widehat{M}_s\widehat{M} + L\langle \boldsymbol{u}\cdot\nabla q \rangle = R_{TOA}\downarrow - Q_{net} \tag{4.21}$$

Because $\widehat{M} \ll 1$, the horizontal advection is as important as the vertical advection, causing the MJO to propagate eastward. Since prognostic water vapor is essential, MJO is considered as a moisture mode (Sobel and Maloney 2013) missing in Matsuno dry wave solutions. Over the Indo-western Pacific warm pool, the moisture mode becomes unstable due to convection-circulation interaction (the top loop of Fig. 3.21), resulting in spontaneous variability internal to the atmosphere. See Jiang et al. (2020) and Adames and Maloney (2021) for recent reviews of the MJO and moisture mode theory.

Review questions

1. The Matsuno theory predicts Kelvin waves with a phase speed of ~50 m/s. Why don't we see eastward propagating waves of such a fast phase speed in OLR measurements? What slows down OLR waves?
2. What is the restoring force for Matsuno dry Kelvin waves? How does the latent heating of water vapor condensation change the restoring force?
3. The $n = 2$ mode of Rossby wave is antisymmetric about the equator. On the frame moving eastward at $c_m = 15$ m/s, calculate the Doppler-shifted phase speed of the antisymmetric Rossby wave. Show the ratio to the dry wave speed $c'_{R2}/c_{R2} = 1 + 5r_c$, where $r_c = c_m/c_d$. Explain why the antisymmetric flow perturbations are not well developed in convectively coupled Kelvin wave even though the convective anomalies are displaced north of the equator along the ITCZ (see Fig. 4.4).
4. Why are convective anomalies of MJO most pronounced over the eastern Indian and western Pacific oceans?
5. Why do we consider the MJO distinct from CC Kelvin waves? Discuss the horizontal structures of circulation and convection.
6. Why did the Matsuno theory "miss" the MJO?
7. Why do meteorologists pay more attention to stream function than velocity potential in the midlatitudes?

8. Why are tropical meteorologists interested in velocity potential?
9. Expand the eastward-propagating wave $\Phi = A \sin[k(x\text{-}ct)]$ into stationary wave modes $\sin kx$ and $\cos kx$. What are the principal components associated with these stationary wave modes? For MJO, what are the typical values of k and c?
10. Identify the MJO phase at which intraseasonal convection is most active during May to September at the following sites: New Delhi, India; Singapore; Jakarta, Indonesia; and Managua, Nicaragua.
11. Estimate the phase speed of the ocean Kelvin wave from the Hovmöller diagram of sea level variability across the equatorial Pacific.
12. Show that the definitions of gross moist stability in Eqs. 4.19 and 2.17 are equivalent.
13. Vary the vertical structure function $\Omega(p)$ and consider its impact on the gross moist stability $\widehat{M}$: $\Omega(p)$ peaks in the upper/lower troposphere above/below the $\overline{m}$ minimum.

CHAPTER 5

Summer Monsoons

Contents

Because the axis of the Earth rotation is titled by about 23.5° from the axis of revolution around the Sun, solar radiation at the top of atmosphere (TOA) in the Northern Hemisphere peaks at the summer solstice (June 21) and reaches the minimum at the winter solstice (December 21). The solar radiation is nearly symmetric about and of a broad peak on the equator on spring (March 21) and fall (September 21) equinoxes when the length of day is 12 hours everywhere on Earth. The annual cycle in downward solar radiation at the TOA increases with latitude but that in the absorbed solar radiation peaks in the subtropics because atmospheric scattering and surface reflection both increase with the solar zenith angle. Because of large heat capacity, the seasonal cycle is generally much smaller in sea surface temperature (SST) than air temperature in the planetary boundary layer (PBL) over land. This is apparent in a map of the summer-winter surface air temperature (SAT) difference.

The differential heating of the atmospheric column between continents and oceans drives planetary-scale circulation patterns that vary from winter to summer. In Arabic, monsoon originally refers to seasonal wind reversals. Such seasonal wind reversals take place on the southeastern part of the Asian continent, south and east of high mountains of Tibet. Over the North Indian Ocean (including the Arabian Sea and Bay of Bengal), India, Indochina Peninsular, and the South China Sea, the southwest winds prevail in

Coupled Atmosphere-Ocean Dynamics: from El Niño to Climate Change
ISBN 978-0-323-95490-7, https://doi.org/10.1016/B978-0-323-95490-7.00005-9

summer while the northeast winds in winter (Fig. 5.1). Over eastern China to Japan, the summer winds are from south to southwest while the winter winds are northwesterly carrying dry cold air from Siberia. Sailors knew the monsoonal winds for more than 2000 years and used them for navigation to trade between China, Malay Archipelago, India, Persia, and East Africa for centuries. The maritime Silk Road became important for trading in the 11th to 13th centuries when the Chinese Song dynasty lost control over the land Silk Road. "Sail out on the northerly winds and come home on the southerlies," Song governor of Quanzhou—a bustling port on the southern Chinese coast where the maritime Silk Road began—wrote in a poem.

During winter, the northeasterly winds push rainfall away from South Asia and blow into the marine intertropical convergence zone (ITCZ) near the equator. Synoptic cold surges from the Siberian intrude as south as the South China Sea. The Tibetan Plateau blocks cold air from intruding southward, making countries neighboring the North Indian Ocean much warmer than at the same latitude in China. For example, the January mean temperature is 20.1°C in Calcutta but only 15.8°C in Hong Kong, both at 22°N.

The southwest monsoon transports moisture from the tropical oceans and produces rainfall over South and East Asia. Monsoon rains drive great migrations of herbivorous mammals and their predators (as beautifully captured by *Nature* television programs) across African savannas. Abundant summer rains are a defining feature of monsoon climate (Fig. 5.2), enabling major population centers of Africa, South Asia, Indochina, East Asia, and the Americas (Section 5.6). Twelve of the world's 15 most populous countries are in monsoon regions, each with 100 million or more and together accounting for half of the world's population. The monsoon rains shape the vegetation, crops, agriculture, culture, and history. For example, people have long developed an empirical knowledge of monsoon climatology to predict the onset of the rainy season. Chinese divided a solar year into 24 terms. One of the solar terms "Grain Rains" (谷雨; April 19–21)

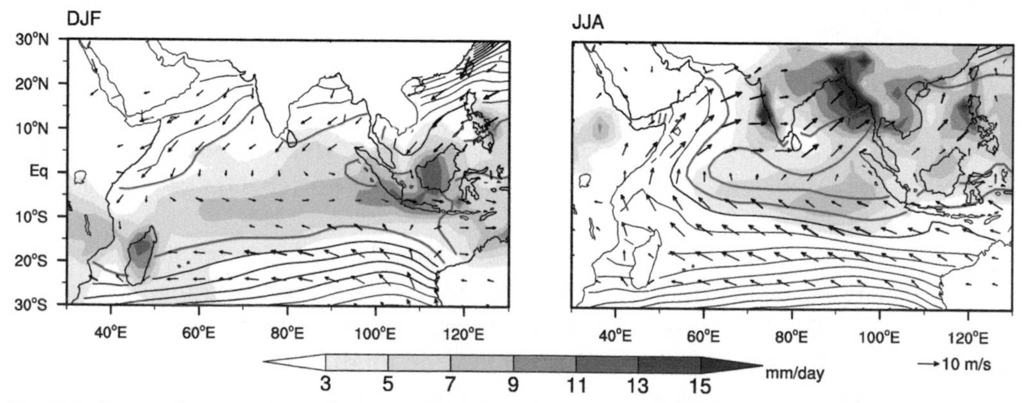

Fig. 5.1 Seasonal-mean sea surface temperature (SST) at 1°C interval (SST ≥28°C thickened in red), precipitation (color shading in mm/day), and surface wind velocity (m/s). *(Courtesy ZQ Zhou.)*

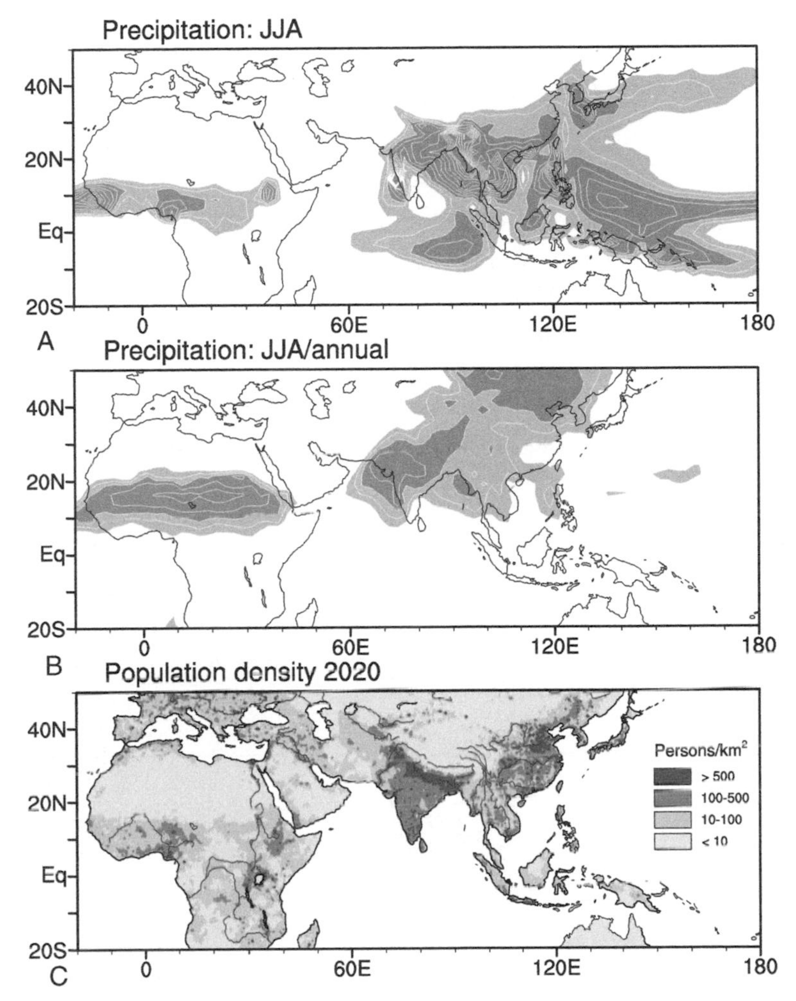

Fig. 5.2 Summer (JJA) Global Precipitation Climatology Project precipitation (A) amount (light gray >400 mm; dark gray >600 mm, white contours 400–1500 with interval of 100 mm) and (B) as a fraction of the annual accumulation (light gray >0.4; dark gray >0.6, white contours 0.4–0.8 with interval of 0.1). Population density in 2020 (color shading: 10, 100, 500 persons/km^2). *(Courtesy ZQ Zhou.)*

literally refers to the seasonal increase in precipitation that marks the time to plant grain crops. Historic conflicts, unrests, and the fall of dynasties were often triggered by major droughts and famines.

We focus on summer monsoons in the Northern Hemisphere. Readers are referred to Chang et al. (2006) for Asian winter/Australian summer monsoons and Vera et al. (2006) for the South American monsoon.

5.1 South Asian monsoon

While the Madden-Julian oscillation (MJO) is the spontaneous mode of tropical convection, summer monsoons are the organization of convection powered by the seasonal increase in solar energy through interactions with land and ocean. In summer (June-July-August [JJA]), convection centers develop off the west coast of India, over the Bay of Bengal and South China Sea. The deepest convection anywhere on Earth at the time is in the northeastern Bay of Bengal with the seasonal mean outgoing longwave radiation (OLR) less than 200 W/m^2. Curiously there remains an Indian Ocean marine ITCZ at the height of the summer monsoon, as a band of enhanced rainfall displaced slightly south of the equator. This south equatorial ITCZ is not that clear in the OLR distribution, however, suggesting shallower convection than in the Bay of Bengal. Broadly, atmospheric connection in summer South Asia is concentrated in a latitudinal band of 10°N to 25°N.

5.1.1 Circulation

Gill (1980) considers a heat source displaced north of the equator to mimic the summer Asian monsoon convection (Fig. 5.3). With the heating centered one Rossby radius north of the equator, the wind and pressure response is zonally confined compared

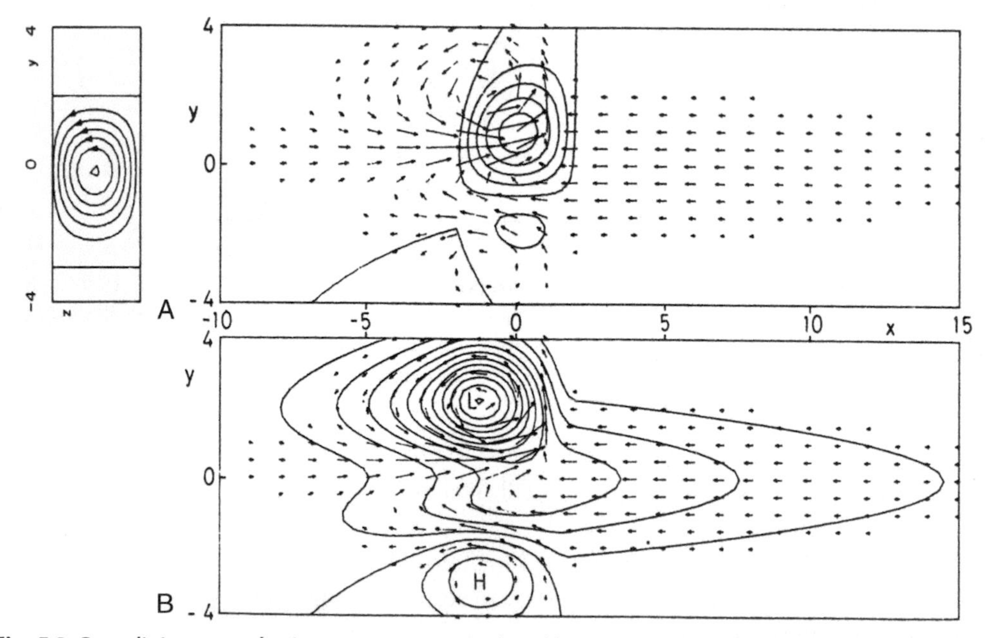

Fig. 5.3 Baroclinic atmospheric response to an isolated heating north of the equator: (A) vertical motion (contours); (B) low-level geopotential height (contours) and velocity. *(top left)* Zonal-mean meridional overturning circulation. *(From Gill [1980].)*

with the case with equatorial-centered heating, but a weak Kelvin wave response is still present to the east, symmetric about the equator. Over and west of the heat source, a strong cyclonic circulation (around a low-pressure center) forms in the lower troposphere with strong westerlies south of the convection center. Southerly cross-equatorial winds converge onto the heat source north of the equator. The response to the west is mixture of the symmetric and antisymmetric Rossby waves. In the zonal-mean meridional section, a one-cell Hadley circulation forms, with rising motions in the prescribed heating and subsidence on the other side of the equator. Such a one-cell Hadley circulation across the equator is observed in the summer.

With the aid of the Gill model, the convective heating over the Asian monsoon region explains much of the circulation at both low and upper levels (Fig. 5.4). In the lower troposphere, strong westerlies are observed along the south coast of Asia, and strong southerly cross-equatorial winds blow in the western Indian Ocean, both features consistent with the Gill model. To the east of the Asian monsoon and western Pacific convection, the easterlies persist in the tropical Pacific. In the upper troposphere, a gigantic anticyclonic circulation called the Tibetan High or South Asian High is observed over

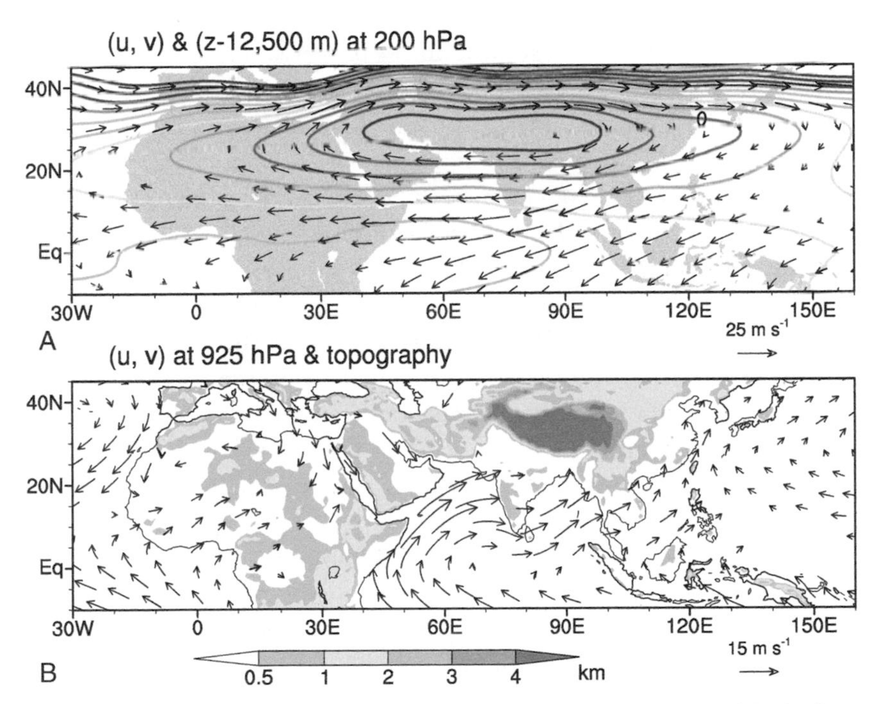

Fig. 5.4 July climatology. (A) Geopotential height (m) and wind velocity at 200 hPa (red >12,530 m, contour interval = 30 m). (B) Wind velocity at 925 hPa superimposed on land topography (shading). *(Courtesy ZQ Zhou.)*

the north Indian Ocean and South Asia, extending westward all the way to north Africa. This upper-level South Asian/Tibetan High is consistent with the Gill solution in response to massive heating in the South Asian monsoon. The upper easterly winds reach 20 to 30 m/s over the vast subtropical region south of the Tibetan High.

The strong vertical shear between the upper easterlies and lower southwesterlies prevents tropical cyclones (TCs) from developing over the North Indian Ocean at the height of summer monsoon despite high SSTs and strong convection. TCs are warm-core vertical vortices. By tearing apart the upright structure, strong vertical shear is unfavorable for TC growth. From the Arabian Sea to Bay of Bengal, TCs develop in the intermonsoon months of April-May and October-November when SSTs are still high but vertical shear weakens.

In the Gill model, the low-level westerlies blow southwest of the monsoon heat sources unimpededly as the Rossby response. In reality, the African highlands interrupt and the east slope of the mountain range (1–2 km high) banks strong pressure gradients (somewhat like a coastal Kelvin wave) as the baroclinic Rossby wave excited by monsoon convection impinges on east Africa. A strong southwesterly jet forms as a result, hugging the Somali coast (Fig. 5.5). This Findlater wind jet commences with the developing Indian monsoon.

Socotra island (54°E, 12°30′N; population ~60,000) stands in the way of the Findlater wind jet at the mouth of the Gulf of Aden. Every spring, island residents are busy fetching supplies from mainland Yemen on the Arabian Peninsula in anticipation of strong winds that will isolate them from the rest of the world from May through September. Summer is dry and cool because of ocean upwelling. Socotra gets most rainfall during the winter months of November-December.

The southwesterly Findlater wind jet drives the offshore Ekman flow at the surface, and the resultant intense ocean upwelling off the Somali and Omani coasts lowers coastal SSTs to as low as 20°C (see Fig. 5.5). The offshore spread of the cold coastal water limits the monsoon convection to the eastern Arabian Sea. The intense coastal wind jets, southwesterly in summer and northeasterly in winter, drive coastal jets in the ocean that reverse the direction and often exceed 1 m/s in each season. Such intense, seasonally reversing jets came as a big surprise to oceanographers, who were used to steady western boundary currents such as the Kuroshio and Gulf Stream,

Summer upwelling is also observed off the southern tips of India and Sri Lanka, and off southern Vietnam, in response to local orographic wind jets. The broad Bay of Bengal is prevented from coastal upwelling cooling possibly by a strong surface halocline that forms due to strong freshwater flux (from rain and river discharge) in an isothermal layer (see Section 4.4.2).

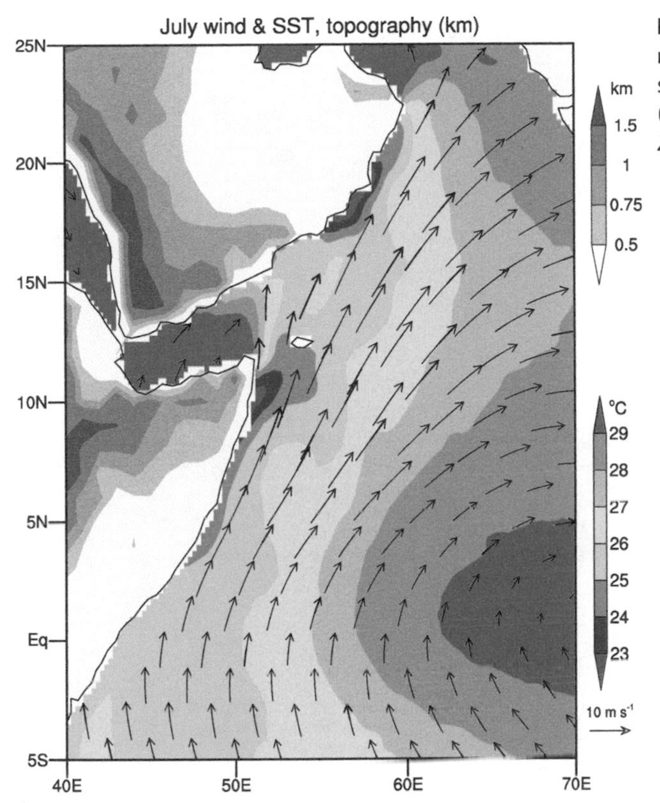

Fig. 5.5 July climatology and topography (grey shading, km). QuikSCAT surface wind velocity (m/s) and SST (color shading, °C). *(Courtesy ZQ Zhou.)*

5.1.2 Orographic effects on convection

Over a large region of the North Indian Ocean south of the Tibetan Plateau, upper tropospheric temperature increases northward (Fig. 5.6C), reversing the thermal gradient. The subtropical maximum of tropospheric temperature is consistent with Gill's response to a northward-displaced heating from India to the Philippines.

The Tibetan Plateau is important for the monsoon convection. The colocation of the tropospheric temperature maximum on the south plateau inspired the hypothesis that the absorption of solar radiation by the elevated plateau surface helps heat the atmosphere above (Yeh 1957; Flohn 1957). Indeed, summer-mean surface temperature in Lhasa (elevation: 3.65 km) is comfortably warm at 15°C, 10°C warmer than over the North Pacific at the same altitude and latitude. The daily maximum at Lhasa is even higher than over the Pacific, which determines the PBL temperature above a shallow nocturnal surface layer. The large Q_1 near the surface indicates that the mountains heat the atmosphere via sensible heat, although the latent heating contribution is significant (Q_2 in Fig. 5.6A), especially in the eastern plateau once summer rains commence. The repeating

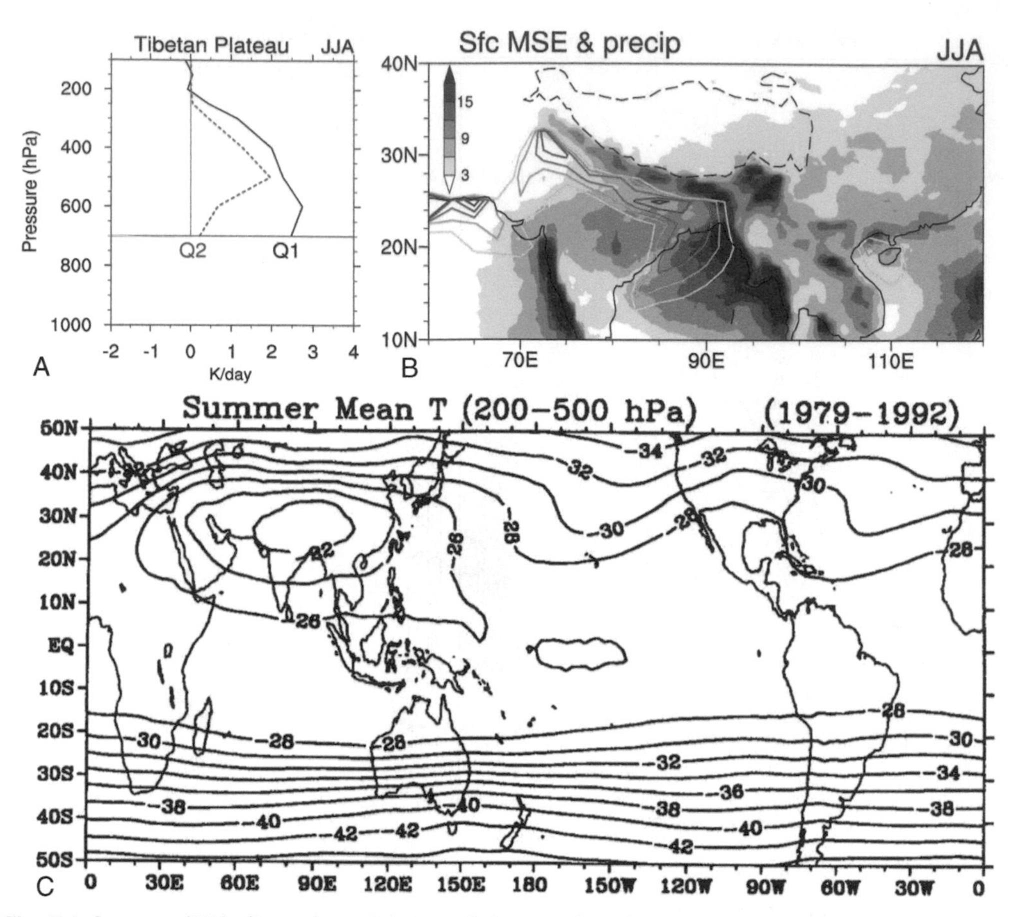

Fig. 5.6 Summer (JJA) climatology. (A) Q_1 and Q_2 over the Tibetan Plateau. (B) Surface moist static energy/C_p (contours >350 at intervals of 1 K), and precipitation (shading, mm/month). (C) Upper tropospheric (500–200 hPa) mean temperature (°C). (A, Based on Yanai and Tomita [1998]; B,). *(courtesy of ZQ Zhou; C, Li and Yanai [1996]. © American Meteorological Society. Used with permission.)*

cycle of strong daytime solar heating and nighttime longwave cooling on the mountain surface produces towering convection on the south slope of the Himalayas during the afternoon/early evening, with suppressed convection at night. Thus with sensible heating on its surface the Tibetan Plateau triggers a strong interaction of circulation and convection, making it a major forcing mechanism for the Asian monsoon. Deep moist convection is an efficient way to collect solar energy over the vast Indian Ocean (in latent heat of water vapor evaporation), concentrate, and release the energy to warm the atmosphere in a small area.

A complementary view of the orographic effect stresses the role of the high Himalayas as a barrier protecting air of high moist static energy (MSE) to the south from mixing with low-MSE air north of the Plateau (Boos and Kuang 2010). South of the Himalayas, surface MSE is indeed high (see Fig. 5.6B) because of solar heating and the moisture transport from the warm Indian Ocean. In convection, undiluted parcels of high MSE reach a high level of neutral buoyancy, and the latent heat release results in high temperature in the upper troposphere. Regions of highest surface MSE do not exactly coincide with regions of strongest convection and are displaced toward dry Pakistan away from the most convective Bay of Bengal. Rising in the dry ambient atmosphere over Pakistan, parcels lose buoyancy through lateral entrainment. The elevated surface heating and barrier effects of the Tibetan Plateau seem complementary in driving monsoon convection.

The orographic effect on the Asian summer monsoon has been demonstrated with atmospheric general circulation models (GCMs) by artificially removing mountains. Without the plateau, monsoon convection retreats southward and weakens over South and East Asia (Fig. 5.7). Without the orographic blockage by the African highlands, the Findlater jet disappears. (In winter, by contrast, the northeast monsoon intensifies over South Asia and the North Indian Ocean as the Siberian high expands and drives cold air southward.)

High-resolution satellite observations reveal striking influences of north-south—oriented mountain ranges as narrow as a few 100s km across and only 1 km high or less. As the southwest monsoon impinges on the narrow mountains, moisture-laden air is forced to rise, causing intense convection on the windward side. Specifically, from the west to east, a sharp rain band each is found windward of the Western Ghats, the mountain ranges of the eastern coast of the Bay of Bengal and off the Cambodian coast,

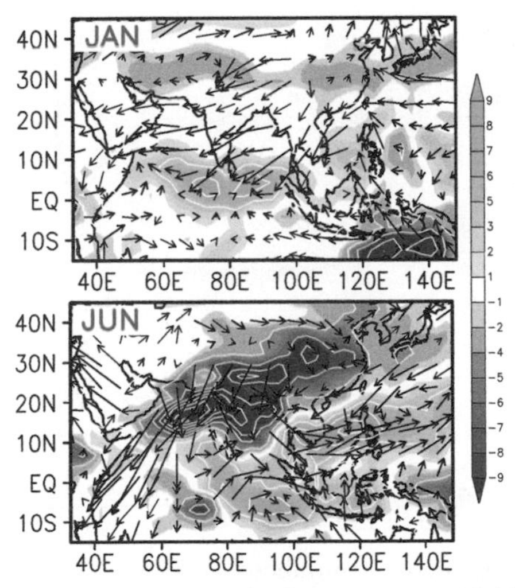

Fig. 5.7 Effects of removing global orography as the flat minus control difference with an atmospheric general circulation model for *(top)* January and *(bottom)* June: precipitation (shading, white contour intervals of 4 mm/day) and 850 hPa wind velocity (m/s). *(From Okajima [2006], PhD thesis.)*

the Annam Cordillera range on the border between Laos and Vietnam, and mountains of the Philippines (Fig. 5.8; see also Fig. 5.6B). Each of these mountain ranges entails a rain shadow on the leeside, another manifestation of the orographic effect. Even the Andaman Islands, a tiny island chain in the southern Bay of Bengal with the highest peak at 738 m, anchor a rain band to the west and a rain shadow to the east. Surprisingly, these orographic rain bands are consistently displaced windward of the mountain ranges, often off the coast over the flat ocean. The offshore rainfall peak off the Indian west coast was first detected in the 1970s in an aircraft field campaign.

In the time-longitude section (see Fig. 5.8), orographic rain bands appear as stationary features. As the Sun moves into the Southern Hemisphere, the first wind reversal takes place in October over the South China Sea and the western Pacific, where the

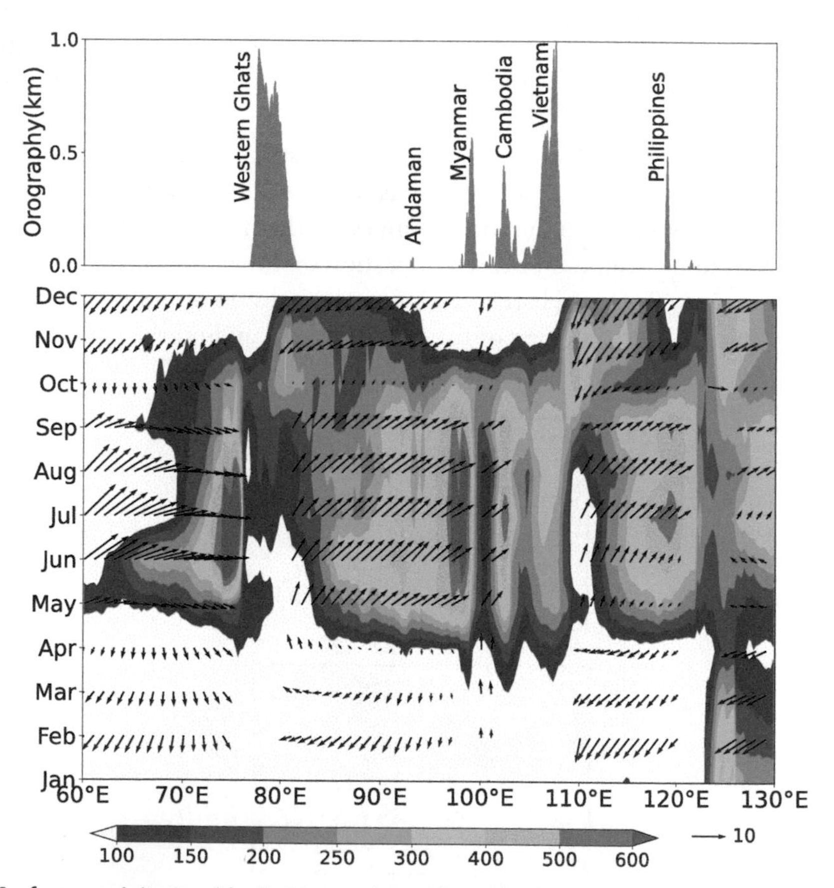

Fig. 5.8 Surface precipitation (shaded in mm/month) and surface wind velocity (m/s) at 12.5°N as a function of longitude and calendar month, along with land orography (*top panel* in km). Mountain ranges on display are western Ghats, Andaman Isle, Tenasserim/Myanmar, Cardamom/Cambodia, Annam Cordillera/Vietnam, and the Philippines. *(Adapted from Xie et al. (2006).)*

orographic rain bands shift from the west to the east side of Annam Cordillera and the Philippines, facing the prevailing northeasterlies. The Vietnam coast rain band persists from October to December and weakens eventually because of the cooling of the South China Sea and advection of dry continental air by the northeast monsoon.

A narrow rain band hugs the south slope of the steep Himalayas and displays a pronounced diurnal cycle. Further to the south in the vast floodplain that stretches through Bangladesh, another hotspot of summer convection develops as the moisture-laden southerlies from the Bay impinge on the Shillong Plateau. The south-facing station of Cherrapunji, India (25.2°N, 91.7°E) is crowned the wettest place on Earth with a whopping annual rainfall of 12,700 mm, most of it during the 5 summer months of May to September. Satellite data suggest even heavier summer rainfall out in the ocean off the coastal ranges of Myanmar (see Fig. 5.6B).

5.1.3 Onset

The arrival or onset of the summer rainy season is of vital importance for people, agriculture, and environment in South and East Asia. Takio Murakami, a pioneer monsoon meteorologist based at the University of Hawaii, was dispatched by the World Meteorological Organization to the Indian Institute of Tropical Meteorology in 1960. The prologue of his book *Monsoon* (Murakami 1986) describes his personal observations in Pune, western central India (18.5°N):

> *In April and May—day after day—temperatures burned above 45°C, earth was parched, no cloud in sight, no water ran in river channels. When one or two cumulus clouds appeared in the sky, I would jump out and admire them. Some days later, Monsoon arrived. In no time, trees turned green, grass grew, temperature dropped to a pleasant level. Children ran around joyfully in the rain and farmers turned busy. Monsoon's drama has opened on the seasons' revolving stage.*

Under intense sunlight, heat waves are common in India during April to May and early June prior to the arrival of monsoon rains. The dry ground can only balance the solar heating by raising temperature to increase sensible heat flux and infrared radiation (Box 5.1). Daily highs often exceed 45°C. Following the onset of the rainy season, daily highs drop 8° to a "pleasant" level of 30°C (Fig. 5.9) as surface evapotranspiration from the wet ground surface helps offset solar radiation. Nighttime minimum also decreases across the monsoon onset, but the drop is only half of that for the daytime maximum.

Prior to the rainy season, the hot land surface drives shallow dry convection and a shallow meridional overturning circulation below 600 hPa (Fig. 5.10A). With the onset of monsoon rains, a deep local Hadley cell emerges with deep upward motions over India and the northern Bay of Bengal and sinking motions over the Indian Ocean to the south (see Fig. 5.10B).

TOA solar radiation is a smooth function of time, but the evolution of monsoon convection displays some discontinuities. Over India, a monsoon rain band appears abruptly in June and stays there until late September (Fig. 5.11). In the tropics, surface MSE is key to convective instability. The shallow overturning transports moisture inland, specific humidity q increases rapidly from April to May, driving the seasonal increase in surface MSE over India (see Fig. 5.9). The MSE over India eventually exceeds that over the northern Bay of Bengal, heralding the monsoon onset. Following the onset, MSE continues to increase because of the feedback from lateral moisture advection and surface evaporation despite the large decrease in surface temperature.

BOX 5.1 Land surface-atmosphere interactions

Compared to the ocean, land surface is of high reflectivity for solar radiation and small heat capacity. The heat capacity effect is apparent in larger diurnal and seasonal cycles in SAT over land than ocean. Unlike the ocean, evapotranspiration on land is often limited by the availability of water in the soil especially during summer daytime when the solar radiation is strong. The regulation of evapotranspiration is often modeled with a bucket model of a finite depth, say 20 cm, which allows runoff into rivers when the holding capacity is exceeded. An evaporation factor is often introduced, varying from unity when the bucket is nearly full to zero when it is nearly empty. Over dry surface, evapotranspiration is strongly limited, and the ground surface heats up in response to solar heating. The heated ground causes dry convection that deepens the daytime boundary layer. In this water-limited regime of evapotranspiration, incoming solar radiation is balanced by infrared radiation and turbulent exchange of sensible heat with the atmosphere aloft. When the surface is wet, say over a swamp, daytime temperature increase is moderate because of evaporative cooling.

The effect of land surface regulation of evapotranspiration is apparent by comparing May 21 and July 21 over central India. Although TOA solar radiation is nearly identical, surface temperature is much higher on May 21 than July 21 (see Fig. 5.9), especially during daytime because of the lack of soil moisture for evapotranspiration. MSE is much higher on July 21 because of higher moisture and latent heat, allowing deep convection.

In the midlatitudes, summer heatwaves are due to adiabatic warming in the downdrafts of persistent atmospheric blocking. (This is akin to the ocean upwelling effect, albeit with reversed sign.) The surface warming promotes evapotranspiration, which dries the topsoil. The reduced soil moisture limits the evapotranspiration, amplifying the surface warming. Decreased cloud cover in downward motions is another positive feedback for surface warming. Central US states and the region surrounding the Mediterranean Sea have been identified as hotspots of the soil moisture-evapotranspiration feedback, where the forementioned positive feedback is responsible for a long tail in the occurrence of extreme high temperatures (Box Fig. 5). In model experiments, the effect of initial soil-moisture anomalies can last for up to 2 months on surface temperature (Koster et al. 2010; Merrifield et al. 2019), while the effect on precipitation is weaker and more variable.

BOX 5.1 Land surface-atmosphere interactions—cont'd

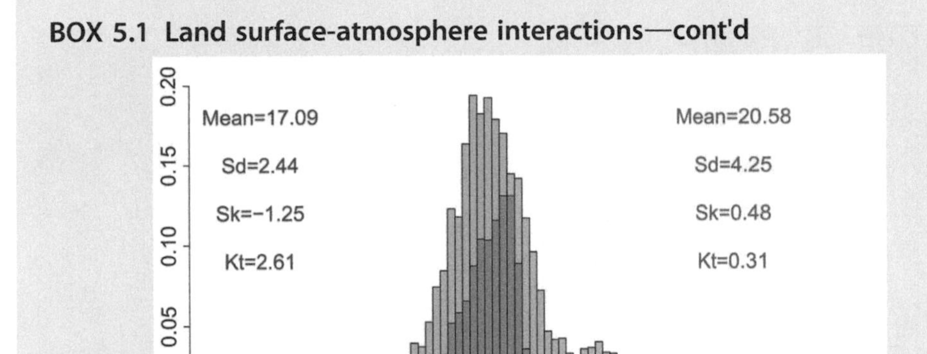

Box Fig. 5 Histogram of daily 2-m temperatures at a central US grid point for June-July-August 1971-2000 in an atmospheric general circulation model simulation with *(red)* interactive soil moisture. When soil moisture is prescribed with the monthly climatology, the occurrence of warm temperature *(blue bars)* is strongly suppressed. The legend lists the first four moments of distributions: mean, standard deviation *(Sd)*, skewness *(Sk)*, and kurtosis *(Kt)*. *(From Berg et al. (2014). © American Meteorological Society. Used with permission.)*

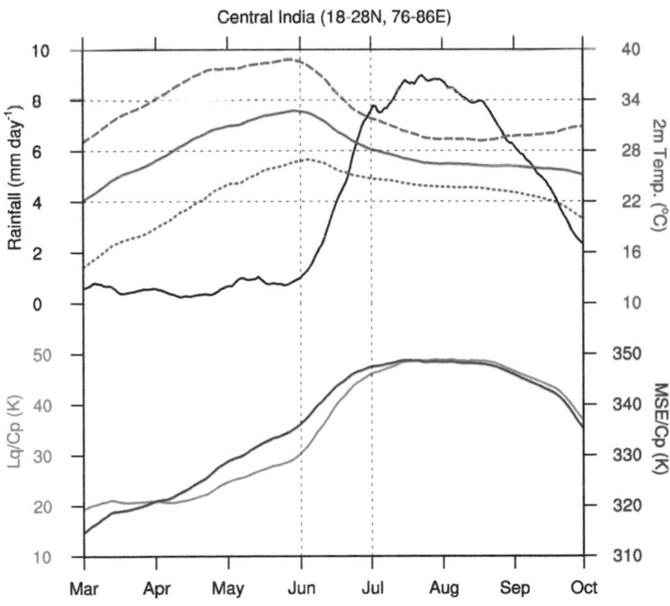

Fig. 5.9 Precipitation in central India (76°–86°E, 18°–28°N), air temperature (solid in °C; daily max/min dashed/dotted), moist static energy (m/cp), and specific humidity (Lq/cp) at 2 m above ground. *(Courtesy ZQ Zhou.)*

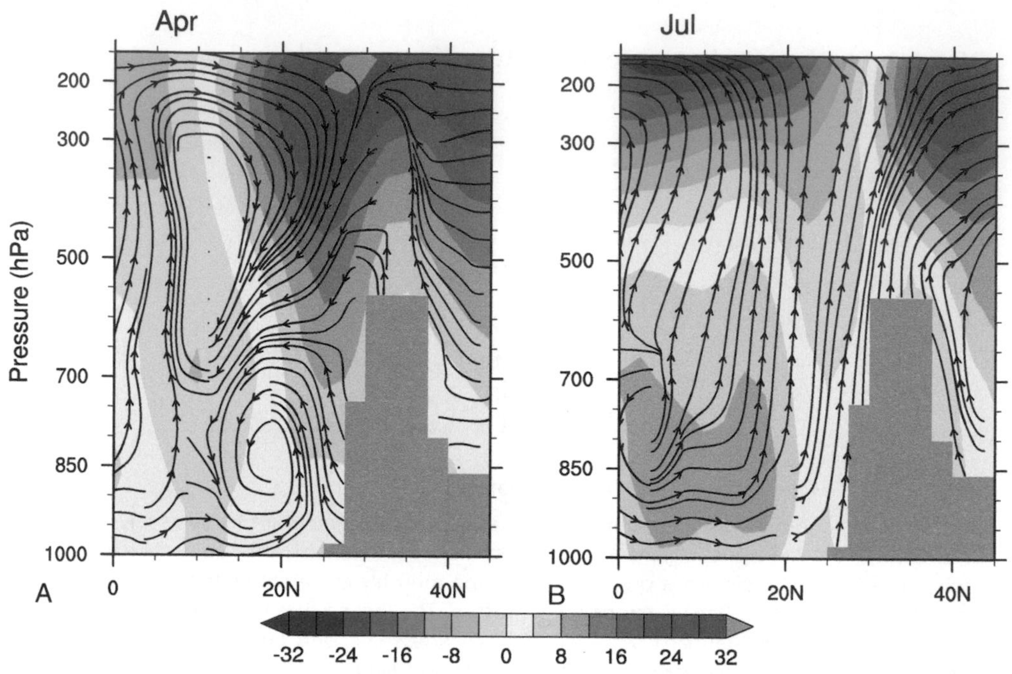

Fig. 5.10 Local Hadley cell (stream function) and zonal wind velocity (selected contours with color shading for westerly/easterly) in 78° to 90°E in (A) April and (B) July. *(Courtesy ZQ Zhou.)*

The Indian Meteorological Department defines the date of Indian summer monsoon onset as the time when the rainy season commences in Kerala State on the west coast near the southern tip of the country. The long-term mean onset date in Kerala is June 1. The rain front then advances north to northwestward and takes 2 or more weeks to cover the entire country.

Climate over India is drastically different between the equinoxes (see Fig. 5.11): The ground and atmosphere are dry at the spring equinox (March 21) but the monsoon is just about to wrap up at the fall equinox (September 21) with wet soil. Summer monsoon is often described as a planetary sea breeze driven by the thermal contrast between the Asian continent and Indian Ocean. Daily mean SAT over India exceeds Indian Ocean SSTs as early as at the spring equinox, but monsoon convection does not commence over India for another 2 months or more. What causes the apparent large delay in the monsoon onset?

Consider two columns of water in a tank, of the same mass but different temperatures and separated by a wall. Surface level is higher above the warm column. The pressure difference drives the warm surface water to flow into the cold column when the separating wall is removed. As a result, a thermal low forms at the bottom of the warm column and a thermal high under the cold column. The pressure gradient drives the cold

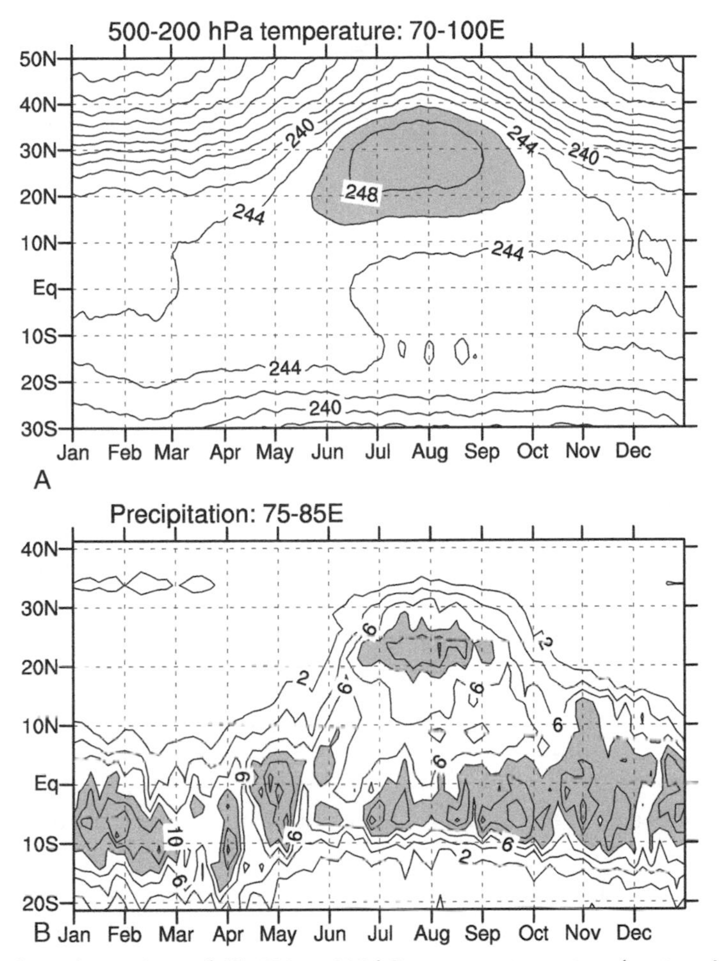

Fig. 5.11 Time-latitude sections of (A) 500 to 200 hPa mean temperature (contour interval of 2 K, shading >246 K), averaged in 70°E to 100°E, and (B) CPC Merged Analysis of Precipitation (contour interval of 2 mm/day, shading >8 mm/day), averaged in 75°–85°E. *(Courtesy ZQ Zhou.)*

water toward the warm column at the bottom. If the Earth rotation is unimportant as in a diurnal sea breeze, the overturning circulation will continue to develop, driving the cold water toward the warm water column near the bottom (Fig. 5.12A). If the Earth rotation is important as in the planetary-scale monsoon, the horizontal temperature gradient can be simply balanced by a vertical shear (see Fig. 5.12B; Eq. 2.22), without necessarily an overturning circulation. Indeed, the vertical shear between low-level westerly and upper easterly winds over India is in thermal wind balance with the temperature gradient between the Asian Continent and Indian Ocean during the summer monsoon. Thus the seasonal monsoon is fundamentally distinct from the diurnal sea breeze in being

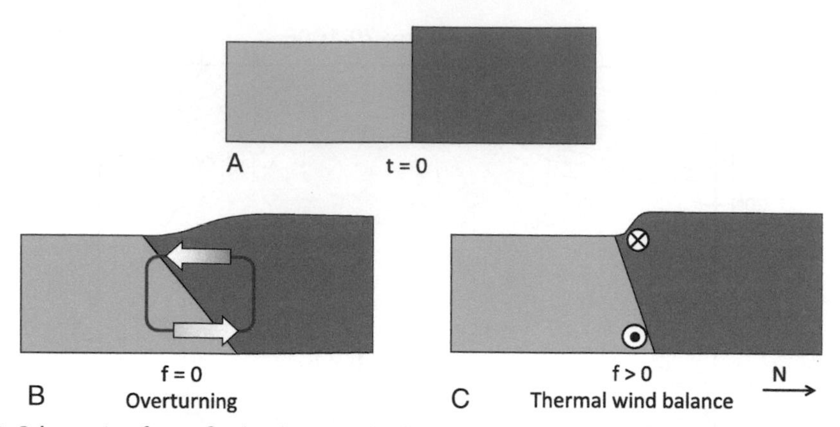

Fig. 5.12 Schematic of two fluid columns of different temperatures initially at rest (A), resulting in (B) vertical overturning without Earth rotation, and (C) vertical shear in thermal wind balance sufficiently away from the equator with finite f.

geostrophic. A geostrophic monsoon resists the development of and delays the onset of a deep overturning circulation.

As deep convection requires high surface MSE, the summer monsoon may be viewed as driven by the land-sea contrast in energy rather than simply in temperature. The premonsoon shallow overturning circulation advects high (low) MSE near the surface (at 600 hPa), causing the column-integrated MSE to increase. The resistance to a deep overturning circulation due to geostrophy probably contributes to the large delay of the deep convective onset despite a northward temperature gradient in the PBL.

During the 4-month summer (June—September), the Indian monsoon is punctuated by break periods with reduced rainfall over India. This 30- to 60-day cycle of active and break monsoon over India is part of the planetary-scale MJO. The MJO rain band is slanted northwestward, causing an apparent northward propagation as the system moves eastward (see Section 4.2.4).

The North Indian Ocean responds strongly to monsoons. With the seasonal increase in solar radiation, SST reaches as high as 30°C in May just before the onset of the southwest monsoon. Following the monsoon onset, active convection increases cloud cover, and the strong southwest winds intensify surface evaporation, both causing SST to drop from 30°C in May to 28°C in August (Fig. 5.13). The monsoon-induced decrease in SST is a negative feedback on the monsoon convection. The most clear example is off the Somali and Omani coast, where the intense ocean upwelling lowers SST well below the convective threshold, driving atmospheric convection away from the western Arabian Sea.

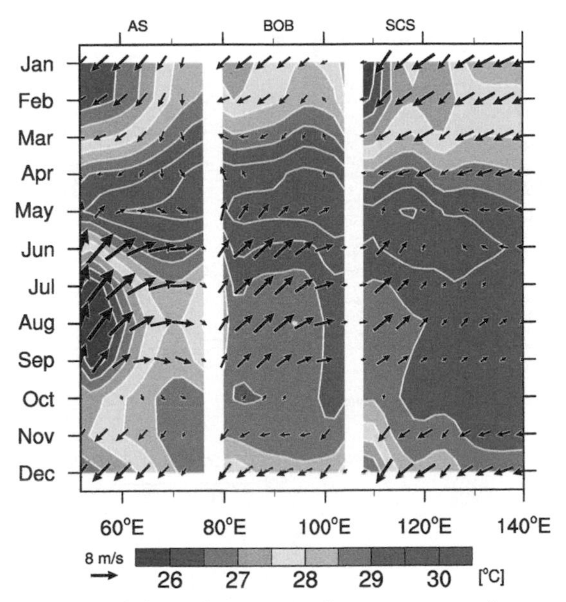

Fig. 5.13 Time-longitude section of climatologic sea surface temperature (contours at 0.5°C interval) and 10-m winds at 11°N: Arabian Sea *(AS)*, Bay of Bengal *(BOB)*, and South China Sea *(SCS)*. *(Courtesy CY Wang.)*

5.2 East Asian monsoon

In midsummer (mid-June to late July), a northeast-slanted rain band develops from the east flank of the Tibetan Plateau through southeastern China and Japan to the North Pacific (Fig. 5.14A). Called *Meiyu* in China, *Changma* in Korea, and *Baiu* in Japan, it is a single most important climate phenomenon for the region that produces the major rainy season. Within this rain band, mesoscale disturbances develop accompanied by intense convection and heavy rains.

At low levels, the southerly winds prevail over East Asia as a result of the land-sea contrast between the heat low over the continent and the subtropical high of the Pacific. The low-level southerlies advect moisture and feed the Meiyu-Baiu rain band. The Mieyu-Baiu rain band appears as a humidity front that separates moist airmass to the south from the dry airmass to the north (see Fig. 5.14A). It is not a temperature front over east China, however, where boundary-layer temperature is nearly uniform in the meridional direction between the sunny and dry north and cloudy/wet south. The meridional temperature gradient builds up over the East China Sea toward the Kuroshio Extension east of Japan, maintained by the underlying SST distribution.

5.2.1 Thermal advection by the westerly jet

Different from the tropical monsoon of South Asia, the westerly jet of Asia is an important factor that steers the Meiyu-Baiu rain band. At the midtroposphere, strong diabatic

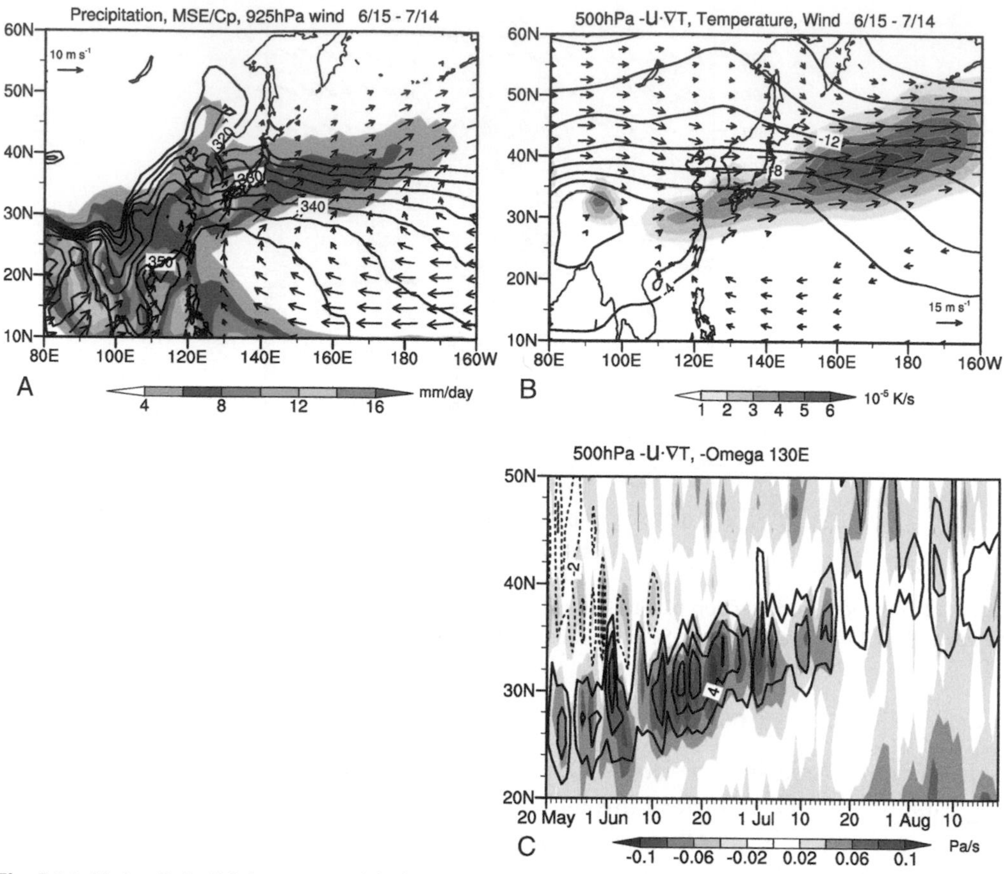

Fig. 5.14 Meiyu-Baiu (15 June—14 July) climatology. (A) Precipitation (color shading, mm/day), surface MSE/c_p (contours, K, from 320—355 at intervals of 5 K), and wind velocity vectors (m/s) at 925 hPa. (B) Advection (color shading, 10^{-5}K/s) of temperature (black contours in °C, from −18 to −2 with intervals of 2°C) by wind velocity (vectors in m/s) at 500 hPa. (C) Time-latitude section of temperature advection (black contours, −4, −2, 2, 4, and 6x10^{-5} K/s) and vertical velocity ($-\omega$, Pa/s; color shading) at 130°E. *(Courtesy ZQ Zhou.)*

heating maintains a global temperature maximum over the southern Tibetan Plateau. The westerly flow as part of the broad Asian jet advects warm air eastward along 30°N over southeastern China, through the East China Sea and southwest Japan (see Fig. 5.14B). Riding on the westerlies, warm air from the eastern flank of the Tibetan Plateau gradually rises on the sloping isentropic surface. Indeed, the warm advection and upward motion are collocated with each other and with the Meiyu-Baiu rain band. The adiabatic induction of ascending motions by the warm advection triggers convection along the westerly jet.

The life cycle of the Meiyu-Baiu season supports the mechanism of adiabatic induction along the midtropospheric westerly jet (see Fig. 5.14C). From May to June, the growth of the Tibetan High displaces the westerly jet north of the Plateau (see Fig. 5.4A). With the buildup of the tropospheric temperature maximum over southern Tibet, the westerly warm advection and the resultant upward motion begin to appear in East Asia (25°–30°N) in late May to early June and intensify while migrating slowly northward (see Fig. 5.14C). They begin to weaken in the second half of July as the jet stream migrates further northward, detached from the tropospheric temperature maximum.

The Tibetan Plateau is an important orographic forcing that steers the westerly jet. In winter, the subtropical jet flows south of the Plateau (Fig. 3.18). During mid-March to mid-May, the jet position is bimodal between south and north of the mountain. The convective heating of the South Asian monsoon pushes the jet to a northerly position during the summer, setting the stage for the East Asian summer monsoon. The Tibetan Plateau contributes to Meiyu-Baiu formation (Fig. 5.15) by (1) strengthening the surface southerly flow over East Asia, (2) anchoring the westerly warm advection in the troposphere, and (3) initiating cyclonic storms—called southwest vortices in China (Ding and Chan 2005)—on the eastern flank that propagate eastward along the westerly jet. Fig. 5.16 shows that a southwest vortex developed on the east flank of the Plateau in June 1998, contributing to heavy rainfall in the Yangtze River basin. In atmospheric

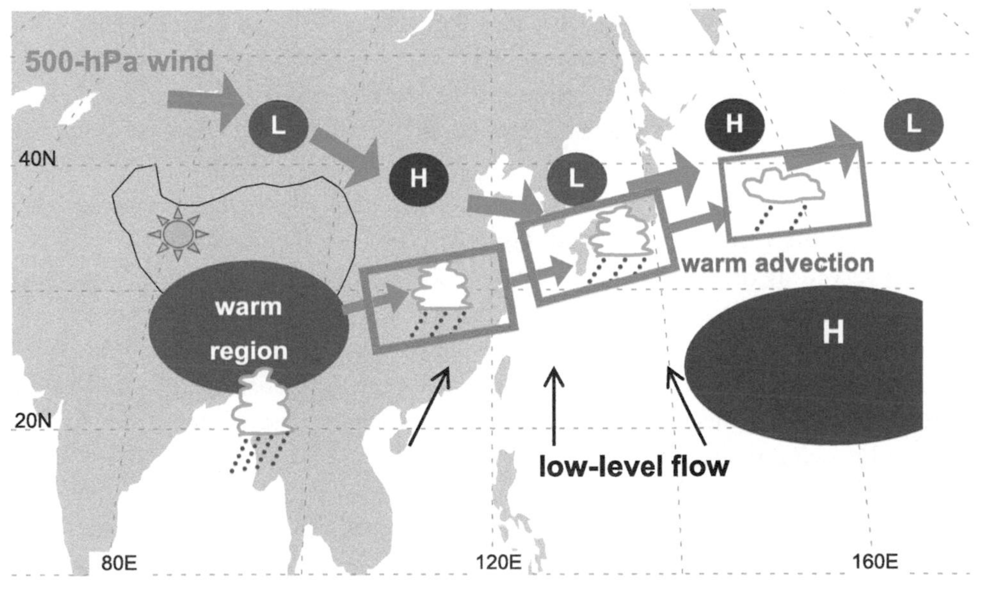

Fig. 5.15 Schematic of the East Asian summer monsoon. *(From Sampe and Xie (2010).)*

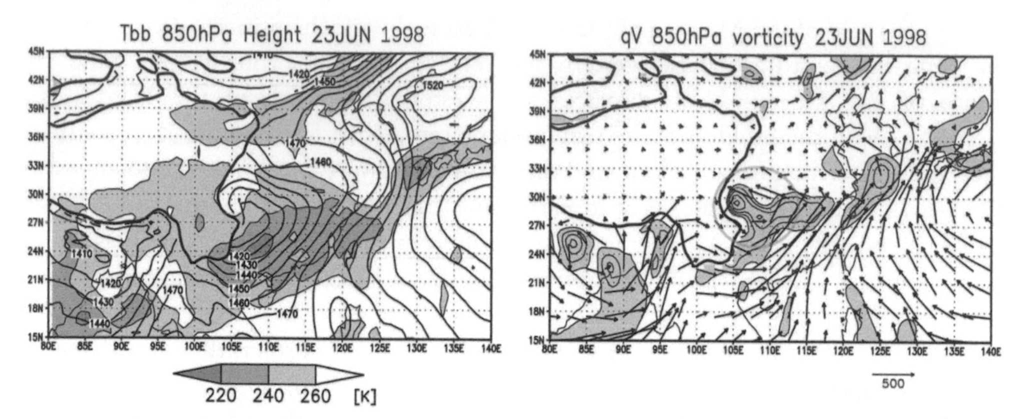

Fig. 5.16 Tbb (shading <260 K), 850 hPa geopotential height *(left)*, moisture transport (g/kg m/s) and 850 hPa vorticity *(right)* over the Asia monsoon region on June 23, 1998. The Southwest vortex is highlighted by the blue circle. Topographic contour for 1500 m is shown in thick solid line. *(From Yasunari and Miwa [2006].)*

GCMs, the removal of mountains substantially weakens summer rainfall over subtropical East Asia (see Fig. 5.7B).

5.2.2 Socioeconomic impacts

The Meiyu-Baiu rain band is perhaps the single most important climate phenomenon for East Asia. It shapes the important north-south gradients across China in food and culture. Rice culture developed along and south of the rain band while the drier north adopted wheat (e.g., noodles) as the staple. Meiyu and Baiu share the same Chinese characters. The current preference is to write the characters of plum rain (梅雨), indicating the coincidence of the rainy season with ripening plums. The term of Meiyu/Baiu is also believed to originate from the Chinese character for mold (霉), an unpleasant effect of prolonged muggy conditions.

The dramatic covariations between Meiyu and winds are widely known among people living under the recurrent rain band. Great poet Shi Su (1037–1101) of Song dynasty, who served as the governor of Hangzhou in Meiyu region, wrote: "As plum rains cease, the breezes bring home sails from far seas." Prevailing winds shift from southerly during the Meiyu to easterly post the rainy season over the East China Sea (Fig. 5.17B). After long voyages from the Indian Ocean, merchant ships waited over the East China Sea for the post-Meiyu southeasterly winds to send them home along the eastward flowing rivers.

Over the East China and Yellow seas—marginal seas divided by the line between Shanghai and Jeju Island—fog forms as the prevailing southerlies advect warm and humid air across northward-decreasing SST gradients. Qingdao (36°N, 120°E) is a major city on the south coast of the Shandong Peninsular north of the Meiyu rain band. It is frequented

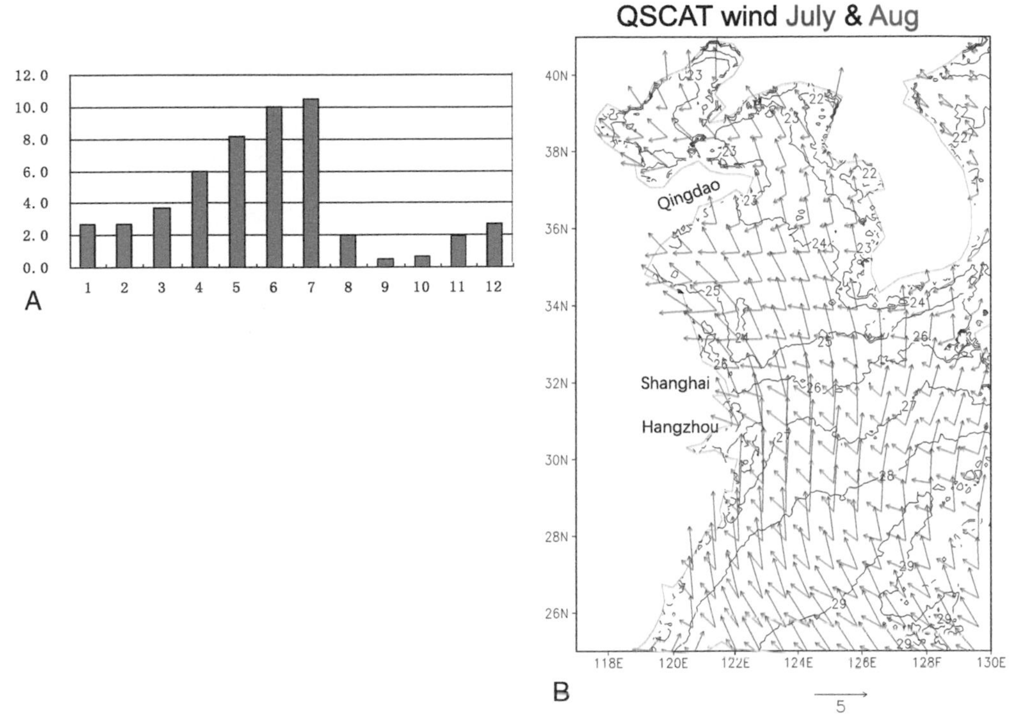

Fig. 5.17 (A) Sea fog days as a function of calendar month in Qingdao, China. (B) July sea surface temperature (*black contours*, °C) and QuikSCAT surface wind velocity in July (*blue arrows*, m/s) and August (*red*). *(Adapted from Zhang et al. (2009).)*

with sea fog (one-third of the days in June–July) under the prevailing southerlies. The end of sea fog season in Qingdao coincides with the Meiyu's termination as the wind turns easterly in parallel with the SST contours in the Yellow Sea (see Fig. 5.17).

5.2.3 Subtropical convection

The termination of the Meiyu-Baiu rain band is part of large-scale changes in the second half of July over the Northwest Pacific. By the mid-July, the 28°C SST isotherm has expanded to 25°N over the Northwest Pacific. Despite a warm bulge at 20°N in the SST distribution, the Pacific ITCZ stays south of 10°N (Fig. 5.18A). In a short time of 2 weeks, deep convection suddenly appears to fill the subtropical warm bulge (see Fig. 5.18B). We call this abrupt onset of subtropical connection the Ueda jump after its lead discoverer (Ueda et al. 1995). The Ueda jump signals the onset of the Northwest Pacific monsoon, where active convection causes an eastward expansion of the monsoon westerlies to its south (~10°N) (see Fig. 5.18C). Meanwhile, it puts an end to the Meiyu rain band over East Asia, and sea fog starts to clear over the Yellow Sea as the winds turn easterly.

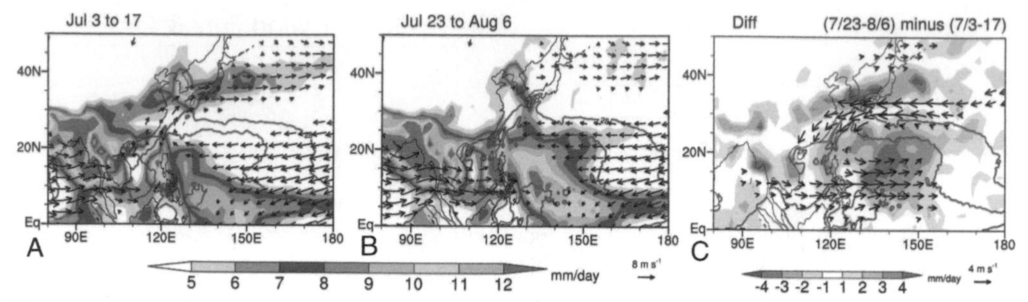

Fig. 5.18 Sea surface temperature (SST) (28° and 29°C contours), precipitation (shading, mm/day), and 850 hPa wind (m/s) for (A) July 3–17, (B) July 23–August 6, and (C) the difference. SST contours are identical between (B) and (C). *(Modified based on Zhou et al. (2016), and courtesy ZQ Zhou.)*

High SSTs are a necessary but not the sufficient condition for the Ueda jump. In the subtropical Northwest Pacific (140–160°E, 15–25°N), SST exceeds 28°C in early June but deep convection (precipitation >6 mm/day) does not start until 1.5 months later following the buildup of the column precipitable water (Fig. 5.19). Prior to the convective jump, a weak temperature inversion caps the atmospheric boundary layer, and the free troposphere is dry. As the capping inversion gradually weakens and the lower troposphere moistens, dry entrainment weakens to allow deep convection to develop. Local SST decreases slowly after the convective onset and stays at ~29°C through September. Deep convection in this subtropical warm pool persists through September. While the onset of the Ueda jump is an abrupt event, the southward withdraw of subtropical convection is smooth in time.

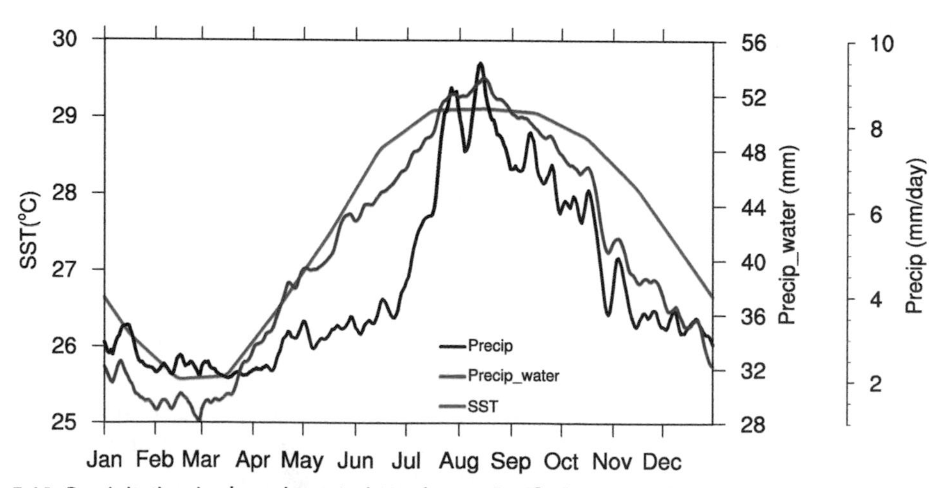

Fig. 5.19 Precipitation in the subtropical Northwest Pacific (140°–160°E, 15°–25°N) as a function of calendar month, along with sea surface temperature *(SST)* and column precipitable water (mm). (Courtesy ZQ Zhou.)

5.3 Asian summer monsoon system

5.3.1 Subsystems

The summer circulation and convection over the broad region from the Arabian Sea to Northwest Pacific may be viewed as an interconnected system. The onset of the rainy season is societally important, and the convective heating drives the large-scale circulation. The distribution of the time of convective onset (Fig. 5.20) is often used to identify the following, dynamically distinct subsystems.

1. The South Asian monsoon is a tropical system, including the traditional Indian monsoon (from the Arabian Sea to Bay of Bengal), Indochina, and South China Sea. The three-way interaction among the ocean, land, and atmosphere creates complex spatiotemporal variations (e.g., in onset). Deep convection sets in mid-May over a broad region from the Bay of Bengal to South China Sea, but the onset is noticeably later (~June) over India (see Fig. 5.20). The mean flow effect (e.g., the midlatitude westerly jet) is not of first-order importance, and the Gill model without a mean flow captures salient features of this tropical monsoon circulation.
2. The East Asian monsoon is a subtropical system characterized by the Meiyu-Baiu rain band. The westerly jet is fundamentally important through the warm advection from the Tibet and as a waveguide for weather perturbations (e.g., southwest vortices).
3. The Northwest Pacific monsoon is tropical and maritime. It is unique in its late and abrupt onset, more than 1 month after local SSTs rising above 28°C. The Ueda jump puts an end to the Meiyu-Baiu rain band but is associated with little coherent change in the South Asian monsoon (from the Arabian to South China Sea).

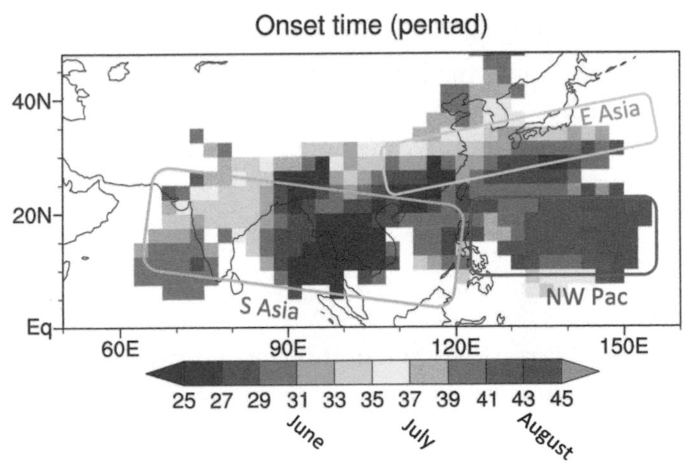

Fig. 5.20 The Asian summer monsoon and its subsystems based on the time (pentad) of the monsoon onset. The onset is defined as the first pentad that relative rainfall rate (removed January mean rainfall) exceeds 5 mm/day during May to September. *(From Sperber et al. (2013); courtesy ZQ Zhou.)*

These subsystems are mutually interactive. The South Asian monsoon sets the stage for the East Asian monsoon. At low levels, the southerlies on the eastern flank of the monsoon low feed the moisture to fuel the Meiyu-Baiu rains, while at upper levels the Tibetan High displaces the Asian jet northward, which steers the northeastward-slanted Meiyu rain band. The midsummer termination of Meiyu-Baiu is tied to the belated onset of the Northwest Pacific maritime monsoon (the Ueda jump).

While the seasonal variation in solar radiation is smooth in time and uniform in longitude, the Asian summer monsoon system in response often displays mysterious jumps. Strong interactions among circulation (moisture supply), convection (heating), and soil moisture often make it difficult to pinpoint the true cause of phenomena of interest (e.g., monsoon onset, intraseasonal to interannual variability). Better dynamic understanding is needed.

5.3.2 Connection to the Sahara Desert

The subsidence with the Hadley circulation is often invoked to explain subtropical deserts. There are two problems with this argument: In summer, (1) the zonal-mean Hadley cell is weak poleward of 15°N (Fig. 5.21), and (2) zonal variations in vertical velocity are much larger than the zonal mean subsidence in the Northern Hemisphere subtropics. In fact, at similar subtropical latitudes (20°−30°N), the Sahara of North Africa is a dry desert, but East Asia is blessed with monsoon rains. A simple atmospheric dynamic argument that follows suggests that summer convection over South and East Asia is an important factor that keeps the Sahara dry.

Let us take another look at Gill's solution with a northward-displaced heat source. The Asian monsoon heating excites a baroclinic Rossby wave response of deep subsidence that propagates far westward. In observations, the upper-level Tibetan High extends westward across the Middle East all the way through North Africa (see Fig. 5.4). Thermodynamically in the nonconvective region west of Tibet, the radiative cooling is balanced by the adiabatic warming associated with the slow subsidence. Alternatively, we can diagnose vertical motion from the potential vorticity conservation, which requires a poleward-moving upper tropospheric column to expand in the vertical (i.e., midlevel subsidence, $w_2<0$) on the western extension of the High,

$$\beta v = f\frac{\partial w}{\partial z} \propto -f w_2$$

where we have assumed zero vertical motion at the tropopause. This is known as the monsoon-desert mechanism,[9] which is equivalent to the Sverdrup relation in physical oceanography (see Eq. 7.26).

Climate model simulations with idealized continental geometry suggest an intrinsic preference for monsoon climate on the southeastern continent and desert on the

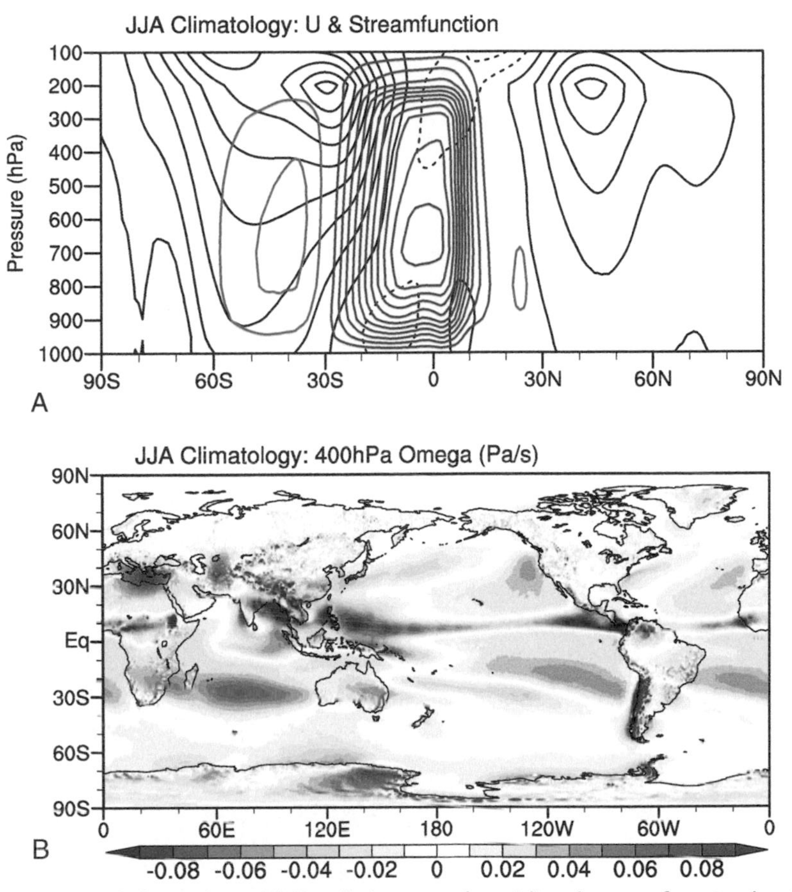

Fig. 5.21 Summer (JJA) climatology. (A) Zonally integrated meridional stream function (positive/negative in red/blue contours, 2×10^{-10} kg/s intervals with zero omitted) and zonal-mean zonal wind (black contours, 5 m/s intervals). (B) Vertical velocity at 400 hPa. *(Courtesy ZQ Zhou.)*

southwestern continent. With a flat Afro-Eurasian continent, the summer monsoon rain band slants in a northeast direction (Fig. 5.22A). This zonal asymmetry emerges as a result of the land-sea contrast in the zonal direction. The rapid warming of the continent forms a heat low, and the relatively cool ocean helps form a thermal high near the surface. The resultant geostrophic wind on the eastern continent advects moist marine air poleward, and the increased MSE is conducive to deep convection. On the western continent, on the other hand, the geostrophic wind between the thermal high over the (Atlantic) ocean and heat low on land advects cool, dry air of low MSE equatorward, creating unfavorable conditions for deep convection. Model experiments show that a zonally larger continent causes a greater poleward penetration of wet climate on the eastern continent and a further equatorward expansion of dry climate on the western continent. Despite the

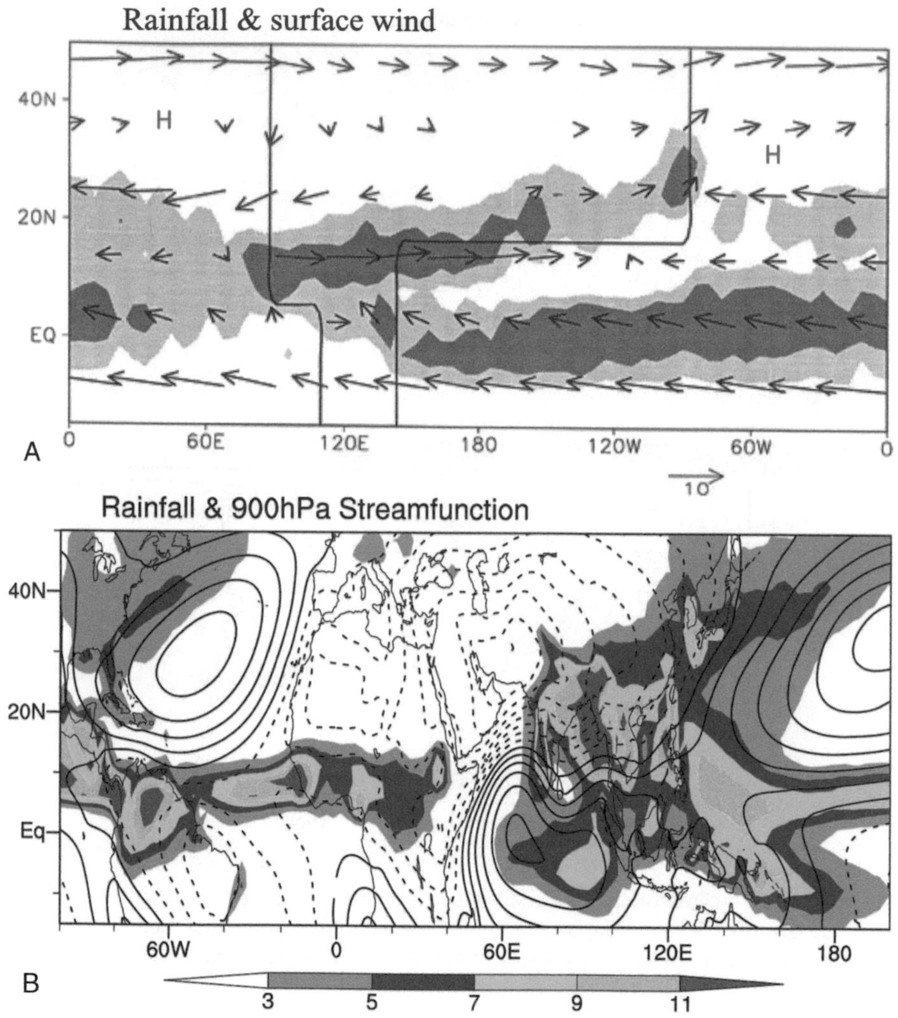

Fig. 5.22 (A) JJA rainfall (shading) and surface wind in an idealized model experiment with an Afro-Eurasian continent. (B) JJA rainfall (shading, mm/day) and 900 hPa stream function (zonal mean removed, black contours, 2×10^6 m^2/s intervals) in observations. *(A, From Xie and Saiki (1999), JMSJ; B, Shaw (2014), JAS. © American Meteorological Society. Used with permission.)*

simplifications in the experimental design, the tilted rain band and surface circulation bear a remarkable resemblance to observations (see Fig. 5.22).

The following factors contribute to the dramatic zonal variation in subtropical climate from the Sahara Desert to lush East Asia.

- Zonal land-sea contrast tilts the summer rain band in favor of wet climate in East Asia as illustrated by idealized model simulations.

- Tibetan Plateau strengthens the low-level cyclonic circulation, increasing rainfall over East Asia. It also protects high MSE air south of the mountain.
- Baroclinic Rossby wave response to convective heating of the Asian summer monsoon is associated with subsidence that keeps the Arabian and Sahara dry.

The Tibetan High, excited by the Asian summer monsoon system and extending far westward due to the Rossby wave dynamics, bridges the east and west of the Afro-Eurasian landmass. The deep subsidence in the far western extension of the Tibetan High keeps the Sahara dry and prevents the summer rainband from advancing northward over West Africa. In addition, the high plateau of Tibet connects the South and East Asian summer monsoons and anchors the Meiyu-Baiu rain band by steering the Asian jet for warm advection and adiabatic ascent.

About 50 million years ago, the Indian plate collided from the south with the Eurasian plate. At the time, a narrow mountain range of 4 km high existed along south Tibet. Complex mountain building ensured, with the northward expansion of the Plateau. The rise of the Tibetan Plateau transformed the Asian monsoon system in important ways (Molnar et al. 2010), which remains an important topic of paleoclimate research.

5.4 West African monsoon

West Africa faces the Atlantic Ocean on the west (15°W) and south coasts (~5°N). The equatorial Atlantic Ocean to the south is often called the Gulf of Guinea. In summer (JJA), a zonal rain band extends from the tropical North Atlantic to the Great Rift Valley of East Africa (see Fig. 5.2A). The Sahel refers to a belt of the seasonal savanna over West Africa in 5° to 15°N between equatorial forest and the Sahara Desert to the north. In the Sahel, winter is dry and summer is the rainy season. Lives flourish with the arrival of monsoon rains, heralding great migrations of animals of all kinds. At the upper level, a high-pressure belt extends over Africa as the subsidence Rossby waves forced by the Asian monsoon heating (see Fig. 5.4A). At the low level, the intense sensible heating over the Sahara Desert forms a heat low, with southwesterly winds blown across the wet Sahel onto the desert (Fig. 5.23). The southwesterly surface flow feeds both the deep overturning cell at the Sahel rainband (10°N) and a shallow overturning cell capped below 600 hPa with the ascent at 20°N (Fig. 5.24), resulting in the apparent displacement of the surface wind convergence and the monsoon rain band.

Over the Sahara, the ground is dry and there is no evaporation to balance the daytime solar radiation. The ground surface heats up with temperature often in excess of 50°C, and dry convection develops in a PBL up to 600 hPa in height. Of high temperatures but depleted of water vapor, air in the Sahara is of lower MSE than in the Sahel (~10°N). With temperature increasing northward and humidity peaking in the rainy zone, the maximum MSE is slightly displaced north of the monsoon rain band, creating a tendency for the couplet to advance northward.

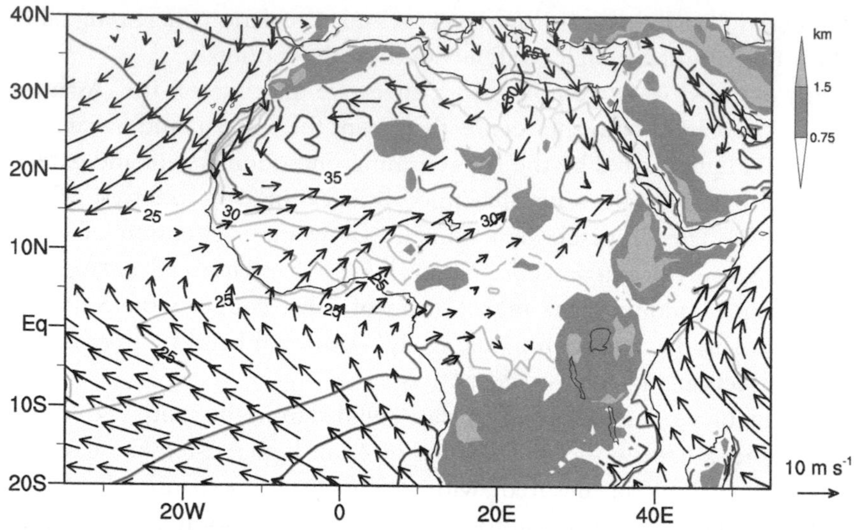

Fig. 5.23 July temperature (°C) and wind (m/s) at 925 hPa, along with topography (gray shading, km). *(Courtesy ZQ Zhou.)*

Immediately following the spring equinox after the Sun marches into the Northern Hemisphere, a rainband develops on the south coast of West Africa and slowly moves inland (Fig. 5.25A). In the West African sector, the maximum of the monsoon rainfall abruptly jumps northward from the coast to 10°N in July. The Sahel rainband stays at 10°N through August and then gradually moves back southward thereafter. The reduced coast rainfall appears due to the coastal ocean cooling during July and August. Coastal SST drops precipitously from above 28°C in April to below 24°C in August as an upwelling Kelvin wave shoals the thermocline on the coast (Section 10.1.2). In the Sahel, rainfall begins in May, increases rapidly thereafter, and peaks in August (see Fig. 5.25A). The evapotranspiration during the rainy season reduces SAT by 6°C from May to August. This evaporative cooling causes a marked semiannual cycle in air temperature over the Sahel (see Fig. 5.25B).

Desert surface is highly reflective of visible light, while the dry atmosphere allows surface infrared radiation to escape into space. The reduced net downward solar radiation and high OLRs cause the atmospheric column over the Sahara to lose energy at the TOA. Over the Sahara, the net radiative cooling is balanced by the adiabatic warming with the subsidence above the warm PBL. In the mid- to upper troposphere (600–200 hPa), the southward temperature gradient between the deep convective Sahel and subsidence over the Sahara creates a westerly wind shear in the vertical, while in the lower layer from the surface to 600 hPa, the northward temperature gradient between the cool Sahel and warm Sahara creates an easterly vertical shear. The reversed temperature gradient

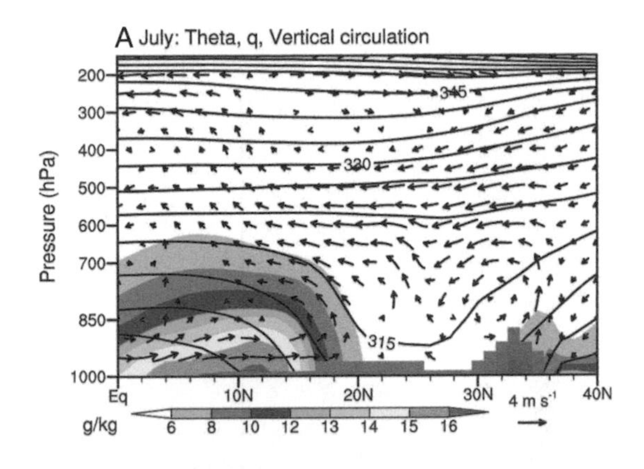

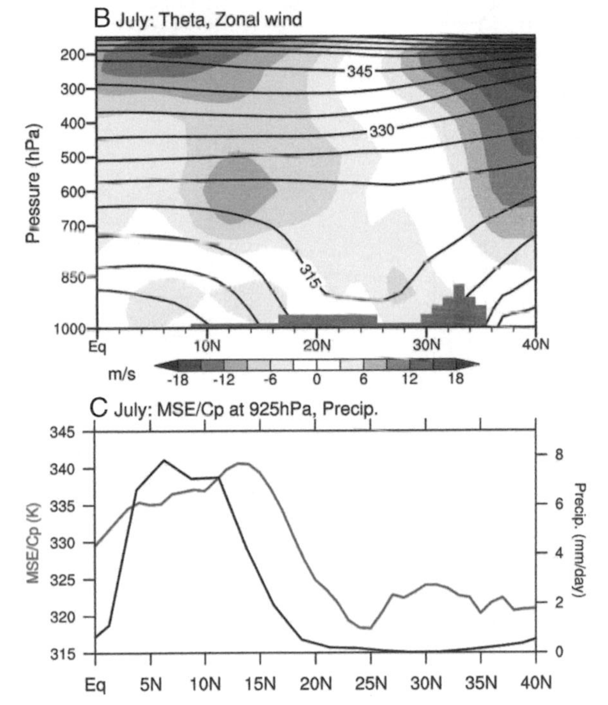

Fig. 5.24 July climatology in 10°W to 10°E as a function of latitude and pressure. (A) Potential temperature (black contours, K), specific humidity (color shading in g/kg), and vertical circulation (v, $-\omega$, vertical velocity $\times 20$). (B) Potential temperature (black contours) and zonal velocity (color shading in m/s). (C) Meridional profiles of MSE/cp at 925 hPa and precipitation. *(Courtesy ZQ Zhou.)*

between the lower and upper troposphere supports an easterly jet core centered at 600 hPa on the boundary (10°–15°N) between the wet Sahel and dry Sahara (see Fig. 5.24A).

The African Easterly Jet, at 12 m/s in core speed, is hydrodynamically unstable due to its lateral and vertical shears, generating weather disturbances—called African easterly waves—that propagate westward into the Atlantic. Of 4000 km in wavelength, they often display a northwest (southwest) phase tilt on the north (south) flank of the mean

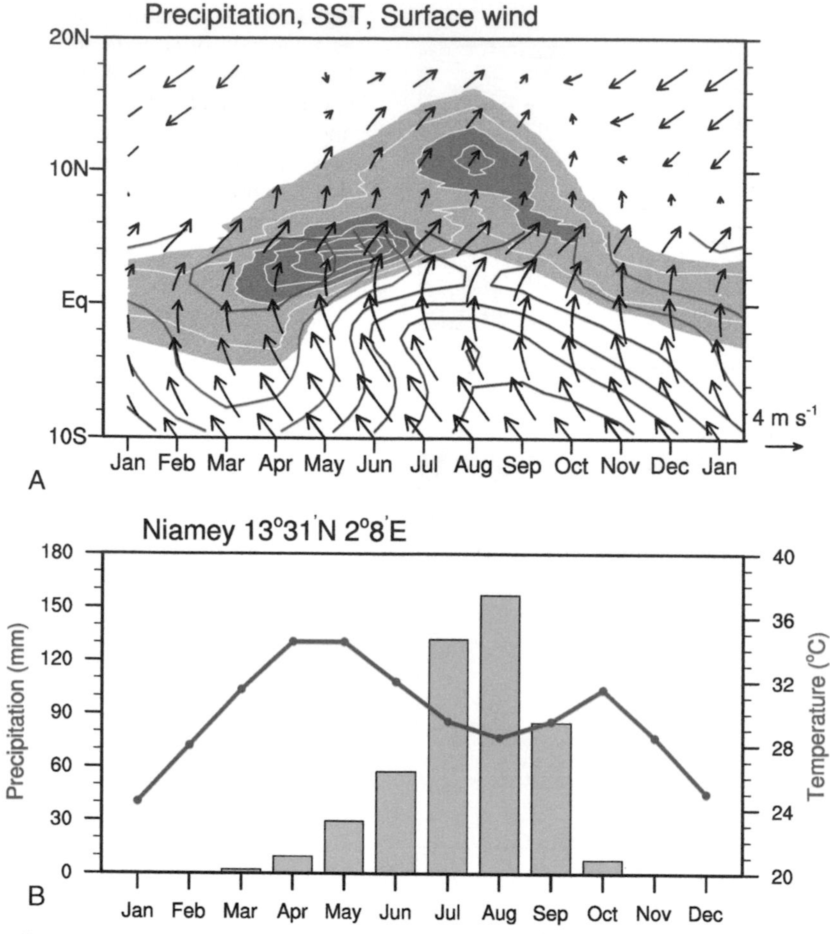

Fig. 5.25 Climatology at 0°E. (A) Time-latitude section of precipitation (light/dark grey shading ≥100/200 mm/month, white contours at 50 mm/month intervals) averaged in 10°W to 10°E, surface temperature *(SST)* (1°C intervals, red ≥27°C, blue <27°C) and sea surface wind velocity (m/s) at 0°E. (B) Surface air temperature (red curves in °C) and precipitation (bars, mm/month) at Niamey of Niger, 13°31′N, 2°8′E. *(Courtesy ZQ Zhou.)*

easterly jet (Fig. 5.26). The phase tilt against the mean lateral shear is characteristic of shear instabilities. Indeed, observations indicate a phase tilt in the vertical suggestive of baroclinic instability on the north flank of the jet. These storms propagate across the Atlantic and often seed hurricanes. It is quite remarkable that the dry Sahara Desert spawns tropical storms that pack violent winds and torrential rains and reach as far west as the Gulf of Mexico.

A wide range of paleoclimate and archaeologic evidence indicates that 5000 to 10,000 years ago, the Sahara was much wetter than at present, with abundant rainfall to fill lakes.

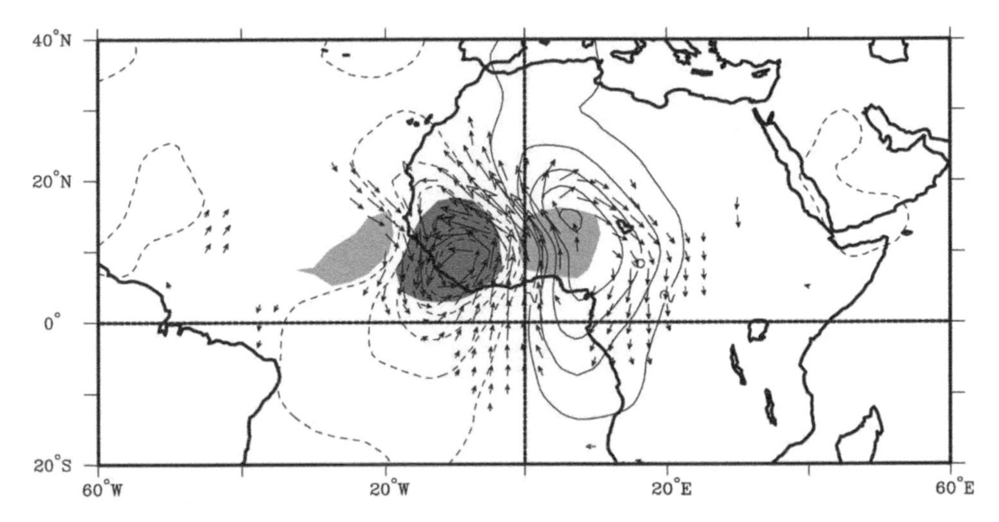

Fig. 5.26 Composite outgoing longwave radiation (OLR) (negative in dark gray), stream function (line contours, negative dashed), and wind anomalies (m/s) at 850 hPa, referenced to synoptic OLR variability at 10°W, 10°N. The mean African easterly jet peaks at 10°N at 700 hPa. *(From Kiladis et al. [2006], JAS. © American Meteorological Society. Used with permission.)*

The green Sahara is considered due to increased summer solar radiation. At present, the Earth is closest to the Sun (perihelion) in December, but 10,000 years ago the perihelion was in Northern Hemisphere summer due to the 26,000-year precessional cycle of the Earth orbit around the Sun. The increased summer solar radiation drives the African monsoon rainband further northward than present (deMenocal and Tierney 2012). Climate models simulate the northward advance of the monsoon rains during the green Sahara but with a reduced magnitude.

5.5 North American monsoon

Mexico is a narrow land bridge that separates the North Pacific and North Atlantic anticyclonic surface circulations in summer (Fig. 5.27A). On the east slope of the Sierra Madre mountain range, the North American low-level jet flows northward on the west flank of the North Atlantic subtropical high. Off the west coast of Baja California, the northwesterly winds induce coastal upwelling, keeping the SST below 25°C on average (see Fig. 6.15). In the narrow Gulf of California between the Peninsular range (~1 km high) on Baja and the Sierra Madre Occidental range (~2 km high), SST is high (~30°C), and the low-level southerly winds transport moisture toward Arizona. Baja California protects the North American monsoon by keeping the ocean upwelling

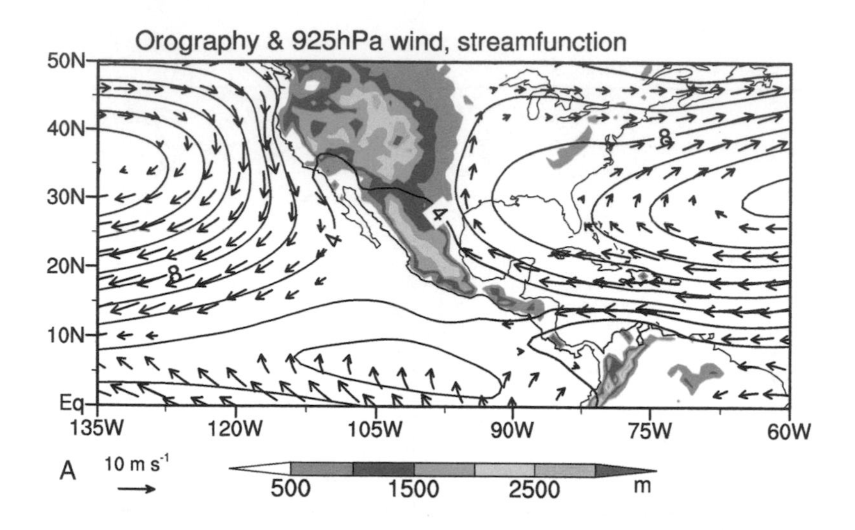

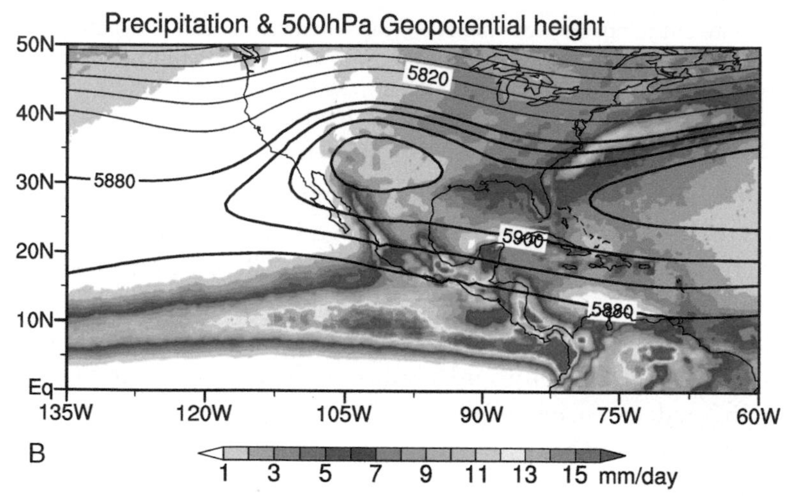

Fig. 5.27 July to August climatology. (A) 925 mb wind, stream function (black contours, 2×10^6 m^2/s intervals), and orography (shading). (B) Rainfall (shading) on 500-mb height (contours, interval of 30 for $z \leq 5850$ m, and interval of 10 for $z \geq 5880$ m). *(Courtesy ZQ Zhou.)*

to the west, and the Peninsular range blocks the cool and dry Pacific air from entering the Gulf of California.

From June to late July, intense solar heating drives rainfall northward along the western slope of the Sierra Madre Occidental range. The latent heating in summer convection induces an upper-tropospheric anticyclone (see Fig. 5.27B), a miniature counterpart of the Tibetan High. The monsoon convection starts in early June in southern Mexico, and thunderstorms reach Arizona and New Mexico in July, bringing in the major rainy

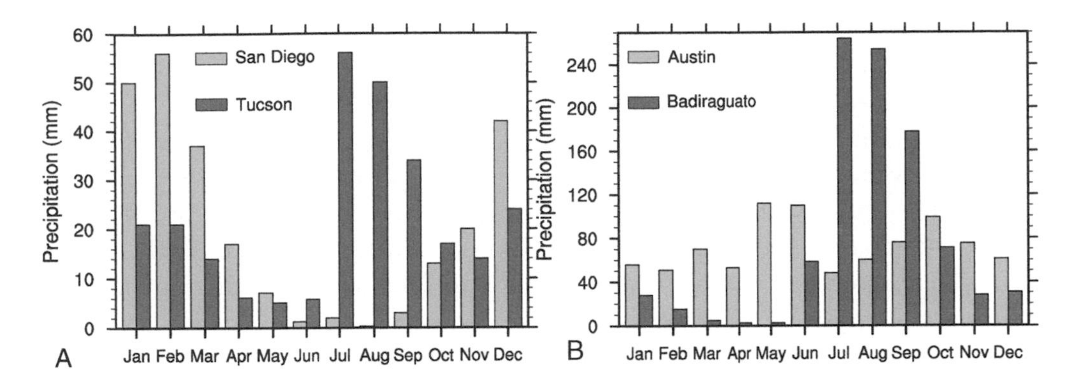

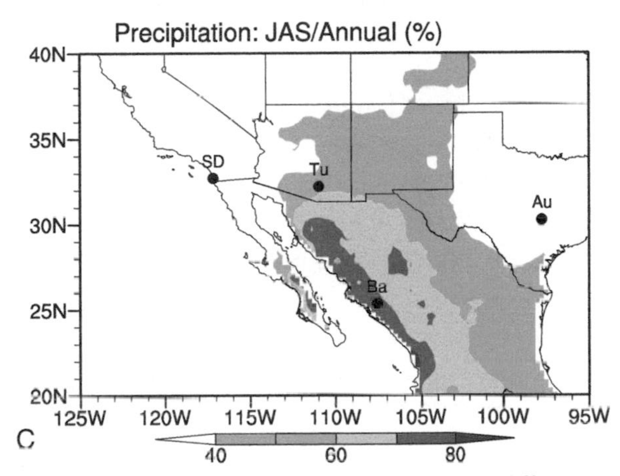

Fig. 5.28 Bar graphs of station precipitation. (A) San Diego and Tucson (different color/shading for each station. (B) Austin and Badiraguato (Mexico). (C) JAS/annual × 100 based on Tropical Rainfall Measuring Mission/Global Precipitation Measurement 3B43. *(Courtesy ZQ Zhou.)*

season for the US states. Fig. 5.28 shows the ratio of summer (July-August-September [JAS]) to annual rainfall. The 40% contour encompasses the entire country of Mexico (except Baja) and the US states of Arizona and New Mexico. This is the North American summer monsoon, also known as the Mexican monsoon. Summer convection is strongly diurnal, peaking in the late afternoon/early evening. The thunderstorms put up impressive shows of lightning in the Arizona desert.

Watch a 6-minute movie of *Monsoon* of Arizona, at https://vimeo.com/106827999. Convection, gust fronts, and gravity waves are in full display. In slow motion, the dynamic, ever-changing clouds, pouring rains and jagged lightning, will change your view of thunderstorms.

In Tucson, JAS months account for 50% of the annual rainfall (see Fig. 5.28), in contrast with the US western coast, which gets rainfall from North Pacific storms in winter/spring (see Fig. 6.17). Across the Afro-Eurasian continent, the transition in climate type from Mediterranean (winter rainy season) in Morocco, Africa to monsoonal (summer rainy season) in southeastern China takes place over a vast distance of 10,000 km (see Fig. 5.22B). Remarkably in North America, a similar transition occurs from San Diego, California, to Tucson, Arizona—only a half-day drive (600 km) apart! The eastern edge of the North Pacific subtropical high and the western edge of the North Atlantic subtropical high are separated by the narrow Sierra ranges (see Fig. 5.27). Unlike rice paddies in southeast China, however, Tucson is in the desert with baking temperatures in summer, and summer precipitation is tiny. In Tucson, the secondary winter rainy season (from the same storms passing through San Diego) is more important for vegetation (e.g., spring flower bloom) than the major summer monsoon rainy season.

East of the Sierra Madre Oriental and over the Gulf of Mexico, rainfall dips in the midsummer months of July and August, with two humps in June and September, respectively (Fig. 5.28B for Austin). This midsummer drought covers a wide region of the Gulf of Mexico and much of its coast (Fig. 5.29; see also Fig. 8.2), associated with the westward expansion of the North Atlantic subtropical high.

During the midsummer despite high SSTs of ~29°C, convection is suppressed over the Gulf of Mexico (see Figs. 5.27 and 5.29). Only in September and October rainfall

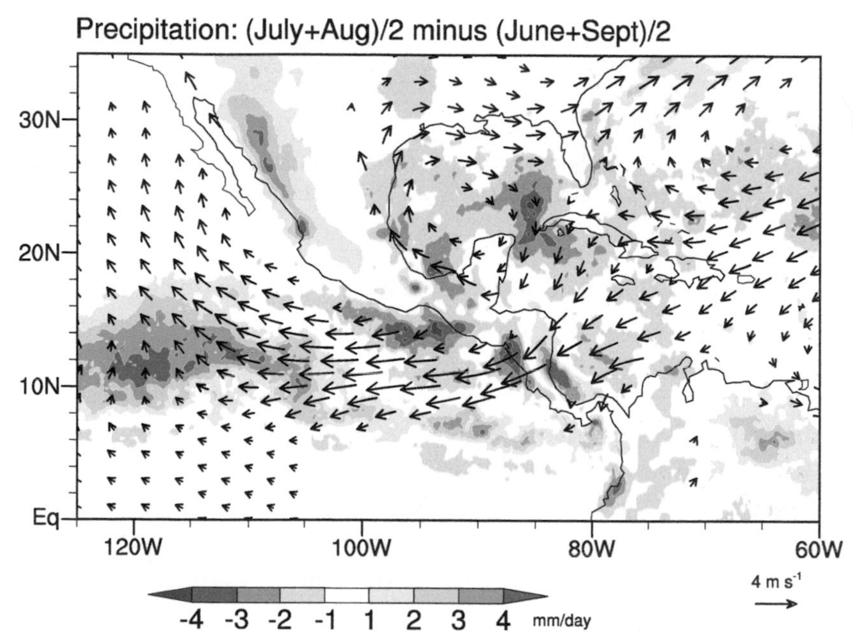

Fig. 5.29 (July + August)/2 − (June + September)/2 difference in climatologic precipitation and 925 hPa wind. *(Courtesy ZQ Zhou.)*

increases as the surrounding land begins to cool. This late-summer/early-fall rainy season coincides with the hurricane season. For comparison, the onset of deep convection over the Gulf of Mexico is more than 1 month later than that in the subtropical Northwest Pacific (the Ueda jump; see Section 5.2.3).

5.6 Global monsoon

In the zonal mean, the tropical rain band migrates across the equator following the seasonal march of the Sun with a delay of 1 to 2 months (see Fig. 3.1). The zonal mean rainfall maximum reaches its northernmost (southernmost) latitude in August (February). The delay in the poleward extension of the zonal mean rainfall behind the summer solstice is due mostly to the thermal inertia of the ocean. In 10°S to 10°N between the southernmost and northernmost swings of the rainfall maximum, the cross-equatorial wind reverses direction, with the easterly trades accelerating (decelerating) on the winter (summer) side the equator. This global seasonal cycle in response to the seasonal solar forcing, including the wind reversal and summer rainy season, meets the general definition of monsoon and is sometimes called the global monsoon.

A precipitation-based definition of a monsoon domain (Wang and Ding 2008) requires the following:

1. The monsoon precipitation index ≡ (annual range)/(annual mean) >0.5, to ensure the summer rainy season
2. The annual range >300 mm to exclude dry regions

Here the annual range refers to local summer minus winter difference. The definition identifies South and East Asia, the subtropical Northwest Pacific, the African Sahel, and Central America to Mexico as the monsoon precipitation domain (Fig. 5.30B) in the Northern Hemisphere. In the Southern Hemisphere, Indonesia to northern Australia, tropical southern Africa, and tropical South America are identified as monsoon regions.

Regions of large JJAS-DJFM rainfall difference coincide with the earlier monsoon definition and are associated with large seasonal variations in low-level winds (see Fig. 5.30A). Over the Indo-western Pacific, a C-shaped low-level wind pattern—southwesterly in the Northern Hemisphere and southeasterly in the Southern Hemisphere—is consistent with the atmospheric response to increased rainfall over the Asian summer monsoon and decreased rainfall south of the equator. Generally, the summer minus winter wind difference is westerly on the equatorward flank of the summer monsoons in the tropics. The subtropical East Asian monsoon is an exception, with a southeasterly summer-winter difference, consistent with the nonlocal effect of the westerly jet in addition to the land-sea thermal contrast.

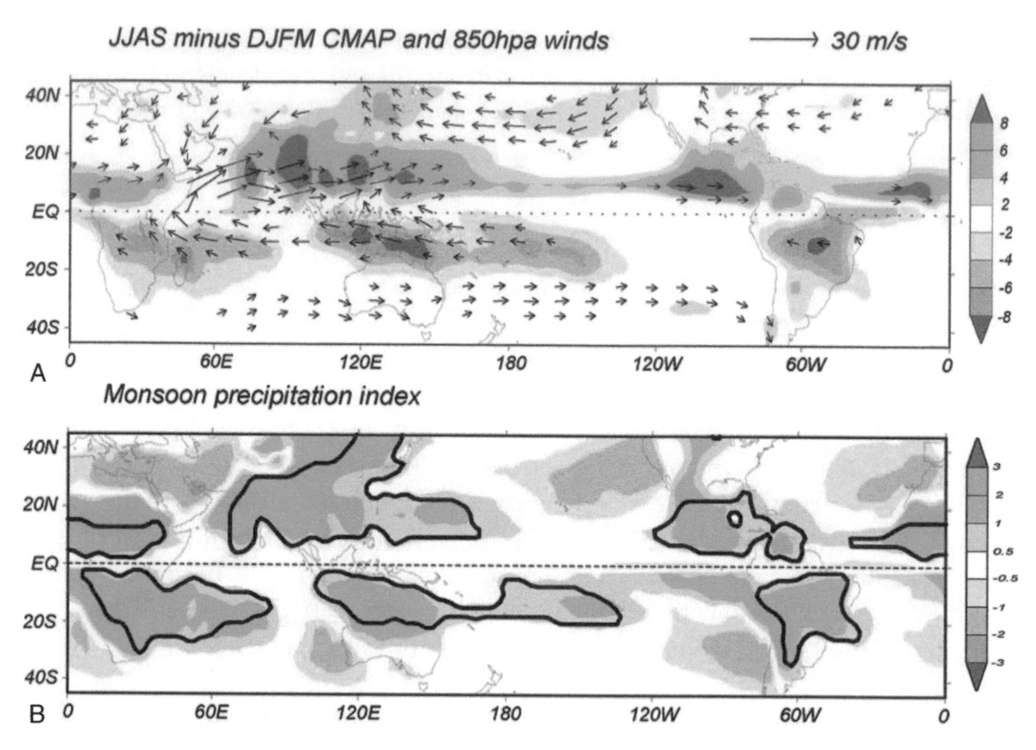

Fig. 5.30 (A) JJAS-DJFM differences in precipitation (mm day−1) and 850 hPa winds. (B) Monsoon precipitation index (color shading) and the monsoon precipitation domain outlined by the black lines. *(From Wang and Ding (2008).)*

5.7 Discussion

Monsoons may be viewed as locally enhanced seasonal variations. While the TOA insolation is a smooth function of time and zonally uniform, the presence of continents and mountains markedly modulates the timing and poleward extension of regional monsoons. For example, the rainband over East Asia reaches as far north as 36°N (the latitude of Tokyo), but on the other side of the Afro-Eurasian continent the African Sahel rain band is kept south of 15°N. The onset of the summer rainy season varies a great deal zonally as well, from May over the Bay of Bengal and Indochina to late July over the subtropical Northwest Pacific (see Fig. 5.20).

Land-sea thermal contrast is often emphasized as the dynamic driver of monsoons, but the Indian monsoon offers a counterexample, where the land-sea temperature contrast is greatest during the pre-monsoon months and decreases sharply following the monsoon onset. Surface MSE is a better predictor of the Indian monsoon onset, while the gradual increase in column-integrated water vapor path seems to trigger the Northwest Pacific monsoon onset. In either case, sufficiently high SSTs above the convective threshold are necessary to sustain deep convection of the monsoon, as illustrated by suppressed rainfall over the western Arabian Sea where the intense monsoon southwesterlies induce ocean upwelling. The peak SST, however, tends to occur just before the monsoon onset (see Fig. 5.13), from the Bay of Bengal to the Northwest Pacific.

Even though the TOA insolation is cyclic from one year to another, monsoons display marked year-to-year variations. The search for a scheme to predict monsoon over India led Walker (1932) to discover the Southern Oscillation. Coupled ocean-atmospheric interactions give rise to recurrent patterns of year-to-year variability, a major conceptual advance that led to climate prediction at seasonal leads. Chapters 9 to 11 present coupled dynamics that drive interannual variability of monsoons.

Review questions

1. Refer to Yanai and Tomita (1998) and discuss the seasonal variations in Q_1 and Q_2 profiles over the Tibetan Plateau and Bay of Bengal (regions G and E).
2. The (*northeast, southwest*) monsoon is wet over India. Why?
3. Calculate and compare the contributions from temperature and humidity to the MSE change in Fig. 5.9 from May to June.
4. How does the upper-level westerly jet vary from winter to summer in the Indian sector? How is this change related to the tropospheric temperature distribution?
5. The Sahara Desert, India, and southeastern China are at about the same latitudes. Why is the Sahara dry while India and southeastern China experience abundant summer rainfall?
6. Consider the thermodynamics of an air column west of Tibet. How does the Asian monsoon heating cause a tropospheric warming there? How does this affect summer convection over the large region from the Mideast to Sahara?
7. Gill's solution for a northward-displaced heat source explains much of the Asian summer monsoon circulation. Discuss the thermodynamic balance and vertical motion in and west of the heat source.
8. Discuss observational evidence and physical mechanisms for orographic effects on the Asian summer monsoon. Consider both the Tibetan Plateau and narrow mountain ranges.
9. When does the SST in the Arabian Sea reach the annual maximum? Why does it not happen in summer, say July?
10. Solar radiation is about the same on May 6 and August 6, each 45 days away from the summer solstice. Compare climate conditions (rainfall, surface and tropospheric temperatures) over India between these dates. Discuss factors that cause these differences.
11. Is the same asymmetry seen between May 6 and August 6 also seen for the North America monsoon?

12. How does the Earth rotation affect the Asian summer monsoon, making it distinct from the diurnal sea breeze? Consider the thermal-wind relationship between temperature and wind at lower and upper levels. How might this affect the timing of the arrival of deep convection over India?
13. How does summer climate (late July) change from San Diego to Tucson? What is the equivalent transition in terms of rainfall seasonal cycle at ~32°N on the Afro-Eurasian continent?
14. Is there evidence for mountain effects on rainfall of the North American summer monsoon?
15. Where is the core of the African Easterly Jet located in terms of latitude and height? Use the thermal wind to explain.

CHAPTER 6

Subtropical climate: Trade winds and low clouds

Contents

In the lower branch of the Hadley circulation, air moves equatorward and gains an easterly wind component because of the Coriolis force. The easterly trade winds are subject to surface friction, do not conserve angular momentum, and are about 7 to 8 m/s in surface speed over the subtropical oceans. A belt of surface high pressure forms in the subtropics as air is being displaced poleward (Fig. 6.1) from the tropical rain bands in the upper troposphere (see Fig. 3.18).

The trade winds are in geostrophic balance with the pressure gradient between the subtropical high and equatorial low along the major rain bands. The easterly trades prevail in all seasons and over all the tropical oceans, except for the summer North Indian Ocean that is dominated by the southwesterly monsoon. The northeast and southeast trades converge onto the intertropical convergence zone (ITCZ) near but not always on the equator. The subtropical high is stronger and more robust in the summer hemisphere, with eastward intensification toward the eastern boundary of the ocean basin. For example, the North Pacific subtropical high shifts northward and intensifies in summer off the west coast of North America. The northeast trades over Hawaii (160°W, 20°N) and the northwesterly winds along the Southern California coast are parts of this subtropical high circulation.

6.1 Trade wind climate

Atmospheric temperature generally decreases with height below the tropopause. In the subtropics, however, the marine boundary layer (MBL) of the atmosphere is often capped

Coupled Atmosphere-Ocean Dynamics: from El Niño to Climate Change
ISBN 978-0-323-95490-7, https://doi.org/10.1016/B978-0-323-95490-7.00006-0

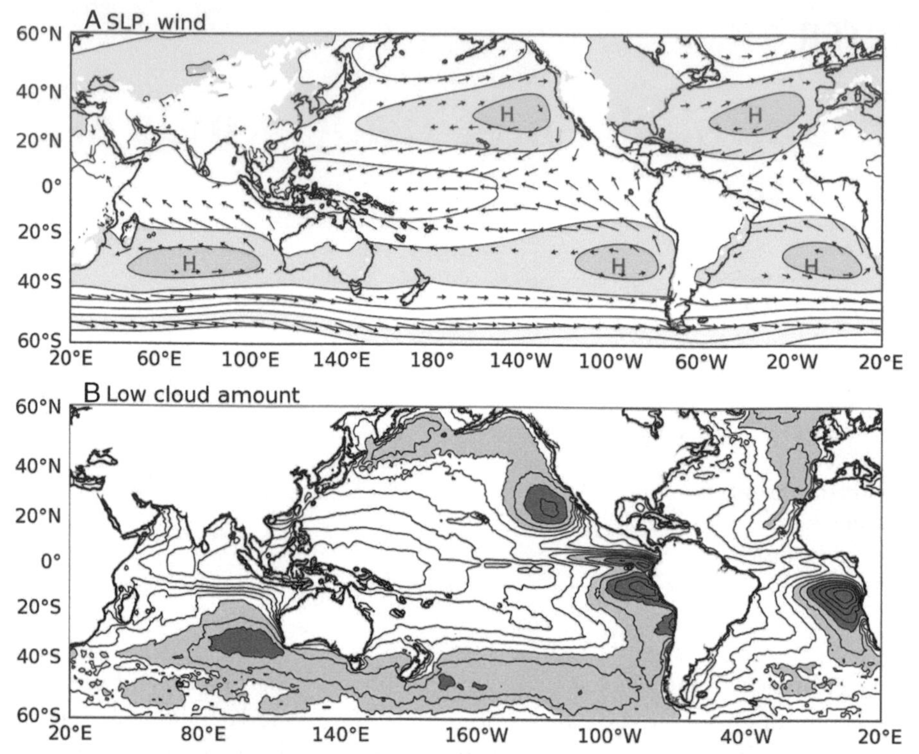

Fig. 6.1 Annual-mean climatology. (A) Surface winds and sea level pressure (*SLP*; blue >1015, yellow <1000; interval of 5 hPa; SLP and wind vectors with elevation >1 km are omitted). (B) Low cloud amount (line contours at interval of 5%; light grey shading >35%, dark >45%) based on International Satellite Cloud Climatology Project during 1984-2018. *(Courtesy L Yang.)*

by a thin layer in which temperature increases with height. This temperature inversion represents a very stable layer of the atmosphere. With a base typically of 1 to 2 km in height, the trade wind temperature inversion separates the moist layer beneath and a very dry free troposphere above (Fig. 6.2). Here dew point temperature T_d refers to the temperature at which water vapor would condense when the air parcel is cooled adiabatically. It is an alternative measure of humidity to vapor pressure and mixing ratio. Sharp temperature and humidity differences indicate that air is of different origins above and below the inversion. The dry air above the inversion originates from the upper troposphere (of low in situ temperature and humidity) and is subject to adiabatic warming in slow descent. With little cloud above the MBL, the thermodynamic balance in the free troposphere is between radiative cooling and adiabatic warming associated with slow downward motion of the Hadley circulation (see Fig. 3.8C).

In a three-dimensional view, equatorial waves flatten horizontal temperature gradients in the tropical free troposphere. As a result, the free tropospheric temperature profile is set to be nearly moist adiabatic by deep convection over warm waters (and in summer

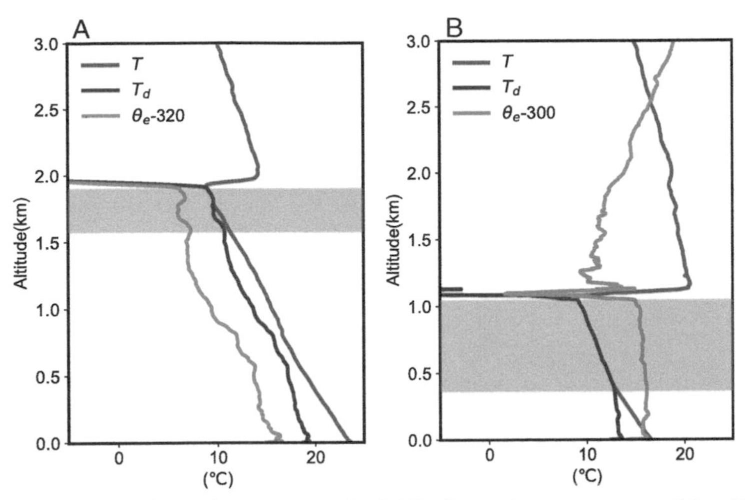

Fig. 6.2 Representative profiles of temperature (red, °C), dew point temperature (blue, °C), and equivalent potential temperature (green with offset, K) near (A) Hawaii (152°W, 23°E, October 24, 2012) and (B) Los Angeles (126°W, 31°E, 21 July 2013). Gray shading indicates the cloud layer. *(Courtesy JW Liu.)*

monsoon; see Section 3.6). Fig. 6.3 compares the potential temperature profiles between the northwestern Pacific warm pool and the cool tropical southeastern Pacific. Above 700 mb, the temperature profiles are nearly identical, set to be moist adiabatic by deep convection over warm oceans (e.g., the tropical western Pacific). Over the Southeast Pacific where the sea surface temperature (SST) is lower by 8°C, a temperature inversion forms in transition from the warm free troposphere to a colder profile in the MBL.

The MBL is moist as sea surface evaporation supplies the moisture, and the strong stratification in the inversion inhibits mixing with the dry free troposphere. In the surface mixed layer (typically 500 m deep), both potential temperature and specific humidity are nearly uniform in the vertical (see Fig. 6.2B). Clouds (stratus, stratocumulus, and shallow cumulus) often form under the inversion as the moist surface air is mixed upward by turbulence and subject to adiabatic cooling. Such extensive low-cloud decks are commonly observed over the subtropical oceans.

The easterly trades, low clouds capped by an inversion, weak precipitation, and suppressed weather disturbances, are characteristic of climate in the subtropical high. Satellite-borne cloud radar (CloudSAT) and lidar (CALIPSO) observe deep convective clouds in the ITCZ as tall as 16 km, prevalence of low clouds in the subtropics (void of middle and high clouds, especially in summer), and storm track clouds further poleward (Fig. 6.4). The tropopause that caps high cloud tops shows discontinuities in the subtropics, dropping from 16 km in the tropics to 12 km at 30°N/S and 9 to 10 km at the poles.

Turbulence lifts moist surface air above the lifting condensation level (LCL, equivalent to cloud base) for cloud formation. Of high liquid water contents and optically thick, low clouds are an effective infrared radiator (close to a black body). The vertical

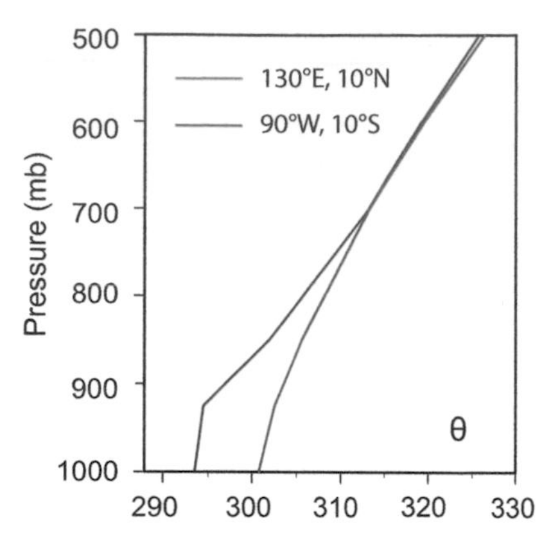

Fig. 6.3 Potential temperature (K) over *(red)* the warm, convective Northwestern and *(blue)* cool Southeast Pacific.

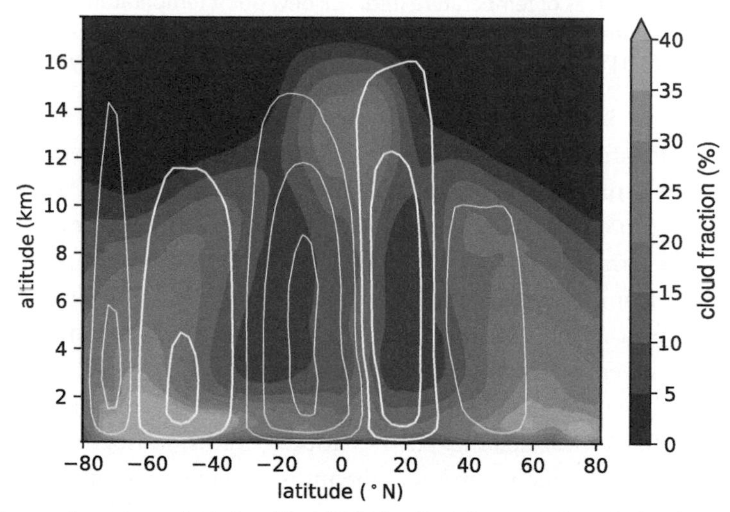

Fig. 6.4 Zonal annual mean vertical cloud fraction (shading) from combined CloudSat-CALIPSO climatology for 2006-2018, with the Hadley circulation streamfunction (white line contours, positive thickened) superimposed. *(Data: W.J. Bertrand and J.E. Kay, courtesy YF Geng.)*

minimum in the coarse-resolution Q_1 profile at 850 hPa in Fig. 3.8C is suggestive of the cloud-top radiative cooling, but the real radiative cooling is much larger at and immediately below the cloud top. The intense cloud-top radiative cooling promotes turbulent mixing in the MBL in favor of cloud formation, and the MBL cooling strengthens the inversion (Fig. 6.5). This implies positive feedback between the inversion, vertical mixing, and low clouds. The large-scale subsidence aids low-cloud formation by drying out the free troposphere and reducing the downward atmospheric radiation, and by strengthening the capping inversion.

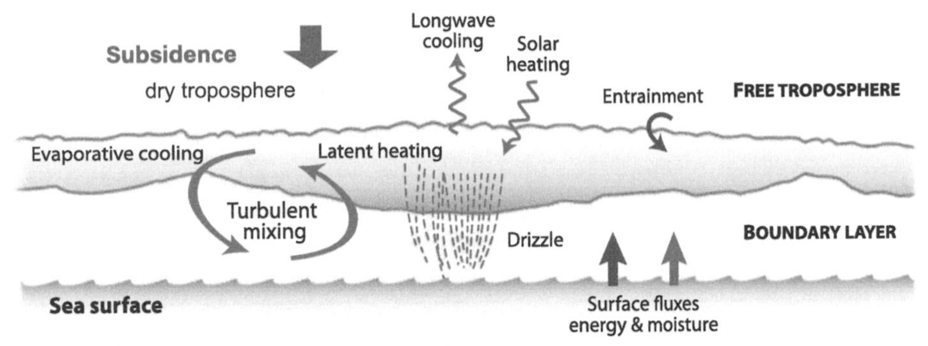

Fig. 6.5 Schematic of important physical processes for marine boundary layer clouds. *(Adapted from Wood [2012]. © American Meteorological Society. Used with permission.)*

Low clouds are important for global energy balance. Compared to the dark ocean surface, low clouds are highly reflective of solar radiation (Fig. 6.6) but emit upward infrared radiation of similar intensity to the ground surface (because of similar temperatures), as illustrated by a comparison of visible and infrared images from satellite. Thus low clouds reduce the net downward radiation at the top of the atmosphere that enters the atmosphere-ocean column. Reflection of solar radiation by bright low clouds is an important contributor to planetary albedo (~ 0.3). Likewise, low clouds are an effective cooling mechanism for the ocean surface: They greatly reduce the downward solar radiation (at the rate of 20 W/m^2 per 10% increase in cloud cover) but increase downward infrared radiation only slightly because of the low cloud base (~ 500 m) and high moisture in the MBL.

A well-mixed MBL consists of a surface mixed layer and a solid stratus cloud layer capped by the inversion (see Fig. 6.2B). In the former, temperature decreases upward at the dry lapse rate ($\Gamma_d = 9.8$ K/km), and specific humidity is constant. In the cloud layer, q decreases with height, and temperature follows the moist adiabat as water vapor condenses to form stratus, releasing latent heat. Equivalent potential temperature (θ_e) is constant in the vertical.

6.2 Cloud-regime transition

Vertical shear of the trades within the MBL is weak except near the surface. Consider an atmospheric column following the MBL-mean wind from the southern California coast (Fig. 6.7). The temperature profile above the inversion does not change much along the path, but the underlying SST increases from 20°C on the coast to 26°C near Hawaii (see Fig. 6.2). The cool MBL column that moves over the increasingly warm ocean surface (known as the cold advection) is gravitationally unstable, promoting turbulence that maintains a surface mixed layer. The radiative cooling at the cloud top generates

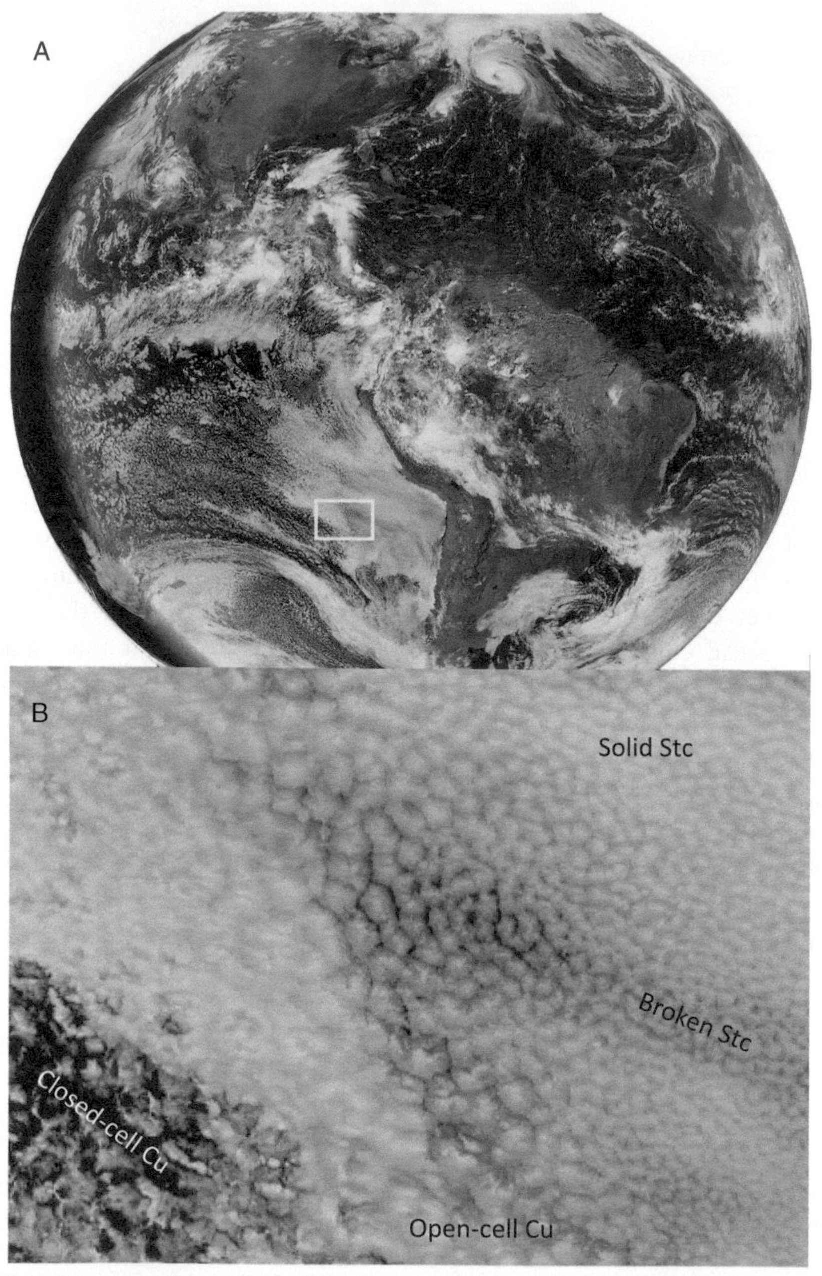

Fig. 6.6 (A) Visible image from NOAA GOES East satellite on September 10, 2021, 1530 UTC. (B) A closeup in a 1400 km × 1000 km region *(yellow box)* of the Southeast Pacific. *(Image: NOAA.)*

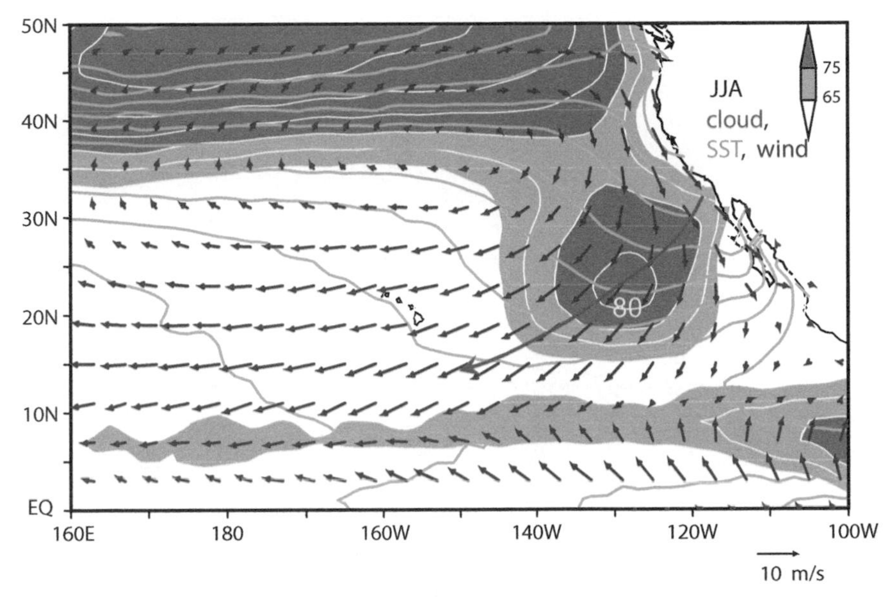

Fig. 6.7 Summer *(JJA)* sea surface temperature *(SST)* (2°C interval), surface wind velocity (m/s), and cloud fraction (light/dark gray shading >65/75%; white contours at 5% interval) based on iCOADS ship observations. Along a trade wind trajectory *(purple line)*, a marine boundary layer column moves over an increasingly warmer ocean surface.

turbulence that entrains warm and dry air from the inversion layer, deepening the MBL. The MBL deepens from 1 km off the coast to 2 km near Hawaii (see Fig. 6.2) with the temperature jump in the inversion layer decreasing from 12°C to 5°C. Off California, the inversion is strong and low in altitude with solid, unbroken stratus (St) decks. Along the trajectory southwestward, MBL clouds change gradually to cumulus under stratocumulus (Stc) to scattered shallow cumulus (Cu) near Hawaii (Fig. 6.8). The cloud type transition in response to the increase in underlying SST corresponds to a decrease in bulk cloudiness from 80% off California to 30% in Hawaii. Fig. 6.6B is an example of rich low-cloud types from satellite. Convective cellular structures 10s to ~100 km across are clear from the above, and the cloud cover decreases with more convective clouds.

Status/stratocumulus clouds are found in a well-mixed MBL over the cool ocean, in which turbulence couples the cloud layer with the sea surface. As the air column moves southwestward and warms, the rising MBL top becomes permanently decoupled from the sea surface as turbulence generated at the cloud top can no longer reach the surface mixed layer (see Fig. 6.8). In the so-called subcloud layer above the surface mixed layer, potential temperature increases and specific humidity decreases with height. The decoupling from the surface mixed layer deprives the moisture supply and starves the cloud layer under the inversion.

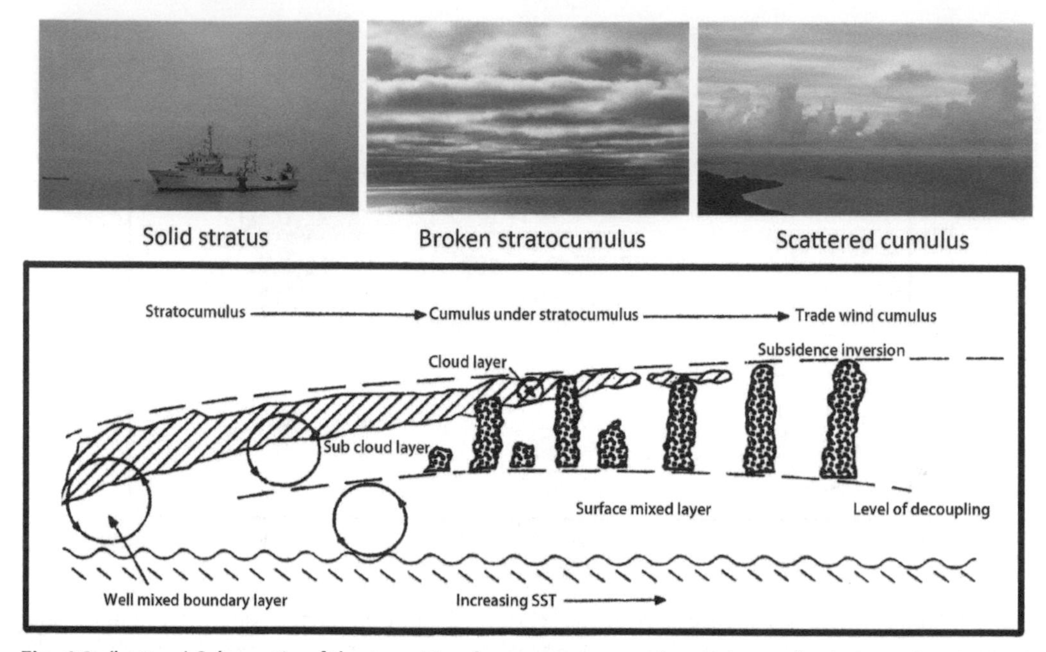

Fig. 6.8 *(bottom)* Schematic of the transition from stratus to trade wind cumulus in inversion-capped marine boundary layer (MBL). With increasing sea surface temperature *(SST)*, the MBL deepens and cloud cover decreases. *(top)* Low clouds seen from below: stratus off Peru, stratocumulus off La Jolla, California, and trade cumulus off Oahu, Hawaii. *(Top, https://www.eol.ucar.edu/content/rv-jose-olaya-lima-peru, QH Peng, and author; bottom: Albrecht et al. (1995). © American Meteorological Society. Used with permission.)*

In the surface mixed layer, the air parcel riding on the northeast trade winds continues to warm and moisten due to the SST increase. The increase in moist static energy in the surface layer eventually makes the air parcel conditionally unstable ($d\theta_e/dz < 0$) (see Fig. 6.2A). Shallow cumulus clouds develop and detrain under the inversion base, forming sheets of stratocumulus above the cumulus. Overshooting cumuli entrain dry air from the inversion layer, evaporating the detrained cloud droplets and breaking up the stratocumulus sheets. Cumuli often organize into mesoscale convective complexes (see Fig. 6.6B), with clear sky in downdraft regions. Precipitation under stratocumulus/cumulus clouds reduces total water content above the surface mixed layer. The latent heating above and evaporation of drizzles in the surface mixed layer stratify the MBL As a result, bulk cloudiness (averaged in space and time) is small for the decoupled, cumulus regime.

CALIPSO is a satellite-borne earth-pointing LIDAR and can measure MBL cloud-top height accurately over the subtropics, where the atmosphere above the trade wind inversion is mostly clear. Fig. 6.9A shows the mean cloud-top height and fraction over the Northeast Pacific, and Fig. 6.9B is a transect of cloud-top occurrence along a great

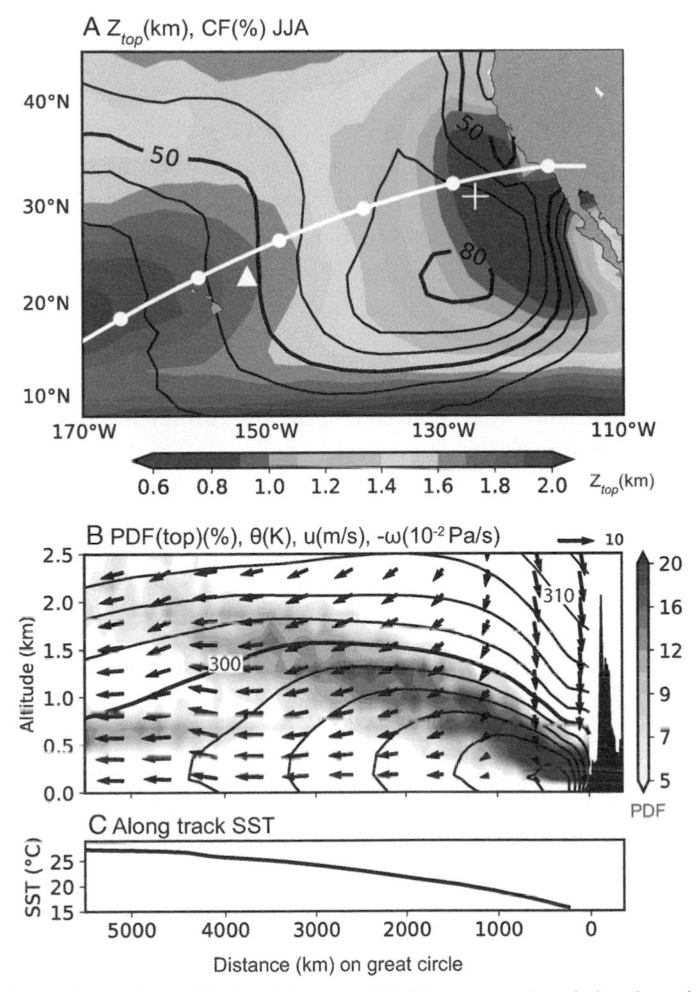

Fig. 6.9 CALIPSO JJA climatology: (A) cloud fraction *(CF)* (%, contours) and cloud-top height (km, color shading). (B) Frequency of cloud-top occurrence as a function of height (blue shading) along a great circle (white curve in A, dots spaced at 1000 km), superimposed on potential temperature (K) and wind velocity (u, $-\omega$). (C) Alongtrack sea surface temperature *(SST)*. The plus and triangle indicate locations for the soundings in Fig. 6.2. *(Adapted from Liu et al. (2022).)*

circle between Los Angeles and Hawaii. In summer, the hot desert air meets and rides above the cool marine air on the California coast. A strong inversion develops with a low base. The inversion and cloud-top rise and the inversion intensity weakens southwestward in this transect. On average, the cloud top deepens from 0.7 km off Los Angeles to 2 km over Hawaii (see Fig. 6.9B), and the cloud type changes from solid stratus to scattered cumulus with the bulk cloudiness decreasing from 70% to 30%. Here the cloud cover is defined as the percentage of the sky covered by clouds. Over warm SSTs near

Hawaii, cloud-top occurrence is bimodal, peaking at the inversion base and the top of the surface mixed layer.

Riding on the trades further south, the MBL continues to deepen and eventually merges into deep convection in the ITCZ over very warm waters with SSTs above the convective threshold. On its way toward the equator along the trade wind trajectory, the air column in the MBL warms and gains moisture from surface evaporation. The trade wind inversion limits the mixing with the dry air above, minimizing moisture loss in the MBL. Thus heavy convective rainfall in the narrow ITCZ is supplied by surface evaporation over the broad subtropical ocean through the inversion-capped MBL. On the global scale, surface evaporation peaks over the subtropical trade wind region due to the strong wind and reduced relative humidity at the sea surface as a result of vertical mixing within the MBL and with the free troposphere under the broad subsidence.

On the poleward flank of the subtropical high (north of 40°N), advective fog frequents the summer subpolar oceans where the poleward flow advects warm moist air down the SST gradient over the increasingly cool ocean (see Figs. 6.1 and 6.7).

6.3 Climate feedback

Globally, low clouds that top the MBL tend to occur over the eastern subtropical oceans (e.g., off California over the North Pacific, off Peru over the South Pacific, and off Namibia over the South Atlantic) (see Fig. 6.1B). Low SSTs over these regions (as much as 10°C lower than in the tropical western Pacific) require a temperature inversion to transition to the much warmer free troposphere. As over the subtropical Northeast Pacific, the cold dry advection by the prevailing winds promotes turbulence that mixes moisture upward to form stratus/stratocumulus clouds.

6.3.1 Cloud-SST feedback

From a meteorologic point of view, low SSTs are favorable for low clouds that top the MBL over the subtropical Northeast Pacific. From an oceanographic point of view, the low SSTs are due, at least in part, to the reduced incoming solar radiation at the surface under stratus/stratocumulus clouds. Coastal upwelling and the offshore Ekman advection also contribute to low SSTs, as does the cold, dry advection by the northeast trades in MBL (see Fig. 6.7), which intensifies surface evaporative cooling by increasing sea minus air temperature difference ($T_s - T_a$) and lowering surface relative humidity (R_H).

The downward solar radiation at the surface may be expressed as

$$Q_{SW} = Q_{S0}(1 - a_c)$$

where Q_{S0} is the clear sky value, and $a_c \sim 0.7C$ is the cloud albedo, with C being the bulk cloud cover. The difference from clear-sky radiation, $-a_c Q_{S0}$, is also known as the cloud radiative effect (CRE). Around the cloud cover maximum (130°W, 25°N), an enhanced cloudiness of 0.7 compared to a background value of 0.3 in the western North Pacific reduces the downward solar radiation by $0.7\delta C Q_{S0} = 90\ \text{W/m}^2$, with $Q_{S0} = 320\ \text{W/m}^2$ the clear sky insolation in the tropics (Section 7.3). Compared to the mean latent heat flux from the ocean $\overline{Q_E} = 150\ \text{W/m}^2$, the CRE on solar radiation is important to keep the subtropical Pacific cooler in the east than in the west. The discussion here implies a positive feedback between SST and MBL cloud: Low SSTs increase cloud cover, and the cloud-induced reduction in solar radiation lowers SST even more.

Seasonal and spatial variations (Fig. 6.10A) suggest an empirical relationship between low-level cloud cover and lower tropospheric stability (LTS),

$$\text{LTS} \equiv \theta_{700} - \theta_s$$

where the subscripts denote 700 hPa and the sea surface. A large LTS corresponds to a strong inversion, which is favorable for solid stratus/stratocumulus with a large cloud cover, while a weak inversion with a deep MBL (small LTS) favors scattered cumuli of a small bulk cloudiness. If one assumes that free tropospheric temperature is horizontally uniform in the tropics (weak temperature gradient; Section 3.6), then the lower-atmosphere stability is determined by surface air temperature or SST. Indeed, positive correlations are found in observed interannual variability between CRE on surface shortwave radiation and SST, over the low cloud-covered subtropical eastern oceans of the Atlantic and Pacific (Fig. 6.11; Norris and Leovy 1994).

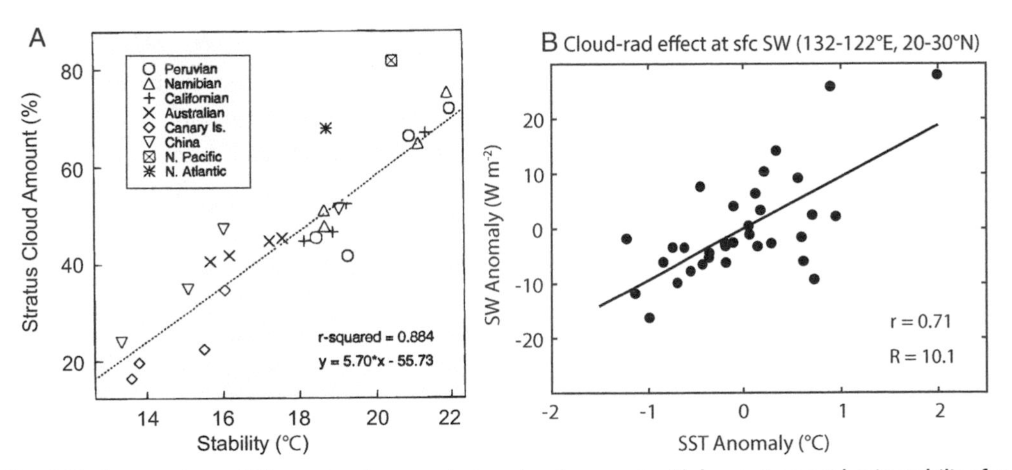

Fig. 6.10 Scatterplots of (A) seasonal-mean stratus cloud amount with lower-tropospheric stability for different regions, and (B) interannual anomalies of JJA sea surface temperature and shortwave cloud radiative effect over the subtropical Northeast Pacific (132–122°E, 20–30°N). *(A, From Klein and Hartmann (1993); B, courtesy L Yang. © American Meteorological Society. Used with permission.)*

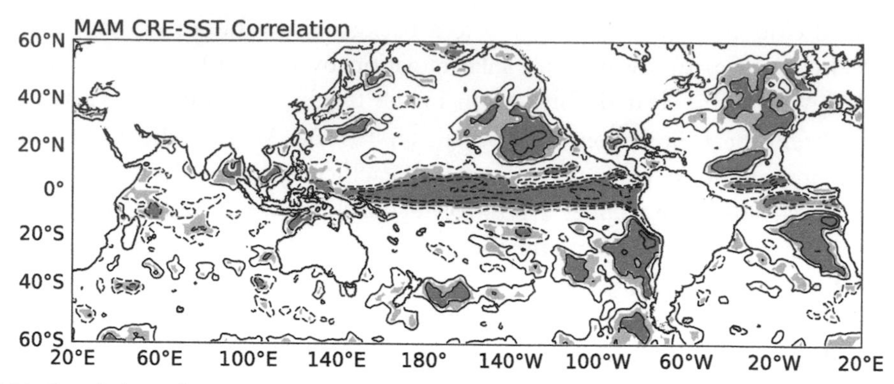

Fig. 6.11 Correlation of interannual variability in March-May shortwave cloud radiative effect *(CRE)* and sea surface temperature (SST) during 1984-2018. Contour interval is 0.2, negative contours dashed, and zero omitted. Light/dark shading indicates 95/99% significance level. *(Courtesy L Yang.)*

The cloud-radiative effect for shortwave can be written as

$$Q'_{SW} = -a_c Q_{S0} = b_c T'$$

where $b_c = 10\ \mathrm{W/m^{-2}/K}$ is estimated from interannual variability on the south flank of the Northeast Pacific stratus cloud deck (see Fig. 6.10B). Likewise, change in surface latent heat flux due to an SST variation can be estimated using the Clausius-Clapeyron (CC) equation,

$$Q'_E = b_E T'$$

where $b_E = \overline{Q}_E \alpha = 10\ \mathrm{W/m^{-2}/K}$, and $\alpha = \frac{1}{q_s}\frac{dq_s}{dT} = 0.065\ \mathrm{K^{-1}}$ is the CC coefficient (see Chapter 8 for details on the evaporative feedback). Note that the latent heat flux is upward (to cool the ocean) and a negative feedback on SST change. Under low-cloud decks, the cloud radiative feedback is positive and comparable in magnitude to the surface evaporative damping on SST. Although the value of $b_c = 11\ \mathrm{W/m^{-2}/K}$ is evaluated near the center of the low-cloud deck for summer and represents an upper limit on the stratus feedback, the net SST damping is markedly reduced over cloud-covered subtropical oceans.

To illustrate the low-cloud effect, we disable the CRE over the Northeast Pacific cloud deck in an otherwise realistic coupled ocean-atmosphere general circulation model (GCM) (see Miyamoto et al. 2021 for details). The comparison with the full-physics control run shows a local SST response of more than 3°C in summer (JJA) (Fig. 6.12A). While the cloud radiative scheme is perturbed in a limited domain, the resultant SST cooling triggers a coupled ocean-atmospheric response that extends well beyond the perturbed region toward the southwest. The pattern resembles the so-called Pacific meridional mode of interannual variability (see Fig. 6.12B) to be discussed in Section 12.4. This

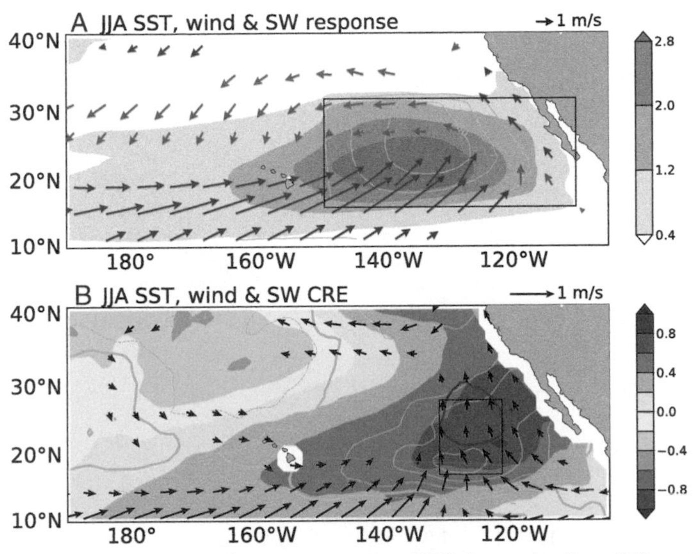

Fig. 6.12 (A) JJA changes in sea surface temperature (SST) (gray shading, °C), surface downward shortwave radiation (green contours at interval of 20 W/m^2; positive values for downward flux), and surface wind (m/s; blue/red arrows for weakened/strengthened background winds) in an experiment where the low-cloud radiative effect is artificially turned off within the black box (150°W–110 °W, 16°N–32°N). (B) Regression coefficients of JJA shortwave cloud radiative effect *(CRE)* (green contours every 3 W/m^2 °C^{-1}; negative dashed), SST (shading, °C), and surface wind (vectors, m/s) onto seasonal-average SST anomaly in the black box. The thick blue contour highlights the stratus deck (mean cloud cover = 0.8). *(A, Courtesy A Miyamoto; B, adapted from Yang et al. (2022).)*

implies that the radiative cooling of the low-cloud deck off California intensifies the northeast trade winds over a broad region to the southwest, including Hawaii.

The scatter diagram of climatologic-mean cloudiness and local SST over the tropical region of 30°S to 30°N suggests two regimes of cloud-SST feedback (Fig. 6.13). Over the cool ocean with SST below the deep convective threshold (~26.5°C), low clouds dominate and the cloudiness increases with decreasing SST (a positive feedback). Over the warm ocean, deep convective clouds prevail, and the cloudiness increases with increasing SST (a negative feedback). Thus the cloud-SST feedback depends on the dominant cloud type and can be either positive (shallow stratiform clouds) or negative (deep convective clouds). There is evidence that SST variability is locally enhanced under stratiform cloud decks while it is suppressed under the ITCZ, consistent with the cloud radiative feedback.

Over the tropical Atlantic, a cross-equatorial dipole pattern in SST arises due to coupled ocean-atmospheric feedback (to be discussed in Chapter 10). In the deep tropics, the ITCZ displaces into the anomalously warm hemisphere, and the resultant cloudiness anomalies are of the same sign as the underlying SST anomalies (Fig. 6.14). In the

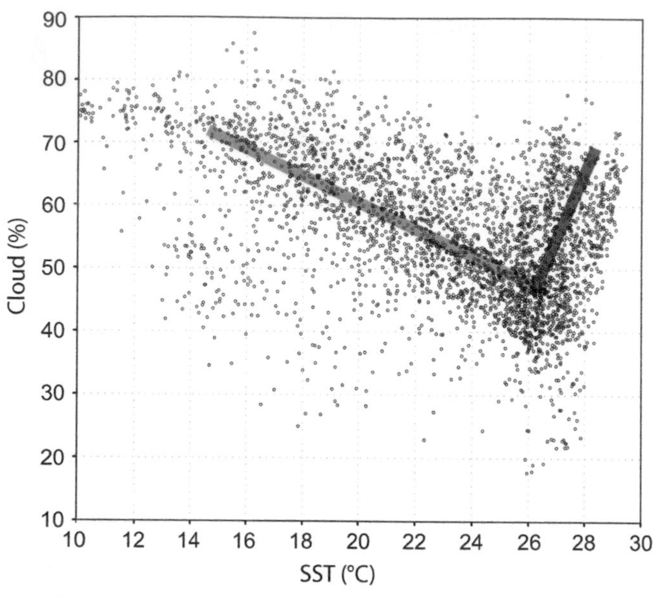

Fig. 6.13 Scatterplot of climatologic-mean sea surface temperature *(SST)* and cloud amount (%) in February over the tropical eastern Pacific and Atlantic (180°W–20°E, 30°S/N).

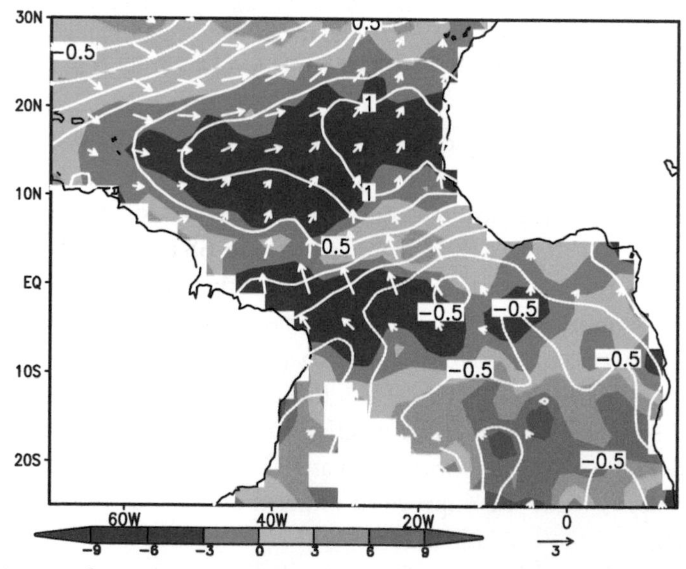

Fig. 6.14 COADS sea surface temperature (SST) (contours in °C, negative dashed), surface wind (m/s), and cloudiness (color shading in %) composites in January to March based on a cross-equatorial SST gradient index. Note that the cloud-SST correlation in the subtropics (stratus) is opposite to that in the deep tropics (deep convective clouds in intertropical convergence zone). *(From Xie (2004a).)*

off-equatorial regions covered by low MBL clouds, cloudiness and SST anomalies are of the opposite signs. The coupled SST-cloud pattern here illustrates the feedback of two different kinds.

6.3.2 Global radiative feedback

The vertical and horizontal scales of low clouds (<1 km) are one order of magnitude smaller than those of deep convection (10 km). It requires 10^3 times more computing power to resolve explicitly the cloud-topped MBL than deep convection. The adequate representation and realistic simulation of low cloud decks remain a major challenge for atmospheric/climate modeling. In fact, low-cloud response (area fraction, thickness, and reflectivity) to global warming dominates the global cloud-radiative feedback on global warming, representing a major source of uncertainty in model projections of future change in global temperature (see Chapter 13).

Recent studies suggest that low-cloud cover is likely to decrease in warmer climate, resulting in a positive feedback on greenhouse warming (Bretherton 2015). In warmer climate, the increased greenhouse gases and water vapor increase the emissivity of the free troposphere above the MBL, thinning low clouds by reducing the cloud-top cooling. The increased jump in specific humidity across the capping inversion intensifies the drying of MBL by the turbulence at the cloud top. Both thermodynamic effects thin MBL clouds. There are competing mechanisms acting to enhance low clouds in warmer climate: (i) The upward amplified tropospheric warming intensifies the inversion, and (ii) the reduced subsidence deepens the MBL in favor of a thicker cloud layer. Large-eddy simulation results indicate, with some observational support, that MBL clouds thin overall in greenhouse warming with a positive global radiative feedback.

Anthropogenic aerosols are often good cloud condensation nuclei. Increased aerosols in MBL clouds increase the number of cloud condensation nuclei (N_d). For a given liquid-water path integrated over the MBL (L), the effective radius of cloud droplets is proportional to $r_d = (L/N_d)^{1/3}$. The total surface area of cloud droplets is proportional to

$$N_d r_d^2 = L^{2/3} N_d^{1/3}$$

A polluted cloud contains more cloud condensation nuclei and is more reflective with a higher albedo. Bright bands are often seen from satellite visible images on stratus cloud decks as a result of sulfur dioxide emissions from ships underneath (known as the Twomey effect). Smaller cloud droplets are harder to precipitate, and the reduced drizzle increases the lifetime of the cloud (known as the Albrecht effect). Through both the Twomey and Albrecht effects, pollutants increase the bulk cloud cover, helping cool the planet.

6.4 California climate

California is classified as Mediterranean climate with winter being the rainy season. The Pacific Ocean to the west, coastal mountains, the broad atmospheric subsidence, and subtropical high are all important in shaping California climate and the year-to-year variability.

6.4.1 Coastal upwelling

Northwesterly alongshore winds prevail off the coast of southern California year round. The alongshore winds drive offshore Ekman flow in the surface ocean, forcing cold, nutrient-rich water to rise. The coastal upwelling keeps coastal water several degrees cooler than offshore, and the nutrients upwelled from beneath enable the growth of California kelp forests, spawn planktons, and sustain rich fisheries.

The atmospheric high-pressure system off California exists throughout the year (see Fig. 6.1). In summer, the high-pressure system occupies the entire North Pacific, causing coastal upwelling off the entire west coast of the United States (see Fig. 6.15A). In winter, the closed anticyclonic circulation is confined to the eastern third of the subtropical North Pacific, with the upwelling favorable northwesterly alongshore winds limited to the south of San Francisco. The mechanism for the subtropical high likely differs between winter and summer. GCM experiments show that the winter high pressure is a remote response to the orographic effect of the Tibetan Plateau acting on the intense westerly jet. In summer this remote orographic effect is weak, and the thermal contrast between the cool Pacific Ocean and warm North American Continent is important.

In summer, coastal SST reaches a local minimum in northern California (14°C). While the upwelling favorable alongshore winds are strongest (Fig. 6.15A) there, the offshore Ekman mass transport $\rho u_E H_E = \tau_y / f$ is nearly constant on the entire California coast, where ρ is the water density, u denotes the cross-shore current velocity, y is the alongshore distance, and τ_y is the alongshore wind stress. The thermocline, say as represented by the 12°C isotherm, shoals sharply northward, by 50 m from San Diego to San Francisco (see Fig. 6.15B), in response to the northwesterly alongshore winds. In a two-layer model of the upper ocean (see Chapter 7), the alongshore momentum equation is

$$fu = -g\Delta\rho \frac{\partial h}{\partial y} + \frac{\tau_y}{\rho H}$$

where f is the Coriolis parameter, g the gravity, $\Delta\rho$ the density difference between the warm upper layer and cold lower layer, and h and H the perturbation and mean thermocline depth, respectively. On the coast, cross-shore flow is not allowed ($u = 0$), and the alongshore wind stress is balanced by the pressure gradient due to the shoaling thermocline. The northward shoaling thermocline and intensified upwelling cooling on the California coast is similar to the eastward intensification of the equatorial cooling across the Pacific and Atlantic (Section 7.2.1). In San Francisco, the summer

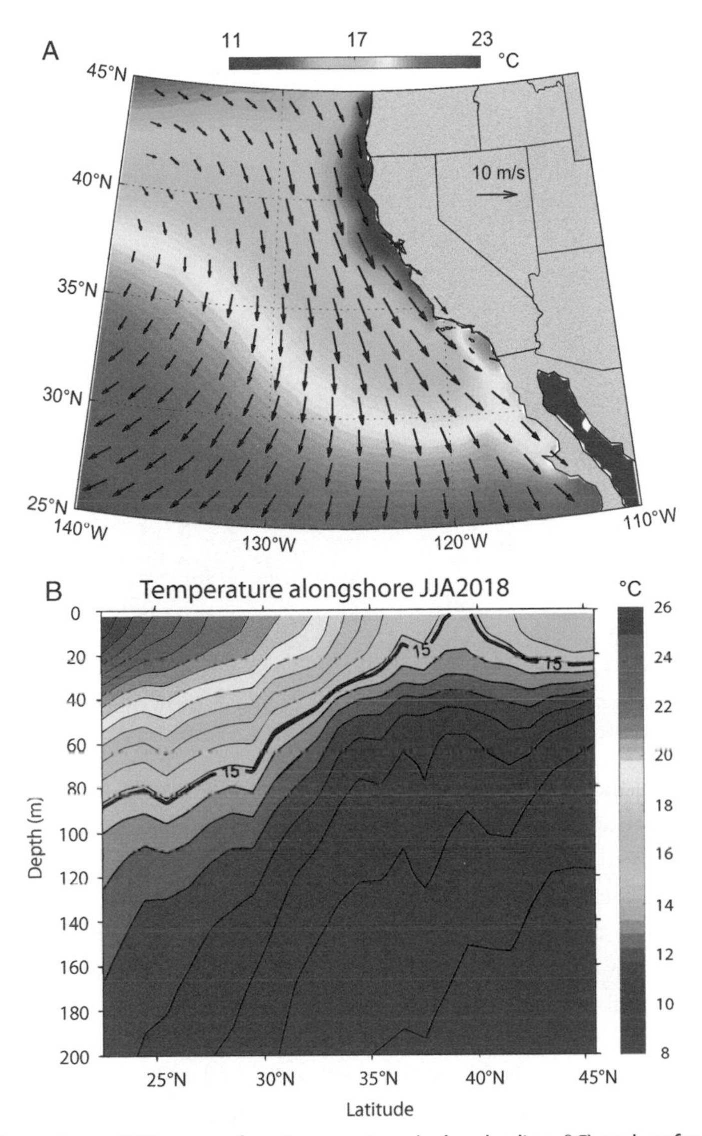

Fig. 6.15 JJA climatology of (A) sea surface temperature (color shading, °C) and surface wind velocity (m/s), and (B) ocean temperature (°C) along the west coast of North America. *(A, courtesy K Li; B, courtesy JR Shi.)*

upwelling is very strong, keeping the annual SST maximum at 15°C (September), only 3°C warmer than the winter minimum (April). In the western Pacific at similar latitudes, the summer SST maximum is about 26°C, with much lower winter minima (15°C in Tokyo [Japan] and 4°C in Qingdao [China]).

South of the Point Conception, the coast curves eastward, forming the Southern California Bight between Point Conception and San Diego. A northwesterly wind jet leaves

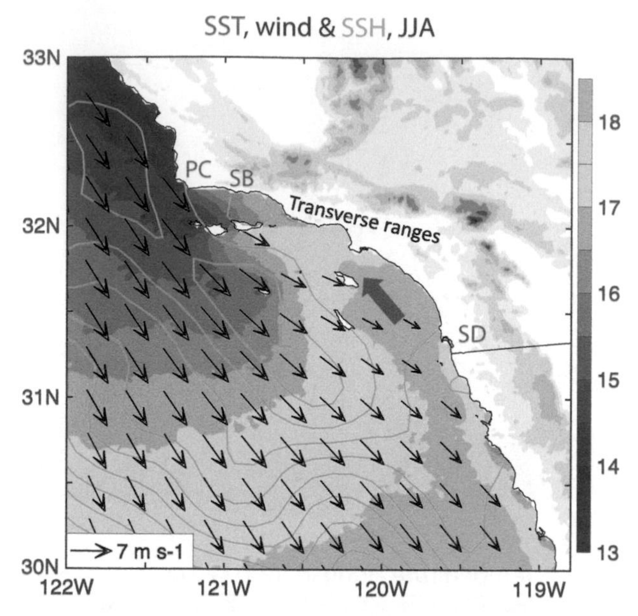

Fig. 6.16 JJA climatology of sea surface temperature *(SST)* (color shading, °C), surface wind (m/s), and sea surface height (SSH, green contours at 2 cm internal), along with land topography (color shading at 500 m interval). The purple block arrow indicates the Southern California Countercurrent. *PC,* Point Conception; *SB,* Santa Barbara; *SD,* San Diego. *(Adapted from Kilpatrick et al. [2018].)*

the coast at Point Conception, while the winds are weak in the Bight (Fig. 6.16), blocked by the transverse mountain range. The southward California Current leaves the coast at Point Conception and becomes an offshore jet, creating a southeastward cold tongue in SST. The cyclonic shear of the offshore wind jet drives a cyclonic circulation within the Bight, and a northwestward coastal current (South California Countercurrent) develops, advecting the warm water from the south. The reduction in alongshore winds and coastal upwelling also contributes to the increase in SST on the coast of the south Bight; the August SST increases from 18°C at Santa Barbara to 20°C at San Diego.

Capped by a temperature inversion that is less than 1 km high, shallow stratus clouds develop over the Southern California Bight. The stratus-topped MBL is called the marine layer locally in California. During the day, solar radiation heats up the ground surface and burns the stratus by reducing the surface relative humidity and lifting the cloud base.

6.4.2 Atmospheric rivers

Winter is the rainy season for the US west coast. A water year is defined as the 12-month period that starts from 1 October, capturing the rainiest months of December to February. In winter, successive storms arrive from the Pacific along the storm track steered by the westerly jet. The storminess, measured by upper-level meridional wind variance, is large over the North Pacific and Atlantic storm tracks but weak over Eurasia

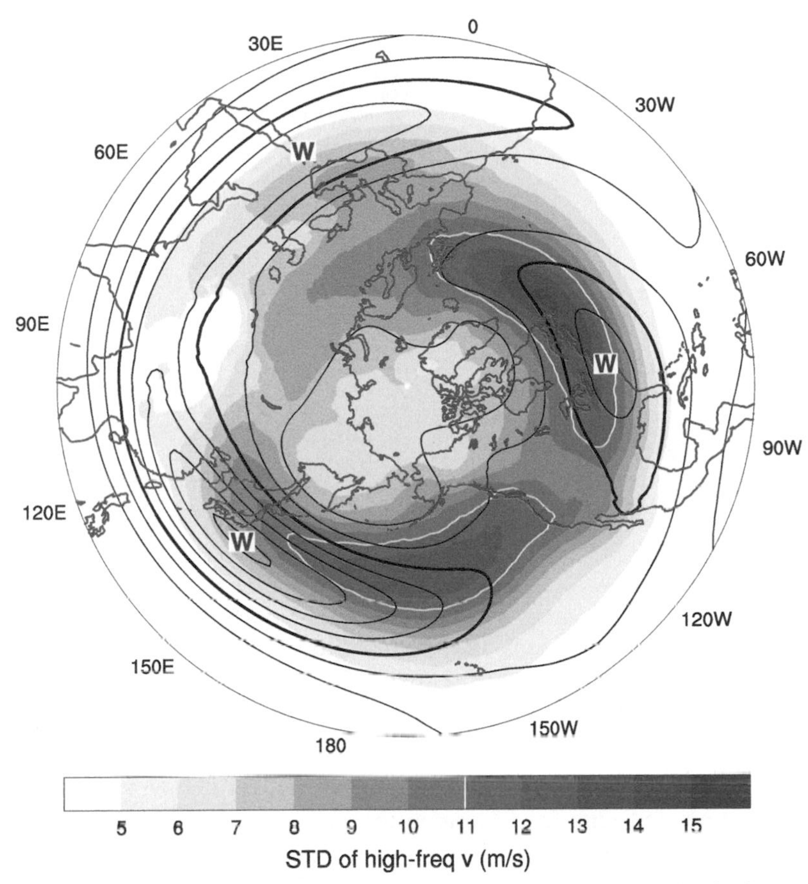

Fig. 6.17 DJF mean flow and storminess at 250 hPa. Standard deviation of 8-day high-pass variability of meridional wind velocity (shading), along with the zonal wind velocity (black contours, interval of 10 m/s, 30 m/s thickened). Based on ERA5 for 1980-2019. *(Courtesy Y Kosaka and S Okajima.)*

east of the Urals (Fig. 6.17). On the west coast of North America, the storminess peaks over the Pacific Northwest.

Pacific storms moderate west coast temperatures and bring rainfall on coastal mountain ranges. In the warmer sector of these storms, the low-level southwesterly flow channels moisture poleward in narrow corridors called atmospheric rivers (Fig. 6.18; Ralph et al. 2020). Globally this is the major mechanism for the poleward transport of atmospheric moisture from the tropics. Impinging on coastal mountains, the southwesterly onshore flow induces orographic lifting, causing precipitation. The large and sustained moisture transport in atmospheric rivers renders intense precipitation of long durations, heightening the risks of flooding and landslides on the windward slope of the mountains. On the US west coast, atmospheric rivers contribute 30% to 50% of annual precipitation and is responsible for 60% to 100% of the most extreme rainstorms (Lamjiri et al. 2017).

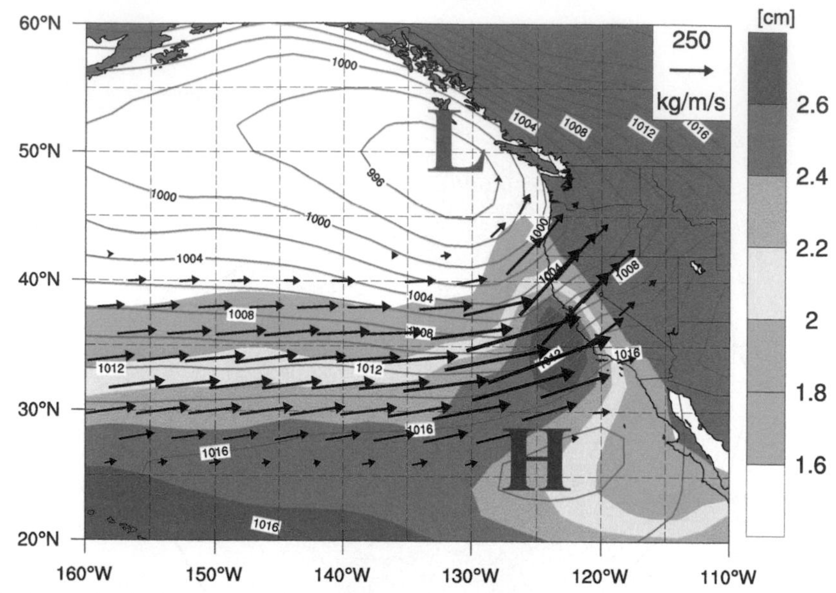

Fig. 6.18 Atmospheric river composites on days of San Francisco Bay Area landslide onset: sea level pressure (line contours, 2 hPa interval), column-integrated water vapor (color shading, cm), and column-integrated vapor transport vectors (>150 kg/m/s). *(From Cordeira et al. (2019). © American Meteorological Society. Used with permission.)*

In California, half or more of the annual precipitation takes place within a total of 5 to 15 days, often associated with landfalling atmospheric rivers. A small difference in the number of storms that come onshore can make a difference between a wet and dry year. As a result, annual precipitation and streamflow in California experience large variations due to atmospheric internal dynamics (see Section 12.1) and/or El Niño and the Southern Oscillation (ENSO) influence from the tropics (see Section 9.5).

6.4.3 Hydroclimate

Rainfall is highly unevenly distributed in California, generally larger in the northern than the southern state. The annual rainfall in southern California is 200 to 300 mm, far from sufficient to sustain major population centers such as Los Angeles and San Diego. At large economic and environmental costs, major engineering projects were undertaken to divert water from rivers of Sierra mountains and the Colorado River to supply Southern California. Large interannual variability in rainfall poses additional challenges for water management.

Much of the winter precipitation is locked in the snowpacks on Sierra mountains above the snowline (1–1.5 km above the sea level). The snowpacks reach the annual maximum in size in April and then start melting. Collecting the melt water, rivers swell. The resultant rapid increase in river flow is called the spring pulse (Fig. 6.19B). The

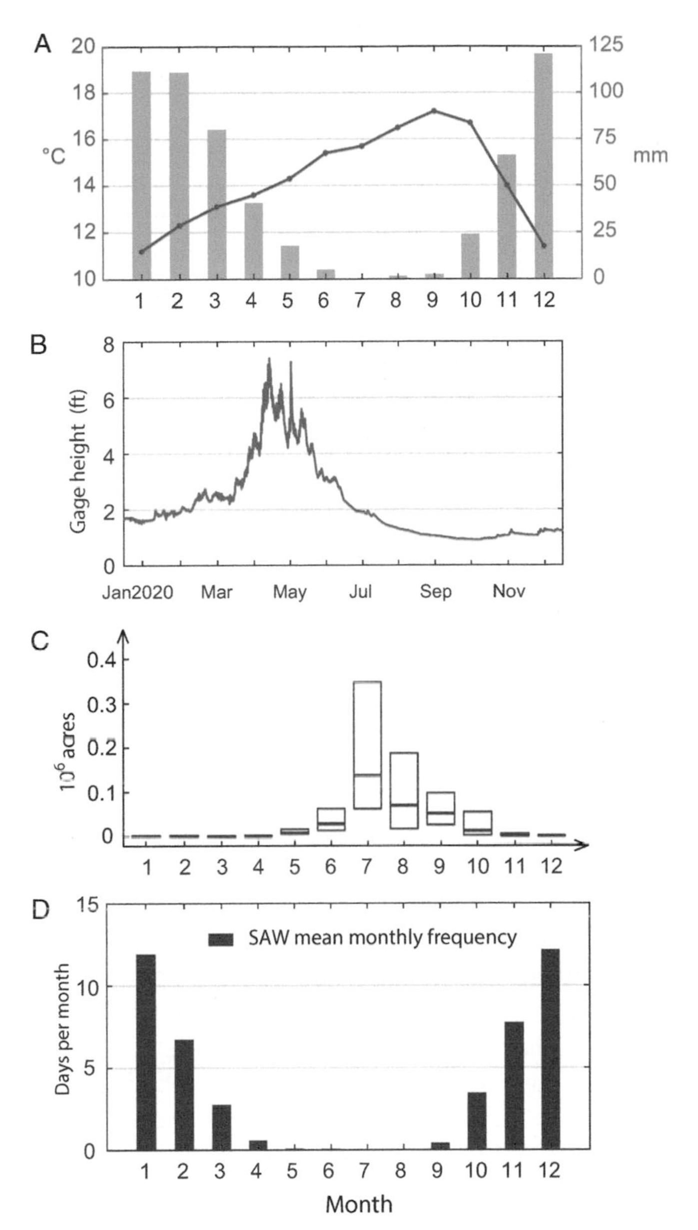

Fig. 6.19 Monthly climatology of (A) temperature and precipitation in San Francisco, (B) daily flows in 2020 of the Merced River (USGS gauge 11264500), (C) burned area in California (10^6 acre), and (D) Santa Ana wind *(SAW)* days per month. *(A-B, Courtesy ZH Song; C, Li and Banerjee (2021); D, Gershunov et al. (2021).)*

seasonal melting of the snowpacks sustains the river flow through summer despite the diminishing precipitation itself. Thus snowpacks act much like a natural water reservoir for California.

California deserts transform into colorful flower beds during the spring, followed by a long dry season of May to September with little precipitation. The vegetations that grow in spring dry and turn into fuels for wildfires when ignited. The fire season peaks in summer (July–August) (see Fig. 6.19C), aided by high temperatures. The vapor pressure deficit, VPD $\equiv e_s - e_a = (1 - R_H)e_s$, is a meteorologic parameter often used to assess fire risks. It increases exponentially with temperature. The fire-burned area in California displays pronounced year-to-year variability, which is highly correlated with VPD ($r =$ 0.72 for 1972–2018). It is also correlated with ENSO, which affects winter storm tracks and precipitation on coastal California (see Section 9.6).

Downslope wind or Foehn is favorable for wildfire because of the adiabatic warming and drying plus high wind. Santa Ana is the important downslope wind for southern California that blows offshore across the coastal mountain ranges. Conditions favorable for Santa Ana wind are:

1. Synoptically, a high-pressure anomaly over the Great Basin to the north causing offshore geostrophic winds roughly perpendicular to the region's mountain ranges (Fig. 6.20A)
2. Thermodynamically, a large temperature gradient between the cold desert surface east of coastal ranges and the warm ocean air on the other side of the mountains (see Fig. 6.20B), say at 1.5 km, that creates an offshore pressure gradient for katabatic-like offshore flow in a thin layer near the surface

The downslope wind causes adiabatic warming and drying, resulting in single-digit relative humidity in severe cases. For this reason, Santa Ana winds often cause wildfires, sometimes ignited by fell power lines, but they occur most frequently during winter associated with synoptic weather variations (see Fig. 6.19D). During the main fire season of May to September, synoptic variations are weak, and the condition for katabatic wind is not met (desert is warmer than the coast in summer).

Greenhouse warming has a strong direct effect on the hydroclimate of California, elevating the risks of heavy rainfall and landslides. In warmer climate, the moisture transport in atmospheric rivers is likely to increase because of the CC equation. The snow line rises, and snowpacks melt earlier, reducing the capacity as water reservoir.

Climate models project a robust increase of VPD with rising temperatures (Fig. 6.21B). Perturbing the expression for VPD and retaining the linear terms yields

$$VPD' \approx (1 - \overline{R}_H)e_s' - R_H'\overline{e}_s \approx \overline{e}_s\left[(1 - \overline{R}_H)\alpha T' - R_H'\right]$$

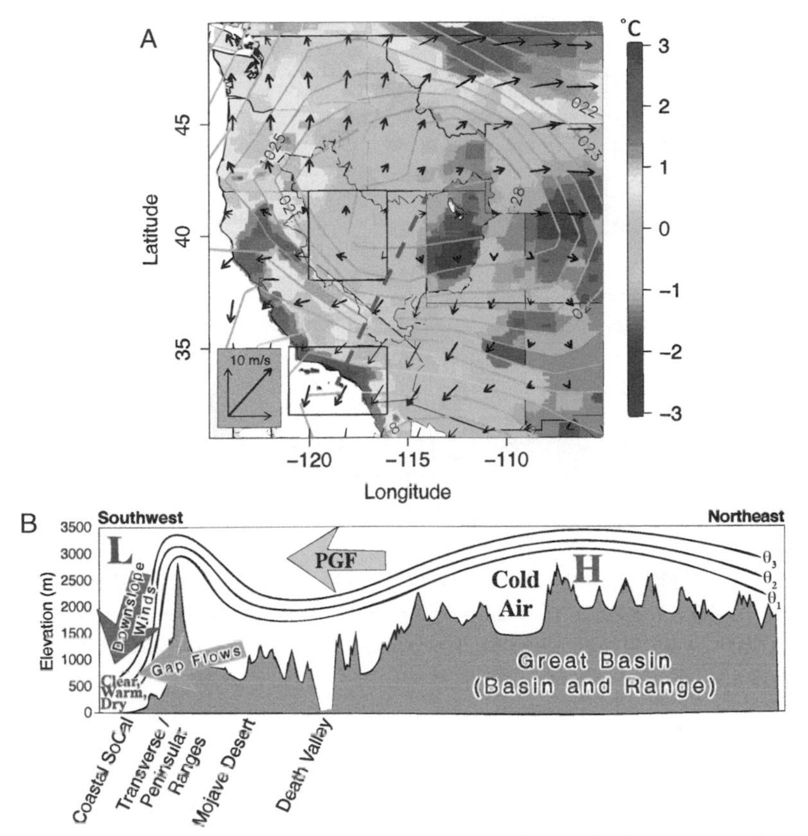

Fig. 6.20 (A) Santa Ana wind composites of Tmax anomaly (color shades), sea level pressure (line contours, 1 hPa interval), and 10 m wind (black arrows). (B) Schematic of Santa Ana winds along a transect (dashed red line in A) from the Great Basin to southern California coast. *(From Gershunov et al. [2021].)*

Fire risks increase because both terms on the right-hand side are positive; surface relative humidity on land decreases in warmer climate. The early snowpack melt also contributes to the expansion of the fire season. The burned area has been increasing in California since the 1970s (see Fig. 6.21A), reaching the record extreme of 4.26 million acres in 2020. Climate warming is a major driver of the rising trend, but fire suppression policy has also contributed to the accumulation of potential fuels in the forest. The pre-1800 natural level of wildfires is estimated at 4.4 million acres per year in California, much higher than the actual yearly burned area with fire suppression policy.

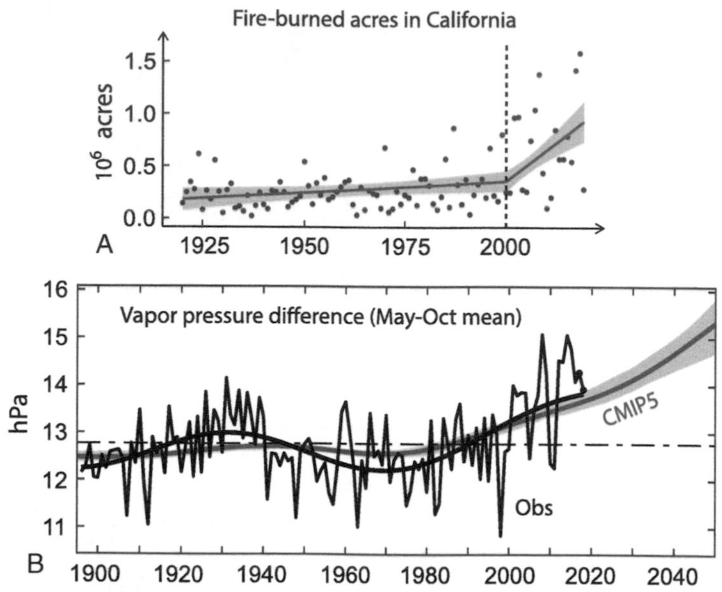

Fig. 6.21 (A) Annual fire-burned area (million acres), and (B) May to October vapor pressure difference in California in observations *(black line)* and climate model projections *(red)*. *(A, Li and Banerjee (2021); B, Williams et al. (2019).)*

Review questions

1. From a typical atmospheric sounding in Hawaii (Fig. 6.2A), how do we know that the air above the trade wind inversion is of a different origin than in the MBL below?
2. How does water vapor or cloud above the trade wind inversion affect the boundary layer cloud underneath? *Hint:* Consider cloud-top radiation.
3. Why does temperature inversion form in the subtropical ocean, especially in the eastern basin?
4. Why is the eastern subtropical ocean covered by low clouds?
5. What are the important factors for cloud types and bulk cloudiness?
6. How does the MBL decoupling affect surface relative humidity? How does shallow cumulus due to the conditional instability affect humidity in the surface mixed layer?
7. Compare the moist static energy $m/c_p = T + Lq/c_p$ between surface air parcels in San Diego ($T = 15°C$) and Hawaii (25°C). Assume $R_H = 0.8$.
8. Discuss the importance of MBL clouds in the global energy budget.
9. Suppose that you can control the amounts of deep convective and low MBL clouds. What would you do if you'd like to cool the planet to counter increasing greenhouse gases?
10. What are the major features of ocean circulation in the Southern California Bight? How do they affect SST? Discuss the orographic effect of coastal mountain ranges.

11. Discuss roles of mountains in the California hydroclimate.

12. Air temperature is known to decrease with height in most places. In monsoon Asia, summer resorts are built in elevated towns to avoid the intense heat. This is different in southern California. On the warm day of 5 September 2021, the daily maximum was 79°F at San Diego Airport (elevation 15 ft) and 98°F at Lake Forest (970 ft). Both stations are near the coast. Explain why.

13. Consider an air parcel at the sea level with a relative humidity R_H. Raising the air parcel in the vertical will cause water vapor inside to condense and form cloud.

a. Show that the cloud base height is $z_c = -\ln(R_H)/(\alpha\Gamma_d)$, where Γ_d is the dry lapse rate (9.8 K/km), and $\alpha \sim 0.06\ K^{-1}$. Use the CC equation $dq_s/dT = \alpha q_s$, where q is the specific humidity and the subscript s denotes saturation. Assume that the parcel does not mix with the environment as it rises from the sea surface to the cloud base (i.e., $R_H\ q_s[z = 0] = q_s[z = z_c]$).

b. Over ocean, relative humidity is about 80%. What is the typical cloud base height for low clouds?

c. Over land, R_H can drop to 50% during the day. What is the cloud base height? Discuss how sunlight can burn stratus clouds in coastal San Diego.

d. Over ocean, the diurnal cycle in surface temperature is negligible but the amount of boundary-layer cloud decreases during the day. Explain why. *Hint:* Consider cloud-top radiation.

CHAPTER 7

Equatorial oceanography

Contents

Incoming solar radiation at the top of the atmosphere is most intense at the equator. Indeed, sea surface temperatures (SSTs) reach 30°C over the western Pacific off Indonesia. We associate the equator with warm ocean beaches, frequent rainstorms, and lush rainforests where African elephants, Indonesian orangutans, and Amazon crocodiles live. The Galapagos are a group of volcanic islands 1100 km west of South America on the equator. The island group is famous for the samples of speciation Charles Darwin collected there that eventually led him to develop the idea of evolution. At the Galapagos, Darwin measured a SST of 20°C, surprisingly low compared to those off Indonesia. The water is cold enough for penguins—known for their preference for freezing weather on Antarctica—to take a foothold on the Galapagos.

The cool eastern equatorial Pacific results from the mechanic upwelling of cold water from beneath, corroborated by satellite observations of high nutrients and primary production (Fig. 7.1B). Upwelling also takes place off the California coast to compensate the offshore flow of the surface warm water driven by the northerly alongshore winds (see Section 6.4.1). In the case of California upwelling, the Coriolis effect and the presence of a solid boundary are necessary. There is not a solid boundary in the interior equatorial Pacific. Then why does upwelling occur in the open ocean on the equator? What

Coupled Atmosphere-Ocean Dynamics: from El Niño to Climate Change
ISBN 978-0-323-95490-7, https://doi.org/10.1016/B978-0-323-95490-7.00007-2

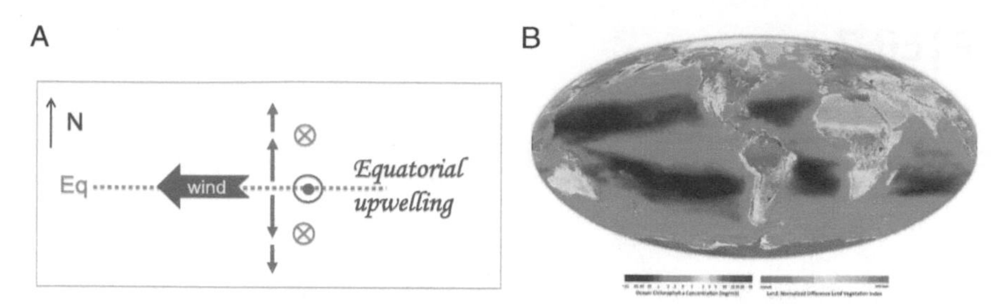

Fig. 7.1 (A) Schematic of equatorial upwelling. (B) Primary production in the marine and terrestrial biosphere. Dark blue and orange indicate low productivity over the ocean and land, respectively. (B, courtesy of NASA.)

controls its intensity and zonal variations? We will explain these and other important features of equatorial oceans.

7.1 Dynamical models

7.1.1 Equatorial upwelling

The classic Ekman pumping (see Eq. 1.6) suffers singularity at the equator ($f = 0$). To overcome the singularity, we include friction due to strong vertical shear across the base of the Ekman layer ($z = H_E$),

$$-fv_E = \frac{\tau_x}{\rho H_E} - \varepsilon u_E, \tag{7.1}$$

$$fu_E = \frac{\tau_y}{\rho H_E} - \varepsilon v_E, \tag{7.2}$$

where ε is the frictional coefficient (typical e-folding timescale: 1–2 days). We have neglected the pressure gradient force as it is smaller than the wind stress term if the horizontal scale of the wind forcing is sufficiently large. The correction due to the pressure gradient effect will be made in the next section. The above friction term becomes negligible 2° to 3° away from the equator ($|f| > \varepsilon$).

Now consider an easterly wind field as observed over the equatorial Pacific and Atlantic oceans. For simplicity, further assume that the wind is spatially uniform. Sufficiently off the equator, the Coriolis effect becomes important, and the Ekman flow is poleward in either hemisphere. On the equator, the Coriolis force vanishes and the surface currents are in the wind direction. As a result of the Ekman flow divergence, cold water rises from beneath on the equator and is then advected poleward (Fig. 7.1A). Ocean upwelling is an effective "natural conditioner" that can cool the surface ocean to as low as 18°C on the equator near the Galapagos Islands in the eastern Pacific.

Photosynthesis by phytoplankton consumes nutrients, while sinking organic matters dissolve to enrich the deep ocean. As a result, a sharp nutricline forms with low nutrients near the surface and high nutrients below. Much of the ocean productivity is limited by the supply of nutrients. In the upwelling zones, rich nutrients enter the surface layer to spawn high productivity and marine life, as captured by ocean color observations from space (see Fig. 7.1B).

The Ekman solution is

$$u_E = \frac{1}{\rho H_E} \frac{\varepsilon \tau_x + f \tau_y}{\varepsilon^2 + f^2}$$

$$v_E = \frac{1}{\rho H_E} \frac{\varepsilon \tau_y - f \tau_x}{\varepsilon^2 + f^2}. \tag{7.3}$$

For uniform easterly winds, the vertical Ekman pumping velocity at the base of the Ekman layer,

$$w_E = -\beta \frac{\varepsilon^2 - f^2}{\left(\varepsilon^2 + f^2\right)^2} \frac{\tau_x}{\rho} \tag{7.4}$$

An upwelling zone forms within $-y_E < y < y_E$, with a peak at the equator. Here, the width of the equatorial upwelling zone $y_E = \varepsilon/\beta \sim 200$ km for $\varepsilon = 1/\text{day}$. Broad downwelling takes place poleward of $y = \pm y_E$ because of the beta effect.

7.1.2 Thermal stratification

In the central equatorial Pacific, water temperature drops from 27°C at the sea surface to 15°C only 200 m below (Fig. 7.2). The tropical ocean response to wind variations is confined above the 12°C isotherm. Colder isotherms stay flat while the warmer isotherms shoal eastward across the equatorial Pacific (Fig. 7.3B). For this reason, the 12°C isotherm is taken as the lower boundary of the main thermocline, while the 20°C isotherm is often used to track the main thermocline. Below the main thermocline, water temperature continues to decrease downward but at slower rates than in the thermocline.

On the poleward flank of the anticyclonic subtropical Pacific gyre, the downward Ekman pumping pushes the winter mixed layer water into the thermocline while the Sverdrup flow carries the subducted water equatorward. The equatorward thermocline flow is manifested as a salty tongue on either side of the equator indicative of the subtropical origin (see Fig. 2.8A). At the equator, the cold thermocline water upwells as is readily seen in the hydrographic transect as the doming and outcropping of warm isotherms (see Fig. 7.2). The easterly winds drive the surface water poleward, completing an overturning circulation between the subtropics and equator. This shallow meridional overturning

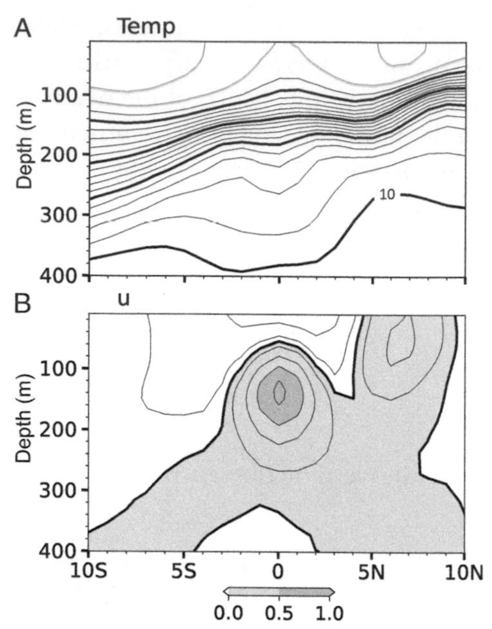

Fig. 7.2 Annual-mean climatology in the central tropical Pacific (155°W). (A) Temperature (contours at intervals of 1°C; contours of 10°C, 15°C, 20°C, 25°C thickened and temperature ≥27°C in red). (B) Zonal current velocity (contours, intervals of 0.2 m/s; positive shaded in grey). *(Courtesy Q. Peng.)*

circulation is a major mechanism for meridional heat transport in the ocean by carrying the warm surface water poleward and the cold thermocline water equatorward (see Section 2.3).

7.1.3 Pressure perturbations due to thermocline displacements

Spatial variations in the thermocline depth cause pressure variations in the warm upper layer. This is illustrated by the following thought experiment (Fig. 7.4). Consider a tank of cold water of density ρ_2. Heat the upper column of depth h to a lighter density ρ_1. Thermal expansion of the heated water raises the water surface slightly by

$$\eta = \frac{\rho_2 - \rho_1}{\rho_1} h$$

where η is the sea surface height. The pressure perturbation in the warm upper layer due to the water surface rise is

$$\delta p / \rho = g\eta = g' h \tag{7.5}$$

where $g' = \frac{\rho_2 - \rho_1}{\rho_1} g$ is called the reduced gravity and only a few thousandths of the gravity.

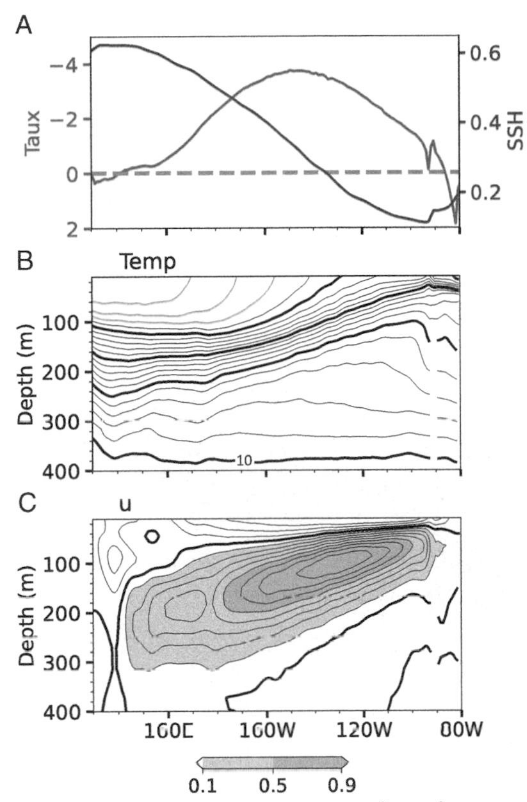

Fig. 7.3 Climatology at the equator. (A) Zonal wind stress (10^{-2} N/m^2) and sea surface height (*SHH*, m). (B) Ocean temperature (contours at intervals of 1°C, contours of 10°C, 15°C, 20°C, and 25°C thickened, and temperature ≥27°C in red). (C) Zonal current velocity (contours at intervals of 0.1 m/s, the zero contour thickened, grey shading ≥0.1 m/s). *(Courtesy Q. Peng.)*

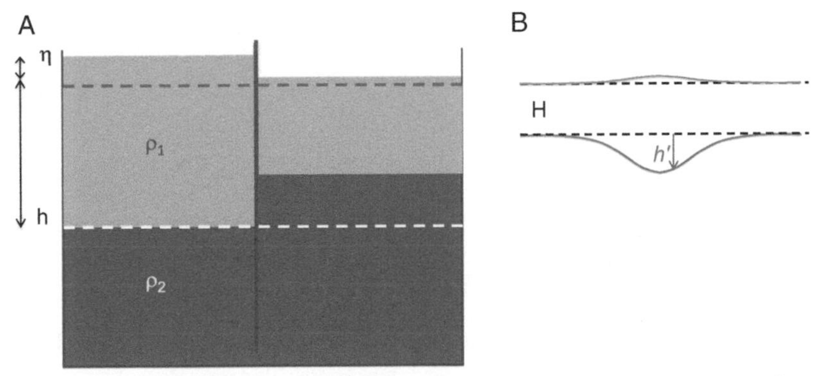

Fig. 7.4 (A) Relationship between the thermocline depth and sea level. (B) An internal wave with a deepened thermocline *(blue)* and rising sea level *(orange)*.

Now divide the tank into two parts with a solid division (see Fig. 7.4). Suppose that the depth of the upper layer that is heated to ρ_1 differs between the left and right columns. At the bottom and in the lower layer, the water pressure is the same between the two columns because the total mass above a given level in the lower layer has not changed. In the warm upper layer, however, a pressure gradient is created by the difference in the rise of the free water surface. Now if we remove the division, the pressure gradient will drive flow in the upper layer. Thus variations in the thermocline depth cause horizontal pressure gradients, which ocean waves will try to flatten.

For surface waves such as swells and wind waves, the free surface displacements (η) are subject to a restoring force of $g\eta$ as the air density is negligibly small. For ocean internal waves, the pressure perturbation arises because of the density difference between the upper and lower layers, and the restoring force is $g'h$ as if the gravity were reduced from g to g' for thermocline displacements h. If salinity is constant, $g' = \alpha(T_1 - T_2)g$, where $\alpha = 2.6 \times 10^{-4}$ K^{-1} is the thermal expansion coefficient. For the tropical thermocline with $T_1 = 27°$C and $T_2 = 12°$C, $g'/g \sim 4 \times 10^{-3}$. While open-ocean surface waves are typically only a few meters tall, it is not uncommon for internal waves to have thermocline displacements of 100 m or more (Alford et al. 2015) because the reduced gravity g' is small.

Thus by measuring sea surface height, one can infer the vertical displacements of the thermocline without even taking a cast of ocean density structure in the vertical. This requires extremely high accuracy in sea level height measurements; a 10-m deepening of the thermocline is associated with a mere 4-cm increase in sea level. NASA's SeaSat satellite proved the concept of measuring sea surface height from space in 1978. In the early 1990s, Topex/Poseidon satellite achieved the amazing feat of measuring sea surface height variations at an accuracy of 3.3 cm from an orbit of 1330 km above. A constellation of satellite altimeters has since been observing the ocean surface topography and its change continually, revolutionizing physical oceanography. Satellite altimetry has become an essential component of climate-observing systems.

7.1.4 1.5-layer reduced-gravity model

Consider a two-layer system of constant density ρ_1 in the upper and ρ_2 in the lower layer, separated by an infinitesimally thin thermocline of a mean depth H. The lower layer is much deeper, where the flow is negligibly slow compared to the upper layer. Assuming that the horizontal pressure gradient vanishes in the motionless abyssal layer, one obtains a relationship between the perturbation sea surface height (η) and thermocline depth (h):

$$g\nabla\eta = g'\nabla h \tag{7.6}$$

Integrating the momentum equations from the sea surface to beneath the bottom of the upper layer yields

$$\frac{\partial u}{\partial t} - fv = -g'\frac{\partial h}{\partial x} + \frac{\tau_x}{\rho H} - \gamma u \quad (7.7)$$

$$\frac{\partial v}{\partial t} + fu = -g'\frac{\partial h}{\partial y} + \frac{\tau_y}{\rho H} - \gamma v \quad (7.8)$$

where (u, v) are the velocities averaged vertically over the upper layer above the thermocline, H is the mean depth of the upper layer, and γ is the Rayleigh friction coefficient for the drag between the moving upper and resting lower layers. The continuity equation is

$$\frac{\partial h}{\partial t} + H\left(\frac{\partial u}{\partial x} + \frac{\partial v}{\partial y}\right) = -\gamma h \quad (7.9)$$

where γ is the dissipation rate due to the mixing with the lower layer. Although we set the Rayleigh friction and interfacial mixing coefficients the same here for convenience, they are generally not the same. For simplicity, all the nonlinear terms have been omitted. Fully nonlinear model calculations show that this is a good approximation near the equator, where ocean waves propagate faster than the layer-mean flow.

Eqs. 7.6, 7.7, and 7.8 are called the 1.5-layer reduced-gravity model, and the motionless lower layer is counted as the half layer. It is a shallow water system on the beta-plane with the long gravity wave speed $c_o = (g'H)^{1/2} \sim 2\text{–}3$ m/s for a mean thermocline depth $H \sim 200$ m. The resemblance to Matsuno's baroclinic model of the tropical atmosphere (see Section 3.4) is obvious. We are interested in basin-scale ocean variations, for which the longwave approximation applies. By dropping small terms, Eq. 7.7 approximates into

$$fu = -g'\frac{\partial h}{\partial y} \quad (7.10)$$

The zonal current velocity is always in geostrophic balance. The longwave approximation filters out all the gravity waves, retaining only the eastward propagating Kelvin wave at the phase speed of c_o and westward propagating nondispersive Rossby waves at phase speeds of $c_n = c_o/(2n + 1)$, where $n = 1, 2, \ldots$ is the number of the meridional mode.

The following is a comparison of the 1.5-layer ocean and 2-level atmospheric models (Table 7.1).

Table 7.1 Comparison of key parameters between the atmosphere and ocean. Phase speed of long gravity wave (c) and equatorial Rossby radius of deformation ($R_e = \sqrt{c/\beta}$).

	c (m/s)	Re (km)	Pressure	Forcing
Ocean	2.5	340	$g'h$	wind stress
Atmosphere	50	1500	Φ	diabatic heating

a. The atmospheric Kelvin wave speed ($c_a = 50$ m/s) is 20 times faster than that in the 1.5-layer ocean. As a result, the atmosphere quickly adjusts to slow ocean variations. This is called the quasi-equilibrium approximation ($\partial/\partial t \sim 0$) for the atmosphere in modeling ocean-atmosphere interaction.

b. The meridional scale of ocean dynamic fields (currents and thermocline depth) is an order of magnitude smaller than that of the atmosphere (Rossby radius: $\sqrt{c/\beta}$ near the equator; c/f in the midlatitudes).

c. The upper ocean circulation is driven by the wind stress while the atmospheric circulation by diabatic heating (sensible heat in the planetary boundary layer and condensational heat in the free troposphere). Pressure perturbations are associated with thermocline displacements ($g'h$) in the 1.5-layer ocean model.

7.1.5 2.5-layer model

While the 1.5-layer model treats only the vertical-mean velocity above the bottom of the thermocline, equatorial ocean currents vary markedly and even reverse the direction within this upper layer above the 12°C isotherm. For example, the easterly winds drive the westward currents near the surface but the equatorial undercurrent flows eastward underneath in the thermocline at 1 m/s speed (see Fig. 7.3). Following Zebiak and Cane (1987), we consider a 2.5-layer system to represent the current shear between the surface Ekman layer and the rest of the upper ocean above the thermocline, with the motionless abyssal ocean further below where the pressure gradient vanishes. Readers who are not interested in the derivations can skip to the end of the section (Eq. 7.20).

For simplicity, we assume a constant density in both the Ekman and lower moving layers. Thus horizontal pressure gradient does not vary between layers 1 and 2 and is due to the sea level variations. The momentum equations for layer 1 are

$$\frac{\partial u_1}{\partial t} - fv_1 = -g'\frac{\partial h}{\partial x} + \frac{\tau_x - a(u_1 - u_2)}{\rho H_1}, \tag{7.11}$$

$$\frac{\partial v_1}{\partial t} + fu_1 = -g'\frac{\partial h}{\partial y} + \frac{\tau_y - a(v_1 - v_2)}{\rho H_1}, \tag{7.12}$$

where a is the drag coefficient at the interface between layers 1 and 2 due to vertical advection and mixing. The momentum equations for layer 2 are

$$\frac{\partial u_2}{\partial t} - fv_2 = -g'\frac{\partial h}{\partial x} - \frac{a(u_2 - u_1)}{\rho H_2} \tag{7.13}$$

$$\frac{\partial v_2}{\partial t} + fu_2 = -g'\frac{\partial h}{\partial y} - \frac{a(v_2 - v_1)}{\rho H_2} \tag{7.14}$$

where the friction at the thermocline is neglected for clarity here. The governing equations for the vertical shear $\boldsymbol{u}_s = \boldsymbol{u}_1 - \boldsymbol{u}_2$ are

$$\frac{\partial u_s}{\partial t} - f v_s = \frac{\tau_x}{\rho H_1} - \varepsilon u_s \tag{7.15}$$

$$\frac{\partial v_s}{\partial t} + f u_s = \frac{\tau_y}{\rho H_1} - \varepsilon v_s \tag{7.16}$$

where $\varepsilon = \frac{a}{\rho}\left(\frac{1}{H_1} + \frac{1}{H_2}\right)$. The vertical mean velocity in layers 1 and 2, $\boldsymbol{u} = \frac{u_1 H_1 + u_2 H_2}{H_1 + H_2}$, follows the 1.5-layer reduced-gravity model (Eqs. 7.6–7.8) with $\gamma = 0$.

Time derivative terms in the shear (Eqs 7.14 and 7.15) are small and can be dropped given the large Rayleigh friction coefficient (1/day). Then the shear equations are identical to the Ekman layer model (Eqs. 7.1 and 7.2). From the vertical shear and mean flow, we obtain the velocity in each layer as

$$\boldsymbol{u}_1 = \boldsymbol{u} + \frac{H_2}{H}\boldsymbol{u}_s \tag{7.17}$$

$$\boldsymbol{u}_2 = \boldsymbol{u} - \frac{H_1}{H}\boldsymbol{u}_s \tag{7.18}$$

At the limit of a thin Ekman layer $H_1/H_2 \ll 1$,

$$\boldsymbol{u}_1 = \boldsymbol{u} + \boldsymbol{u}_E \tag{7.19}$$

$$\boldsymbol{u}_2 = \boldsymbol{u} \tag{7.20}$$

The surface flow shear is governed by Ekman dynamics Eqs. 7.1 and 7.2, while the flow in the second layer by the shallow-water dynamics Eqs. 7.6 through 7.8.

Here we have neglected the difference in the pressure gradient force between layers 1 and 2. By considering the density difference between the layers, the Ekman layer model (Eqs. 7.14 and 7.15) is replaced with a shallow water model for the second baroclinic mode, with a smaller longwave phase speed and much stronger damping than the first mode. As a result, pressure perturbations of the second baroclinic mode are not well developed, and the Ekman layer model (Eqs. 7.1 and 7.2) is a good approximation.

7.2 Ocean response to wind stress forcing

In the above 2.5-layer model, the upper layer flow is governed by local quasi-steady Ekman flow and nonlocal wave dynamics of the 1.5-layer reduced gravity model. This section focuses on the latter.

7.2.1 Currents at the equator

We first consider the steady-state solution to the 1.5-layer system in response to zonal wind stress forcing. On the equator, the wind stress is balanced by the zonal pressure gradient force associated with the tilt of the thermocline,

$$0 = -g'\frac{\partial h}{\partial x} + \frac{\tau_x}{\rho H} \tag{7.21}$$

which is obtained from the zonal momentum equation (Eq. 7.6) by set $f = 0$ and $\gamma = 0$ (the friction with the abyssal ocean is small). This is an important balance for slow variations with timescales much longer than the wave transit time, which is 12 months for the Pacific (3 months for the Kelvin wave to cross the basin, and 9 months for the first Rossby waves reflected on the eastern boundary to reach the western boundary).

In the equatorial Pacific, the thermocline shoals eastward to balance the prevailing easterly winds. This explains the observed shoaling of the 20°C isotherm, from 200 m in the west to less than 50 m in the east (see Fig. 7.4A). Upon a close look, the thermocline shoals most steeply in the central equatorial Pacific consistent with the zonal maximum of the easterly trades. The thermocline depth change is weak in the western equatorial Pacific where the zonal winds are weak. The lower thermocline with temperatures at or lower than 12°C stays flat across the basin on the equator, indicating that the local wind effect is confined to warmer isotherms.

Equatorial currents feature strong vertical shear. Near the surface, the frictional Ekman flow dominates and is westward under the easterly wind stress force. The westward surface flow, called the South Equatorial Current (SEC), piles warm water toward the Asian coast and deepens the thermocline there while shoaling it in the east. This sets up an eastward pressure gradient force, with a higher sea level on the Asian than South American coast. Underneath the surface Ekman layer, the direct wind forcing is weak, and this eastward pressure gradient drives an intense eastward equatorial undercurrent (EUC) in speeds often exceeding 1 m/s (see Fig. 7.3B). The strong vertical shear between the westward SEC (-30 cm/s) and eastward EUC (100 cm/s) is conducive to pronounced turbulent mixing on the equator. The strong equatorial upwelling advects the eastward momentum from the EUC up to slow the surface SEC. This advection of the EUC momentum produces an equatorial minimum in the SEC speed as observed by surface drifters (Fig. 7.5).

Below the surface Ekman flow, zonal currents are nearly in geostrophic balance with the meridional pressure gradient force associated with the thermocline topography (Eq. 7.9). This is true even on the equator. By taking the meridional derivative of Eq. 7.9, we obtain

$$\beta u = -g'\frac{\partial^2 h}{\partial y^2} \text{ at } y = 0 \tag{7.22}$$

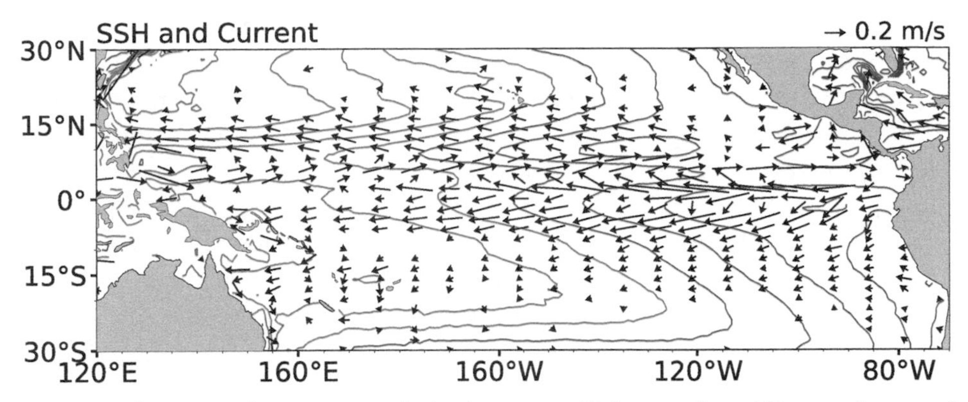

Fig. 7.5 Annual-mean surface current velocity (vectors, m/s) from surface drifters and sea surface height (SSH) contours (intervals of 0.1 m, red ≥1 m). *(Courtesy Q. Peng.)*

which is called the equatorial geostrophic balance. In a meridional section (see Fig. 7.2), the depressed lower thermocline (e.g., Z_{14}) implies an eastward EUC, while the shoaling of the upper thermocline (e.g., Z_{25}) supports the westward SEC.

7.2.2 Yoshida jet

Yoshida (1959) considered the response of an unbounded, 1.5-layer equatorial ocean to the sudden onset of easterly winds uniform in both the zonal and meridional directions. The response is thus zonally uniform ($\partial/\partial x = 0$). The poleward Ekman flow in the upper layer shoals the thermocline on the equator. The thermocline displacement calls for an intense westward jet on the equator, in geostrophic balance (Eq. 7.22). The upwelling zone spans ~1.5 Rossby radius of deformation on either side of the equator (Fig. 7.6). As $|f|$ increases poleward, the poleward Ekman flow decelerates, and the convergence causes an Ekman downwelling that depresses the thermocline. It can be

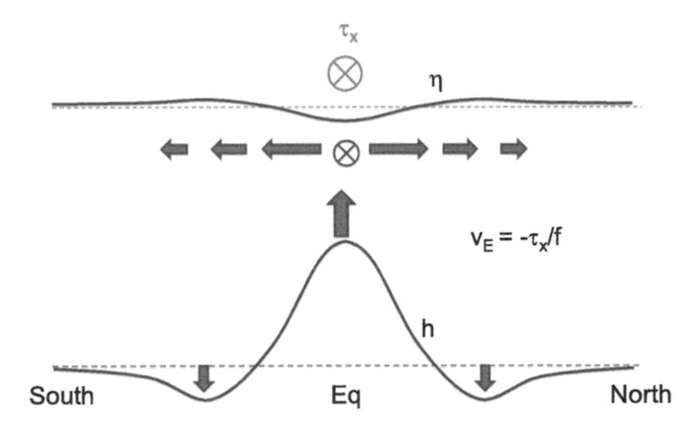

Fig. 7.6 The thermocline response to a uniform easterly wind forcing. Zonal variations are neglected. The thermocline shoals with a westward jet on the equator.

shown that the meridional flow does not change with time, while both the zonal flow and thermocline displacement increase linearly with time. Wyrtki (1973) observed such an eastward jet in the central equatorial Indian Ocean in response to the abrupt onset of westerly winds during the intermonsoon seasons (April-May and October-November), thereby verifying the Yoshida theory.

In the zonally uniform Yoshida solution, the zonal wind stress is balanced by the acceleration of the zonal current (Eq. 7.6). With zonal variations either in wind distribution or due to the presence of meridional boundaries, the wind stress term can be balanced, in part or full, by the zonal pressure gradient force. Eq. 7.21 is an example.

7.2.3 Wave adjustments in an unbounded ocean

We continue to consider a simple case of an unbounded 1.5-layer ocean, but forced by a patch of easterly wind stress in $x_W < x < x_E$. The wind is uniform in both the zonal and meridional directions but vanishes outside. In response to easterly winds (τ_x <0), the poleward Ekman flow that drains the warm upper-layer water away from the equator, and the equatorial upwelling lifts the thermocline in the wind patch. Maximum on the equator, the shoaling of thermocline projects on the Kelvin wave and propagates eastward. At the same time, the poleward Ekman flow deepens the thermocline on either side of the equator. The off-equatorial deepening of the thermocline projects strongly on Rossby waves and propagates westward.

We can cast the solution into the sum of the Kelvin and Rossby waves. Consider the Kelvin wave first. The thermocline displacement on the equator is given by the following one-dimensional wave solution,

$$\frac{\partial h}{\partial t} + c_o \frac{\partial h}{\partial x} = F_0 \tag{7.23}$$

where F_0 is the projection of the wind forcing on the Kelvin wave structure (F_0 <0 for the easterly wind). The steady-state solution is

$$h = \begin{cases} 0, \ x < x_W \\ \dfrac{(x - x_W)}{L_F} h_0, x_W < x < x_E \\ h_0, \ x > x_E \end{cases}$$

where $L_F = x_W - x_E$ is the width of the wind patch, and $h_0 = \frac{L_E}{c_o} F_0$ is the maximum displacement in response to the wind patch. The boundary condition is $h = 0$ at $x = x_W$ because the Kelvin wave propagates only eastward. The thermocline shoals eastward within the wind patch to balance the easterly wind stress.

The transient adjustment to an abrupt switch-on of the wind stress forcing is helpful to see the Kelvin wave in action. We consider the following three stations on the equator.

- West of the wind patch ($x < x_W$). No response because of the eastward propagation of the Kelvin wave.
- Within the wind patch ($x_W < x < x_E$). The response takes place in two stages: (1) the Yoshida solution with the thermocline shoaling with time, $h = F_0 t$, initially for $t < \frac{(x-x_W)}{c_o}$ before the Kelvin wave from the western edge of the wind patch arrives; (2) the steady-state solution, $h = \frac{(x-x_W)}{L_F} h_0$ thereafter.
- East of the wind patch ($x > x_E$). The response consists of three stages: (1) $h = 0$ for $t < t_E = \frac{(x-x_E)}{c_o}$ before the arrival of the Kelvin wave from the eastern edge of the wind patch; (2) the shoaling thermocline with time, $h = \frac{t-t_E}{t_W-t_E} h_0$ for $t_E < t < t_W = \frac{(x-x_W)}{c_o}$ before the arrival of the Kelvin wave from the western edge of the wind patch; and (3) the steady-state solution, $h = h_0$ for $t > t_W$ after the Kelvin wave from the western edge of the wind patch has passed.

The two Kelvin waves emanated, respectively, from the western and eastern edges of the wind patch are evident in the Hovmoller diagram of the thermocline depth at the equator (Fig. 7.7A).

The same applies to the Rossby wave solution, which is downwelling in response to Ekman convergence with the maximum thermocline displacement off the equator. Like the Kelvin wave, the thermocline displacement at the latitude of peak h perturbations is given:

$$\frac{\partial h}{\partial t} - c_n \frac{\partial h}{\partial x} = F_n \tag{7.24}$$

where $c_n = c_o/(2n + 1)$ is the Rossby wave phase speed ($n = 1, 3, 5\ldots$ for symmetric perturbations), and F_n is the Rossby wave projection of the wind forcing ($F_n > 0$ for the easterly wind). The slow westward propagating Rossby waves are also evident in the Hovmoller diagram of the thermocline depth at the equator (see Fig. 7.7A).

Because of different meridional wave structures between the Kelvin and Rossby waves, the steady-state solution in the unbounded, zonally infinite domain features the shoaled thermocline with the maximum displacement on the equator east of the wind patch, and the depressed thermocline with the maximum displacement off the equator west of the wind forcing (see Fig. 7.7A for large t). Within the wind patch, the solution is a combination of both.

7.2.4 Ocean adjustments with meridional boundaries

The Pacific Ocean is bounded by Asia/Australia on the west and the Americas on the east. The abovementioned unbounded solution needs to be modified to meet the condition of no normal flow on these boundaries. Upon arrival at the eastern boundary, the upwelling Kelvin wave reflects into equatorial Rossby waves and a coastal Kelvin wave of

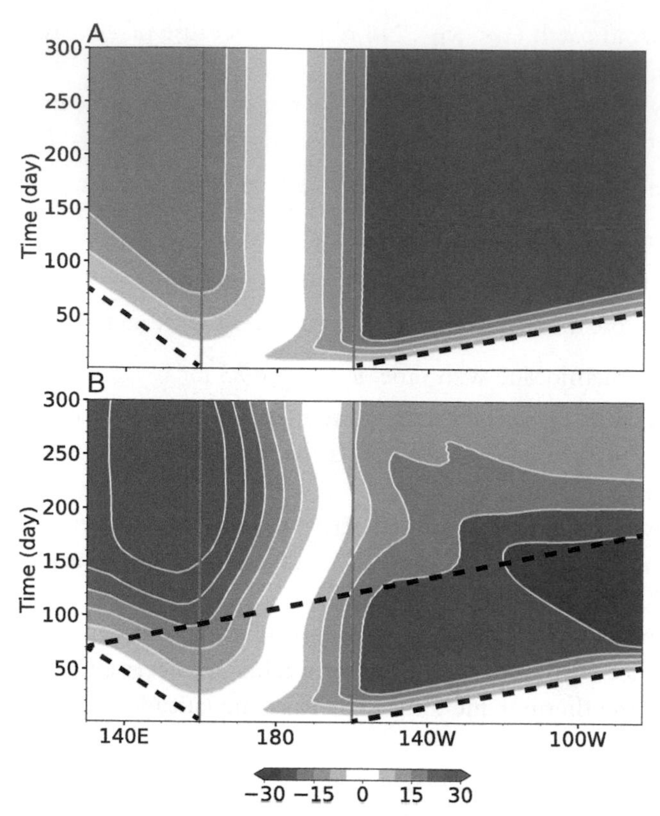

Fig. 7.7 Hovmoller diagrams of the thermocline displacement (color shading, m) in response to an easterly wind forcing (Gaussian meridional distribution) over the western equatorial Pacific Ocean (160°E–160°W) in (A) a zonally unbounded, and (B) bounded ocean, based on a 1.5-layer reduced-gravity model. *(Courtesy Q. Peng.)*

upwelling type (i.e., with the shoaling thermocline) (Cane and Sarachik 1977; Clarke 1983). The principle of the reflection is to conserve the volume of the upper-layer warm water because of the continuity. The upwelling coastal Kelvin wave, with the equatorward alongshore flow, radiates Rossby waves as it propagates poleward. The upwelling Rossby waves reflected from the eastern boundary propagate slowly westward. The upwelling Kelvin wave emanated from the wind patch and the upwelling Rossby waves reflected from the eastern boundary are both evident in the Hovmoller diagram (see Fig. 7.7B). Here the meridional boundaries are at 130°E and 80°W.

On the western boundary, the arriving downwelling Rossby waves are reflected into downwelling equatorial Kelvin waves that propagate toward the east along the equator. The downwelling Rossby waves are evident in the thermocline deepening that starts progressively later west of the wind patch.

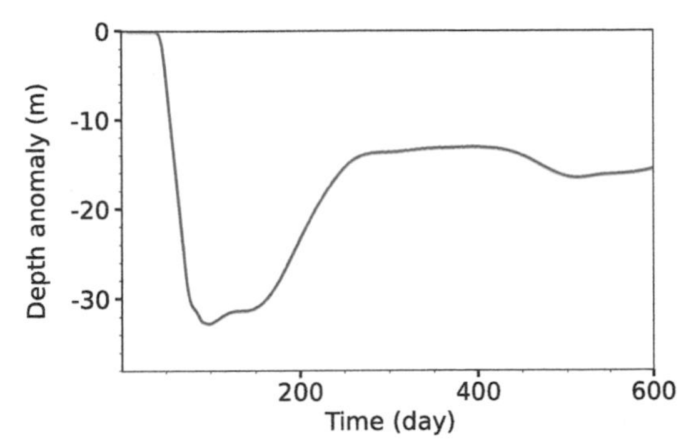

Fig. 7.8 Thermocline displacement (m) on the equatorial eastern boundary in response to an easterly wind forcing over the western equatorial Pacific Ocean (160°E—160°W, meridionally Gaussian distribution with e-folding scale of 5° latitude). *(Courtesy Q. Peng.)*

Consider an observer at the eastern boundary on the equator (Fig. 7.8). Initially, no change is observed in the thermocline depth. The thermocline begins to shoal as the upwelling Kelvin wave arrives. The reflected upwelling Rossby waves from the eastern boundary also make minor contributions to the thermocline shoaling on the equator. The arrival of the downwelling Kelvin wave reflected from the western boundary begins to bring back the shoaling thermocline. The bouncing back and forth of the equatorial waves causes the thermocline depth at the eastern boundary to oscillate at a period of four times of the Kelvin wave transit time, 4 L/c_0, where L is the zonal width of the ocean basin. This oscillation is called the basin mode. The wave propagation and reflection can be seen in a movie of thermocline perturbations (online supplement).

The dissipation damps slow Rossby waves reflected from the eastern boundary. In the steady state, the e-folding zonal scale of the damped Rossby wave is given by

$$L_n = c_n/\gamma$$

The lower the meridional mode, the longer the Rossby wave extends. As the Rossby wave of a lower meridional mode features the peak thermocline displacements closer to the equator with a longer damping length scale, a wedgelike pattern of the thermocline shoaling forms near the eastern boundary as a manifestation of the frozen Rossby waves (Fig. 7.9C).

7.2.5 Long Rossby waves and Sverdrup balance

Sufficiently away from the equator, ocean currents below the Ekman layer are nearly in geostrophic balance (see Eqs. 1.1—1.2). Vertical integration of the continuity equation (see Eq. 1.5) from the base of the Ekman layer to below the thermocline yields the equation for long, nondispersive Rossby waves forced by Ekman pumping (upward positive),

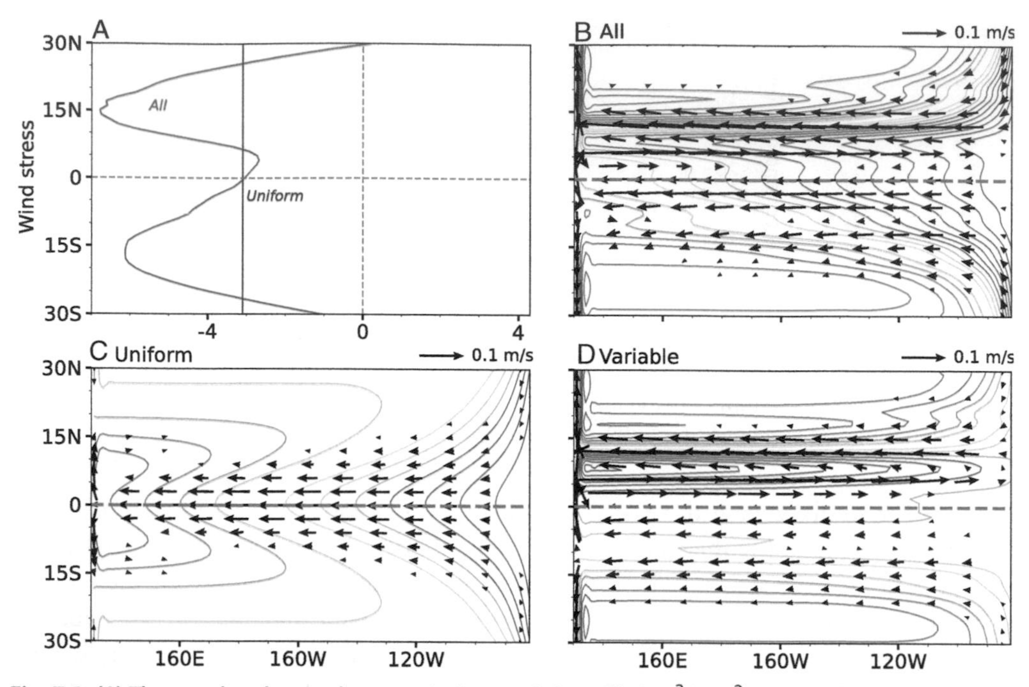

Fig. 7.9 (A) The zonal and annual mean wind stress (τ_x) profile (10^{-2} N/m^2) derived from GODAS data. The steady-state response of the thermocline depth (contours at intervals of 10 m, positive red and negative blue, zero contour omitted) and currents (vectors, m/s) to (B) the full τ_x profile, (C) the meridionally uniform, and (D) variable components. The meridionally uniform component is of a magnitude of the 3°S–3°N mean. *(Courtesy Q. Peng.)*

$$\frac{\partial h}{\partial t} + c_R \frac{\partial h}{\partial x} = -w_E \tag{7.25}$$

where $c_R = -\beta R_f^2$ is the longwave Rossby wave phase velocity with the negative sign indicating the westward propagation, and $R_f = c_o/f = \sqrt{g'H}/f$ is the Rossby radius of deformation.

In steady state, we obtain

$$\beta H v = f w_E \tag{7.26}$$

Hv is called the Sverdrup transport, which is southward under the downward Ekman pumping. The thermocline slopes downward from the eastern boundary is

$$h = h_E - \frac{1}{c_R} \int_x^{x_E} w_E dx \tag{7.27}$$

The zonal geostrophic current is given as

$$u = \frac{g'}{c_R f} \int_{x}^{x_E} \frac{\partial w_E}{\partial y} dx \tag{7.28}$$

where we have required no normal flow into the eastern boundary at $x = x_E$. In the classic subtropical gyre under the downward Ekman pumping, an anticyclonic gyre forms, with an intense norward western boundary current that returns the zonally integrated southward Sverdrup flow in the interior. The western boundary current is about 100 km wide (1% of the Pacific basin width) and hence 100 times stronger in speed than the interior Sverdrup flow. The formation of the intense western boundary current is a result of the beta effect due to the poleward increase in the Coriolis parameter. See Vallis (2017; chap. 17) for a more detailed treatment of the subtropical gyre.

7.2.6 Equatorial current system

A cyclonic wind curl drives upward Ekman pumping at the base of the ocean Ekman layer (w_E) while surface friction induces upward motion at the top of the atmospheric boundary (w_B). It can be shown (as a homework) that the mass flux at the base of the ocean Ekman layer equals that at the top of the atmospheric boundary layer,

$$\rho_o w_E = \rho_a w_B = curl\left(\frac{\tau}{f}\right)$$

where subscripts a and o denote the air and water densities, respectively.

The Pacific intertropical convergence zone (ITCZ) is displaced north of the equator. Convective heating in the ITCZ drives cyclonic low-level winds (see Fig. 2.13) with upward Ekman pumping at the bottom of the ocean Ekman layer. The thermocline depth reaches a meridional maximum at 5°N (see Figs. 7.2 and 7.5). The thermocline then shoals toward the ITCZ at ~9°N in geostrophic balance with the North Equatorial Countercurrent (NECC), which flows eastward against the prevailing easterly trade winds. Further to the north, the thermocline slopes downward in response to the Ekman downwelling, in association with the broad westward North Equatorial Current (NEC). The NEC is both part of the anticyclonic subtropical gyre with downward Ekman pumping ($w_E < 0$) to the north and part of the cyclonic tropical gyre with $w_E > 0$ to the south. The northward Kuroshio is the western boundary current of the subtropical gyre while the Mindanao Current flows south off the Philippines and connects to the NECC to form the tropical gyre. In the Southern Hemisphere, the surface SEC occupies the broad tropical to subtropical South Pacific. This asymmetry in equatorial currents is a result of the basin-scale climatic asymmetry in the ITCZ and surface winds (see Section 8.1.3).

The 1.5-layer model reproduces the observed currents in the tropical Pacific remarkably well (cf. Figs. 7.5 and 7.9B). The tropical circulation may be viewed as the sum of the equatorial circulation driven by the tropical easterly trade winds and the off-equatorial circulation forced by wind curls. We decompose the zonal wind stress into a meridionally uniform easterly component ($\tau_x|_{y=0}$) and the residual (τ_x^*) that vanishes on the equator,

$$\tau_x(y) = \tau_x|_{y=0} + \tau_x^*(y)$$

For simplicity, we have zonally averaged the observed zonal stress across the Pacific and modeled the Pacific as a rectangular basin. The upwelling equatorial Kelvin wave driven by the broad easterlies reflects into upwelling Rossby waves that travel westward and the upwelling coastal Kelvin wave on the eastern coast that propagates poleward. The dissipation damps the westward traveling Rossby waves more strongly on higher meridional modes, producing the wedgelike shoaling of the thermocline toward the east (i.e., the frozen upwelling Rossby waves reflected from the eastern boundary) (see Fig. 7.9C). The easterly winds also excite downwelling Rossby waves in the west, with the deepened thermocline off the equator. The basinwide, westward intensified NECC (~7°N) forms with wind curls added (see Fig. 7.9B). Along the thermocline ridge under the ITCZ (10°N), the thermocline depth is nearly constant in the zonal direction because the eastward shoaling in response to the broad easterly winds opposes the wind curl effect. The positive wind curls in the ITCZ drive a westward intensified cyclonic tropical cell with the thermocline shoaling toward the west (see Fig. 7.9D).

7.3 Mixed-layer heat budget

7.3.1 Governing equation

A well-mixed layer develops in the top 20 to 100 m of the ocean due to turbulent mixing by wind, current shear in the vertical, and buoyancy (e.g., nocturnal cooling). Integrating the thermodynamic equation of the ocean from the surface to the bottom of the mixed layer yields

$$\frac{\partial T_s}{\partial t} = -\boldsymbol{u}_m \cdot \nabla T_s - w_m \frac{T_s - T_e}{H_m} H(w_m) + \frac{Q}{\rho c_p H_m} \tag{7.29}$$

Here T_s is the mixed layer temperature (equated as SST here), H_m is the mixed layer depth, $\boldsymbol{u}_m$ is the horizontal velocity averaged in the mixed layer, w_m is the vertical velocity at the bottom of the mixed layer, which advects the subsurface water of temperature (T_e)

at $z = -2\ H_m$. Only the upward advection affects the mixed layer temperature, as represented by the Heaviside function, $H(x) = \{1 \text{ for } x > 0; 0 \text{ otherwise}\}$. The subsurface temperature gradient in the upwelling term on the right-hand side is evaluated with a finite difference below the mixed layer.

In the eastern equatorial Pacific, the surface mixed layer is shallow and sits atop the shoaled thermocline. Here upwelling brings cold water from the thermocline to keep SSTs much lower than those in the western Pacific. Thus the easterly winds are important for the equatorial cold tongue in two different but complementary ways: by shoaling the thermocline in the east and upwelling cold water from a shallow thermocline. The thermocline depth effect is obvious from the following comparison: While the easterly winds and hence upwelling are strongest in the central Pacific, SST decreases eastward as the thermocline continues to shoal, exposing colder subsurface temperatures to upwelling.

The mixed layer depth and Ekman layer are often assumed of the same depth, say at 50 m, for simplicity (Zebiak and Cane 1987). From the 2.5-layer model, $H_m = H_1$, $\boldsymbol{u}_m = \boldsymbol{u}_1$, and

$$w_m = H_1 \nabla \cdot \boldsymbol{u}_1 = H_1 \nabla \cdot \boldsymbol{u} + w_E \tag{7.30}$$

where $w_E = H_1 \nabla \cdot \boldsymbol{u}_E$ is the frictional Ekman pumping velocity calculated from Eqs. 7.1 and 7.2. Near the equator, currents in the thermocline are strongly sheared in the vertical (e.g., between SEC and EUC), and the upwelling velocity at the bottom of the mixed layer is often dominated by the second term in Eq. 7.30. The entrainment temperature T_e is a function of the thermocline depth, the latter obtained by solving the shallow-water, reduced-gravity model (Eqs. 7.6–7.8). Both upwelling velocity and the thermocline depth display large interannual variability as the equatorial Pacific oscillates between warm El Niño and cool La Niña states (see Chapter 9). From an oceanographic point of view, the trade wind intensity is an effective regulator of equatorial SST by its effect on the upwelling and entrainment temperature.

7.3.2 Surface heat flux

The equatorial ocean is heated from the sea surface by heat flux Q. In the third term of the mixed layer temperature equation, $\rho = 1025\ \text{kgm}^{-3}$ is the water density, and $c_p = 4000\ \text{Jkg}^{-1}\text{K}^{-1}$ is the specific heat of water. The surface heat flux is made up of four distinct components: the downward shortwave (solar) radiation (Q_{SW}), the net upward longwave radiation (Q_{LW}), the latent heat of evaporation (Q_E), and sensible heat through conduction and turbulence (Q_H):

$$Q = Q_{SW} - Q_{LW} - Q_E - Q_H \tag{7.31}$$

Both latent and sensible heat fluxes are due to turbulent exchange between the sea surface and atmosphere, and the sum is often called the turbulent heat flux. Here we briefly discuss each flux component based on empirical formula (e.g., as summarized in Rosati and Miyakoda 1988) and ship reports of meteorologic conditions nominally at 10 m above the surface, as archived in the Comprehensive Ocean-Atmospheric Data Set (COADS).

Downward solar radiation is a function of cloud cover,

$$Q_{SW} = Q_{S0}(1 - a_s)(1 - 0.7C) \tag{7.32}$$

where Q_{S0} is the clear-sky solar radiation that varies with latitude ($\sim$320 W/m^{-2} on the equator), a_s is the sea surface albedo $\sim$0.04, and C is the fraction of sky covered by cloud (0$\sim$1). Surface solar radiation is a strong function of cloud cover; a 0.3 difference in cloud cover causes a change of as much as 65 W/m^{-2} in solar flux into the ocean. A local minimum is found under the ITCZ (Fig. 7.10).

Net upward longwave radiation is given by

$$Q_{LW} = 0.985\sigma T_S^4\left(0.39 - 0.05e_a^{\frac{1}{2}}\right)\left(1 - 0.6C^2\right) \tag{7.33}$$

where 0.985 is a correction factor for the departure of ocean surface from blackbody, σT_S^4 is the blackbody flux emitted by the ocean surface with $\sigma = 5.7 \times 10^{-8}$ Wm^{-2}K^{-4}, and e_a is the 10-m vapor pressure in mb. The net longwave flux is always upward because SST is higher than atmospheric temperature, which decreases with height. Q_{LW} is nearly constant across much of the world ocean within a small range of 30 to 50 Wm^{-2} (see Fig. 7.10) because high water vapor content makes the lower atmosphere opaque and the effective radiative temperature is not far from SST. As clouds are close to blackbody for longwave radiation, the atmospheric downward radiation increases with cloud cover. The cloud effect on the net longwave opposes but is generally only

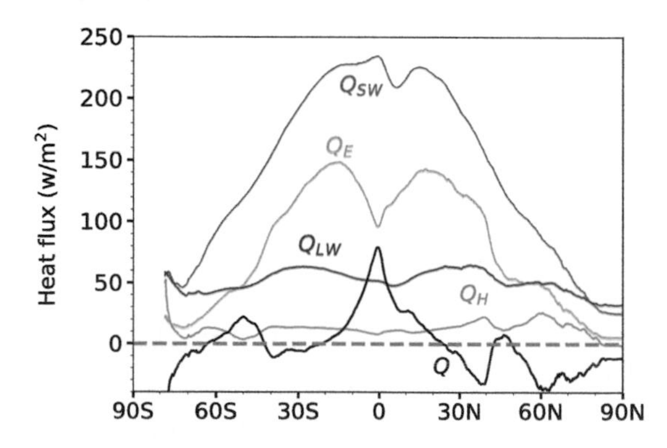

Fig. 7.10 Zonal and annual mean ocean surface heat flux (W/m^2): downward shortwave (Q_{SW}), net upward longwave radiation (Q_{LW}); upward latent (Q_E) and sensible heat flux (Q_H). The net surface heat flux $Q = Q_{SW} - Q_{LW} - Q_E - Q_H$. *(Courtesy Q. Peng.)*

~20% of that on the downward solar radiation. Thus the cloud radiative effect at the sea surface is dominated by the shortwave component.

For these reasons, the net longwave radiation at the sea surface varies little either in space or time and is often neglected in studying spatiotemporal variations in SST. This is in sharp contrast with outgoing longwave radiation at the top of the atmosphere, which varies from more than 300 Wm^{-2} in dry subsidence regions (e.g., the Sahara Desert) to less than 200 Wm^{-2} in deep convective regions (e.g., the summer Bay of Bengal) (see Fig. 3.3).

Latent heat flux due to surface evaporation E is the major mechanism for the tropical ocean to balance the net downward radiation (~150 Wm^{-2}):

$$Q_E = LE = \rho_a L C_E W(q_s - q_a) \tag{7.34}$$

where $\rho_a = 1.3$ kgm^{-3} is the surface air density, $L = 2.5 \times 10^6$ Jkg^{-1} is the latent of evaporation, $C_E \sim 1.3 \times 10^{-3}$ is the exchange coefficient that varies with stability ($T_a - T_s$) and wind speed, W is the wind speed at 10 m above the surface, q_s is the saturated specific humidity at the surface, $q_a \equiv R_H \cdot q_s(T_a)$ is air specific humidity at 10 m, and R_H is relative humidity. Surface evaporation increases with wind speed and specific humidity difference. This is consistent with our experience on the beach. Coming out of the water, we feel that the air is "cooler" on windy and/or dry days even with the same air temperature.

Typically over tropical oceans, $R_H \sim 0.8$ and 10-m air temperature is about 1°C cooler than SST, $\Delta T = T_s - T_a \sim 1°C$. The latent heat flux can be approximated as

$$Q_E = \rho_a L C_E (1 - R_H) W q_s(T_s) \tag{7.35}$$

if we neglect the small air-sea temperature difference. Evaporative cooling increases with q_s and hence SST. This is a damping effect on SST variations, say caused by changing equatorial upwelling or wind speed.

Eq. 7.34 may be used to calculate monthly-mean flux as long as the mean scalar wind speed is used for W. The mean scalar wind speed $\overline{|\boldsymbol{u}|} = \overline{(u^2 + v^2)^{1/2}}$ is generally greater than the scalar speed of the mean vector wind $|\overline{\boldsymbol{u}}| = \left(\overline{u}^2 + \overline{v}^2\right)^{1/2}$, with the overbar denoting the time mean. This is especially so where the mean vector wind is weak (e.g., in the western Pacific warm pool, ITCZ, and boundaries between the westerly and easterly trade wind regimes [~30°N/S]). There, high-frequency weather variability generates turbulence in the atmospheric boundary layer for surface evaporation.

Latent heat flux (see Fig. 7.10) has a meridional minimum on the equator (50~100 Wm^{-2}) because of low SSTs in the cold tongue and relatively weak winds. It reaches the meridional maximum in the subtropics (15–20°) in either hemisphere (~150 Wm^{-2})

because of the strong trades and low relative humidity, and then decreases poleward with the rapid decrease in temperature.

Sensible heat flux is given by

$$Q_H = \rho_a c_p^a C_H W(T_s - T_a) \tag{7.36}$$

where $c_p^a = 1030\,\mathrm{Jkg^{-1}K^{-1}}$ is the specific heat of air at constant pressure, and $C_H \sim C_E$ is the exchange coefficient. Sensible heat flux is small in the tropics ($\sim$10 Wm^{-2}) but increases in the extratropics (see Fig. 7.10), especially near major continents and ocean fronts where air-sea temperature difference is large.

Over tropical oceans, the sensible component of surface heat flux is negligibly small. While not negligible in absolute magnitude, the net longwave radiation does not vary much either in space or time. To first order, we only need to consider the shortwave and latent heat components of surface heat flux when studying spatiotemporal deviations (denoted with a prime) from the mean

$$Q' \approx Q'_{SW} - LE' \tag{7.37}$$

In earlier studies of El Niño (e.g., Zebiak and Cane 1987), sea surface flux was often simplified as a linear Newtonian cooling term, $Q' = -\lambda T'$. Linearizing the latent heat flux term with regard to SST in Eq. 7.35 yields such a linear damping. In reality, cloud and surface wind also affect the surface flux and hence SST in important ways as will be discussed in the next chapter.

Review questions

1. How does the ocean reduced gravity vary with latitude? Assume that the abyssal temperature is nearly constant.
2. How does upwelling occur on the equator without a solid boundary to force it?
3. Why is net longwave radiation at the sea surface often neglected in studying climate variability despite its significant value ($\sim$50 W/m^2 upward) in the mean?
4. What causes the net downward heat flux ($\sim$70 W/m^2 in zonal mean) at the surface and equator (see Fig. 2.9A)? Where is this downward flux largest over the equatorial Pacific: west near Indonesia or east off Ecuador? Why?
5. Where does the water in the equatorial thermocline come from? How does the shallow overturning circulation work?
6. During El Niño, sea level rises as much as 30 cm on the coast of Ecuador. How much does this translate into the thermocline displacements?
7. What is the pressure gradient force below the thermocline in the reduced-gravity model?
8. Why does the thermocline shoal toward the east in the equatorial Pacific?

9. Over the equatorial Pacific, zonal wind stress peaks around 140°W. Why is the lowest SST found well to the east?
10. In the equatorial Pacific, why is the current westward near the surface (SEC) but eastward in the thermocline (EUC)?
11. The sea surface height distribution has a broad structure in the tropics (see Fig. 7.5). Explain why the EUC is narrowly trapped on the equator (see Fig. 7.2).
12. Apply the equatorial geostrophic balance to show that the Yoshida equatorial jet flows in the direction of the wind forcing.
13. How does a westerly wind patch centered on the equator excite a downwelling Kelvin wave but upwelling Rossby waves at the same time?
14. Draw schematics (x-y) to illustrate the reflection of a Kelvin wave on the eastern boundary, and reflection of Rossby waves on the western boundary of an equatorial ocean basin.
15. Read McCreary and Anderson (1984; only paragraphs 2–3, including figs. 1–2, section 2c). In their fig. 1, explain why the thermocline in the eastern equatorial ocean shoots up for the first 3 months and why it deepens afterwards. In their fig. 2 bottom panel ($t = 5$ years), use the balance between pressure and wind stress to explain why the thermocline tilt on the equator is confined to $2500 < x < 7500$ km where the wind forcing is applied.
16. Examine the longitude-depth section of temperature in the equatorial Pacific, and read the thermocline depth, say as represented by the 20°C isotherm ($Z20$) in the west and east. Use the 1.5-layer reduced-gravity model to estimate the west-east (Indonesia-Ecuador) sea level height difference across the equatorial Pacific.
17. Use $Z20$ in the western Pacific to estimate the phase speed of internal Kelvin wave propagating along the thermocline. Calculate the ocean radius of deformation on the equatorial beta-plane. Compare with the atmospheric radius of deformation (assume a wave speed of 50 m/s).
18. Justify that in coupled ocean-atmospheric theory, the atmospheric state (i.e., surface wind) can be assumed to be in steady state with the ocean and that we only need to retain the time derivative terms in the ocean model.

CHAPTER 8

Coupled feedbacks and tropical climatology

Contents

The tropical Pacific and Atlantic oceans share many common characteristics: a northward-displaced intertropical convergence zone (ITCZ) and an equatorial cold tongue that features a marked annual cycle and year-to-year variability. The rich structures of the mean climate are not only interesting by their own rights but also key to interannual variability such as El Niño.

8.1 Meridional asymmetry

This section draws in part from Xie (2004b).

Before the invention of steam engines, information on wind direction and speed was of vital importance for navigating sailing boats. By the late 17th century, the traffic between Europe and the New World had grown to such a level that Edmund Halley (1686) of comet fame was able to compile a quite accurate map of surface-wind streamlines for the tropical Atlantic and Indian oceans by gathering information from navigators. Fig. 8.1 reproduces the Atlantic portion of Halley's wind map that depicts the trade winds in the Northern and Southern Hemispheres. Remarkably, the southeast and northeast trades meet north of, instead of on, the equator as one might expect from equatorial symmetry. In the ITCZ where the trade winds meet, Halley wrote: "it were improper to say there is any Trade Winds, or yet a Variable; for it seems condemned to perpetual Calms, attended with terrible Thunder and Lightning, and Rains."

Coupled Atmosphere-Ocean Dynamics: from El Niño to Climate Change
ISBN 978-0-323-95490-7, https://doi.org/10.1016/B978-0-323-95490-7.00008-4

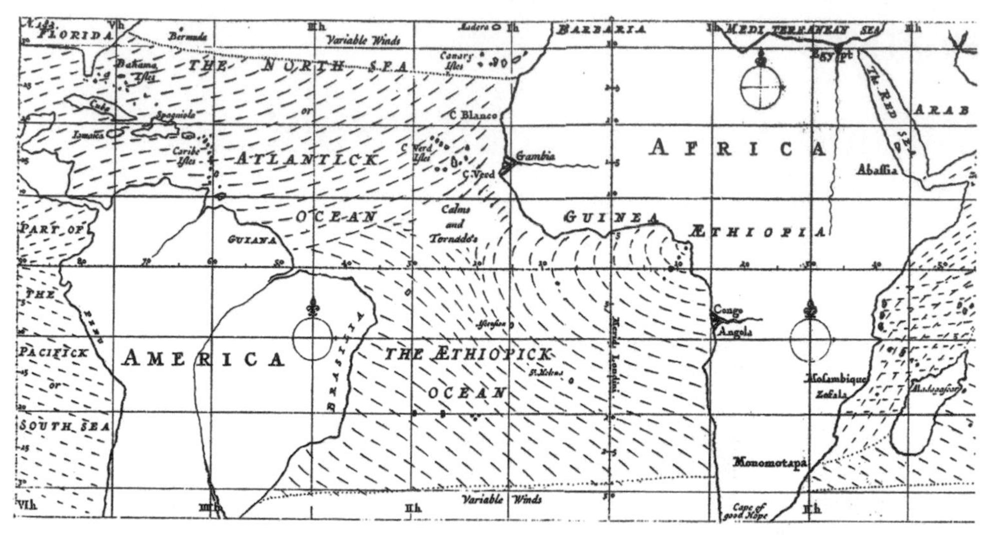

Fig. 8.1 Halley's (1686) trade wind map for the Atlantic.

The easterly winds aided trading boats to sail from Europe to the Americas, while captains avoided the equatorial doldrums as slow, and variable winds could keep a sailing boat stuck for a long time. As will become clear, the collocation of the Doldrums with the ITCZ is the key to solving the mystery of their northward displacement from the equator. As air rises in the ITCZ, the water vapor it carries condenses, resulting in intense rains and releasing latent heat that drives the global circulation of the troposphere.

At the time of Halley, the vast Pacific Ocean was much less navigated than the Atlantic. For lack of information and out of consideration of equatorial symmetry, the Pacific trades were drawn to converge on the geographic equator on Halley's map. Had the relationship between deep convection and surface wind convergence been known, the rainfall distribution on the Pacific coast of the Americas would have suggested a northward displaced ITCZ over the eastern Pacific, much as in the Atlantic. Abundant rainfall supports lush forecasts in warm Panama and Nicaragua, while dry desert extends on the Pacific coast of Peru on the other side of the equator (Fig. 8.2). In Lima, Peru, the annual rainfall is only 11 mm with the annual mean temperature of 19°C. The cool temperatures result from the coastal ocean upwelling induced by the southeast trades and preclude deep convection.

Fig. 3.3A shows the annual-mean precipitation climatology based on satellite observations. Over the continents and the Indo-western Pacific sector, the annual-mean precipitation distribution in the tropics is more or less symmetric about the equator, consistent with solar radiation distribution. This solar control of tropical convection breaks down over the eastern half of the Pacific and entire Atlantic, where deep

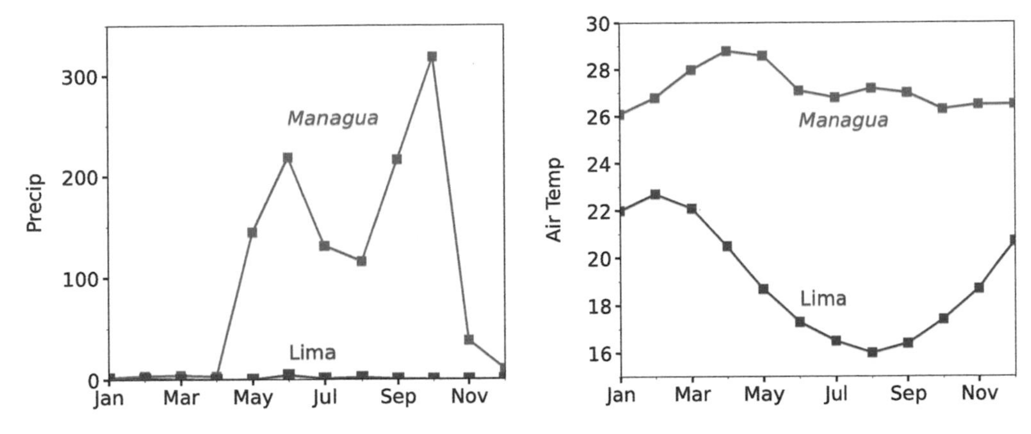

Fig. 8.2 North-south asymmetries on the Pacific coast of the Americas. Monthly climatology at Managua, Nicaragua, and Lima, Peru: *(left)* rainfall (mm/month) and *(right)* surface air temperature (°C) based on in situ observations. *(Courtesy Q. Peng.)*

convection is confined to the north of the equator. The ITCZ is sometimes called the thermal equator, where a great amount of latent heat is released into the atmosphere. The rainfall distribution over the eastern Pacific and Atlantic begs answers to the following questions: why the ITCZ is not on the equator where the annual-mean solar radiation is the maximum, and why it is displaced north of the equator.

8.1.1 Wind-evaporation-sea surface temperature feedback

Major tropical precipitation is confined within the 27°C sea surface temperature (SST) contours (see Fig. 3.3A). From a meteorologic point of view, the ITCZ remains north of the equator over the eastern Pacific and Atlantic because SST is higher north than south of the equator. From an oceanographic point of view, on the other hand, SST is higher north of the equator because the ITCZ stays in the Northern Hemisphere. This circular argument suggests that the northward-displaced ITCZ and high SST band are just two sides of the same coin and a coupled ocean-atmosphere approach is needed to explain them as a pair.

Surface evaporation is the major means for tropical oceans to balance incoming solar radiation. Surface wind speed reaches a minimum at the ITCZ in both the eastern Pacific and Atlantic (Fig. 8.3). Surface evaporation is a function of both SST and wind speed (see Eq. 7.30). If one assumes that everything else is the same between 10°N and 10°S, a 25% wind speed difference leads to an SST difference of 3°C according to the Clausius-Clapeyron equation for saturated water vapor content (for a typical wind speed of 7–8 m/s). To balance the net radiative flux, SST must rise (fall) under weak (strong) winds at 10°N (S). The simple calculation suggests that wind-induced evaporation change is a plausible mechanism for meridional asymmetry in SST.

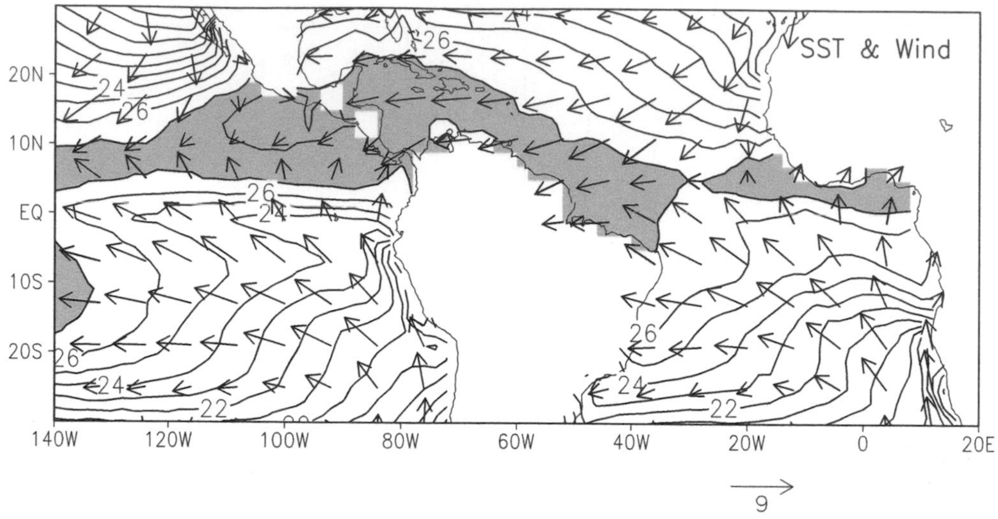

Fig. 8.3 Annual-mean sea surface temperature (SST) (shading $\geq$27°C, black contours at 1°C intervals) and surface wind velocity (m/s).

We introduce an antisymmetric SST perturbation and analyze how it might evolve due to ocean-atmosphere interaction. Suppose that somehow SST becomes slightly warmer north than south of the equator (Fig. 8.4). The resultant sea level pressure (SLP) gradient will drive southerly winds across the equator. The Coriolis force acts to turn these southerlies westward south and eastward north of the equator. Superimposed on the background easterly trades, these anomalous southeasterlies increase surface wind speed and hence evaporative cooling south of the equator. Conversely, north of the equator wind speed and surface evaporation decrease, amplifying the initial northward SST gradient. This implies a positive feedback from wind-evaporation-SST (WES) interaction (Xie and Philander 1994). This WES feedback is in favor of antisymmetric disturbances and helps break the equatorial symmetry set by the annual mean solar radiation.

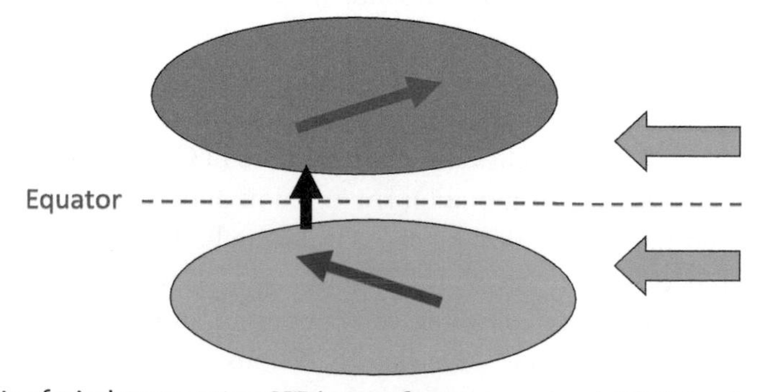

Fig. 8.4 Schematic of wind-evaporation-SST (sea surface temperature) (WES) feedback. Anomalies of SST (*positive shaded orange and negative blue*), and surface wind velocity (*westerly in red and easterly in blue*), on the background of mean easterly wind (*green block arrows*).

8.1.2 Coupled model

We build a simple model to illustrate the WES feedback and the concept of ocean-atmosphere interaction. Neglect air-sea surface temperature difference. Meridional wind is generally much weaker than the zonal wind, and we approximate the scalar wind speed $W = |\overline{U}+U'|$, where the overbar denotes the mean and the prime perturbation zonal wind velocity. The wind speed term is piecewise linear, $W = -(\overline{U}+U')$ for the easterly background wind that prevails in the tropics. Linearizing surface latent heat flux (Eq. 7.30) gives rise to

$$Q_E' = \overline{Q}_E\left(\frac{U'}{\overline{U}}+\alpha T'\right) \tag{8.1}$$

where we have used the Clausius-Clapeyron equation $dq_s/dT = \alpha q_s$, where $\alpha = L\big/R_v\overline{T}^2 \sim 0.06\ \mathrm{K}^{-1}$. By further neglecting perturbation radiative fluxes at the sea surface, we obtain the governing equation for perturbation SST

$$\frac{\partial T}{\partial t} = aU - bT \tag{8.2}$$

where $a = \frac{\overline{Q}_E}{\rho c_p H_m}\left(\frac{1}{-\overline{U}}\right)$, and $b = \alpha\frac{\overline{Q}_E}{\rho c_p H_m}$ is the evaporative damping rate. For simplicity, we drop the prime in the rest of this chapter. The first term reflects its wind-speed dependence and represents the WES effect: westerly (easterly) wind anomalies weaken (strengthen) the background easterlies and evaporation. The second term on the right-hand side (RHS) reflects the SST dependence of surface evaporation and acts as Newtonian damping.

For the atmosphere, the cross-equatorial wind is driven by cross-equatorial SST gradient,

$$V = \gamma \cdot \delta T \tag{8.3}$$

where γ is a constant, and δ denotes the difference between the northern and southern boxes (say at 10° latitude). The cross-equatorial wind induces a zonal wind component due to the Coriolis force (cf. the Matsuno-Gill model, Eq. 3.32),

$$\delta U = \frac{2}{\varepsilon} f_N V \tag{8.4}$$

where f_N is the Coriolis parameter in the northern box. Note that we have dropped time tendency terms in the atmospheric model. This assumption of atmospheric "quasi-equilibrium" is based on the fact that the atmosphere adjusts to SST changes in a matter of a week to a month, an order of magnitude shorter than oceanic adjustments to changes in the atmosphere (say winds). For a 50-m mixed layer and a typical trade wind speed of 7 to 8 m/s, the Newtonian damping rate b has an e-folding timescale of about 6 months.

Coupling the ocean model (Eq. 8.2) and atmospheric model (Eqs. 8.3 and 8.4) leads to a *coupled model* for the north-south SST difference:

$$\frac{\partial}{\partial t}\delta T = (\sigma - b)\delta T \tag{8.5}$$

where

$$\sigma = 2f_N(a\gamma)/\varepsilon \tag{8.6}$$

is the WES *coupling coefficient*, which is the product of oceanic (a) and atmospheric (γ/ε) coupling coefficients. Without the coupling, the initial SST dipole δT would decay at the rate of b. With the WES feedback, the decay slows down at the reduced rate of ($b-\sigma$). If the WES feedback is strong enough ($\sigma > b$), the meridional dipole can overcome the Newtonian damping (b), become unstable, and grow in time.

The meridional shear of zonal wind across the equator due to the beta effect is essential to the WES feedback (Fig. 8.4). The WES coupling coefficient is proportional to the distance of the ITCZ from the equator ($f_N = \beta y_N$). The cold tongue due to equatorial upwelling precludes deep convection on the equator, giving rise to a large f_N. This explains why strong meridional asymmetry in the ITCZ develops only north of the equatorial cold tongue over the eastern Pacific and Atlantic. Without the equatorial upwelling, the ITCZ can simply sit on the equator. With $f_N = 0$, WES feedback vanishes, stabilizing the symmetric climate, as over the Indo-western Pacific warm pool.

The abovementioned positive WES feedback is conditioned on the easterly trade wind background (as over the tropical Pacific and Atlantic). WES feedback turns negative with a westerly wind background (as over the North Indian Ocean in summer).

8.1.3 Continental forcing and the westward control

Air-sea feedbacks are important in keeping the ITCZ north of the equator. They do not fully explain, however, why the Northern but not the Southern Hemisphere is favored in the Pacific and Atlantic. It is reasonable to assume that hemispheric asymmetry in continents ultimately gives rise to the climatic asymmetry. Which continental features, and how do they move the ITCZ away from the equator? For the Pacific, is the shape of the Asian-Australian continents on the west or that of Americas on the east more important? Here we extend the coupled WES model to include zonal variations.

Meridional wind velocity at the equator is a good measure of equatorial asymmetry, vanishing if the climate were perfectly symmetric between the hemispheres. Following Xie (1996), we model it with a quasi-steady Rossby wave equation that is forced by the meridional SST gradient,

$$\left(1 - L_a \frac{\partial}{\partial x}\right) V = \gamma \cdot \delta T \tag{8.7}$$

where $L_a = \frac{c_R}{\varepsilon} = \frac{c_a}{5\varepsilon}$ is the e-folding scale of the damped, antisymmetric long Rossby wave that propagates westward at the phase speed of $c_a/5$. Combining Eqs. 8.2, 8.4, and 8.7 yields an equation for cross-equatorial wind velocity:

$$\left(\frac{\partial}{\partial t} + b\right)\left(1 - L_a \frac{\partial}{\partial x}\right) V = \sigma V \tag{8.8}$$

In general, Eq. 8.8 may be solved by imposing an eastern boundary condition:

$$V|_{x=0} = V_E \tag{8.9}$$

The cross-equatorial wind on the eastern boundary V_E is positive in both the Pacific and Atlantic and may be viewed as induced by the cross-equatorial asymmetry in the continents to the east.

For the Pacific, consider an infinitesimally thin continent slanted northwestward under the uniform easterly trades. While the thin continent does not have any atmospheric effect, the background easterlies induce coastal upwelling south of the equator but cause downwelling north of the equator (Fig. 8.5). The asymmetry in coastal upwelling drives the southerly cross-equatorial wind on the eastern boundary.

The steady-state solution to Eq. 8.8 that satisfies the eastern boundary condition (Eq. 8.9) is

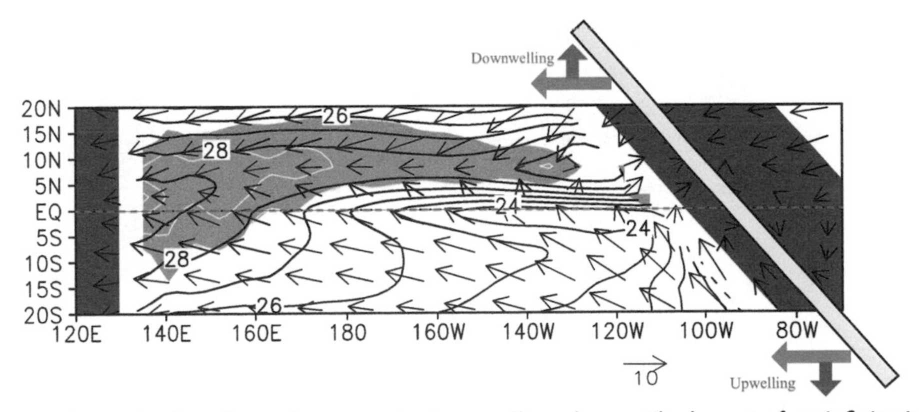

Fig. 8.5 Schematic of north-south asymmetry in upwelling along a tilted coast of an infinitesimally thin continent in response to uniform easterly background wind *(green block arrows)*. Climatology of sea surface temperature (contour interval 1°C), surface wind (vectors; scaled to 10 m/s), and precipitation (shade >4 mm/day with white contours at intervals of 4 mm/day) in a coupled ocean-atmosphere model with a tilted east coast. *(From Okajima et al. (2003).)*

$$V = V_E e^{x/L_c} \tag{8.10}$$

where $L_c = L_a \Big/ \left(1 - \frac{\sigma}{b}\right)$ is the e-folding zonal scale with WES feedback. Without coupling, the atmospheric response to the eastern boundary forcing decays rapidly toward the interior ocean with a small e-folding scale of L_a (~1700 km for $\varepsilon^{-1} =$ 2 days and $c_a = 50$ m/s). WES feedback increases the e-folding zonal scale by a factor of $\left(1 - \frac{\sigma}{b}\right)^{-1}$, allowing the influence of continental asymmetry to penetrate far into the west over the Pacific. Observed meridional wind on the equator peaks on the South American coast and decays westward (e-folding scale L_c ~9000 km) (see Fig. 8.3).

For the Atlantic, the bulge of West Africa is hotter than the ocean to the south, inducing the southerly winds across the equator. The winds in turn cause upwelling both off the southern African coast and in the open ocean south of the equator. This initiates ocean-atmosphere feedback that displaces the ITCZ north of the equator (Fig. 8.6A). Over the warm waters where the southeast and northeast trade-wind systems meet, deep convection takes place, producing the thunders and rains Halley noted. Remarkably, Halley's wind map contains key elements to solving the age-old riddle of

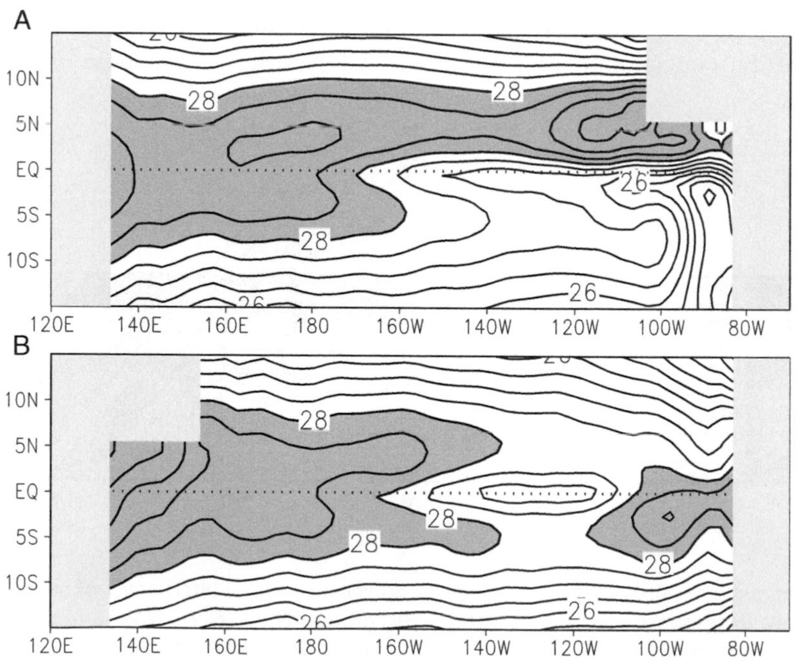

Fig. 8.6 Time-mean sea surface temperature (°C) in a coupled model where a northern land bulge is added to (A) the eastern and (B) the western continent. Land is shaded yellow-green. *(From Xie and Saito (2001).)*

climatic asymmetry; winds are calm in the doldrums of the ITCZ, allowing water to stay warm there.

The continental forcing is distinct between the east and west. The interior asymmetry is sensitive only to the continental asymmetry on the eastern boundary (Eq. 8.10). The continent's westward control over oceanic climate stems from the fact that under the longwave approximation, asymmetric perturbations in the ocean and atmosphere propagate only westward as Rossby waves (Eq. 8.7). (The Kelvin wave propagates eastward but is symmetric about the equator.) Fig. 8.6 is an explicit demonstration of this westward control in a coupled model in which a northern land bulge creates the latitudinal asymmetry. Basin-wide northward displacement of the ITCZ occurs only when the continental forcing is placed on the eastern side of the ocean basin. This westward control mechanism implies that the symmetry-breaking forcing resides on the eastern continent—the Americas for the Pacific and Africa for the Atlantic.

8.1.4 Tropical basin view vs. global zonal-mean theory

It is not the atmosphere, nor the ocean alone, not even their coupling, but the collective effect of the ocean, atmosphere, and land geometry that gives rise to the long silver band of convective clouds, stretching north of the equator over the half the globe in the Pacific and Atlantic (see Fig. 3.3). This climatic asymmetry illustrates the importance of air-sea interaction in driving Earth's climate to deviate markedly in space and time from the pattern of solar forcing. Our simple coupled model suggests that the cause of the climatic asymmetry over the Pacific and Atlantic lies in the continents to the east (e.g., the coastal line).

Global continental geometry (e.g., the circumpolar nature of the Southern Ocean) locks the sinking branch of the deep meridional overturning circulation (MOC) in the subpolar North Atlantic. A complementary global view (to be presented in Section 12.4.2), emphasizes the effect of cross-equatorial ocean heat transport by the deep MOC. The energy transport theory explains the global zonal-mean structure but faces difficulties in the zonal variations of the ITCZ asymmetry. The tropical view presented here focuses on the eastern Pacific and Atlantic where the ITCZ asymmetry is most pronounced. The global energy transport and tropical views await further integration.

Tropical cyclones are organized deep convection and among the most intense storms on Earth. The meridional asymmetry limits warm waters conducive to deep convection to the north of the equator from the eastern Pacific through the Atlantic. Half of the Southern Hemisphere from 140°W to 40°E is devoid of tropical cyclones. Globally, 60 tropical cyclones form each year in the Northern Hemisphere but only 26 in the Southern Hemisphere.

The Southern and Northern Hemisphere trade wind systems meet at the ITCZ. In the Pacific, the upper-ocean circulation develops strong hemispheric asymmetries in response to wind stress curls associated with the northward-displaced ITCZ (see Figs. 7.2 and 7.5). The westward South Equatorial Current intrudes north of the equator and is separated from the North Equatorial Current by the North Equatorial Countercurrent (NECC) on the south flank of the ITCZ. Notably, the NECC flows eastward against the easterly trades and in geostrophic balance with a thermocline ridge beneath the ITCZ. Such an eastward countercurrent is not observed south of the equator. The equatorial undercurrent is an interesting exception to the meridional asymmetry; it is the vanishing Coriolis effect that allows the zonal pressure gradient to drive this current at the equator (see Fig. 7.2).

8.2 Equatorial cold tongue and Walker circulation

The equatorial cold tongue is another example that tropical climate can deviate markedly from the top of atmosphere (TOA) solar radiation. From an oceanographic point of the view, SSTs in the eastern equatorial Pacific are low because the prevailing easterly winds shoal the thermocline and upwell the cold water from the shoaled thermocline into the surface mixed layer.

From a meteorologic point of view, on the other hand, the prevailing easterly winds result from the SLP gradient set up by the SST difference between the warm west and cool east equatorial Pacific Ocean. Active deep convection in the west and the lack thereof in the east on the equator contributes to the SLP gradient that drives the equatorial easterlies as part of the zonal overturning Walker circulation.

The abovementioned circular argument suggests a positive feedback from the ocean-atmosphere interaction. Suppose that SST is somehow cooler in the east than west. The zonal SST gradient drives an easterly wind anomaly, which shoals the thermocline in the east. The equatorial upwelling induced by the easterly wind anomaly amplifies the initial negative SST anomaly in the east going through the loop of ocean-atmosphere interaction. This is known as Bjerknes feedback, a concept we will further develop in the next chapter on El Niño and the Southern Oscillation. Bjerknes and WES feedbacks are distinct; the former operates in the east-west direction and involves ocean dynamic adjustments of the thermocline depth and upwelling to wind changes, while the latter operates in the meridional direction. Surface heat flux is mostly a damping for Bjerknes feedback on the equator while it gives rise to WES feedback for antisymmetric perturbations across the equator.

The background wind on the equator is easterly. With zonally uniform SST, idealized atmospheric model experiments on a water-covered earth yield a background easterly surface wind of typically $\sim$2 m/s on the equator (Fig. 8.7). This background easterly wind breaks the east-west symmetry with equatorial upwelling and the eastward shoaled thermocline. Bjerknes feedback then acts to amplify the eastern cooling in the equatorial Pacific.

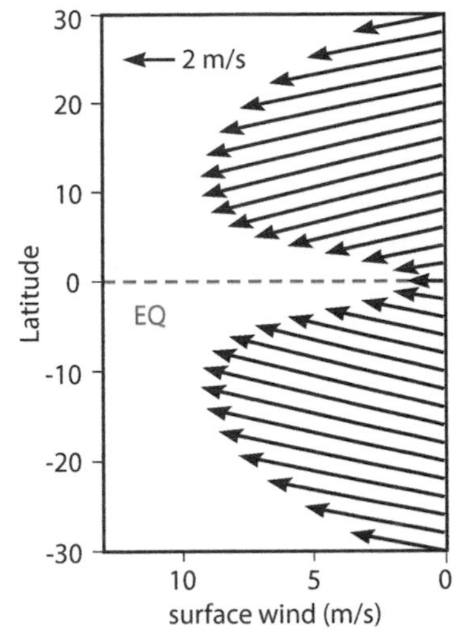

Fig. 8.7 Surface zonal wind as a function of latitude in an aquaplanet atmospheric general circulation model run (Song et al. 2023) with a zonally uniform, meridionally symmetric sea surface temperature distribution. *(Courtesy Z. Song.)*

In the eastern equatorial Pacific, the easterly winds weaken gradually toward the South American coast (Fig. 8.3) due to the blockage of the low-level flow by the steep meridional mountain range of the Andes. Despite this eastward reduction in the easterly wind, SST continues to decrease toward Ecuador as the cross-equatorial southerlies strengthen and induce ocean upwelling south of the equator. Fig. 8.8 is a schematic of how a spatially uniform meridional wind causes open-ocean upwelling. Sufficiently

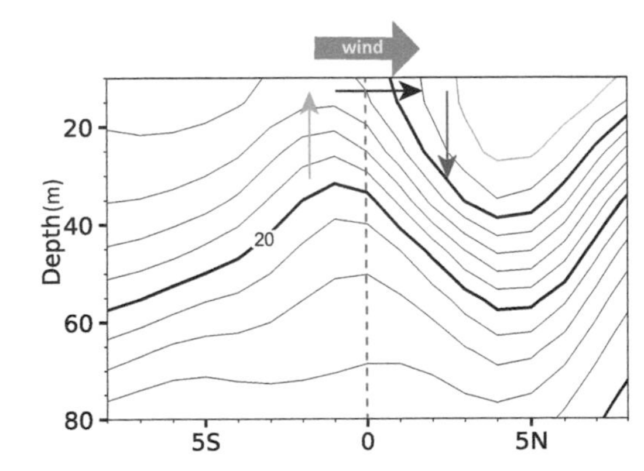

Fig. 8.8 Schematic of south-equatorial upwelling *(blue arrow)* induced by southerly cross-equatorial wind *(green block arrow)*, on the background of annual-mean ocean temperature (contours at intervals of 1°C, thickened every 5°C and temperature ≥27°C red) at 90°W. *(Courtesy Q. Peng.)*

away from the equator with a large Coriolis effect, the surface Ekman flow is deflected eastward (westward) north (south) of the equator. On the equator, the Coriolis effect vanishes, and the warm surface water flows in the direction of the wind. As a result, upwelling takes place south of the equator while downwelling north of the equator (Philander and Pacanowski 1981). In the eastern Pacific Ocean, the meridional minimum of SST, indicative of the peak upwelling, is indeed displaced slightly south of the equator (~1°S).

From the surface Ekman flow (see Eq. 7.3), the vertical Ekman pumping velocity for spatially uniform wind stress is

$$w_E = -\frac{\beta}{\rho}\frac{\left(\varepsilon^2 - f^2\right)\tau_x + 2\varepsilon f \tau_y}{\left(\varepsilon^2 + f^2\right)^2} \tag{8.11}$$

An easterly wind stress induces upwelling that peaks at the equator, while a southerly wind stress induces upwelling (downwelling) south (north) of the equator. It can be shown that the latitude of the peak south-equatorial upwelling under the southerly wind is $y = -y_E/2 = -\varepsilon/(2\beta) \sim -100$ km, in broad agreement with the observed SST pattern with a southward displaced SST minimum. Since the thermocline shoaling induced by the basin-scale easterlies is meridionally broad near the eastern boundary (Fig. 7.9; see also Fig. 7.5), the slightly southward displaced upwelling is effective in lowering SST. The Ekman currents induced by the southeast trade winds advect the upwelled water away from the equator.

As the southerly cross-equatorial wind in the eastern Pacific is associated with the northward displaced ITCZ, the ITCZ asymmetry strengthens the equatorial cold tongue. The cold tongue, by pushing the ITCZ away from the equator, strengthens WES feedback (larger f_N in Eq. 8.6) and hence meridional asymmetry. This implies a coupling between the cold tongue and ITCZ asymmetry across the equator.

8.3 Equatorial annual cycle

In most places, summer is warmer than winter because solar radiation is higher. At the equator, TOA solar radiation hardly varies throughout the year, with weak peaks at the equinoxes of March 21 and September 21. Indeed, the seasonal range of air temperature variation is small in Singapore (1°N), from 26°C in December-January to 27.7°C in May-June.

What does the seasonal cycle look like on the Galapagos in the eastern equatorial Pacific? Darwin's observations there (Box 8.1) offer an important hint: The standing leafless trees imply a savanna climate. There must be a rainy season to keep these trees alive, but rain-producing deep convection requires much higher SSTs than 20°C Darwin measured in September 1835. It turns out that, on average, SST around the Galapagos varies from a

BOX 8.1 Climate on the Galapagos

The Galapagos are situated in the equatorial Pacific cold tongue (90°W). Low SSTs prevent deep convection from occurring for much of the year. Condensation of cloud droplets on tree leaves is important for precipitation on volcanic mountains on the islands. According to Wikipedia:

> *During the season known as the 'Garua' (June to November) the temperature by the sea is 22°C, a steady and cold wind blows from South and Southeast, and frequent drizzles (Garuas) last most of the day, along with dense fog which conceals the islands. During the warm season (December to May) the average sea and air temperature rises to 25°C, there is no wind at all, there are sporadic though strong rains and the sun shines.*

Charles Darwin set foot on the Galapagos as the naturalist on *HMS Beagle* in September 1835. In his diary (Keynes 2001), he described the southeasterly trades during the cool season: "The weather, now & during the passage, has continued as on the coast of Peru, a steady, gentle breeze of wind & gloomy sky." Darwin reported on the dry condition: "In all these Islds the dry parts" feature "an arid Volcanic soil, a flowering leafless Vegetation in an Intertropical region, but without the beauty which generally accompanies such a position." He also noted the cloud forest effect on mountains hugged by low clouds. If he had come to the islands during the warm season, Darwin would have seen a quite different climate with rain showers and trees with green leaves. Leafless trees Darwin noted are also commonly seen on the leeside of Hawaii (e.g., Diamond Head Mountain) and in Southern California during much of the year. These trees turn green during the brief winter rainy season.

September low of 21vC to a March high of 26vC (see Fig. 8.9). As a result, it never rains in cold September, but moderate rainfall is observed in the warm season of February-April. The annual range of 5vC in SST is much larger than the 3°C range in Hawaii, 20° to the north. Unlike Hawaii, TOA insolation is nearly identical between March and September on the equator, but March is much warmer than September on the Galapagos.

The annual cycle in equatorial SST is weak in the western Pacific and intensifies eastward, with a coherent westward propagation in the eastern half of the basin (Fig. 8.10). SST peaks in March at 90°W and in May at 150°W, 60° to the west. Important questions arise: What causes the large difference in SST between March and September? Why is this annual cycle large in the east but diminishes toward the west? What causes the apparent westward propagation of the SST annual cycle in the eastern half of the equatorial Pacific?

8.3.1 Annual frequency

In the far eastern Pacific, the southerly winds drive an upwelling slightly south of the equator. In March, the seasonal cooling/warming in the Northern/Southern Hemisphere weakens the southerly winds, reducing the south-equatorial upwelling and

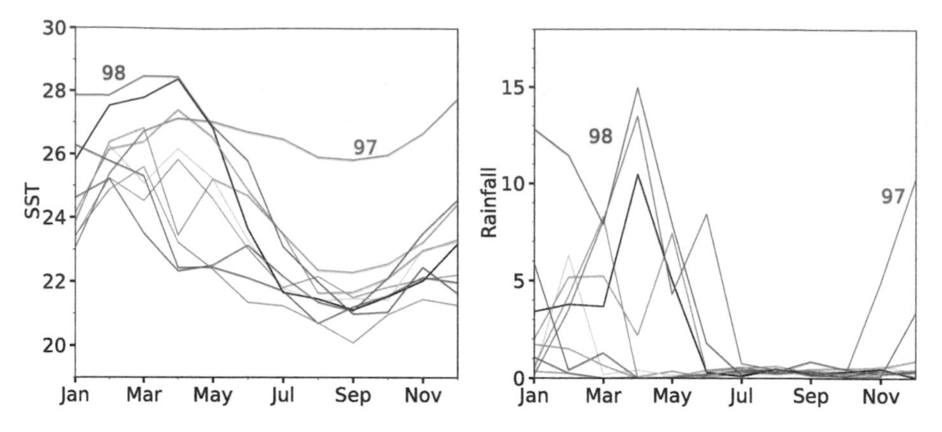

Fig. 8.9 Sea surface temperature *(SST)* (°C) and precipitation (mm/day) at the Galapagos as a function of calendar month for a 10-year period of 1989-1998. Data from Galapagos in situ data. *(Courtesy Q. Peng.)*

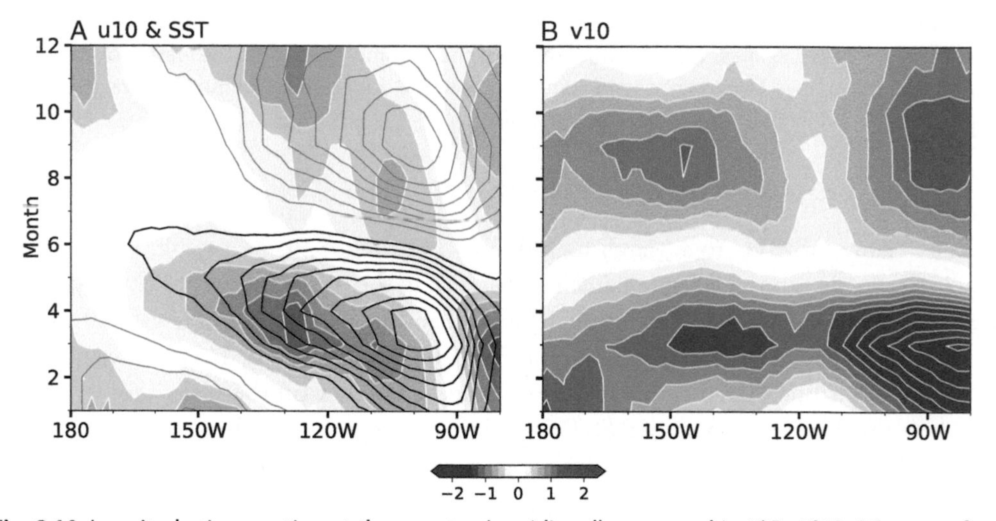

Fig. 8.10 Longitude-time section at the equator (meridionally averaged in 2°S–2°N): (A) sea surface temperature (SST) (contours at intervals of 0.4°C, positive in black and negative in grey) and surface zonal wind velocity (color shading and white contours, 0.2 m/s intervals); and (B) surface meridional wind velocity (color shading, m/s). all the deviations from the annual mean, with the zero contours omitted. *(Courtesy Q. Peng.)*

causing the surface water to warm (Fig. 8.11). Thus March and September are made different by the intensity of the southeast trades at the equator due to the northward-displaced ITCZ. (What would happen if the annual mean climate were symmetric about the equator?)

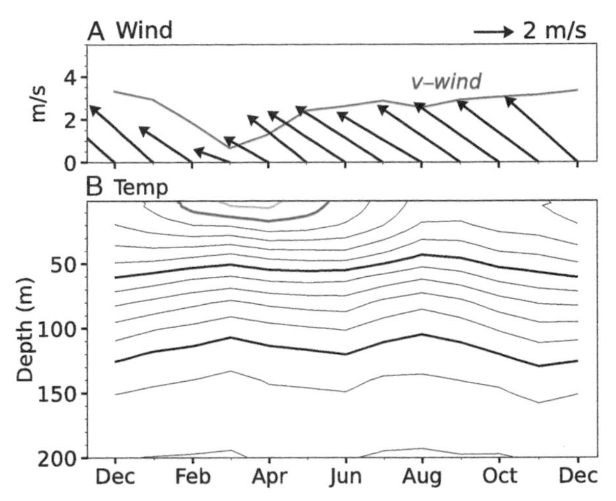

Fig. 8.11 Monthly climatology at 110°W, equator: (a) wind velocity vector (m/s) and meridional wind velocity (v-wind, curve); and (b) ocean temperature (black contours at 1°C interval, 15, 20 & 25°C thickened, red ≥ 25°C), based on TAO data. Courtesy Q. Peng.

Consider a simple case relevant to the far eastern Pacific with $\tau_x = 0$. From Eq. 8.11, the southerly wind stress drives upwelling,

$$w_E = a_y\left(\overline{\tau}_y + \tau'_y\right) \tag{8.12}$$

which peaks south of the equator at $y = -y_E/2$, where $a_y = \frac{16}{25}\frac{\beta}{\rho\varepsilon^3}$. At the peak latitude of the off-equatorial upwelling, SST reaches the meridional minimum and the meridional advection vanishes. We also neglect zonal advection because $u' = 0$ at the equator under the meridional wind forcing τ'_y. Here the overbar denotes the annual mean and the prime the deviation. From Eq. 7.29, the linearized equation for perturbation mixed layer temperature at the latitude of maximum upwelling is

$$\frac{\partial T'}{\partial t} = -a_y\overline{T}_z\tau'_y - bT' \tag{8.13}$$

where the first term on the RHS is the anomalous upwelling effect, and $\overline{T}_z > 0$ is the vertical gradient of mean temperature between the surface and thermocline and increases eastward with the shoaling thermocline. Here we have neglected the seasonal variation in the thermocline depth based on observations ($T'_e = 0$) (see Fig. 8.11).

The surface heat flux tries to restore SST toward a thermal equilibrium, while upwelling restores it toward the subsurface temperature in the shoaled thermocline. Both effects damp SST perturbations, $b = b_E + b_W$, where b_E denotes the evaporative damping (b in Eq. 8.2), and $b_W = \overline{w}_E/H_m$ the damping due to the mean upwelling. The wind speed effect on surface latent heat flux can be absorbed in the coefficient a_y and is smaller than the variable upwelling effect.

We have assumed in Eq. 8.13 that the mean southerly wind is greater than the perturbation, namely the cross-equatorial wind remains southerly throughout the year. By modulating the intensity of the south-equatorial upwelling, the annual cycle in the southerly wind forces an annual cycle in SST. As $\overline{\tau}_y$ weakens westward, the total cross-equatorial wind reverses direction into northerly in boreal winter, reducing the projection onto the annual harmonic in upwelling intensity (proportional to $\left|\overline{\tau}_y + \tau'_y\right|$). This westward decreasing projection onto the annual harmonic can be absorbed in $a_y(x)$.

The perturbation meridional wind at the equator may be considered due to the seasonal solar forcing, southerly and at the maximum in September when the thermal inertia of the ocean mixed layer is factored in. To first order, τ'_y is annual and does not vary much zonally (see Fig. 8.10B), but its effect on the annual cycle in SST decays rapidly westward because of the decrease in $a_y \overline{T}_z$. This explains why the SST annual cycle is large in the eastern but weak in the western equatorial Pacific.

8.3.2 Westward phase propagation

Now consider the part of the eastern Pacific sufficiently away from the South American coast—say, west of 110°W—so that zonal wind effect on horizontal current and upwelling dominates. At the equator, perturbation upwelling and zonal current are both proportional to perturbation zonal wind:

$$w'_E = -\frac{\beta}{\rho}\frac{1}{\varepsilon^2}\tau'_x, \; u'_E = \frac{1}{\rho H_E}\frac{1}{\varepsilon}\tau'_x \tag{8.14}$$

From Eq. 7.29, the perturbation SST equation is

$$\frac{\partial T}{\partial t} + \overline{u}\frac{\partial T}{\partial x} + a_x \tau'_x = -bT \tag{8.15}$$

where $a_x = \frac{\beta}{\rho}\frac{1}{\varepsilon^2}\left(-\overline{T}_z + \frac{1}{H_E}\frac{\varepsilon}{\beta}\overline{T}_x\right)$ is generally negative for the equatorial Pacific. Assuming that perturbation atmospheric pressure is proportional to perturbation SST, we obtain a simple atmospheric model:

$$\tau'_x = \gamma_\tau \frac{\partial T}{\partial x} \tag{8.16}$$

Fig. 8.10A supports this relationship. Substituting into the ocean model Eq. 8.15 yields a coupled model:

$$\frac{\partial T}{\partial t} + (\overline{u} + a_x \gamma_\tau) \frac{\partial T}{\partial x} = -bT \tag{8.17}$$

The coupling of the zonal wind and zonal SST gradient causes SST perturbations to propagate westward (the second term in the parentheses on the LHS).

In March, the southerly cross-equatorial wind weakens, reducing the south-equatorial upwelling and causing equatorial SST to rise. The March warming intensifies eastward because of the intensification of the ITCZ asymmetry and the shoaling thermocline toward the east (Eq. 8.13). The eastward gradient of the seasonal SST warming reduces the easterly trade winds to the west (Eq. 8.16) (see Fig. 8.10A). The westward displaced SST tendency causes the coupled SST-zonal wind pattern to propagate westward as in observations. Thus the mean state of eastern Pacific climate is essential for the SST annual cycle, the mean southerly cross-equatorial wind and eastward shoaling thermocline in particular.

Fig. 8.12 shows the results from a simple coupled model similar to the one derived here. It captures major characteristics of the observed annual SST cycle in the eastern Pacific, including the eastward increasing amplitude and westward phase propagation. In the western Pacific, the semiannual cycle in local solar radiation forces a weak semiannual oscillation in SST.

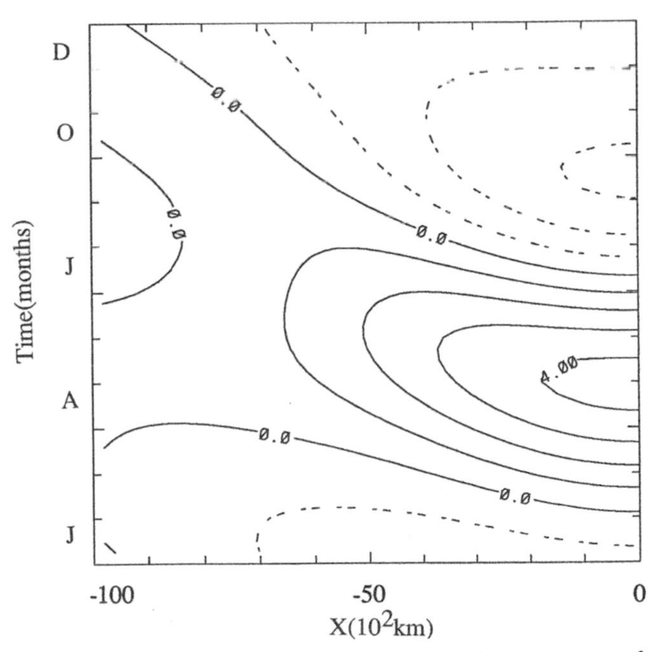

Fig. 8.12 The seasonal sea surface temperature variation (°C) on the equator as a function of distance from the eastern boundary and calendar month in a simple coupled model. *(From Xie (1994).)*

8.3.3 Broad seasonal variations

Over the Indo-western Pacific warm pool and on tropical continents, annual mean climate is nearly symmetric about the equator. On the seasonal timescale, the maximum rainfall in these regions moves back and forth across the equator following the Sun and the meridional maximum of SST. Over the eastern Pacific and Atlantic, the annual-mean ITCZ is displaced in the Northern Hemisphere and behaves like a climatic equator. The distribution of the SST annual harmonic (phase and amplitude) supports this view. South of the ITCZ the seasonal cycle bears the Southern Hemisphere characteristics of being warm in March and cold in September. North of the ITCZ, SST peaks in September and is coldest in March. Fig. 8.13 shows the annual harmonic of observed SST, with shaded areas indicating regions where the annual-mean SST exceeds 27°C. The annual SST harmonic is at its minimum along the annual-mean ITCZ, consistent with the notion of its being the climatic equator. (The collocation of the annual harmonic minimum and ITCZ is particularly conspicuous in the Atlantic.) The negative feedback between convective clouds and local SST could also reduce the annual variations in SST under the ITCZ.

The annual harmonic is large on/south of the equator and on the eastern boundary because of the upwelling and the shoaling thermocline. The seasonal variation in solar radiation is antisymmetric about the equator and drives a pronounced annual cycle in meridional wind across the equator. Over most of the Atlantic and east of 140°W in the Pacific, the ITCZ stays north of the equator, and the annual cycle in the intensity of the cross-equatorial southerlies regulates the amount of cold water upwelled into the ocean mixed layer. Low cloud is another factor for the enhanced annual cycle in the southeastern Pacific and Atlantic. The boundary-layer cloud cover increases in September as the northern ITCZ and hence the subsidence south of the equator are most intense. The positive feedback between SST and low cloud helps amplify the SST annual harmonic south of the equator.

Fig. 8.14 shows the seasonal cycle over the eastern tropical Pacific (120°W), where the ITCZ is displaced north of the equator most of the time. From April to September,

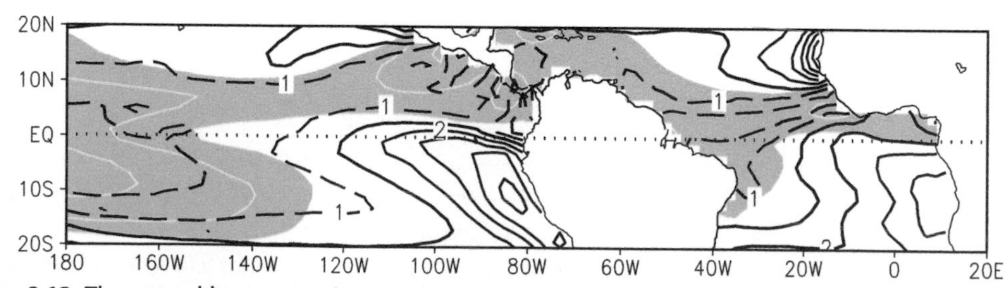

Fig. 8.13 The annual harmonic of sea surface temperature (SST; black line contours at 0.5°C interval; dashed ≤1°C), along with the annual mean SST (shading >27°C and white contours at 1°C intervals).

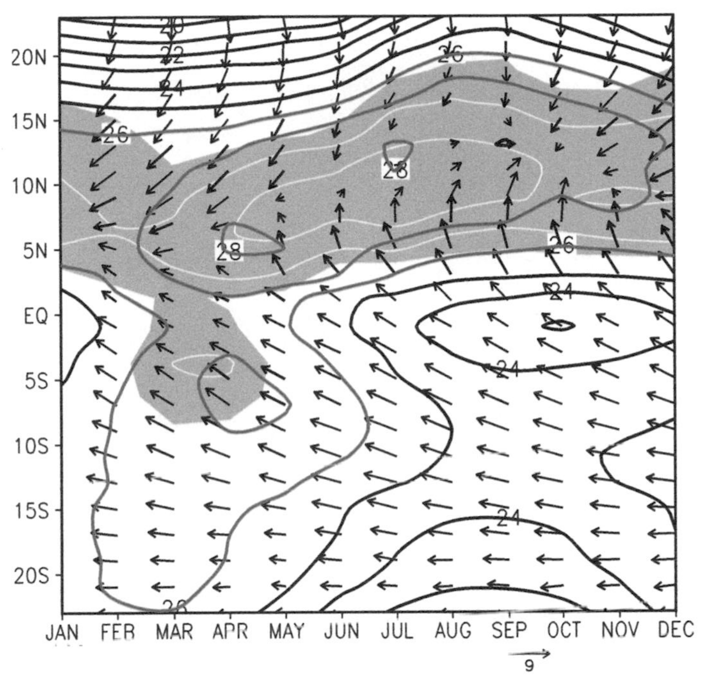

Fig. 8.14 Eastern tropical Pacific climatology (zonally averaged in 120–115°W) as a function of calendar month and latitude: sea surface temperature (line contours in °C, red ≥26°C), surface wind velocity (m/s), and precipitation (shading ≥2.5 mm/day, white contours at 5 mm/day intervals).

the precipitation maximum moves northward from 5°N to 12°N, following the northward march of the SST maximum. From September to March, the SST asymmetry gradually weakens as the solar radiation change increases (decreases) SST south (north) of the equator. Briefly in March and April, SST south of the equator rises above the convective threshold, and a symmetric double ITCZ forms over the eastern half of the tropical Pacific astride the equator (Fig. 8.15A) as the seasonal cycle removes much of the annual-mean meridional asymmetry. Surface wind convergence follows the same seasonal cycle and is tightly coupled with SST and precipitation. Focusing on the equator, the southeast trades weaken during the brief period of the double ITCZ, causing the equatorial Pacific to warm. March and April are the time when equatorial Pacific climate is nearly symmetric about the equator with little meridional SST gradient between 15°S and 15°N. This is in sharp contrast with September when both the meridional asymmetry and equatorial cold tongue are strongest (see Fig. 8.15B).

In the equatorial Pacific, the SST annual cycle represents a westward expansion in September and contraction in March of the cold tongue (see Fig. 8.15). It might be

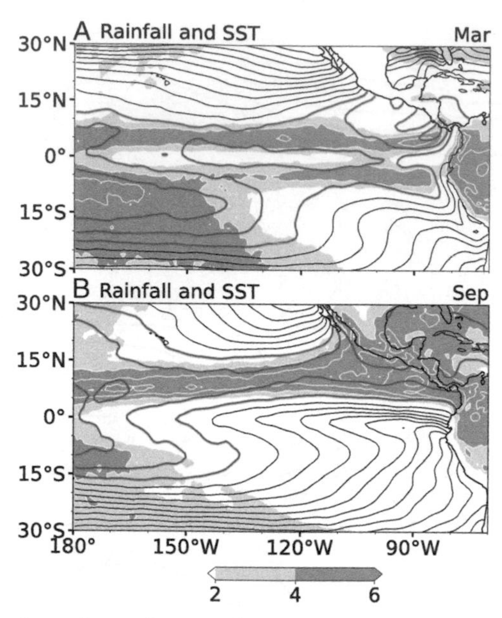

Fig. 8.15 Tropical Pacific climatology of sea surface temperature (SST; contours at intervals of 1°C, red ≥27°C) and precipitation (light/dark gray shading ≥2/4 mm/day, white contours at 4 mm/day intervals) in (A) March and (B) September. *(Courtesy Q. Peng.)*

tempting to draw an analogy with the interannual expansion and contraction of the cold tongue associated, respectively, with El Niño and La Niña, for which Bjerknes feedback is important (see Chapter 9). This analogy is invalid. First, while surface warming of El Niño is due to the deepened thermocline, the thermocline displacements are negligible in a typical seasonal cycle across the equatorial Pacific (see Fig. 8.11). Second, while zonal wind variability drives El Niño warming, it is weak over the far eastern equatorial Pacific, where the meridional wind variability is instead important for SST seasonal cycle (see Fig. 8.8) (Mitchell and Wallace 1992).

On the equator, the zonal current velocity shows a marked annual cycle with conspicuous westward phase propagation (Fig. 8.16), as part of the annual Rossby waves forced by the seasonal variations in the wind curls under the ITCZ. In September, the Pacific ITCZ reaches the northernmost latitude of 12°N (see Fig. 8.14). At 5°N, the anomalous downwelling curls deepen the thermocline. This represents the seasonal intensification of the meridional shear between the westward flow on the equator and the eastward NECC (Fig. 8.17A). The intensified lateral shear generates tropical instability waves—energetic anticyclonic eddies of 1000 km wavelength centered at 6°N (see Fig. 8.17B). The resultant quasi-monthly meanders of the equatorial cold tongue are one of the most notable features from satellite imageries.

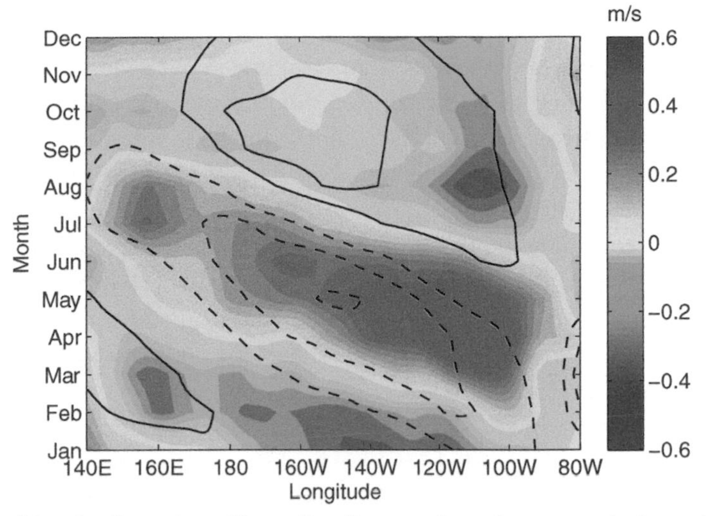

Fig. 8.16 Tropical Pacific climatology. Hovmoller diagram of zonal current velocity at the equator (color shading, m/s) and sea surface height (SSH) at 5°N with the annual mean removed (line contours at 4-cm interval, negative dashed, zero contour omitted). AVISO SSH and OSCAR current velocity are used. *(Adapted from Wang et al. (2017); courtesy M. Wang.)*

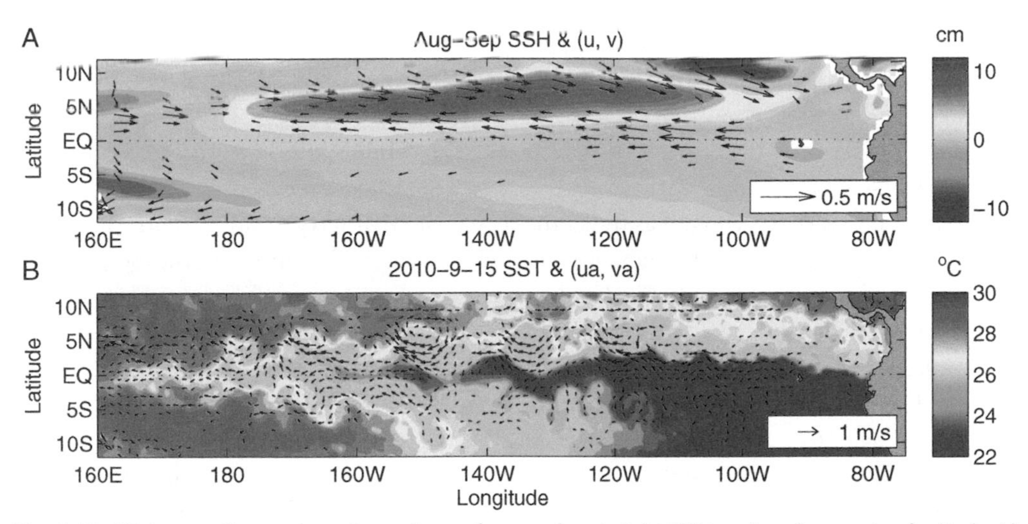

Fig. 8.17 (A) August-September climatology of sea surface height (*SSH*, cm) and current velocity (m/s) with the annual mean removed. (B) SST *(color shading)* and anomalous current velocity (vectors, m/s) on September 15, 2010. AVISO SSH, OSCAR current velocity, and OISST datasets are used. *(Courtesy M. Wang.)*

Review questions

1. Based on the simulation with an early atmospheric general circulation model, Manabe et al. (1974) argued that SST controls the northward-displaced ITCZ. Support this idea based on the seasonal cycles in rainfall and SST over the eastern tropical Pacific. Specifically, when does the double ITCZ occur, and how does the meridional profile of SST look at the time?
2. How does the continent control the climatic asymmetry over the ocean to the west? What is the key argument behind this westward control theory?
3. Consider a symmetric SST perturbation that peaks on the equator. Will the WES feedback help the perturbation to grow as it does for the antisymmetric one? Assume that everything is uniform in the east-west direction.
4. Show that the wind curl is positive in the northward-displaced ITCZ. This band of positive wind curls is bounded by Ekman downwelling on either side. Use the Sverdrup transport to show a NECC as a result, in balance with a thermocline ridge under the ITCZ. When is the NECC stronger, March or September?
5. A friend is going to visit the Galapagos Islands. How are you going to advise your friend about the weather and climate there?
6. Penguins are found on the Galapagos Islands and the coast of Peru. What does this fact imply for the local climate (temperature and rainfall)?
7. At 140°W, what causes the SST minimum at the equator? Why is the SST minimum displaced slightly south of the equator at 90°W?
8. Examine Fig. 8.15A. Is the March SST at 120°W, equator above or below 27°C? Explain.
9. If the ITCZ were displaced south of the equator, when would the rainy season be on the Galapagos Islands?
10. The ITCZ is often viewed as the climatic axis of symmetry. This explains the phase of the annual cycle on the equator, but why does SST's annual harmonic peak on the equator in the eastern Pacific? In other words, what amplifies the SST annual cycle on the equator?
11. Make a case for coupled feedback in creating a cold tongue in the eastern equatorial Pacific. Why is an equatorial warm tongue not possible? El Niño is the best Nature can manage to create, with nearly uniform SST across the equatorial Pacific.
12. Seasonal variations in the thermocline depth are much larger north than south of the equator, say 5°N vs. 5°N. Explain why. Hint: Wind curls of the ITCZ are the major forcing (cf. Fig. 8.14).

CHAPTER 9

El Niño, the Southern Oscillation, and the global influence

Contents

Summer is always warmer than winter, but each summer is different in temperature, precipitation, or both. At the Galapagos, sea surface temperature (SST) varies as much as 4°C to 5°C from one year to another. Remarkably, this variation in the ocean is in lockstep with atmospheric surface pressure at Darwin (Fig. 9.1), a northern Australian port 15,000 km away on the other side of the Pacific Ocean. The discovery and explanation of this remarkable correlation led to a great scientific revolution in the development of coupled ocean-atmospheric dynamics and climate prediction (Box 9.1).

El Niño originally referred among local fishermen to anomalous warming that occurred every few years off the Pacific coast of Peru around the time of Christmas. Now it is widely recognized that the coastal warming is not locally generated but part

Coupled Atmosphere-Ocean Dynamics: from El Niño to Climate Change
ISBN 978-0-323-95490-7, https://doi.org/10.1016/B978-0-323-95490-7.00009-6

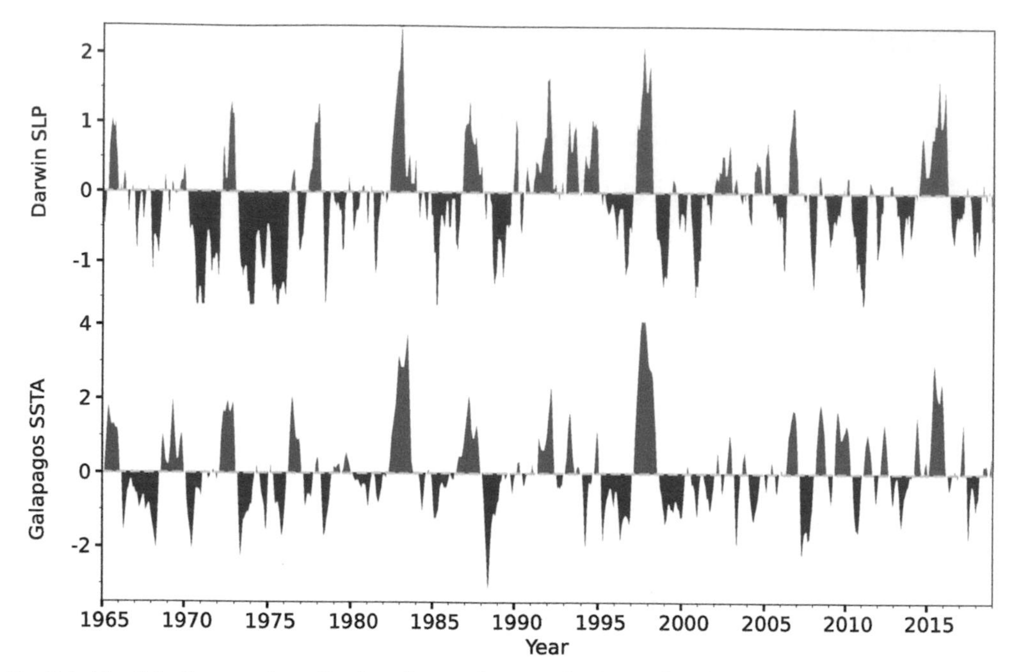

Fig. 9.1 Monthly times series of in situ observed atmospheric sea-level pressure (SLP) anomalies (hPa) at Darwin, Australia and sea surface temperature anomalies (SSTA,°C) at the Galapagos. The cross-correlation is 0.83. *(Courtesy Q. Peng.)*

of the basin-scale warming across the equatorial Pacific Ocean (Fig. 9.2A). Likewise, atmospheric pressure variability at Darwin of Australia is part of a global seesaw G. Walker called the Southern Oscillation between the Indo-western Pacific warm pool and eastern tropical Pacific (see Fig. 9.2B). The high correlation between Galapagos SST and Darwin atmospheric pressure implies that they are different aspects of a coupled ocean-atmospheric phenomenon called El Niño and the Southern Oscillation (ENSO). During El Niño, sea level pressure (SLP) drops in the eastern tropical Pacific (e.g., Tahiti) and rises over the Indo-western Pacific region (e.g., Darwin). The SST warming during an El Niño is centered on the equator and decays rapidly poleward while the SLP correlation with Darwin displays a global pattern, illustrating the global reach of El Niño's influence.

SSTs averaged in the following regions are often used to track ENSO (see Fig. 9.2A): Nino4 (160°E–150°W, 5°S–5°N, central Pacific), Nino3 (150°W–90°W, 5°S–5°N, eastern Pacific), and Nino1+2 (90°W–80°W, 0–10°S, coastal). Each El Niño is different, say in the relative magnitude of Nino3 and Nino4 SST anomalies (see Section 9.4.3). SST in the central-eastern Pacific region of Nino3.4 (170°W–120°W, 5°S–5°N) proves to track most ENSO events well.

BOX 9.1 Road to coupled dynamics

Following several poor monsoons that resulted in devastating famines, the British Raj in India appointed Gilbert Walker in 1904 as the director general of the Indian Meteorological Department with the charge to predict monsoon rainfall. Without prior meteorologic training, Walker viewed year-to-year variability in India as part of what he called "World Weather," or global teleconnection in modern jargon. Using Pearson's correlation and multidecadal observations around the globe, Walker (1933) set out to search for global teleconnections and discovered the Southern Oscillation, a dipole in which the fall of atmospheric pressure over the Indian Ocean to Maritime Continent is associated with a rise in the eastern tropical Pacific.

Bjerknes (1969) made the crucial connection that the equatorial cold tongue drives the Walker circulation, while the surface easterlies of the Walker circulation drive the equatorial upwelling that cools the eastern equatorial Pacific. He suggested that the same feedback can work to amplify El Niño. He marveled at the "remarkable teleconnection" that temperature of "the upwelling water at the coast of Peru is correlated to Djakarta pressure anomalies" halfway around Earth.

Wyrtki (1975) was puzzled by basinwide El Niño warming despite a lack of coherent wind changes over the eastern Pacific. Noting that the 15°C isotherm deepened by as much as 100 m off the coast of South America during El Niño, he suggested "upwelling still takes place but as the warm surface layer is so thick, upward movements will no longer bring cool water to the sea surface." For the mechanism bridging wind changes over the central equatorial Pacific and thermocline displacements off the Peru coast, Wyrtki cited Godfrey (1975) who suggested "reduction of easterly winds will cause a (Kelvin wave) surge of downwelling…to enhance the warming…noted by Bjerknes."

Busalacchi et al. (1983) ran a reduced-gravity ocean model with crude analysis of monthly wind stress atlas. Their hindcast matches sea level observations at the Galapagos and elsewhere in the tropical Pacific. Marine winds were poorly measured on sparse ship tracks over the vast tropical Pacific, but fortunately equatorial waves of long zonal wavelengths average out errors in wind analysis.

The discovery of the equatorial undercurrent (Cromwell et al. 1954) inspired leading dynamists like Stommel and Charney to develop competing theories. "Oceanographers were looking to explain the EUC but they found something far more interesting: El Nino," commented George Philander. They discovered that equatorial waves, first detected in the stratosphere (see Box 3.1), communicate wind variations to faraway places in the ocean, setting the pace of ENSO. The study of ENSO forged the triumphant cross-disciplinary integration of physical oceanography and dynamic meteorology. Philander et al. (1984) showed that the coupling with the atmosphere through the thermocline feedback destabilizes the oceanic Kelvin wave, introducing coupled instability/modes in the glossary of climate dynamics.

Bjerknes (1969) suggested that "dynamic oceanography" holds the key to "the turnabout" between El Niño and La Niña. McCreary and Anderson (1984) showed that in addition to the downwelling Kelvin wave that causes El Niño, the wind change excites upwelling Rossby waves that would return to cause the phase transition upon reflection on the western boundary. This idea that the ocean is not in equilibrium with the wind and the disequilibrium causes the Pacific to oscillate between El Niño and La Niña was incorporated into ENSO theory in various forms (Suarez and Schopf 1988; Battisti and Hirst 1989; Jin 1997; Wang and Picaut 2004).

(Continued)

BOX 9.1 Road to coupled dynamics—cont'd

El Niño would have remained an academic curiosity for a small group of equatorial oceanographers and tropical meteorologists if it had no effect beyond the equatorial Pacific. The global influence of ENSO, including those on the Indian monsoon (Rasmusson and Carpenter 1982) and North America (Horel and Wallace 1981), inspired the international community to embark on the Tropical Ocean and Global Atmosphere (TOGA) Project (Rasmusson 2015). The legacy of TOGA (1985-95) includes climate prediction at seasonal leads (Cane et al. 1986) and the Tropical Atmosphere Ocean array of moored buoys across the equatorial Pacific Ocean (McPhaden et al. 1998).

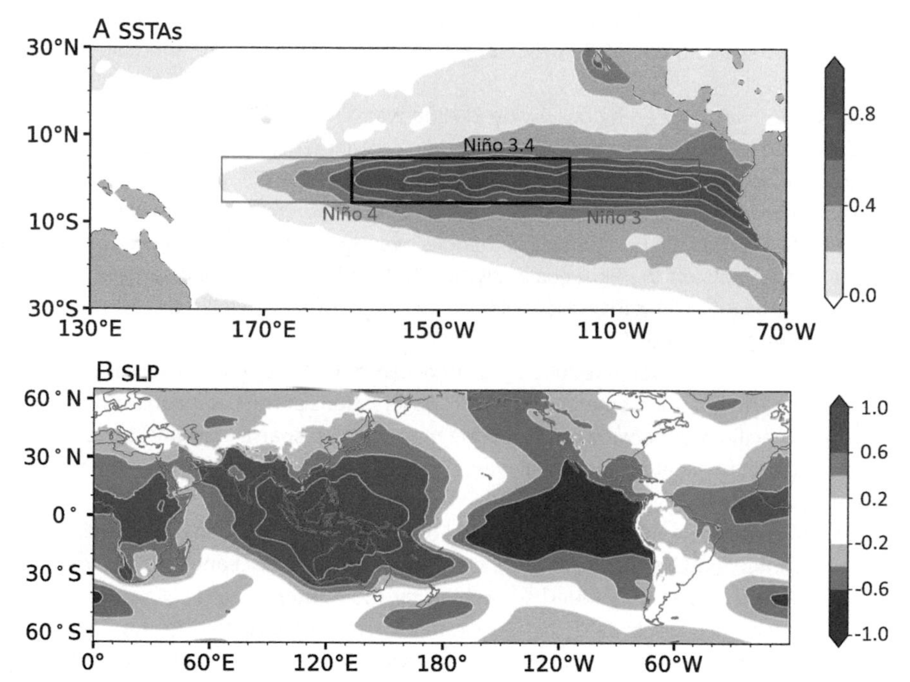

Fig. 9.2 (A) Sea surface temperature *(SST)* regression onto Galapagos SST variability with Niño regions marked. (B) Sea level pressure *(SLP)* correlation with Darwin SLP variability. *(Courtesy Q. Peng.)*

The Southern Oscillation Index (SOI) is defined as the Tahiti-Darwin difference in SLP normalized by the standard deviation at each station. The SOI is highly correlated with Nino3.4 SST. While containing high-frequency weather noise, it has the advantage as a long, uniform record going back to 1882 when SLP measurements at both stations are available.

9.1 1997-1998 El Niño

By some measures (i.e., Nino3 SST anomalies), the 1997-98 El Niño is the strongest on the instrumental record, causing abnormal climatic conditions throughout the world. It was the dramatic debut of El Niño on the public stage, making El Niño a household name and poster boy of climate variability. At the peak of this mega El Niño (~December 1997), the equatorial cold tongue disappeared in the Pacific, the trade winds near the dateline reversed to westerly, and deep convection shifted east of the dateline, leaving the maritime continent in severe drought. Because of the dry conditions, wildfires in Indonesia raged out of control, burning nearly 20 million acres. The smoke and haze reached as far away as Thailand and the Philippines. On the other side of the Pacific, major flooding took place in coastal Peru, and a great lake of 10,000 km^2 emerged, with partially submerged sand dunes reminding us that all this took place in the high desert (Fig. 9.3). It took another 2 years for this Lagoon La Niña to recede into the Peruvian desert. Elsewhere, winter storms buffeted California, while major floods threatened the levees of the Yangtze River in the following summer. The great Yangtze River floods of 1998 caused much soul searching in China, leading to the restoration of wetlands and flood lakes that had given way to economic development.

The changes in the tropical Pacific are dramatic during this mega El Niño. The equatorial cold tongue recedes with warm surface water of SST greater than 29°C spreading across the great expanse of the Pacific from Indonesia to Ecuador (Fig. 9.4). Correspondingly, deep convection fills the entire warm equatorial Pacific. Compared to the climatology, convection moves onto the equator in the eastern Pacific while shifting eastward along the equator. The zonal dipole of rainfall anomalies drives strong westerly wind anomalies between reduced convection over the Indo-western Pacific warm pool and enhanced convection east of the dateline (Fig. 9.5). In the eastern equatorial Pacific, the zonal wind anomalies are weak because the atmospheric Kelvin wave response to the dipole heating cancels each other.

Fig. 9.3 *(left)* Smoke from wildfire in Indonesia 2015. *(right)* A 90-mile-long lake that emerged in Peru's Sechura Desert following the 1997-98 El Niño. *(left, https://news.mongabay.com/2016/09/se-asian-governments-dismiss-finding-that-2015-haze-killed-100300/; right,* National Geographic *March 1999.)*

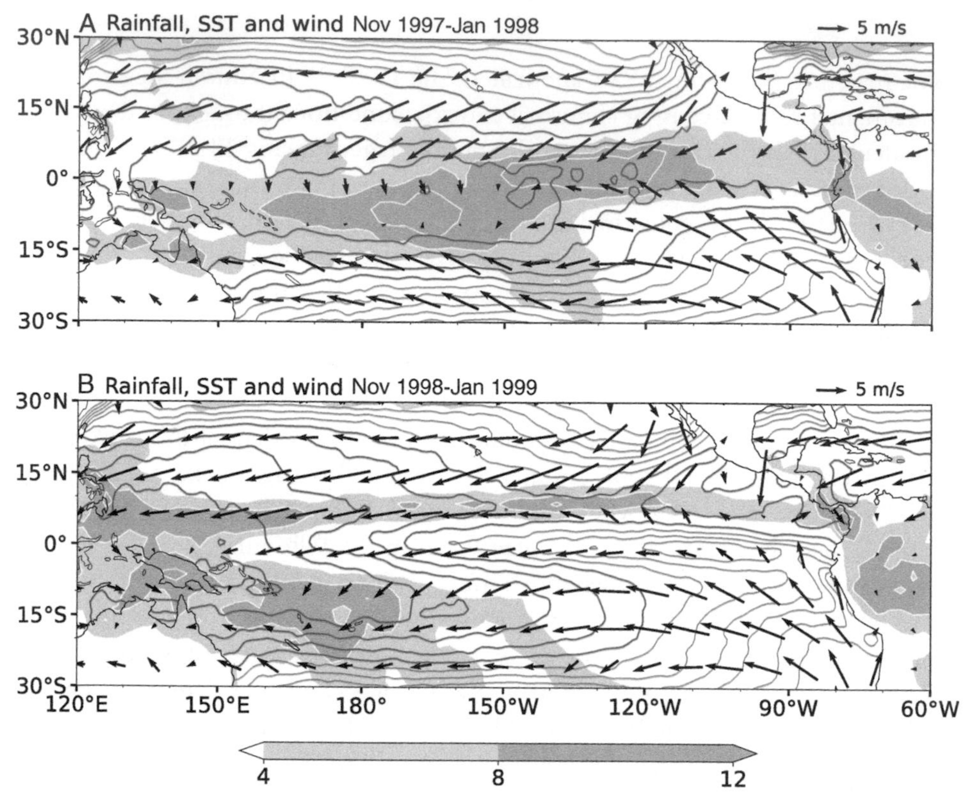

Fig. 9.4 Sea surface temperature (*SST*; ≥26°C in red contours), precipitation (grey shading and white contours at intervals of 4 mm), and surface wind (vectors, m/s) during (A) Nov 1997-Jan 1998 and (B) Nov 1998-Jan 1999. *(Courtesy Q. Peng.)*

While zonal wind anomalies are mostly confined to the western half of the basin, the SST increase of El Niño begins near the dateline and intensifies eastward (see Fig. 9.5A). The westerly wind anomalies flatten the thermocline that usually shoals eastward. Wyrtki (1975) showed that in the eastern Pacific, the deepened thermocline is the direct cause of SST increase as Ekman upwelling brings much warmer water from beneath (Fig. 9.6B) in response to the relaxation of the easterly trades near the dateline. This effect of vertical displacements of the thermocline on SST through upwelling is called the thermocline feedback.

Fig. 9.6 compares transects of ocean temperature along the equator between December 1997 and 1998 (a La Niña). In the eastern Pacific, say at 110°W, the 20°C isotherm (*Z*20) deepens from less than 50 m in December 1998 to more than 100 m in December 1997. This flattened thermocline in 1997 is manifested as a subsurface warming of more than 10°C in the east compared to 1998. This subsurface warming induces a pronounced SST increase of ~5°C, indicative of a strong

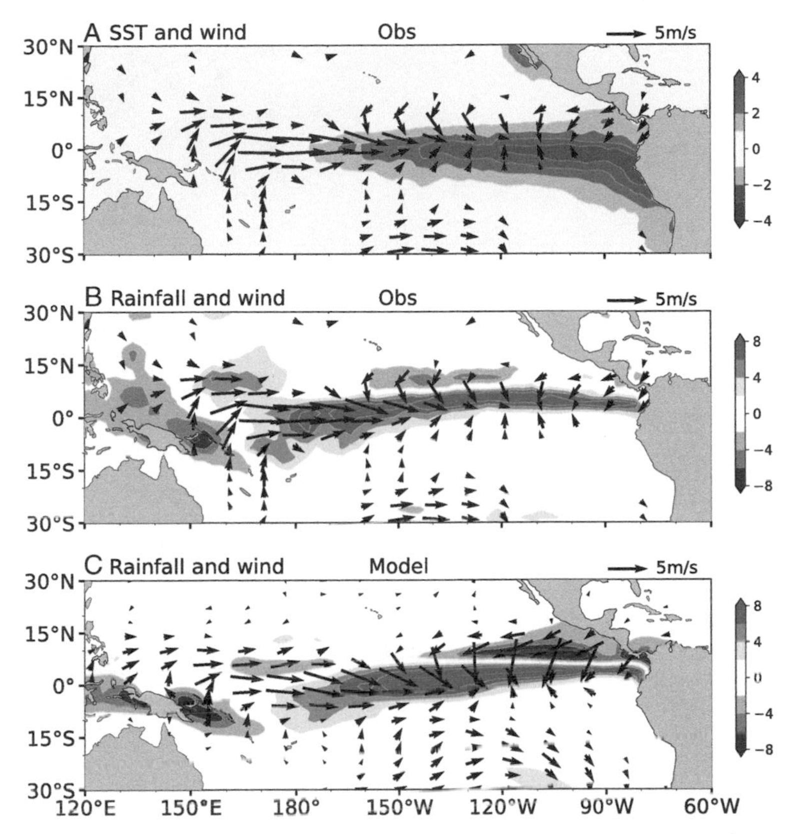

Fig. 9.5 Ocean-atmospheric anomalies in August-October (ASO) 1997: (A) sea surface temperature *(SST)* (color shading, oC) and (B) precipitation (mm/day), along with surface wind (vectors, m/s). (C) Same as (B) but simulated by the Community Atmospheric Model version 5 (CAM5) forced with observed SST. *(Courtesy Q. Peng.)*

thermocline feedback. In the western Pacific, in contrast, the mean thermocline is deep and SST is close to the thermodynamic equilibrium with the atmosphere (SST ~ 30°C). In December 1997, the thermocline shoals in the west, manifested as a subsurface cooling of ~5°C. Unlike the eastern Pacific, the shoaling thermocline is not associated with a robust SST change in the west, both because the mean thermocline is deep and the westerly wind anomalies suppress local upwelling. As a result, thermocline feedback is weak in the west and ocean thermocline variability has limited effects on SST. In the central Pacific (Nino4), the mixed layer warming floats above a subsurface cooling. The SST increase results from the warm surface advection by the anomalous westward currents and reduced upwelling, due to the reduction in local easterly trades and equatorial wave adjustments. Thus leading mechanisms for SST change vary from the western, central, to eastern Pacific.

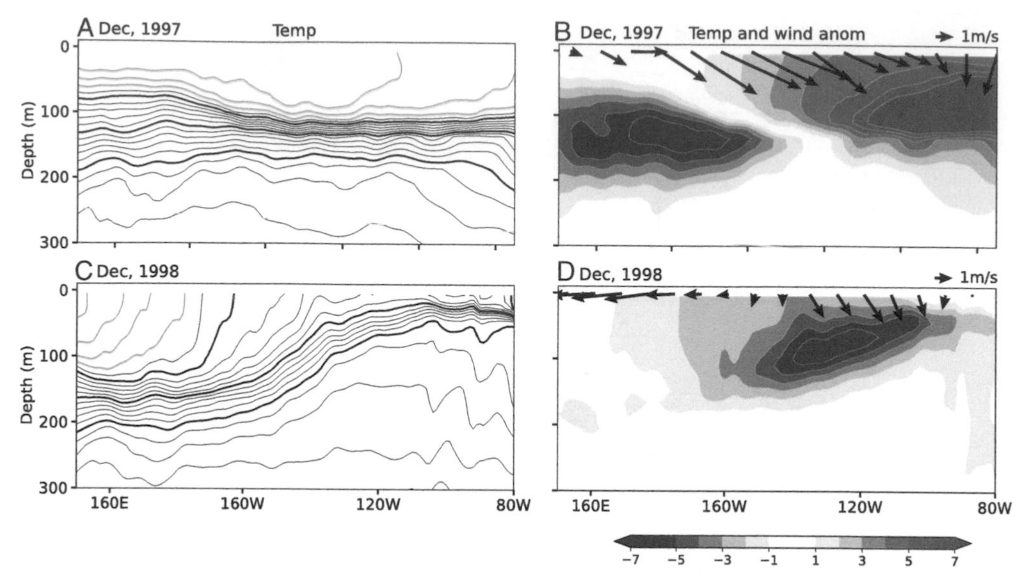

Fig. 9.6 (A) Ocean temperature (contours at intervals of 1°C, thickened every 5°C and temperature ≥27°C red), and (B) temperature anomalies (color shading, °C) along with surface wind anomalies (vectors, m/s) at the equator in December 1997. (C-D) Same as (A-B) but for December 1998. *(Courtesy Q. Peng.)*

The coupled nature of ENSO is illustrated by copropagation of ocean-atmospheric anomalies toward the east in longitude-time sections along the equator (Fig. 9.7). In early 1997, the eastern two-thirds of the equatorial Pacific began warming up, with the eastern

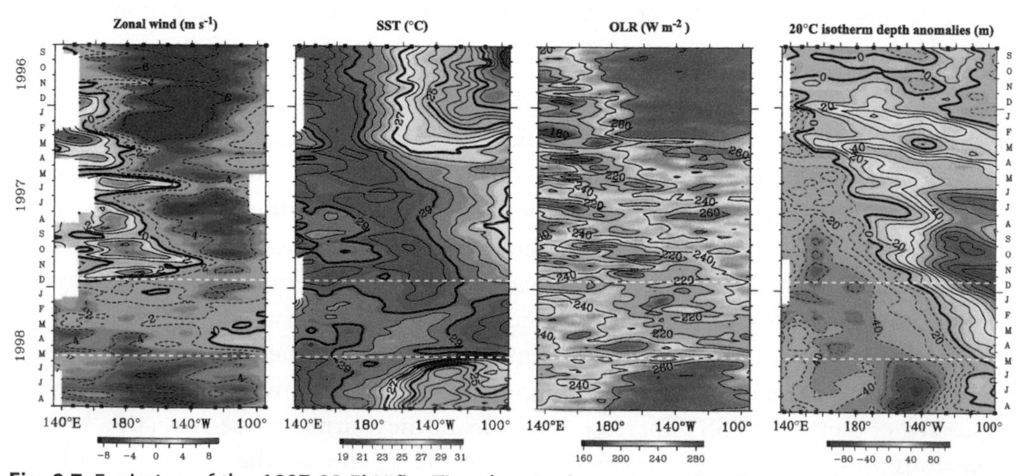

Fig. 9.7 Evolution of the 1997-98 El Niño. Time-longitude sections of surface zonal wind, sea surface temperature *(SST)*, outgoing longwave radiation *(OLR)*, and Z20 anomalies along the equator. *(From McPhaden (1999).)*

edge of the warm pool moving eastward. The eastward expansion is also found in deep convection (as represented by low outgoing longwave radiation) and westerly winds. The westerly wind anomalies suppress the upwelling and deepen the thermocline in the east by as much as 80 m. Embedded in this slow propagation of thermocline depth anomalies are intraseasonal pulses of free Kelvin waves at the beginning of 1997 that travel much faster eastward (forced by Madden-Julian oscillation [MJO] wind bursts in the western Pacific; see Fig. 4.16). The slower eastward expansion of the deepened thermocline that brings about El Niño is not due to the free ocean Kelvin wave but should be considered as a coupled ocean-atmosphere wave (see Section 9.2).

9.2 Bjerknes feedback

Intense currents with strong vertical shear make it challenging to deploy moored buoys on the equator. Slack mooring technology finally enabled long-term (> several months) deployment of equatorial buoys in the early 1980s, just on time to capture the full evolution of the mega El Niño of 1982-83. At 95°W, Eq., the 20°C isotherm dived down by 100 m in December 1982, and SST rose to 30°C just 3 months later in April 1983. What was truly astonishing to many oceanographers is that the Equatorial Undercurrent (EUC) disappeared altogether during these 5 months (Fig. 9.8A); EUC was then considered a permanent fixture of the equatorial Pacific, with a typical core speed of 1 m/s at this site. At the height of the 1982-83 El Niño, the equatorial Pacific Ocean was in a state completely different from the norm; uniformly warm surface water spreads across the basin, with a flat thermocline and without the EUC.

From an oceanographic point of view, El Niño is caused by the relaxation of the easterly trades on the equator; an ocean general circulation model (GCM) forced with the observed evolution of surface winds successfully simulates the dramatic changes observed in the eastern Pacific during 1982-83, including the deepened thermocline, extreme increase in SST, and cessation of the EUC (see Fig. 9.8, lower panels). From a meteorologic point of view, on the other hand, the Southern Oscillation is forced by the El Niño warming across the equatorial Pacific; atmospheric GCMs forced by the observed evolution of SST reproduces the SLP seesaw between the increase at Darwin and decrease at Tahiti as well as the decelerated trade winds near the dateline during an El Niño (see Fig. 9.5C).

The abovementioned circular argument implies a positive feedback from the mutual interaction of El Niño and the Southern Oscillation. Without losing generality, let's start with westerly wind perturbations, say as part of internal atmospheric variability such as the MJO. The wind anomalies deepen the thermocline in the east, causing SST to increase through thermocline feedback. The ocean surface warming from the central to eastern Pacific shifts deep convection eastward, amplifying the initial westerly wind perturbations near the dateline. Based on limited observations, Bjerknes (1969) envisioned

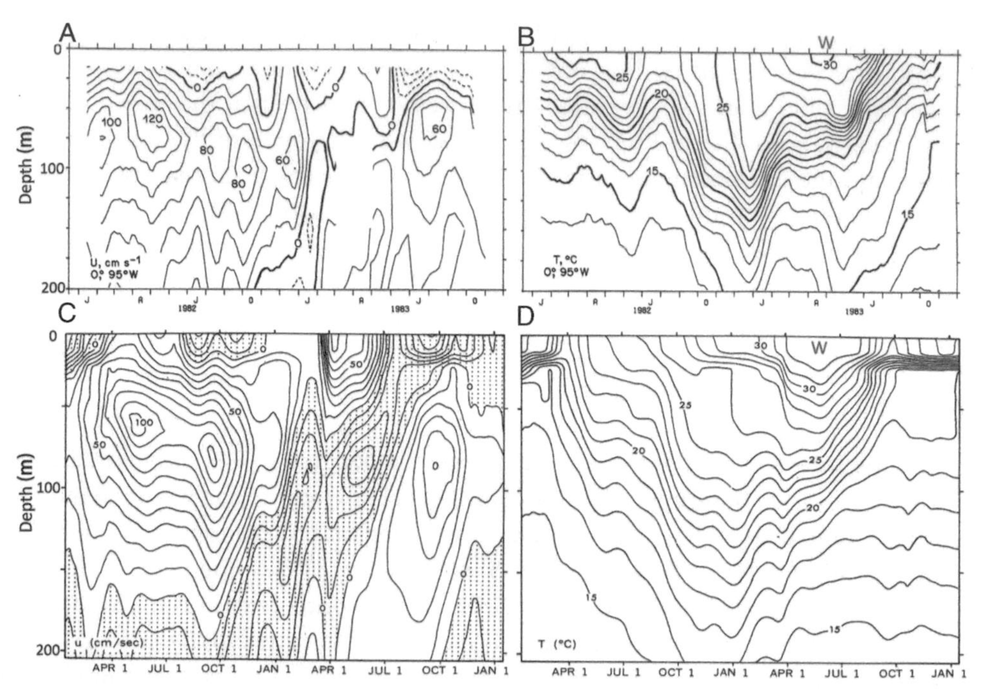

Fig. 9.8 Time-depth sections of (A) zonal current velocity (cm/s) and (B) temperature (oC) at 95°W, equator as measured by Halpern (19873). (C-D): Same as (A-B) but from an ocean GCM simulation forced by observed wind variability. *(From Philander and Seigel (1985).)*

the positive feedback loop, while Wyrtki (1975) added the crucial piece of thermocline adjustments to remote wind forcing. This feedback operates along the equatorial waveguide.

9.2.1 Effect of Earth rotation

Yamagata (1985) proposed the following thought experiments to illustrate the Coriolis effect on Bjerknes feedback. Assume that a deepened thermocline causes SST to rise, much as observed in the eastern equatorial Pacific. First consider a nonrotating frame (as on the equator) (Fig. 9.9A). Suppose that there is a small depression in the thermocline depth. It induces an SST warming, which enhances deep convection and drives winds to converge. Without the Coriolis force, the convergent winds drive surface ocean currents to converge, depressing the thermocline even more. This amplification of the initial thermocline perturbation indicates an unstable air-sea interaction without earth rotation.

On an f-plane, however, the interaction behaves very differently (see Fig. 9.9B). The initial thermocline depression still causes the SST warming and enhanced convection, but the low pressure under intensified convection now drives a cyclonic circulation at the

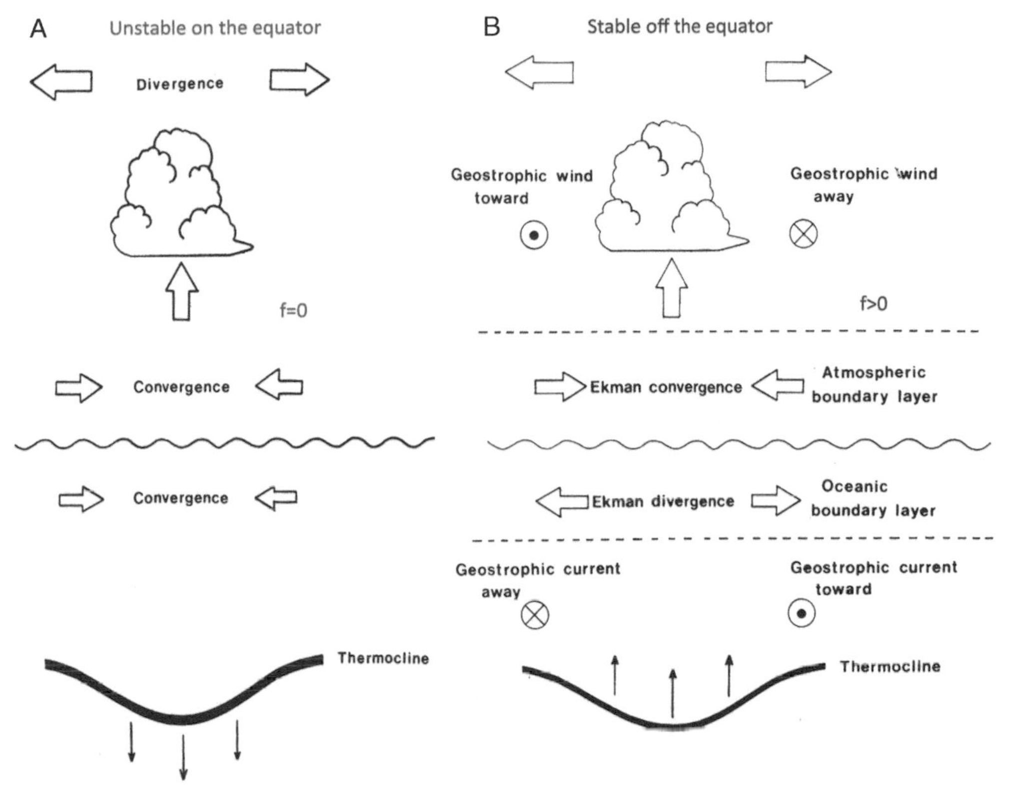

Fig. 9.9 Schematics of ocean-atmosphere interaction (A) without and (B) with the Earth rotation. *(From Yamagata (1985).)*

surface by geostrophy (the atmospheric Ekman flow in the boundary layer converges onto the low, helping supply the moisture for convection). The ocean surface Ekman current driven by the cyclonic winds, however, diverges away from the initial thermocline depression and causes it to decay. Thus the result of the air-sea interaction is to dampen the initial thermocline perturbation when Earth rotation is strong. In conclusion, the Bjerknes feedback is unstable on the equator where the Coriolis force vanishes while it is stable off the equator where geostrophy prevails. This character of the Bjerknes feedback explains why El Niño is observed on the equator.

9.2.2 Ocean heat budget and coupled instability

From Eq. 7.29, the heat budget of the ocean surface mixed layer can be cast as

$$\frac{\partial T}{\partial t} = -u_m\frac{\partial T}{\partial x} - w_m\frac{T - T_e}{H_m} + \frac{Q}{\rho c_p H_m} \tag{9.1}$$

where we have assumed mean upwelling and neglected the meridional advection term, which is small at the equator for symmetric El Niño warming. Linearizing the SST equation about a mean state denoted by the overbar yields

$$\frac{\partial T}{\partial t} = -\overline{u}_m \frac{\partial T}{\partial x} - \left(b_E + \frac{\overline{w}_m}{H_m}\right) T - u_m \frac{\partial \overline{T}}{\partial x} - w_m \frac{\overline{T} - \overline{T}_e}{H_m} + \frac{\overline{w}_m}{H_m} \frac{\partial \overline{T}_e}{\partial h} h, \tag{9.2}$$

where the prime for perturbations has been omitted for clarity, and the surface heat flux term has been parameterized as a linear damping ($-b_E T$; see Section 8.1.2). In the second term on the right-hand side (RHS), mean upwelling provides an extra damping on SST variability toward the climatologic subsurface temperature.

Perturbation currents in the mixed layer are made up of the Ekman component (with subscript E) and shallow water dynamics (without subscript). Eq. 9.2 can be cast as

$$\begin{aligned} \frac{\partial T}{\partial t} = & -\overline{u}_m \frac{\partial T}{\partial x} - \left(b_E + \frac{\overline{w}_m}{H_m}\right) T \\ & -u_E \frac{\partial \overline{T}}{\partial x} - w_E \frac{\overline{T} - \overline{T}_e}{H_m} \quad \text{Ekman dynamics} \\ & -u \frac{\partial \overline{T}}{\partial x} - w \frac{\overline{T} - \overline{T}_e}{H_m} + \frac{\overline{w}_m}{H_m} \frac{d\overline{T}_e}{dh} h. \quad \text{wave dynamics} \end{aligned} \tag{9.3}$$

In the central Pacific where perturbation thermocline displacements (h) are small (see Fig. 9.7), advection by anomalous zonal and vertical velocities is important, called the zonal advection and upwelling feedbacks, respectively. The perturbation velocities include the wave and Ekman components on the RHS. Both the Ekman advection and upwelling feedbacks (row 2) are proportional to the local zonal wind anomalies, in favor of westward phase propagation (see Section 8.3.2).

Thermocline displacements (h) cause SST to change by varying subsurface temperature (T_e) the upwelling brings into the ocean mixed layer. The thermocline feedback depends on the mean thermocline depth; $\frac{d\overline{T}_e}{dh}$ is large in the east where the mean thermocline is close to the surface.

The linearized SST equation can be solved with a reduced-gravity ocean model, with variable winds as forcing. In the Matsuno-Gill atmospheric model, on the other hand, SST variations drive wind changes by modulating diabatic heating to the atmosphere. Stability analysis of such a linear coupled model shows sensitivity to dominant terms in the SST equation (Hirst 1986). With only the thermocline feedback, the ocean-atmosphere coupling destabilizes the ocean Kelvin wave and slows the eastward phase propagation (Philander et al. 1984).

9.3 Mechanisms for oscillation

ENSO is a recurring cycle with a typical period of 2 to 7 years. The Bjerknes feedback explains its growth and coherent structures centered on the equator. It does not explain, however, why El Niño decays eventually and transitions into La Niña, the cold phase of ENSO. Several theories exist, but they all cast transient thermocline depth adjustments as the key to the cyclic nature of ENSO. Here we use the simple delayed oscillator theory to illustrate the ENSO cycle.

The westerly wind anomalies near the Dateline lower the thermocline in the eastern Pacific, causing SST warming there. The volume of the warm upper ocean is roughly conserved in the tropical Pacific basin in the initial adjustment: The warm water that flushes to the eastern equatorial Pacific is drawn from off the equator by the equatorward Ekman flow under the anomalous westerlies (Fig. 9.10A). The loss of the warm upper water shoals the thermocline off the equator, forcing upwelling Rossby waves that propagate westward. Upon reaching the western boundary (Asia and Australia), these upwelling Rossby waves reflect into the upwelling Kelvin wave, which propagates eastward on the equator to shoal the thermocline in the eastern Pacific (see Fig. 9.10B). The thermocline shoaling due to the arrival of the upwelling Kelvin wave eventually causes the El Niño warming to wane, initiating the phase transition toward La Niña. Thus the growth of El Niño seeds its own demise as the anomalous westerly winds excite upwelling

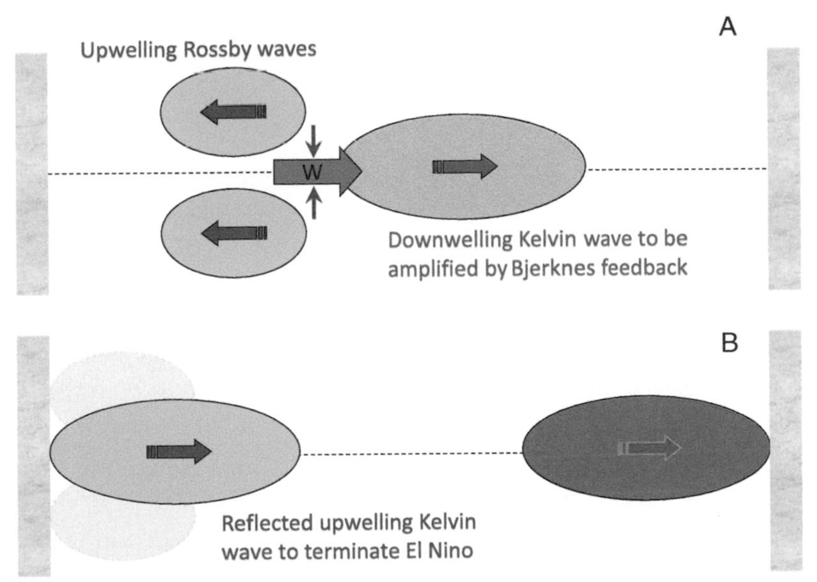

Fig. 9.10 Schematic of a delayed oscillator: (A) westerly wind anomalies excite downwelling equatorial Kelvin wave and off-equatorial upwelling Rossby waves; (B) the reflected upwelling Kelvin wave terminates El Niño, causing the Pacific to transition toward La Niña.

Rossby waves that return several months later to shoal the equatorial thermocline upon reflection on the western boundary.

The ocean response to an abrupt onset of a westerly wind patch centered on the equator (see Section 7.2.4) illustrates the delayed wave action that opposes the initial deepening of the thermocline in the east. On the eastern boundary, the arrival of the downwelling Kelvin wave and the reflection into downwelling Rossby waves deepen the thermocline (see Fig. 7.7B). Then the thermocline displacement stops temporarily until the arrival of the upwelling Kelvin wave reflected from the western boundary drives the thermocline to shoal slowly on the eastern boundary (see Fig. 7.8). McCreary and Anderson (1984) suggested that the slow recovery from the overshoot thermocline depth response causes the coupled Pacific ocean-atmosphere system to oscillate.

In reality, zonal wind in the central Pacific $\tau_x(t)$ is coupled with eastern Pacific SST and varies slowly,

$$\tau_x = \beta T \tag{9.4}$$

The wind variation forces the Kelvin wave and Rossby waves of the opposite sign. The thermocline displacement at the eastern boundary is the sum of the directly forced Kelvin wave (h_K), and a delayed response from the western boundary reflection of Rossby waves forced at an earlier time τ (h_R):

$$h = h_K + h_R = a_K \tau_x - a_R \tau_x(t - \tau) \tag{9.5}$$

where we have neglected the time delay (<100 days; see Fig. 7.8) in the directly forced Kelvin wave and reflection on the eastern boundary, and τ is the time delay for the Rossby waves of the opposite sign to travel to the western boundary and the reflected Kelvin wave to cross the ocean basin.

In the eastern equatorial Pacific, local wind variability is weak and zonal advection is small compared to thermocline feedback. The SST equation may be simplified as

$$\frac{\partial T}{\partial t} = Kh - \varepsilon T = -bT(t - \tau) + cT \tag{9.6}$$

where $c = a_K\beta - \varepsilon$ is the effective growth rate due to Bjerknes feedback, and $b = a_R\beta$ measures the delayed thermocline feedback due to the western boundary reflection.

This delayed action oscillator contains an unstable oscillatory solution, $T = \widetilde{T}e^{-i(\sigma_R + i\sigma_I)t}$, where σ_R and σ_I are the frequency and growth rate, respectively. The wave transition time (τ) is a key parameter: At the limit of $\tau = 0$, the solution is damped for the realistic regime of $b > c$ while at the limit of $\tau \rightarrow \infty$ or without the western boundary reflection, the solution is stationary (but unstable). Battisti and Hirst (1989) estimated that appropriate values for the equatorial Pacific are $b = 3.9$ year^{-1}, $\tau = 180$ days, and $c = 2.2$ year^{-1}, for which the solution is unstable with a period of 3.5 years (Fig. 9.11).

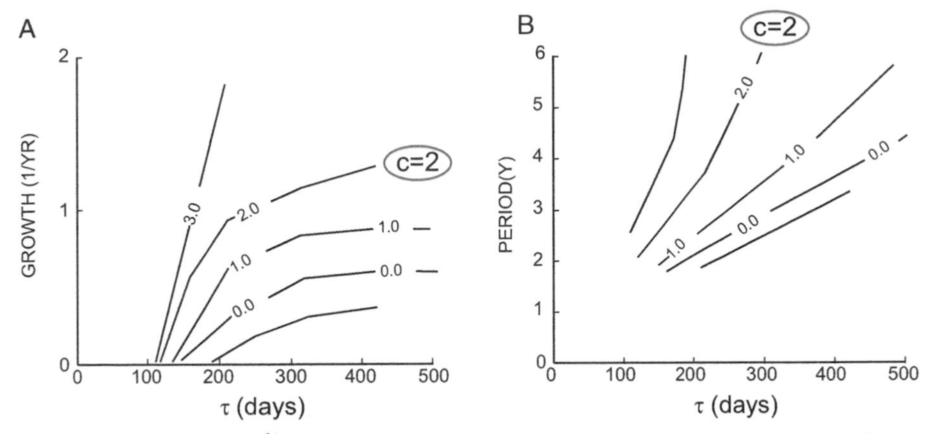

Fig. 9.11 (A) Growth rate (yr^{-1}) and (B) period (yr) of the delayed oscillator solution as a function of time lag τ. Each curve is for a different value of c (yr^{-1}). *(From Battisti and Hirst (1989). © American Meteorological Society. Used with permission.)*

Further analysis of Eq. 9.6 shows the following results (see Fig. 9.11):

- The period increases with the wave delay τ.
- The growth rate increases with τ or the zonal extent of the basin. This explains why equatorial SST variability is more pronounced in the wide basin of the Pacific than in a narrow basin of the Atlantic or Indian Ocean.

Thermocline displacements are related to sea surface height (SSH) variations; $\eta = (g'/g)h$ in a 1.5-layer reduced gravity model. Satellite altimeters capture the full evolution of the 1997-98 El Niño (Fig. 9.12). The eastern Pacific thermocline clearly deepened in October 1997 as indicated by a wedgelike sea level rise along the equator that fans out on the American coast. This wedge of increased sea level grows through December 1997, while negative sea level anomalies develop off the equator in the western Pacific as the westerly wind anomalies drain the warm water from either side of the equator to feed the growing El Niño. The upwelling Kelvin wave reflected from the western boundary becomes apparent by February, eroding the positive sea level anomalies on the equator and initiating the decay of the El Niño. The thermocline depth in the eastern equatorial Pacific returns close to normal in March 1998 (see Fig. 9.7).

9.4 Life cycle

9.4.1 Seasonal phase locking

ENSO shows a strong seasonal dependency. SST variance in the equatorial Pacific peaks in the 3 months of November-January (NDJ) (Fig. 9.13A). Indeed the evolution of a typical El Niño event is phase-locked onto the calendar season, growing in boreal summer to fall, peaking in NDJ, and decaying in the following spring (see Fig. 9.13B). Because an El Niño event typically spans two calendar years and peaks in NDJ, it is convenient to define an

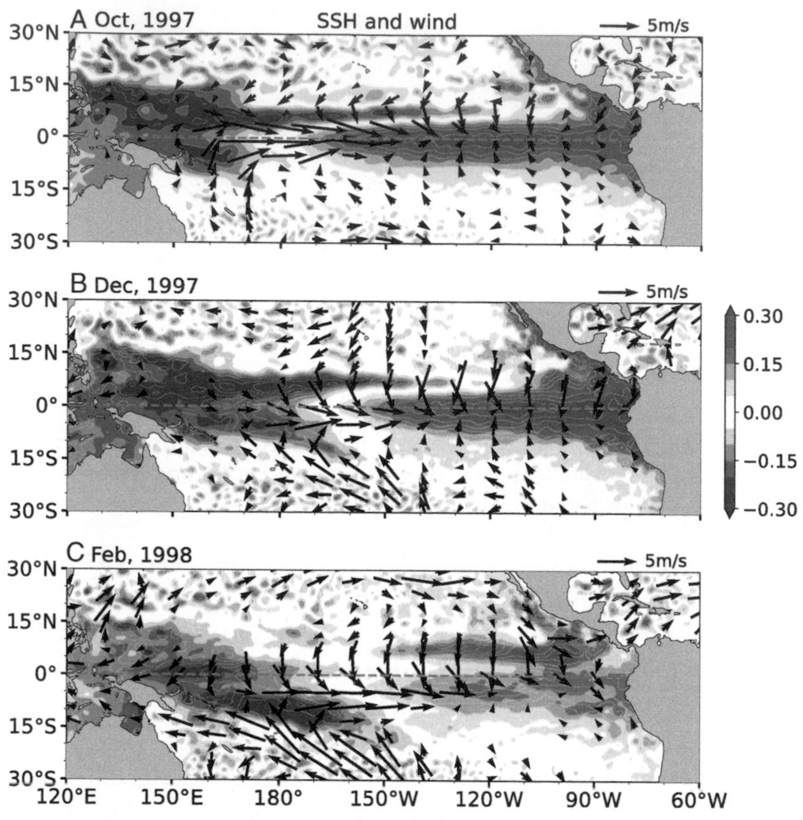

Fig. 9.12 Sea surface height (*SSH*; color shading, m) and surface wind anomalies (vectors, m/s) during the 1997-98 El Niño. (A) October, (B) November 1997, and (C) February 1998. *(Courtesy Q. Peng.)*

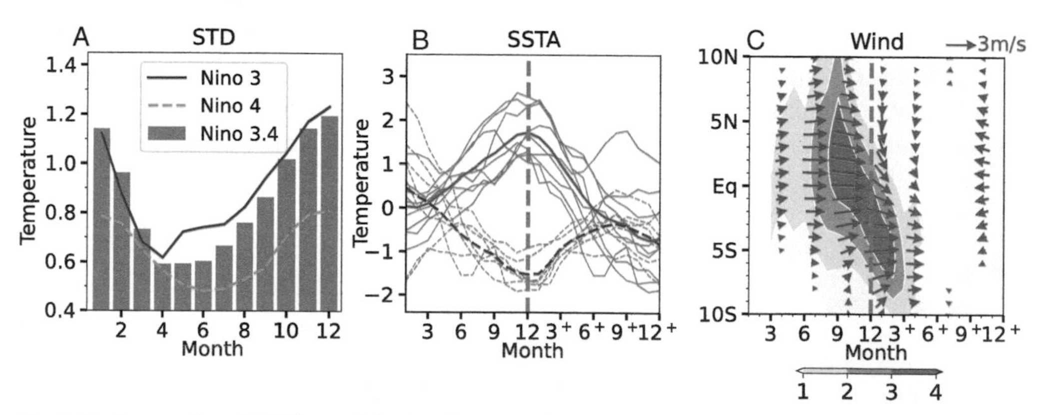

Fig. 9.13 Seasonality of El Niño and the Southern Oscillation. (A) standard deviations (°C) of Nino3.4 *(gray bar)*, Nino3 *(black line)*, and Nino4 *(black dashed)* sea surface temperature anomalies. (B) Nino3.4 (°C) evolution for NDJ Nino3.4 exceeding 0.75σ during 1979-2019; solid *(dashed)* black line indicates the El Niño (La Niña) composite. (C) Corresponding El Niño composite of wind velocity anomalies (vectors, m/s) in the central Pacific (160°E—180°), with zonal wind anomalies in grey shading. *(Courtesy Q. Peng.)*

ENSO year as the 12 months between July when an ENSO event grows [July(0)] and the following June [June(1)]. Here the numerals 0 and 1 in the parentheses denote the ENSO developing and decay years, respectively. Annual averages over an ENSO year capture ENSO effects—e.g., global mean surface temperature (GMST), which increases following an El Niño—more fully than those over a calendar year.

ENSO's phase-locking results from the interactions between Bjerknes feedback and the seasonally varying background state. The mean upwelling and SST gradients are both stronger in the cold (September) than warm (March) season, strengthening the advection and thermocline feedbacks (Eq. 9.3), while the upper thermal stratification increases in the warm season (March), strengthening the vertical advection by anomalous upwelling. Zonal wind variability in the central Pacific is the key atmospheric feedback for unstable ocean-atmosphere interactions giving rise to ENSO. This zonal wind feedback is strong in boreal summer through fall but weakens in December onwards as the zonal wind anomalies begin to displace southward away from the equator (see Figs. 9.12, 9.13C) following the seasonal migration of atmospheric convection (Harrison and Vecchi 1999). Losing the zonal wind anomalies on the equator, the anomalous thermocline displacements that cause SST variability in the eastern Pacific decay rapidly (Fig. 9.14).

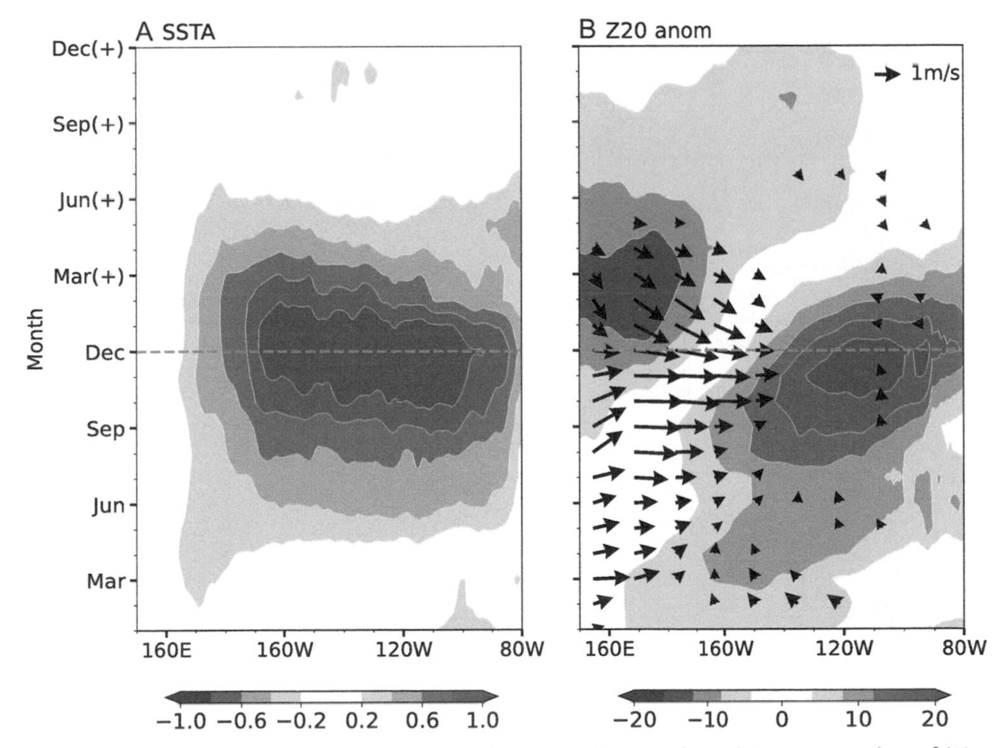

Fig. 9.14 Regressions against NDJ Nino3.4 as a function of longitude and time: anomalies of (A) sea surface temperature (color shading, °C), and (B) *Z*20 (color shading, m) and wind (vectors, m/s), averaged in 2°S–2°N. *(Courtesy Q. Peng.)*

While all these mechanisms contribute, a definitive/quantitative theory for the seasonal phase locking is yet to emerge.

9.4.2 Triggering mechanisms

The delayed oscillator theory illustrates that thermocline displacements on and off the equator are important for the growth and phase transition of ENSO. Fig. 9.15 shows the results of empirical orthogonal function (EOF) analysis on 20°C isotherm depth (*Z*20) variability in the tropical Pacific. EOF1—the tilt mode—represents the zonal tilt of the equatorial thermocline in balance with equatorial zonal wind anomalies during El Niño. Indeed, principal component 1 (PC1) is highly correlated with Nino3 and the correlation peaks with PC1 leading by 2 months. EOF2 features the overall increase/decrease in the equatorial warm water volume, representing the zonal-mean thermocline depth. The two modes explain nearly equal amounts of variance, and their PCs are highly correlated (r=0.77) with PC2 leading PC1 by 9 months (Meinen and McPhaden 2000). In the longitude-time section at the equator, the lagged correlation between the PCs is manifested as an eastward phase propagation of thermocline depth anomalies in the eastern two-thirds of the Pacific basin where EOF2's loading is large (see Fig. 9.14B).

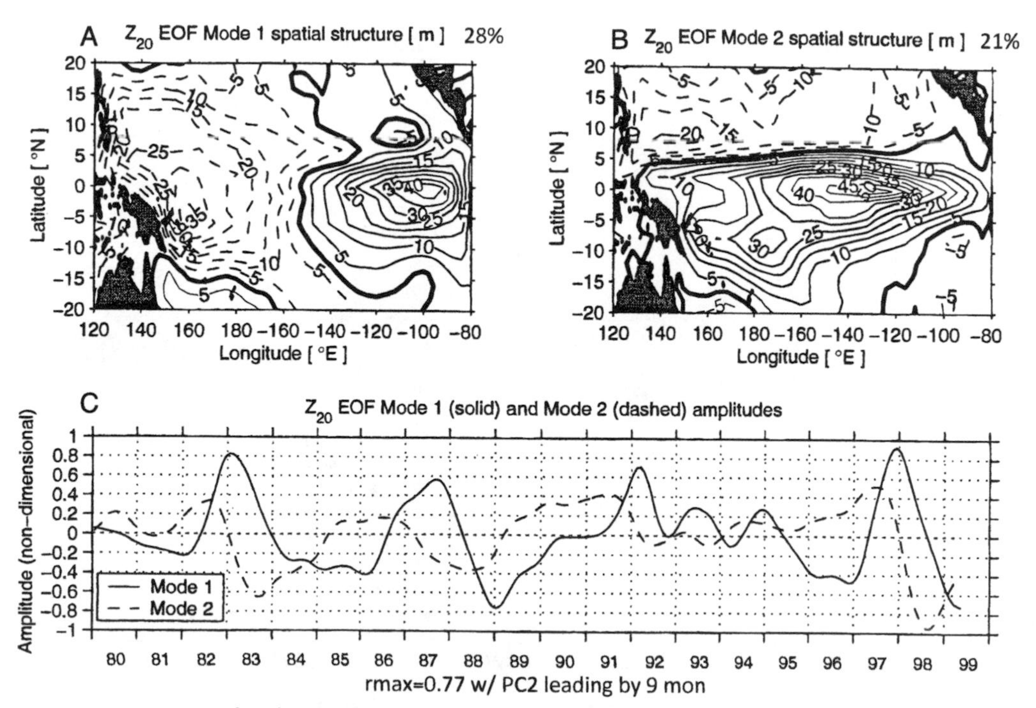

Fig. 9.15 (A) Empirical orthogonal function 1 *(EOF1)* and (B) EOF2 of *Z*20 (m) in the tropical Pacific Ocean. (C) Standardized PCs 1–2, with max cross-correlation and the lag. *(From Meinen and McPhaden (2000) © American Meteorological Society. Used with permission.)*

At the interannual timescale of ENSO ($\sim$4 years), the thermocline tilt is nearly in balance with zonal wind variability (see Eq. 7.21). This is confirmed in the ocean response to an abrupt onset of a patch of westerly wind stress (see Fig. 7.7B); the balance in the zonal tilt (the tilt mode of EOF1) is established quickly in about 2 months, while the zonal mean thermocline depth (the warm water volume mode of EOF2) slowly decreases for 2 to 3 years as a result of the slow warm water export out of the equator in geostrophic balance with the anomalous eastward deepening thermocline. Thus in the literature, the slow discharge of equatorial warm water volume is considered to cause the phase transition from El Niño to La Niña (Jin 1997). To the extent that the tilt and warm water volume modes are distinct in adjustment mechanism, the latter might be considered as a precursor of the former. Indeed, the increased sea level in the equatorial Pacific (especially in the west) in early 1997 preceded the mega El Niño that followed, while the decreased equatorial sea level in early 1998 (see Fig. 9.15C) is followed by a rapid transition into La Niña.

Not all ENSO events are preceded by a robust anomaly of the warm water volume. Stochastic forcing, chiefly of the atmosphere, is important in exciting ENSO. MJO is such a stochastic forcing. Oscillatory zonal wind variability with the zero mean rectifies into eastward mean currents on the equator, with a warm advection effect on SST. At the easterly phase of the MJO, the equatorial upwelling of a westward vertical shear (the Yoshida solution) yields a nonlinear eastward acceleration on the surface current $\left(-w'\frac{\partial u'}{\partial z}>0\right)$. Likewise, the mean surface current acceleration is still eastward at the westerly phase of MJO due to the downward advection of an eastward shear. This nonlinear rectification implies that elevated MJO variance in the zonal wind can cause an anomalous eastward current and warm advection, in favor of El Niño development. There is evidence in support of elevated MJO variance prior to El Niño (Kessler and Kleeman 2000) as in early 1997 (see Fig. 9.7). A strong westerly wind burst can push the warm pool to expand eastward and cause the eastern Pacific to warm by eastward advection and the downwelling Kelvin wave, respectively, setting the stage for subsequent high-frequency wind bursts (Lengaigne et al. 2004). This positive feedback between the slow development of El Niño and stochastic noise helps destabilize the ENSO mode.

Additional stochastic forcing that influences ENSO originates from tropical Atlantic variability (see Section 10.4.2) and the extratropical atmosphere (see Section 12.4). Despite the stochastic forcing outside the tropical Pacific, coupled feedback within the basin is still important by reducing the damping rate of the oscillatory ENSO mode. The least-damped modes are preferentially excited, explaining the recurring coherent structures. Recent studies suggest that the strong discharge of the warm water volume during a major El Niño event makes the subsequent La Niña more deterministic and more predictable than moderate ENSO events (Wu et al. 2021).

9.4.3 ENSO diversity

Earlier studies focus on common characteristics of the few ENSO events for which sufficient data are available. The composites of six El Niños (1951, 53, 57, 65, 69, and 72) by Rasmusson and Carpenter (1982) reveal important features of ENSO, including the seasonal phase locking of eastern Pacific SST variability that has withstood the test of time. Right after the Rasmusson-Carpenter composite was published, a meeting was held in Princeton, New Jersey, in October 1982 to discuss El Niño. Ironically, El Niño researchers at the meeting failed to recognize that a major El Niño was underway (Nino3.4 SST anomaly = 1.95°C) as they spoke. Unlike the Rasmusson-Carpenter composite that identifies the South American coastal warming as a precursor, the 1982 El Niño developed first in the central Pacific and propagated eastward, causing a large coastal warming only in late 1982 and early 1983. The year of 1982 opened with Nino1+2 SST = −0.2°C in January but finished with a strong coastal warming of 3.3°C in December.

The 1982 example illustrates that each El Niño is different in timing, magnitude, and spatial pattern. It is now widely recognized that ENSO events come in different flavors. The zonal distribution of SST anomalies is a popular metric for ENSO diversity (Fig. 9.16); SST anomalies intensify eastward in some El Niños (eastern Pacific [EP] type, e.g., 1982 and 1997) while peaking in the central Pacific in some others (CP type, e.g., 2006 and 2009). Thermocline feedback due to the deepened thermocline is important for EP El Niños, which tend to be preceded by an increase in the equatorial

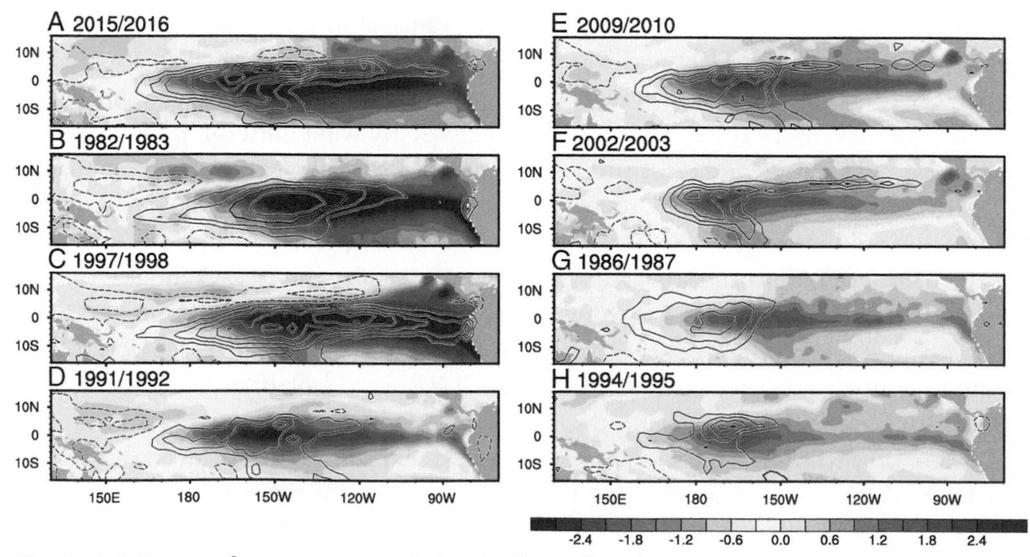

Fig. 9.16 DJF sea surface temperature (color shading, oC) and precipitation (contours at intervals of 2.5 mm/day) anomalies of El Niños ranked by Nino3.4 for 1982–2019. *(From Okumura (2019).)*

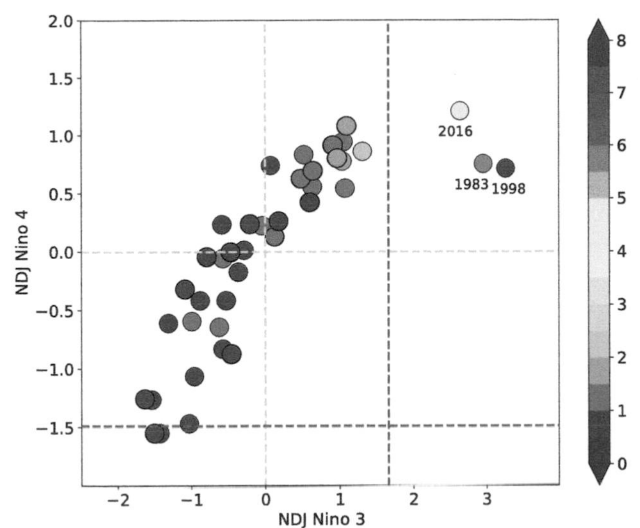

Fig. 9.17 Scatter diagram of NDJ Nino3 vs. Nino4 sea surface temperature *(SST)* anomalies (°C) for 1979−2018. The vertical *(horizontal)* dashed red line denotes Nino3 (Nino4) SST = 27°C. The color shading denotes NDJ rainfall (mm/day) over the Nino3 region. *(Courtesy Q. Peng.)*

warm water volume (see Fig. 9.15). By contrast, the westward advection is important for CP events, which tend to be preceded by the Pacific meridional mode, a key conduit for the atmospheric variability in the midlatitudes to affect the tropics (see Section 12.4).

Fig. 9.17 shows the relationship between CP/Nino4 and EP/Nino3 SST anomalies at the peak phase of ENSO (NDJ) in a scatter diagram for a 46-year period of 1982-2017. Excluding major El Niños of 1983, 1998, and 2016, a tight relationship of $T_3 = \gamma T_4$ exists. The ratio γ is larger for medium El Niños than weak El Niños and all La Niñas, in support of the CP vs. EP classification for ENSO diversity. Alternatively, one may overlook the slight variation in γ, and conclude instead that ENSO mostly oscillates between El Niño and La Niña nearly symmetrically. Occasionally El Niño grows so much that it breaks the convective threshold in the Nino3 region, expanding Bjerknes feedback into the eastern Pacific (see the next subsection). This EP Bjerknes feedback is reserved only for extreme El Niños.

9.4.4 Spring convective view

Nino3 SST anomalies start to decay in boreal spring (FMA), a season when the eastern equatorial Pacific (Nino3) climate is warmest and nearly symmetric about the equator with a double ITCZ (see Section 8.3.3). EOF analysis of FMA rainfall variability over the eastern tropical Pacific offers a convective view of ENSO diversity.

EOF1, explaining 56% of the FMA variance, captures intensified deep convection in the Nino3 region associated with extreme El Niño (Fig. 9.18, upper panels). Highly correlated with Nino3 SST, PC1 is strongly positively skewed, dominated by extreme El Niños of 1983 and 1998 and to a lesser degree by the 1987 and 1992 events. Over

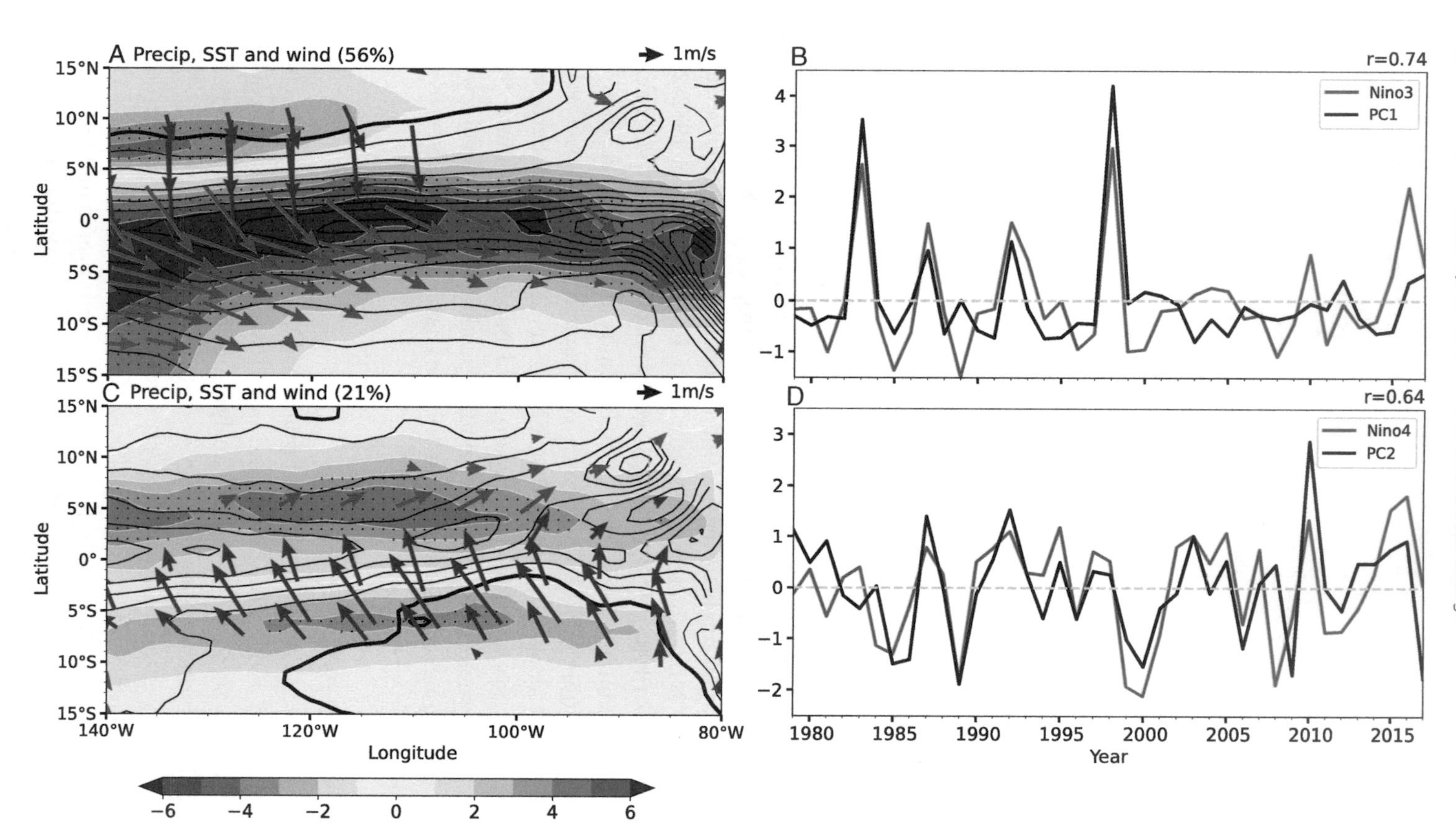

Fig. 9.18 Empirical orthogonal function (EOF) modes of FMA rainfall variability over the eastern tropical Pacific. *(Left panels)* Regressions of rainfall *(color shading)*, sea surface temperature *(SST)* (contours at intervals of 0.3oC, zero thickened and negative dashed), and surface wind (m/s; vectors) against *(right)* the standardized PCs *(blue)*. Also plotted in red are concurrent (B) eastern (Niño3) and (D) central (Niño4) equatorial Pacific SST indices (standardized), with the cross-correlations (*r*) noted at the top right. Blue (red) arrows in the left panels indicate wind anomalies that strengthen (weaken) climatological winds. *(From Xie (2018 https://doi.org/10.1175/JCLI-D-17-0905.1))*

the Nino3 region, active convection drives large westerly wind anomalies. The anomalous westerlies suppress equatorial upwelling and help sustain Nino3 warming through the FMA season, dubbed EP Bjerknes feedback to distinguish from the classic CP Bjerknes feedback where anomalous convection and zonal wind are confined to the central Pacific (Wyrtki 1975). At the mooring of 125°W, Eq., the thermocline shoals sharply from December 1997 and returns to normal by February 1998 (Fig. 9.19). Despite the shoaled thermocline, SST remains high at ~28.5°C through April because the diminished easterly trades shut off the upwelling on the equator. In early May, the easterly trades return at this site, and the recovered upwelling quickly drives SST down by 5°C within a month on the background of a slow shoaling of the thermocline. This example shows that the shoaling thermocline is a necessary condition for the return of the equatorial cold tongue, but the local EP Bjerknes feedback between SST, deep convection, and zonal wind in the Nino3 region determines the timing of the transition from extreme El Niño to La Niña.

EOF2, explaining 21% of the rainfall variance, features a convective dipole across the equator. This represents year-to-year variability in the relative intensity of the northern

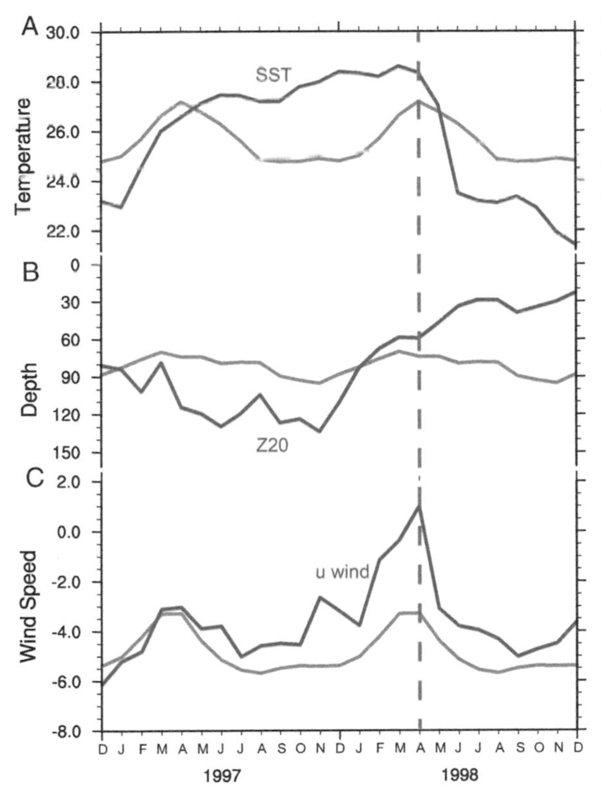

Fig. 9.19 Evolution of the 1997-98 El Niño at 125°W, the equator. (A) Sea surface temperature (*SST*; °C), (B) the 20°C isotherm depth (m), and (C) zonal wind velocity (m/s). The climatology is plotted in thin grey as a reference. Analyses are based on TAO data averaged in 2°N–2°S. *(Courtesy Q. Peng.)*

and southern ITCZs. The associated wind and SST anomalies form WES feedback that amplifies this antisymmetric mode (see Fig. 9.18, lower panels). PC2 is significantly correlated with Nino4 SST and nearly symmetric between positive and negative phases. It represents moderate ENSO that fits the classic view of Wyrtki (1975) with little zonal wind change in the eastern equatorial Pacific. In FMA, the symmetric mean state allows WES feedback to grow the meridional dipole mode. The convective dipole drives anomalous southerly cross-equatorial winds, which dissipates El Niño warming by intensifying the south equatorial upwelling.

While Nino3 SST anomalies peak in DJF, the total Nino3.4 SST is highest in FMA (Fig. 9.20), the season of the greatest possibility of local deep convection. Deep convective state sets El Niño on distinct trajectories: extreme El Niño decays slowly by activating Nino3 convection and the EP Bjerknes feedback, while moderate ENSO decays rapidly by triggering WES feedback that modulates cross-equatorial wind instead. The contrast between extreme El Niño and moderate ENSO in FMA highlights the importance of deep convective adjustment on the equator for ENSO evolution. This convective view supports classifying ENSO into moderate type and extreme El Niño, instead of the classification based on the zonal SST pattern. From May onward, the seasonal cooling precludes atmospheric convection over the equatorial EP, and the extreme El Niño no longer enjoys the EP Bjerknes feedback and eventually decays.

ENSO is asymmetric between the warm and cold phases; the events of extreme amplitudes are limited to El Niño, without extreme La Niña as the counterpart. The above-mentioned results suggest that the asymmetry in atmospheric convection might be the key: Only El Niño has a chance to activate Nino3 convection and boost the growth.

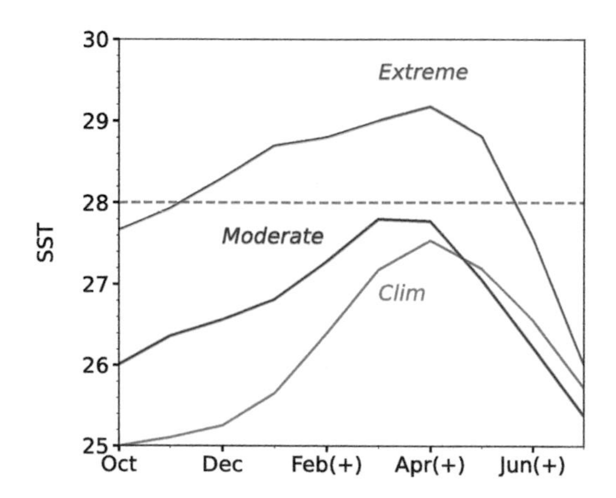

Fig. 9.20 Observed evolution of Nino3 SST (°C) for extreme *(red)* and moderate *(magenta)* El Niños, along with the climatology *(grey)*. *(Based on Peng et al. 2020; courtesy Q. Peng.)*

9.5 Global influences

9.5.1 Tropics

During El Niño, the tropical Pacific warming increases tropospheric temperature. Equatorial waves spread the tropospheric warming throughout the tropical belt, increasing static instability and weakening convection/precipitation in the rest of the tropics, especially over land. Specifically, deep convection moves away from the maritime continent eastward into the central Pacific, and rainfall decreases over the tropical Americas (except the Pacific coast west of the Andes; see Figs. 9.5, 10.11). This general decrease in land rainfall in the tropics during El Niño has important effects on land ecology.

The reorganization of deep convection over the tropical Pacific is associated with changes in the atmospheric circulation, causing, among others, major shifts of tropical cyclogenesis and tracks over the Northwest Pacific and a reduction of hurricane activity over the North Atlantic (see Section 10.5.3). The tropical Indian and North Atlantic oceans warm following El Niño, and such SST changes play important roles in shaping regional climate responses, as discussed further in Chapters 10 and 11.

9.5.2 Pacific North American pattern

In winter, El Niño excites an arch of wave train from the tropical Pacific toward North America: high above Hawaii, low near the Aleutian Islands, high over northwestern Canada, and low centered over Florida in upper tropospheric geopotential height (Fig. 9.21). This wave train is called the Pacific North American (PNA) pattern. In the

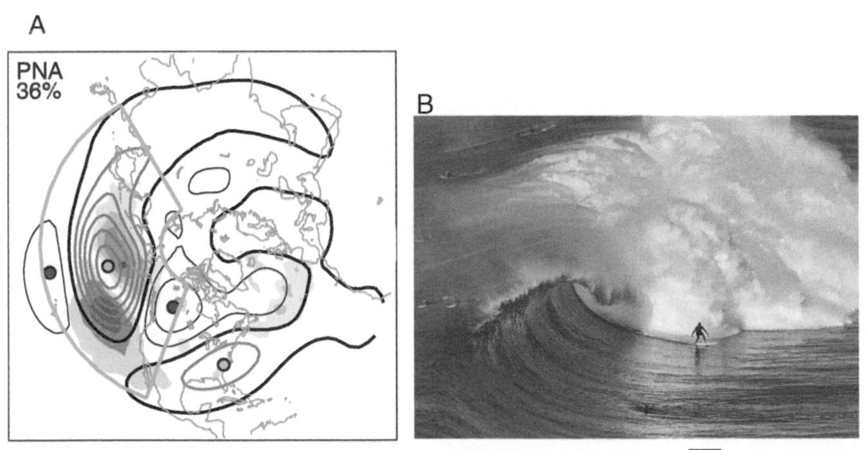

Fig. 9.21 *(left)* Regressions of monthly averaged winter (DJFM) storminess ($\overline{v'^2}$, color shading) and z (contours) at 300 hPa onto standardized Pacific-North American (PNA) index based on sea-level pressure principal component within the Pacific section indicated by a green outline. Prominent maxima and minima in the z_{300} regressions are noted with magenta and cyan markers, respectively. Contour intervals for the $\overline{v'^2}$ and z_{300} regressions are 10 m^2 s^{-2} and 20 m per standard deviation, respectively. Positive (negative) z_{300} regression contours are black *(gray)* and the zero contour is bold. *(right)* Surfs in December 2015 off San Diego, California. *(Left, From Wettstein and Wallace (2010; https://doi.org/10.1175/2009JAS3194.1). © American Meteorological Society. Used with permission.)*

tropics, the baroclinic structure dominates because of the interaction with deep convective heating that peaks at the midtroposphere. Specifically, the Hawaii high is part of a symmetric dumbbell pattern with a maximum on either side of the equator in the central Pacific (see Fig. 10.12), broadly consistent with the Matsuno-Gill model. In the extratropics, the stationary anomalous circulation is barotropic, of the same sign from the surface through the entire troposphere.

Between the anomalous Hawaii high and Aleutian low of the PNA, the anomalous westerlies extend the subtropical jet core, which is normally confined over the western North Pacific, all the way toward California. The intensified subtropical jet steers storms toward California (see Fig. 9.21A), increasing winter precipitation. In California, much of winter precipitation is frozen in snowpacks on the Sierra Mountains, which supply rivers through summer (see Section 6.4.2). Sierra Mountain snowpack tends to grow larger in El Niño winter. The ENSO influence on US Southwest hydroclimate is strong enough to allow the use of tree rings to reconstruct ENSO variability (Li et al. 2013). Remarkably, the ENSO reconstructions are highly correlated with those based on fossil corals on Palmyra Island in the central equatorial Pacific (Cobb et al. 2013), where ENSO-induced rainfall variability is large.

Winter storms generate big swells, and breaking waves erode sand beaches, building up offshore sand bars. Sand beaches recover in summer when small surface waves return sand from offshore sand bars. Californian sand beaches suffer severe erosions in El Niño winter, battered by swelling seas with intensified storm activity offshore (see Fig. 9.21B). During the 2015-16 El Niño winter, the sand surface of La Jolla beach by Scripps Institution of Oceanography dropped by 2 m (Young et al. 2018), exposing rocks that are usually hidden beneath smooth sand.

The south/southeasterlies on the northeast flank of the intensified Aleutian low advect warm air from lower latitudes, causing welcome warmth from Washington State to Alaska. The high pressure centered at Hudson Bay advects warm air from the south, with warm conditions over the northern central United States. The anomalous onshore winds associated with the same high pressure also warm the northeastern United States. The PNA's tail forms a low-pressure band from the Gulf of Mexico through Florida into the North Atlantic. This belt of anomalous low pressure is associated with increased rainfall in the Gulf States. To the extent ENSO is predictable, the PNA pattern is a source of extended range prediction over North America.

9.5.3 Ocean waveguide

ENSO teleconnections also exist in the ocean through the equatorial and coastal waveguide. The thermocline deepens in the eastern equatorial Pacific during El Niño. The increased sea level propagates poleward on the west coast of the Americas (Fig. 9.22A; see also Fig. 9.12). In the case of the 1997-98 El Niño, sea level rises by 30 to 40 cm

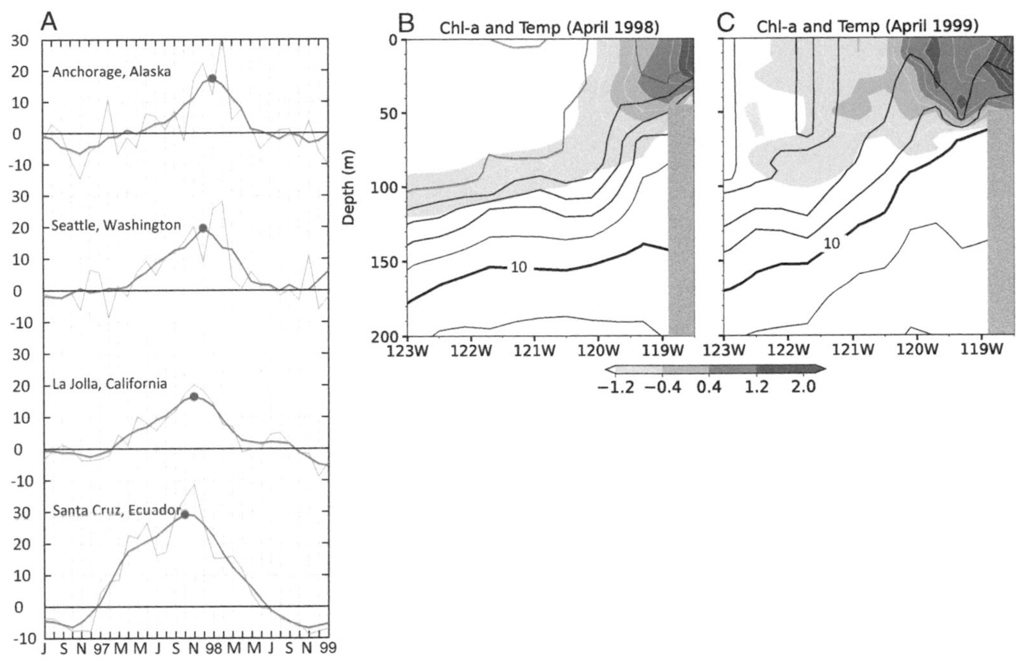

Fig. 9.22 (A) Sea level anomalies (cm) at the Galapagos and three stations on the west coast of North America. Cross sections of temperature (line contours; 15°C in red) and log chlorophyll-a concentration (mg/L; shaded) off Los Angeles following (B) El Niño (April 1998) and (C) La Niña (April 1999), based on CalCOFI. During La Niña, the thermocline shoals, surface water cools, and the intensified upwelling increases biological activity. *(A, Courtesy D.J. Amaya; B,C, courtesy Q. Peng.)*

at the Galapagos. The peak sea level is observed at the Galapagos in November 1997 and arrives progressively later at La Jolla (California), Seattle (Washington), and finally Anchorage (Alaska) in February 1998 (see Fig. 9.22A). The coastal Kelvin wave deepens the thermocline in this waveguide. Off the coast of California, the 10°C isotherm lowers by 80 m in April 1998 compared to a year later in April 1998, reducing the nutrient supplies from the coastal upwelling with decreased chlorophyll concentrations in the mixed layer (see Fig. 9.22B-C). The large thermocline displacements dramatically change the marine ecosystem, including the distribution of fish species (Rykaczewski and Checkley 2008). A general poleward shift of fish species is observed during El Niño.

9.6 Barotropic stationary waves in the westerlies

The PNA wave train follows a great circle of the spherical Earth (see Fig. 9.21A). The pronounced meridional energy propagation (e.g., from the Hawaiian high to Aleutian low) contrasts with the zonal propagation of baroclinic (Kelvin and Rossby) waves in the equatorial waveguide. This section first presents a linear theory for stationary Rossby waves in a uniform westerly flow (see Section 11.5.1, Holton 2004) and then discusses the effects of the meridional shear and zonal variations in the background flow. Readers not interested in detailed wave dynamics may skip to Section 9.6.3.

9.6.1 Energy dispersion

Consider nondivergent barotropic Rossby waves in a zonal flow $\overline{u}$. The vorticity equation on the midlatitude beta-plane is

$$\left(\frac{\partial}{\partial t}+\overline{u}\frac{\partial}{\partial x}\right)\nabla^2\psi'+\widehat{\beta}\frac{\partial \psi'}{\partial x}=0, \tag{9.7}$$

where $\psi = \Phi/f_0$ is the geostrophic stream function with $\boldsymbol{u}_\psi \equiv \left(-\frac{\partial \psi}{\partial y}, \frac{\partial \psi}{\partial x}\right)$, and

$$\widehat{\beta}=\frac{\partial \overline{\eta}}{\partial y}=\beta-\frac{\partial^2 \overline{u}}{\partial y^2} \tag{9.8}$$

is the meridional gradient of the mean absolute vorticity $\overline{\eta}$. The effective beta $\widehat{\beta}$ is locally enhanced in a westerly jet with $\overline{u}_{yy} < 0$. Seeking wave solutions $\psi' = Ae^{i(kx+ly+\omega t)}$ yields the dispersion relation of Rossby wave propagation,

$$\omega=\overline{u}k-\widehat{\beta}k/K^2 \tag{9.9}$$

where ω is the frequency, $K^2 = k^2 + l^2$, (k, l) are the zonal and meridional wavenumbers, respectively.

Monthly anomalies correspond to stationary waves. By setting $\omega = 0$, we obtain

$$K^2=K_s^2\equiv\widehat{\beta}/\overline{u} \tag{9.10}$$

where K_s is called the stationary wavenumber. This means that the stationary wave response is possible only in a westerly mean flow to counter the westward phase propagation of Rossby waves due to the mean vorticity gradient (second term on RHS of Eq. 9.9). For typical mean westerly wind speed of 20 m/s, only barotropic waves (of the same sign between the lower and upper troposphere) can be arrested to become stationary. Baroclinic waves have too slow westward phase speeds to balance the strong eastward advection by the westerly jet. Indeed, disturbances of baroclinic structures (e.g., extratropical storms) are blown eastward.

The wave energy propagates at the group velocity,

$$c_{gx}\equiv\frac{\partial \omega}{\partial k}=\overline{u}+\beta\frac{\left(k^2-l^2\right)}{\left(k^2+l^2\right)^2}$$

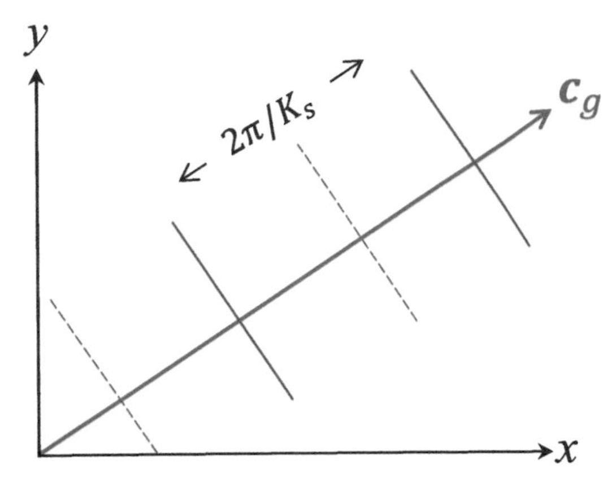

Fig. 9.23 Stationary plane Rossby wave in a westerly flow. The group velocity is perpendicular to the constant phase lines of ridges *(solid red lines)* and troughs *(dashed lines)*, with poleward energy propagation. *(Adapted from Holton 2004.)*

$$c_{gy} \equiv \frac{\partial \omega}{\partial l} = \frac{2\beta kl}{\left(k^2 + l^2\right)^2}$$

For stationary waves, we have the group velocity vector,

$$\mathbf{c}_g = \frac{2\overline{u}k}{\left(k^2 + l^2\right)}(k, l) \tag{9.11}$$

Thus the energy propagates eastward and either poleward or equatorward, depending on the sign of l, from the source region in the subtropics. Specifically, the poleward energy propagation ($l > 0$) (Fig. 9.23) allows tropical convective variability, by extending the divergent flow into the mean westerlies, to excite barotropic Rossby waves with poleward group velocity to affect the extratropics (Fig. 9.24). It can be shown that on the sphere, the group velocity of energy propagation follows a great circle (Hoskins and Karoly 1981).

The group velocity vector for stationary Rossby waves (Eq. 9.11) is in the direction of the wavenumber vector or perpendicular to the wave crests/troughs (constant phase lines). This is indeed the case for the PNA pattern. The arc connecting the four centers of action (see Fig. 9.21) represents the wave energy propagation ($\mathbf{c}_g$), while peak geopotential anomalies tend to be oriented normally against the great circle of energy propagation.

9.6.2 Rossby wave source

(Nat Johnson and Xichen Li contributed to this section.)

The dispersion of Rossby waves is the primary mechanism responsible for tropical forcing of extratropical teleconnection patterns. Consider a stationary convective heating

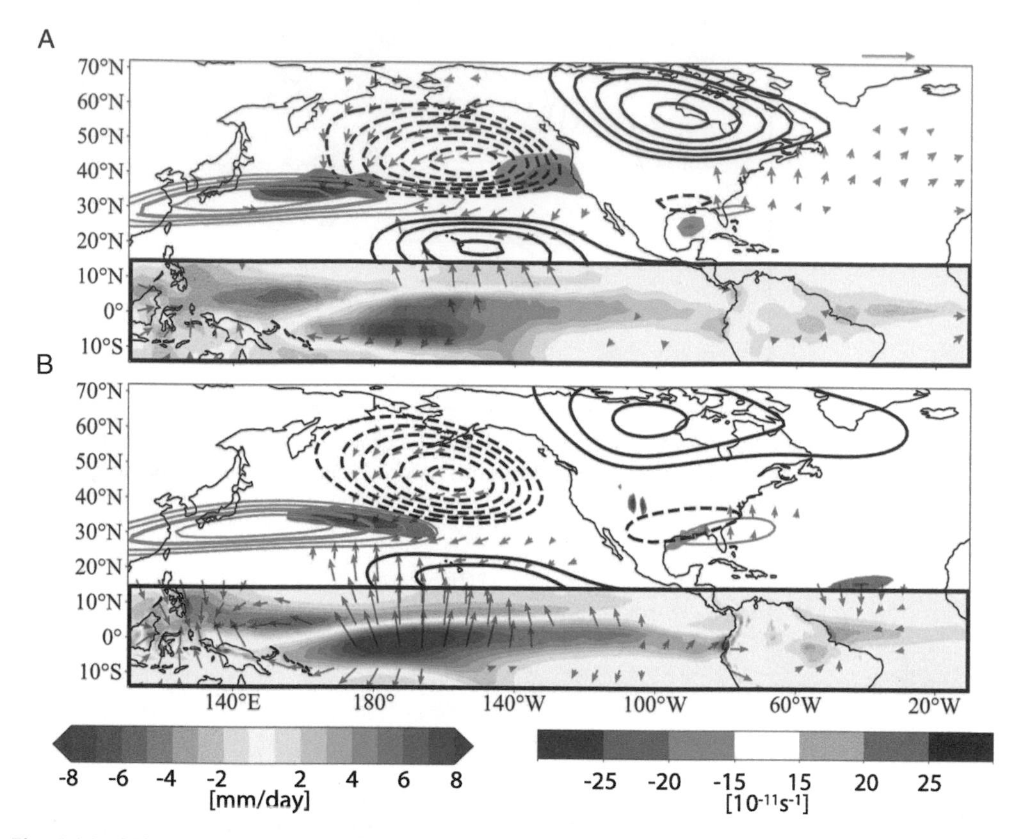

Fig. 9.24 JFM composite anomalies at 200 hPa of El Niño (1979–2010): (A) observations based on NCEP-NCAR Reanalysis and (B) 100-member atmospheric GCM simulation. Geopotential height anomaly (black line contours at ∓[45, 60, 75, 90, 105, 120] m), anomalous Rossby wave source (color shading, color bar at the bottom right), and anomalous divergent wind vectors (scale at the top right: 5 m/s), along with mean zonal wind ([50, 55, 60, 65] m/s green contours, 60 m/s thickened). Inset in 15°S–15°N shows anomalous tropical precipitation (color bar at the bottom left). *(Based on Chapman et al. 2021; courtesy W. Chapman.)*

in the deep tropics that accompanies positive SST anomalies during an El Niño episode. This convective heating induces deep vertical motions, and the resulting upper-level divergence (see Fig. 9.24) provides a substantial forcing on the absolute vorticity tendency equation through vortex stretching.

Generally, the barotropic vorticity equation that allows horizontal divergence is

$$\frac{\partial \eta}{\partial t} + \boldsymbol{u} \cdot \nabla \eta = -\eta \nabla \cdot \boldsymbol{u} \tag{9.12}$$

where $\eta = f + \zeta$ is the absolute vorticity. The horizontal wind velocity field can be decomposed into

$$\boldsymbol{u} = \boldsymbol{u}_{\psi} + \boldsymbol{u}_{\chi}$$

where $\boldsymbol{u}_{\psi}$ is the rotational/nondivergent component of the flow, and $\boldsymbol{u}_{\chi}$ the irrotational/divergent component (see Section 2.5.1). Consider perturbations around a mean state denoted by the overbar. Collecting the advective terms by the rotational/geostrophic flow on the left-hand side (LHS) (barotropic Rossby wave dynamics) yields

$$\frac{\partial \zeta'}{\partial t} + \overline{\boldsymbol{u}}_{\psi} \cdot \nabla \zeta' + \boldsymbol{u}'_{\psi} \cdot \nabla \overline{\eta} = S' \tag{9.13}$$

The RHS,

$$S' = -\nabla \cdot \left(\boldsymbol{u}'_{\chi} \overline{\eta}\right) - \nabla \cdot \left(\overline{\boldsymbol{u}}_{\chi} \zeta'\right) \tag{9.14}$$

is the Rossby wave source (RWS; Sardeshmukh and Hoskins 1988) due to perturbations in absolute vorticity flux by the divergent flow. In the upper troposphere, the Rossby waves are generated from the subtropical region: The downwelling branch of the Hadley cell provides the mean divergent flow, while the strong shear around the subtropical jet induces a strong mean vorticity. We can recover the LHS of Eq. 9.7 by noting $\zeta' = \nabla^2 \psi'$.

Fig. 9.24 shows the RWS and PNA pattern during El Niño winter at 200 hPa. Atmospheric convection over the equatorial Pacific forces the baroclinic Gill pattern (in the equatorial waveguide). The upper tropospheric high of the Gill pattern has a broad meridional structure that is close to the westerly jet waveguide enough to excite a stationary Rossby wave pattern that propagates poleward and eastward and arcs back toward low latitudes. A strong positive RWS is found on the poleward flank of the Asian jet due to the enhanced mean vorticity $\overline{\eta}$ and convergence of anomalous flow $\boldsymbol{u}'_{\chi}$ from enhanced convection over the central equatorial Pacific. The vortex stretching term due to the divergence of anomalous flow, $-\overline{\eta}\nabla \cdot \boldsymbol{v}'_{\chi}$, dominates the Rossby wave source in Eq. 9.14. The positive RWS excites an anomalous cyclone downstream over the Aleutians and the rest of the PNA wave train.

9.6.3 Geographic anchor

The abovementioned theory assumes a zonally uniform mean flow, where the wave solution moves with the longitude of the tropical forcing. The observed PNA pattern is geographically stationary, rather insensitive to slight differences in tropical heating, say between El Niño and La Niña events. Zonal variations in the mean wind are important to anchor these recurrent patterns geographically. The RWS has a geographic preference near the jet core regions (see Fig. 9.24) where vorticity gradients are strongest, rendering the extratropical response less sensitive to the precise location of the tropical convection. In addition, disturbances can gain energy in the jet exit regions through barotropic energy conversions (CK) from mean to eddy kinetic energy:

$$CK \equiv \left(\frac{\overline{v'^2} - \overline{u'^2}}{2}\right)\left(\frac{\partial \overline{u}}{\partial x} - \frac{\partial \overline{v}}{\partial y}\right) - \overline{u'v'}\left(\frac{\partial \overline{u}}{\partial y} + \frac{\partial \overline{v}}{\partial x}\right) \approx -\overline{u'^2}\frac{\partial \overline{u}}{\partial x} \tag{9.15}$$

if we consider long stationary waves ($\overline{u'^2} \gg \overline{v'^2}$) trapped at the jet axis $\left(\frac{\partial \overline{u}}{\partial y} = 0\right)$. Here we assume that the mean flow is nondivergent, $\frac{\partial \overline{u}}{\partial x} + \frac{\partial \overline{v}}{\partial y} = 0$. In the Pacific jet exit region ($\frac{\partial \overline{u}}{\partial x} < 0$; 30°N east of the Dateline), zonal wind anomalies of the PNA peak between the anomalous Aleutian low and Hawaiian high (see Fig. 9.24). This pattern is optimal for the efficient barotropic conversion and is favored for growth (Simmons et al. 1983). The anomalous circulation during El Niño is equivalent to extending the Pacific jet eastward.

The geographically stationary nature of the PNA pattern is best illustrated by revisiting the MJO. In the tropics, the convective heating and wind anomalies show coherent eastward phase propagation (see Chapter 4). From the tropical forcing, one may expect a propagating response in the extratropics. Instead, the geopotential response in the extratropical Northern Hemisphere is strongly locked geographically (Fig. 9.25), peaking over the Aleutians (170°E, 50°N) and northwestern Canada (130°W, 60°N), which are the PNA's centers of action. When the MJO convection is over the maritime continent, the PNA wave train is well developed because the anomalous zonal wind is over in the North Pacific jet exit region to tap into the barotropic conversion (see Fig. 4.13).

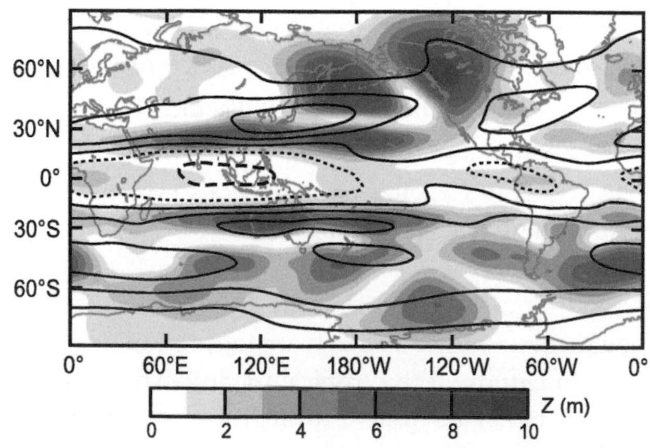

Fig. 9.25 Root-mean-squared (rms) amplitude of geopotential height variability over the MJO cycle superimposed upon the annual-mean zonal flow (black contours, interval 10 m/s, and the zero contour is dotted) at 200 hPa. *(From Adames and Wallace 2014. © American Meteorological Society. Used with permission.)*

9.6.4 Seasonality

Generally, a time series at a weather station can be decomposed into $u = \overline{u} + u' + u''$, where the overbar denotes the long-term mean, the prime the stationary (e.g., monthly mean) anomaly, and the double prime transient (timescale <7 days) eddies (traveling storms of baroclinic vertical structure). The PNA at the positive phase extends the westerly jet eastward, steering transient storms away from the Pacific Northwest toward California (see Fig. 9.21). Transient eddy fluxes are generally a positive feedback amplifying stationary waves.

The PNA pattern is most pronounced in DJF season because of the peak subtropical westerly jet, which enables the generation (RWS) and poleward energy propagation (Eq. 9.11) of barotropic Rossby waves from the tropics. The peak SST forcing in NDJ also helps. Likewise, the Southern Hemisphere counterpart of the PNA, called the Pacific South American (PSA) pattern, is most pronounced in austral winter (JJA) when the subtropical jet there is strong, despite moderate ENSO SST variability.

Although the ENSO-induced PNA pattern is strongest in winter, the internal variability is also strongest in winter. Thus the signal-to-noise ratio may not be highest in winter. For PNA, predictability is high in March, especially following El Niño when the zonal variations in the Pacific jet deceleration and hence the barotropic energy conversion for internal variability are weak (Chapman et al. 2021).

9.7 Seasonal prediction

The growth, propagation, and decay of transient waves in the atmosphere give rise to weather variations. Initialized properly from adequate analysis of current observations, atmospheric GCMs show skills in predicting subsequent weather variations. By the 1980s, numerical weather prediction is well established and operational in weather services worldwide. The skill of weather prediction has since improved steadily and is considered high (see Fig. 1.7). Weather forecast predicts the time of arrival, location, and intensity of a winter storm. The chaotic nature of the atmosphere—the so-called butterfly effect—limits the useful skill of such deterministic weather forecast to 2 weeks. Beyond this limit, the prediction of weather conditions—when and how much precipitation to be expected—becomes unreliable.

While the precise state of weather is no longer predictable beyond 2 weeks, atmospheric GCM simulations, forced with observed SST, show skills in reproducing monthly and seasonal mean statistics of weather (e.g., the intensity and extent of the North Pacific storm track during El Niño winter). This hints at the possibility of predicting seasonal-mean statistics of weather—which are important for beach erosion and hydroclimate—provided that tropical SST variations are predictable. For the atmosphere, weather prediction is an initial value problem while climate prediction is a boundary value problem (SST). In the tropical Pacific, SST and atmospheric variations are closely coupled, giving rise to ENSO.

This implies a scheme for climate prediction: by initializing ocean conditions that include thermocline displacements (hence the warm water volume mode) and SST anomalies (hence the Pacific meridional mode), coupled models can be used to predict ENSO variations, along with the teleconnections such as the PNA pattern. Thus for the coupled ocean-atmosphere system, climate prediction is also an initial value problem, but the initialization of ocean state is the key, instead of the atmospheric state as in the case of weather forecast.

Cane et al. (1986) succeeded in the first proof-of-concept prediction by coupling an ocean model linearized about a realistic mean state and the Matsuno-Gill atmospheric model. Because of the lack of real-time ocean subsurface observations, they drive the ocean model with observed large-scale winds to initialize the coupled model. They show that the forward integration of the initialized coupled model has skills in predicting observed variations in SST and thermocline depth in the equatorial Pacific.

Climate prediction at seasonal leads has since improved. The use of comprehensive coupled GCMs extends SST prediction beyond the tropical Pacific and enables the prediction of such important quantities as rainfall and storm tracks over both ocean and land. Climate prediction benefits from expanding ocean observing systems such as an array of moored buoys across the tropical Pacific, satellite altimeters, and Argo profiling floats. The observing systems provide data for real-time analysis of ocean states such as the thermocline depth variations. The use of atmospheric GCMs improves the representation of SST-forced and internal variability, both affecting ENSO evolution.

A seasonal forecast system is typically initialized each month with ocean state estimates and run for 12 months. An ensemble of model runs is made, each with slightly different initial conditions to sample the range of uncertainty. The ensemble mean represents the deterministic component of the forecast, while the ensemble spread represents the uncertainty in the forecast. When the ensemble mean is much larger than the spread, the confidence is high in the forecast. Conversely, if the ensemble spread is much larger than the ensemble mean, low confidence is assigned.

The skill in predicting equatorial Pacific SST is quite high, especially when a major ENSO event takes place (Fig. 9.26). There is a spring barrier for ENSO prediction: Skill is high for predictions initialized after boreal spring but low initialized before. This may be because equatorial SST variance (signals to predict) is low in boreal spring (see Fig. 9.13A), while uncertainties due to the arrival of the Pacific meridional mode are high (see Section 12.4.1).

Errors and biases of coupled GCMs remain stubbornly large (Planton et al. 2021) and are a limiting factor on the accuracy of seasonal prediction. Seasonal prediction is now operational at about a dozen centers around the world, each producing a perturbed initial-condition ensemble of its own. These models form a multimodel ensemble

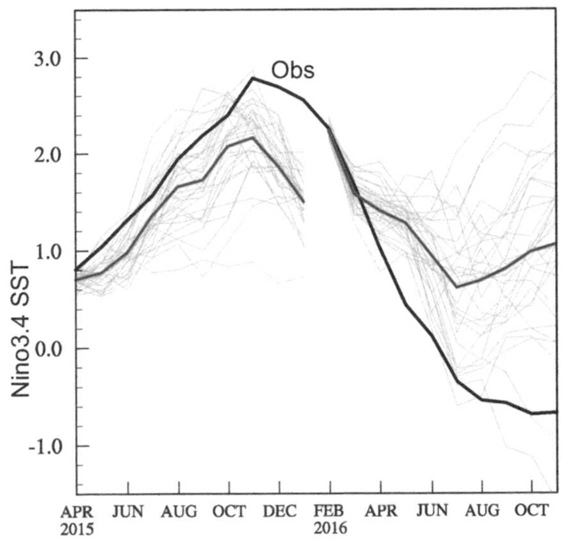

Fig. 9.26 Forecast plumes for Nino3.4 sea surface temperature (*SST*; °C) that start in April 2015 and February 2016: observations *(black)*, 32 member runs *(thin grey)*, and the ensemble mean (magenta). NCEP CFSv2 is used. *(Courtesy J. Ma.)*

(MME), sampling uncertainties due to errors and biases in model physics, data assimilation, and initialization. The MME mean often improves the forecast by reducing errors and biases among different forecast systems.

Fig. 9.27 compares the MME mean forecast with observations, in the composite of three major El Niños of 1982-83, 1991-92, and 1997-98 in the peak winter season. The comparison is generally good in the tropics, including the eastward shift of deep

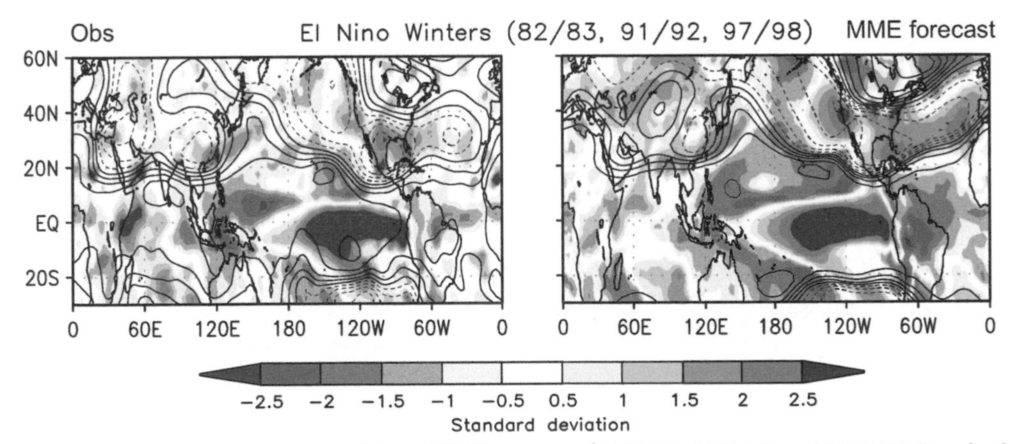

Fig. 9.27 Winter (DJF) composite of three El Niño events of 1982-83, 1991-92, and 1997-98. Standardized anomalies of precipitation *(color shading)* and 500 hPa geopotential height (line contours at intervals of 0.5, negative dashed): *(left)* observations and *(right)* 14-model ensemble mean forecast at 1-month lead. *(From Wang et al.(2009).)*

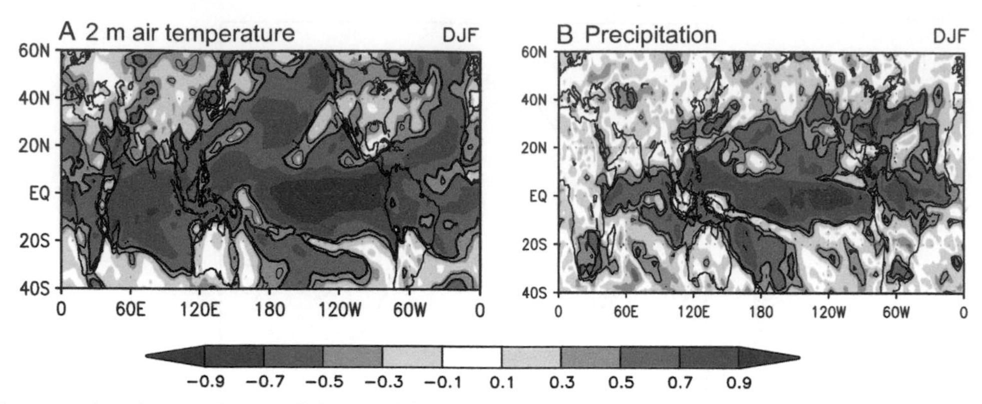

Fig. 9.28 Correlation of 14-model ensemble mean forecast at 1-month lead with observations for DJF 1981-2004. *(left)* Surface air temperature and *(right)* precipitation. *(From Wang et al. (2009).)*

convection over the Pacific and decreased rainfall over South America and the tropical North Atlantic. The MME forecast captures the PNA pattern, and increased rainfall on the US west coast, northern Mexico, and the southern United States.

Fig. 9.28 evaluates forecast skills in terms of anomaly correlation with observations. The skill is generally high in the tropics (20°S–20°N) and over the ocean. Skill in extratropical SST is also high because the initial condition decays slowly in the ocean mixed layer of a large heat capacity. The skill over extratropical land is generally low, except for winter rainfall in southern China and the southern United States (the latter resulting from the ENSO-induced PNA pattern).

Weather forecasting focuses on daily conditions, while seasonal forecasting on monthly or seasonal statistics. Subseasonal-to-seasonal (S2S) prediction aims to fill the gaps of weather and climate forecasting, between week 3 to month 2 following the initialization. Potential sources of S2S predictability include the MJO, soil moisture, snow cover, and sea ice. Intermittent windows of opportunity for skillful S2S prediction exist when these processes contribute constructively (Mariotti et al. 2020).

9.8 Summary remarks

El Niño is not merely an oceanic phenomenon, nor is the Southern Oscillation an atmospheric one. They emerge as a coupled ocean-atmospheric phenomenon, ENSO. In the coupling, the atmosphere adjusts quickly to the slowly evolving ocean state. Ocean wave adjustments are key to ENSO: The equatorial Kelvin wave, in response to zonal wind variability in the central Pacific, causes SST in the remote eastern Pacific to vary through thermocline displacements. Thermocline waves are also important for the phase transition and hence the prediction of ENSO.

Just as the coupling of the atmospheric circulation and deep convection gives rise to the MJO, the coupling of the ocean and atmosphere creates new unstable modes beyond the original family of Matsuno's wave solutions. The mean state of the equatorial Pacific—the prevailing easterly winds, equatorial upwelling, eastward shoaling of the thermocline—sets up the necessary condition for the Bjerknes/thermocline feedback.

Few people live in the equatorial belt, much of which is occupied by ocean. Judging from the geography, the equatorial Pacific would not seem of great importance from the outset. Several dynamic factors elevate it to global prominence. First, deep convection in the tropics is the major driver of the global atmospheric circulation, and both are highly sensitive to tropical SST changes as attested by the Southern Oscillation and PNA teleconnections. Second, Bjerknes feedback favors the equator, and the coupled instability further favors the largest tropical basin of the Pacific. ENSO transforms our view of ocean and atmospheric variability: What happens in the faraway equatorial Pacific can be more influential on North American climate than storms just to the west over the midlatitude North Pacific (see Chapter 12). Tropical meteorology and equatorial oceanography, both on the margins of mainstream quasi-geostrophic framework in the 1970s, now occupy the center stage of climate research.

The study of ENSO provides a template to study other climate phenomena:

- Recurrent atmospheric patterns (e.g., the Southern Oscillation) on seasonal and longer timescales are anchored by recurrent SST patterns (e.g., El Niño). Use atmospheric GCMs to identify the SST forcing pattern(s).
- Pursue the reverse oceanographic problem to identify relevant processes and mechanisms for the pattern and evolution of the SST anomalies.
- Integrate the results from the above ocean/atmospheric studies to identify positive feedback for the growth of coupled anomalies and mechanism for phase transition. Explore precursors.
- Evaluate the predictability of the coupled phenomenon (e.g., by analyzing results of operational climate prediction). Examine whether models adequately represent key physical processes.
- Explore teleconnection patterns and mechanisms to exploit predictability.

The success of coupled dynamics in unraveling the mystery of ENSO inspires studies of coupled phenomena and feedback beyond ENSO. The following chapters showcase the expanding legacy of coupled approach that originated from ENSO research. See also recent edited monographs of McPhaden et al. (2020), Mechoso (2020), and Behera (2021).

Review questions

1. During El Niño, what causes the large SST increase in the eastern equatorial Pacific (e.g., near the Galapagos) even though local trade winds hardly change? Compare with the warming mechanism near the international dateline where wind change is large.

2. What happens to the Equatorial Undercurrent in the eastern Pacific during El Niño? Explain briefly.
3. How does sea surface height change in the western and eastern equatorial Pacific during El Niño? What about the sea level at San Diego?
4. During El Niño, what are the major atmospheric anomalies in the tropics?
5. How does rainfall change during El Niño on the Galapagos and over the Maritime Continent?
6. Why don't winds change much in the eastern equatorial Pacific during ENSO?
7. What is the thermocline feedback? Why is it large in the eastern and small in the western equatorial Pacific?
8. The typical mean upwelling velocity is 1.2×10^{-5} m/s at 50 m in the Nino3 region. Calculate the upwelling damping rate on SST perturbations (Eq. 9.3). Compare it to the evaporative damping rate in Eq. 8.2.
9. What is the typical Rossby radius of deformation for the ocean on the equatorial beta-plane? For ENSO, do you care the most about wind variations within 3°S–3°N, north of 3°N, or south of 3°S? Explain briefly.
10. What causes phase transitions between El Niño and La Niña? How does the basin size matter for the linear growth rate of this oscillation? Why is the Pacific favored for ENSO?
11. The time delay τ in the delayed oscillator theory of ENSO refers to the time for the $n = 1$ Rossby wave to reach the western boundary and the reflected Kelvin wave to travel back to the forcing longitude x. Calculate the delay τ for two cases with the anomalous wind patch centered at $x = 180$ and $x = 155°W$, respectively. Assume the western boundary of the Pacific $x_W = 130°E$, and $c = 2$ m/s. According to the delayed oscillator theory, which case gives rise to a larger growth rate?
12. In which season does ENSO tend to peak? Does this guarantee that ENSO influence peaks in this season everywhere (e.g., India rainfall)? If not, why?
13. How does an extreme El Niño differ in atmospheric anomalies from the canonical/moderate El Niño Wyrtki (1975) described? Explain briefly why.
14. Rainfall averaged in the Nino3 (150°W–90°W, 5°S–5°N) region fails to track the EP meridional dipole mode during the FMA season (see Fig. 9.18C). Design a simple index using gridded rainfall or wind data to track the mode.
15. What are the leading modes of tropical thermocline depth variability? How are they related to ENSO in phase?
16. What gives rise to the predictability for ENSO events at seasonal leads?
17. Why is it possible to predict ENSO? Atmospheric initial conditions are important for weather forecasts. Are they important for ENSO prediction, too?
18. Why is surface air temperature more predictable over ocean than on land, say at 1-month lead? Examine the evidence for PNA effect on the skill in predicting precipitation and surface air temperature.

19. Why is the winter climate of North America more predictable than those of other extratropical continents? During which year do you think NOAA has a better chance of predicting North American climate anomalies: year A with an eastern Pacific warming of 4°C or year B with an eastern Pacific cooling of −0.5°C?

20. During El Niño winter, western Canada and Alaska tend to warm while the Pacific Northwest tends to be dry. Use teleconnection maps in pressure/circulation fields to explain these anomalies. Why does the North American climate response to ENSO tend to be large in boreal winter?

21. Why are ENSO-induced atmospheric circulation anomalies baroclinic in the tropics while barotropic in the extratropics?

22. Despite ENSO diversity, why is the ENSO-induced PNA pattern largely fixed geographically? Which feature of the mean climate is most responsible for this geographically stationary pattern?

23. With the zonal-mean $\overline{u} \sim 20$ m/s, estimate the stationary wavelength at 45°N. Assume for simplicity that both the mean flow and perturbations are uniform in the meridional.

24. The observed westerly profile has a well-defined peak in the meridional direction. The winter westerly jet above Japan is about 50 m/s at 300 hPa. Show that the effective beta ($\widehat{\beta}$) is locally enhanced at the jet core. Estimate the contribution due to the meridional shear and compare it with the planetary beta. Assume that the jet is 20o wide in latitude on either side.

25. How does the sea level change during El Niño winter on the California coast? What effect does this have on beach erosions?

CHAPTER 10

Tropical Atlantic variability

Contents

The tropical Atlantic Ocean is flanked by two large continents, each hosting a major center of atmospheric convection. More than 300 years ago, Halley (1686) recognized the important influence of continents on climate in the Atlantic sector and suggested that the intense surface heating over North Africa drives the southerly winds in the Gulf of Guinea (a geographic name referring to the eastern tropical Atlantic facing the south coast of West Africa). The Atlantic intertropical convergence zone (ITCZ) is displaced north of the equator, onto which the southeast trade winds converge (Fig. 10.1). On the equator, the southeast trade winds shoal the thermocline in the east and induce upwelling. The upwelling of cold thermocline water keeps sea surface temperature (SST) on the equator low in the eastern Atlantic despite strong solar heating.

10.1 Seasonal cycle

10.1.1 Intertropical convergence zone

The ITCZ and its extension of continental convection display large seasonal excursions over the Atlantic sector. Over the continents, the rain band largely follows the seasonal march of the Sun, reaching its northernmost (southernmost) position in July–September (December–February). Equatorial Africa features two rain seasons, one at each equinox

Coupled Atmosphere-Ocean Dynamics: from El Niño to Climate Change
ISBN 978-0-323-95490-7, https://doi.org/10.1016/B978-0-323-95490-7.00010-2

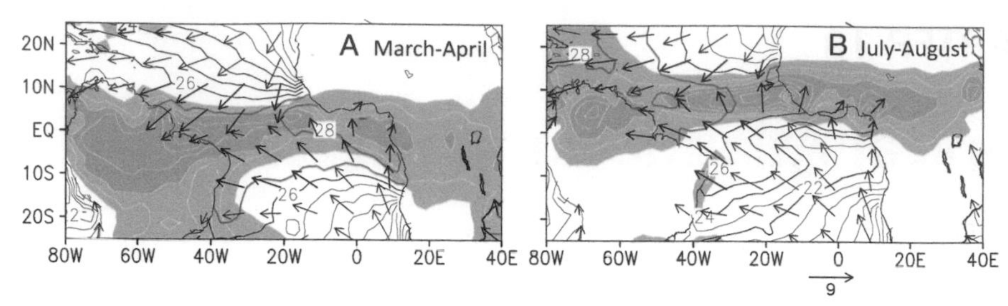

Fig. 10.1 Climatologic distributions of rainfall (light/dark shade >2/6 mm/day, white contours at 2 mm/day interval), sea surface temperature (line contours, red ≥27°C), and surface wind velocity (vectors in m/s) for (A) March-April and (B) July-August. *(From Xie (2004).)*

(Fig. 10.2C). Over the equatorial Amazon, by contrast, convection peaks only once a year, centered in April. At the autumn equinox (September) when the cold tongue develops over the Atlantic, the advection of cool and dry air from the cold ocean suppresses deep convection over the equatorial Amazon (see Figs. 10.1B and 10.2A).

Over the ocean, patterns of SST and the position of the ITCZ are tightly coupled, with major rainfall confined to a band of high SSTs above 27°C. In March-April, the rain band is located nearly on the equator, onto which the trades converge from both hemispheres. SST is uniformly warm in the equatorial zone of 10°S−5°N, making March-April the time when the Atlantic ITCZ is very sensitive to even small changes in interhemispheric SST gradient (see Section 10.3). As the equatorial cold tongue develops in June and persists through September, the ITCZ is kept north of the equator coupled with the northward movement of the high-SST band. The oceanic ITCZ reaches its northernmost position in September, lagging its northward movement over the continents because of the larger heat capacity of the ocean mixed layer. In July-August, rainfall in the ITCZ is considerably stronger than in March-April, despite a 1°C decrease

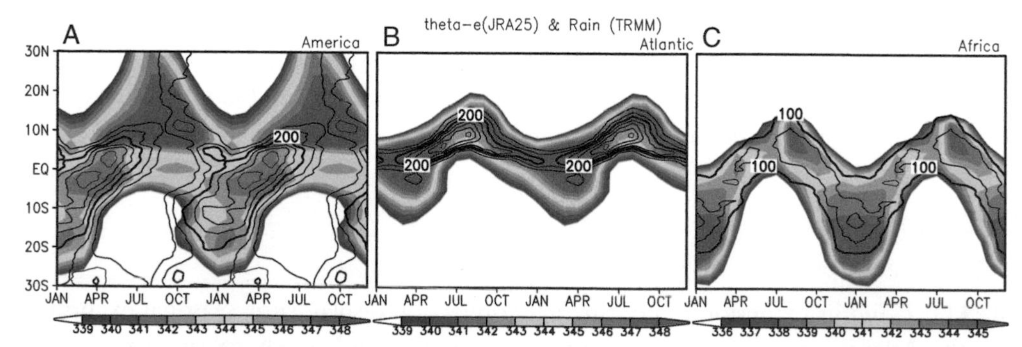

Fig. 10.2 Time-latitude sections of precipitation (black line contours at 50 mm/month intervals) and surface θe (color shading in K) in (A) the Americas (60−50°W), (B) Atlantic Ocean (20−10°W), and (C) Africa (20-30°E). *(From Tanimoto (2010).)*

in SST beneath the ITCZ (see Fig. 10.1). This strengthening of ITCZ convection may be due to the abundance of strong westward propagating easterly wave disturbances that help trigger convection over the ocean. Of a timescale of 3 to 9 days, these disturbances originate from the African rain band and grow on the meridional and vertical shear of the African easterly jet (see Section 5.4). These easterly African waves further amplify in the tropical Atlantic, some into tropical storms and hurricanes that make landfall in the Caribbean and southern United States.

10.1.2 Equatorial cold tongue

Along the equator, SST varies with a strong annual cycle. As in the Pacific, the northward displacement of the mean ITCZ is the ultimate cause by maintaining southerly cross-equatorial winds that relax in boreal spring and intensify in boreal summer (see Fig. 10.1). Equatorial SSTs reach their annual maximum in boreal spring when the southeast trades weaken. As the year progresses the trade winds along the equator intensify, and a distinctive cold tongue forms across the basin, centered slightly south of the equator.

The annual cycle of the equatorial cold tongue differs between the Pacific and Atlantic in a few important aspects. Compared to the Pacific, the annual cycle of eastern equatorial Atlantic SST is sawtooth-like, asymmetric between the rapid seasonal cooling and slow warming (which last 3 and 7 months, respectively) (Fig. 10.3). At 0°W, Eq., SST reaches 29.5°C in April and drops to 25°C in July. From the oceanic point of view, the rapid cooling is attributed to the sudden onset of the West African monsoon and the rapid intensification of the southerly winds in May-June in the Gulf of Guinea. These southerly winds cause upwelling slightly south and downwelling slightly north of the equator, and this upwelling cools the equatorial ocean (see Section 8.2).

Unlike the equatorial Pacific where the thermocline depth (tracked by the 20°C isotherm) is nearly constant through the annual cycle (see Fig. 8.11), the vertical displacements of the thermocline are important for SST in the Atlantic equatorial cold tongue. From April to July, the intensified easterly winds dramatically increase the zonal slope of the thermocline (Fig. 10.4), and the thermocline shoals by ~30 m in the equatorial Gulf of Guinea (see Fig. 10.3), causing SST to decrease together with the intensification of wind-induced upwelling. The seasonal shoaling of the thermocline in the equatorial Gulf of Guinea resembles the oceanic Kelvin wave, with a meridional structure that peaks on the equator and is evanescent poleward (Fig. 10.5B). Upon reaching the African coast, the upwelling Kelvin wave reflects into the coastal Kelvin wave, manifested as a delayed thermocline shoaling on the Guinea coast (5.5°N) 1 month later in July-August (see Fig. 10.5B). The shoaling thermocline off the coast of Guinea causes the coastal ocean to cool (see Fig. 10.5A), contributing to the abrupt northward jump of the West African monsoon rainband (see Section 5.4).

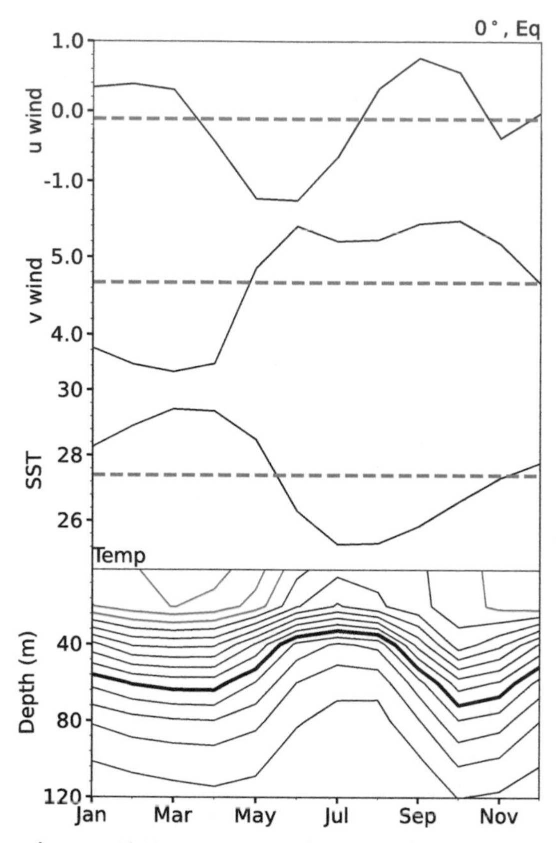

Fig. 10.3 Monthly climatology at 0°W, equator as a function of time: zonal and meridional wind velocity (m/s), sea surface temperature (*SST*; °C), and ocean temperature (1°C interval; 20°C contour thickened, red ≥27°C). The dashed line indicates the annual mean value. Based on PIRATA buoy data. *(Courtesy Q. Peng.)*

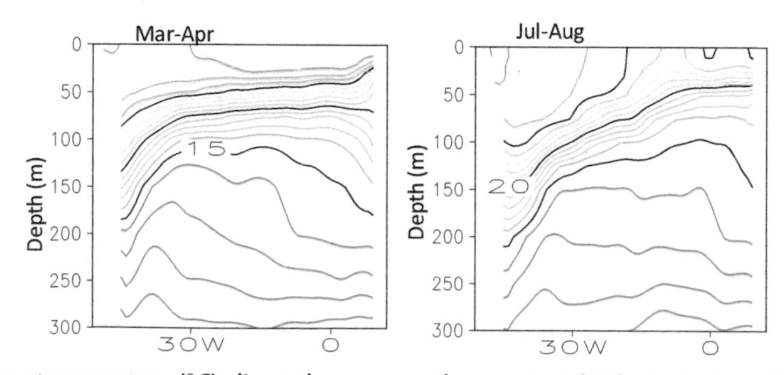

Fig. 10.4 Ocean temperature (°C) climatology across the equatorial Atlantic in *(top)* March-April and *(bottom)* July-August.

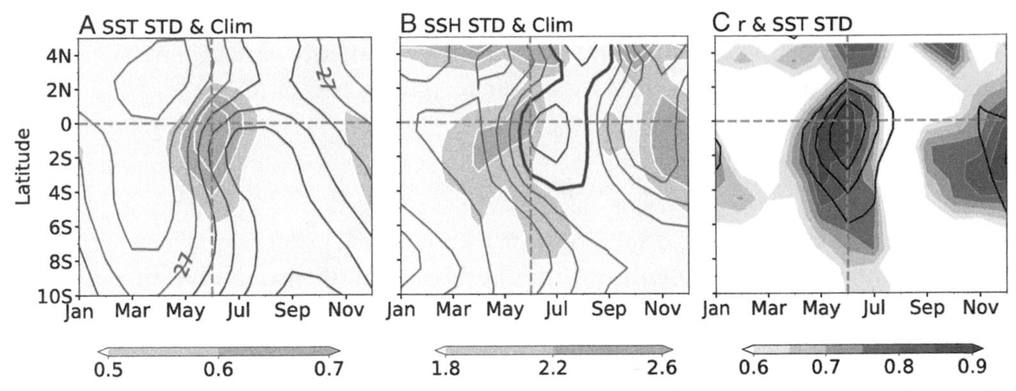

Fig. 10.5 10° W–0° zonal mean Hovmöller diagrams. Standard deviation of interannual variability (IAV): (A) sea surface temperature (*SST*; gray shading, only STD ≥0.5°C shown), (B) SSH (gray shading, only STD >1.8 cm shown), and (C) SST (black contours at 0.1°C interval, starting at 0.5°C). Also plotted in (A) is mean SST (line contours at intervals of 1°C, SST ≤26°C in blue and ≥27°C in red), in (B) is mean SSH (contours at intervals of 2 cm, positive in red, zero thickened and negative in blue; annual mean removed), and in (C) is the SSH-SST correlation (color shading). *(Courtesy Q. Peng.)*

10.1.3 Easterly winds over the equatorial Gulf of Guinea

By causing the thermocline depth to vary in the east, the zonal wind variations near the equator are important for the SST annual cycle. The annual cycle in equatorial zonal wind is driven by the interaction with equatorial SST—as in the Pacific—and by the continental monsoon. In an atmospheric model experiment where the SST field is kept at the April values over the equatorial Atlantic, the easterly acceleration is largely reproduced over the eastern equatorial basin south of West Africa (Okumura and Xie 2004). Because of the northward displaced ITCZ, the easterly winds peak around 10°S and weaken toward the ITCZ ($\frac{\partial \bar{u}}{\partial y}$ >0 from 10°S–5°N, the overbar denotes the annual mean) (see Fig. 10.1). As West Africa heats up from April on, the southerly winds intensify over the Gulf of Guinea, advecting the mean easterly momentum from the south to accelerate the equatorial easterlies ($-v'\frac{\partial \bar{u}}{\partial y}$ <0; the prime denotes the deviation from the annual mean). Thus the West African monsoon southerlies are a cause of the easterly intensification from March to May in the equatorial Gulf of Guinea, another example of continental influence on tropical Atlantic climate.

Here we present a quantitative analysis of the advection of easterly momentum onto the equator in the atmospheric boundary layer (ABL). For simplicity, we neglect zonal variation ($\frac{\partial}{\partial x} = 0$). The zonal momentum equation may be written as

$$v\frac{\partial u}{\partial y} - fv = -\varepsilon u \tag{10.1}$$

where $f=\beta y$ is the Coriolis parameter and ε the drag coefficient (10^{-5} s^{-1} or 1/day for an ABL of 1 km deep). We consider only the antisymmetric wind perturbations as meridional advection vanishes at the equator for symmetric disturbances ($v = 0$ and $\frac{\partial u}{\partial y} = 0$). The first-order linear solution is

$$u_0 = fv/\varepsilon \tag{10.2}$$

The linear zonal flow vanishes on the equator without zonal pressure gradient. Using the linear solution (Eq. 10.2) to estimate the meridional momentum advection yields a higher-order correction,

$$u_1 = \left(fv - v\frac{\partial u_0}{\partial y}\right)/\varepsilon = \left[fv - \frac{v}{\varepsilon}\frac{\partial}{\partial y}(fv)\right]/\varepsilon$$

On the equator, it simplifies into

$$u_1 = -\frac{\beta}{\varepsilon^2}v^2 = -\frac{\beta}{\varepsilon^2}\left(\overline{v}^2 + 2\overline{v}v' + v'^2\right) \tag{10.3}$$

where $v = \overline{v} + v'$, with $\overline{v} > 0$ due to the northward displaced ITCZ in the annual mean and v' an annual cycle driven by the seasonal variation in solar radiation. Let $v' = \widetilde{v}_1 \sin(\omega_1 t + \varphi)$, where $\widetilde{v}_1$ is the annual harmonic of meridional wind on the equator, ω_1 the annual frequency, and φ the phase. Eq. 10.3 becomes

$$u_1 = -\frac{\beta}{\varepsilon^2}\left[\left(\overline{v}^2 + \widetilde{v}_1^2/2\right) + 2\overline{v}\widetilde{v}_1 \sin(\omega_1 t + \varphi) - \left(\widetilde{v}_1^2/2\right)\cos 2(\omega_1 t + \varphi)\right] \tag{10.4}$$

In the angular brackets on the right-hand side, the terms in the first parentheses represent the easterly acceleration in the annual mean due to the mean ($\overline{v}$) and seasonal variation ($\widetilde{v}_1$) of the cross-equatorial wind, respectively; the middle term drives an annual cycle in the equatorial easterly acceleration that peaks in boreal summer; and the last term generates a semiannual cycle in the equatorial easterlies. In the equatorial Atlantic, the cross-equatorial wind is always from the south ($\overline{v} \geq \widetilde{v}_1$) (see Fig. 10.3), the annual cycle (the middle term) dominates the equatorial easterly variations due to the meridional advection of the easterly momentum (by a factor of ≥ 4 compared to the last term). In other words, the northward displaced ITCZ is in favor of an annual cycle in the easterly wind-induced equatorial upwelling and hence SST, in addition to the south-equatorial upwelling in response to the meridional wind stress itself (see Section 8.3). In the equatorial Indian Ocean, on the other hand, the cross-equatorial wind is monsoonal ($\overline{v} \approx 0$), and the semiannual cycle dominates the zonal wind on the equator (see Section 11.1).

10.2 Zonal mode: Atlantic Niño

The Atlantic equatorial cold tongue exhibits considerable interannual variability. In fact, the interannual variance of SST peaks in the cold tongue and on the African coast south of the equator (Fig. 12.8). The upwelling-favorable southeast trades and the eastward shoaling of the thermocline are conducive to thermocline feedback in the eastern equatorial Atlantic. SST anomalies in the eastern equatorial Atlantic (Atl3 region: 5°S–5°N, 20°W–0°) are often used to track interannual variability of the cold tongue. As in the case of El Niño and the Southern Oscillation (ENSO), Atl3 SST is correlated not with the local wind variability but with zonal wind variability in the western equatorial Atlantic (wAtl region: 5°S–5°N, 40°–20°W). This, together with the thermocline feedback in Atl3, suggests coupled Bjerknes feedback at work in the equatorial Atlantic (Zebiak 1993). For this reason, anomalous warming of the Atlantic cold tongue is dubbed "the Atlantic Niño" while the cooling events "Atlantic Niñas." Lübbecke et al. (2018) is a recent review on the topic.

The Atlantic Niño mode is strongly phase-locked onto May-August, the time when the equatorial cold tongue develops (see Fig. 10.5A). This phase locking coincides with the seasonal shoaling of the thermocline and hence enhanced thermocline feedback in the east. At the peak phase (June-July) of the Atlantic Niño, the entire equatorial Atlantic warms up, with the strongest warming located in the eastern half of the basin and extending southward along the African coast (Fig. 10.6). Although the SST warming is slightly displaced south the equator, rainfall anomalies are confined mostly north of the equator, an apparent effect of the northward-displaced mean ITCZ. During Atlantic Niño, the ITCZ is drawn southward toward the anomalously warm equator. The anomalous intensification and southward displacement of the ITCZ cause an anomalous increase in rainfall on the Guinea coast.

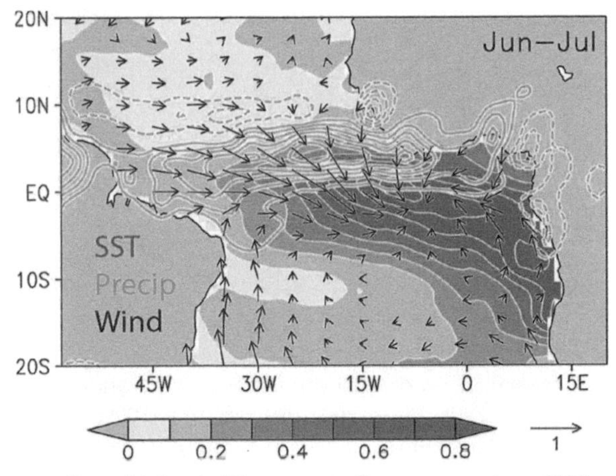

Fig. 10.6 June-July anomalies of Atlantic Nino: sea surface temperature (*SST*; color shading, °C), precipitation (green contours at intervals of 0.3 mm/day, zero omitted and negative dashed), and surface wind velocity (m/s). *(From Okumura and Xie (2006).)*

In the delayed oscillation theory (see Section 9.3), the period and growth rate both decrease with the wave delay τ or the zonal basin width. This is consistent with the reduced SST variance and short duration of Atlantic Niño events compared with ENSO in the much larger Pacific basin. As in the Pacific (see Fig. 9.13C), the equatorial zonal wind response to Atl3 SST anomalies is strongly modulated by the seasonal migration of the ITCZ. In an Atlantic Niño composite (Fig. 10.7A), the wAtl wind anomalies peak in May and decay rapidly thereafter as the mean ITCZ moves northward away from the equator, even though the Atl3 SST warming peaks 1 month later in June. The reduced projection of zonal wind anomalies on the ocean equatorial waveguide partly explains the rapid decay of the Atlantic Niño in SST after July. The 1-month lead of the wAtl wind over Atl3 SST anomalies implies atmospheric stochastic forcing (see Section 12.2). Results from a large-ensemble atmospheric model simulation indicate that much of the wAtl wind variability is indeed due to internal variability unrelated to SST anomalies (cf. green solid vs. dotted line in Fig. 10.7B).

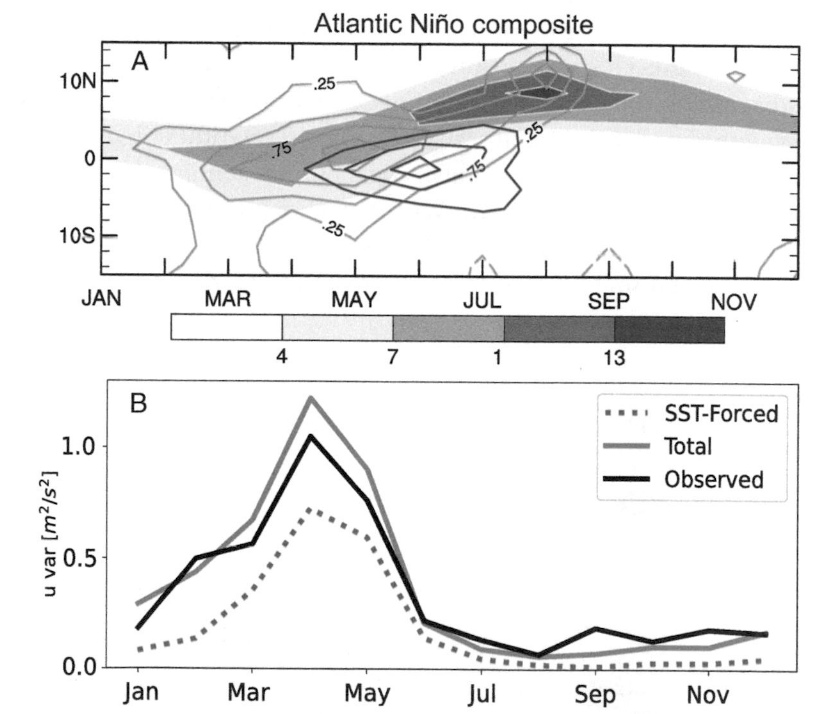

Fig. 10.7 (A) Composite Atlantic Niño: sea surface temperature (*SST*; 20°W–0 mean; purple contours at 0.25 K intervals, starting from 0.5 K) and zonal wind (40–10°W; orange contours at 0.25 m/s intervals, zero omitted) anomalies, with mean precipitation (40–10°W; gray shading with white contours, mm/day). (B) Zonal wind variance (m^2/s^2) in the wAtl region in observations *(solid black line)*, along with the total variance in a 100-member atmospheric general circulation model ensemble (*green solid*; Mei et al. 2019) and SST-forced variance as represented by the ensemble mean *(dotted)*. *(A, Adapted from Richter et al. 2017; B, inspired by Richter et al. 2014; courtesy W. Chapman.)*

The seasonal deepening of the mean thermocline in Atl3 after July also weakens the thermocline feedback on SST (as measured by SST-SSH correlation) (see Fig. 10.5C). Because the seasonal windows for zonal-wind and thermocline feedbacks overlap over a very short period of time (mostly May), Bjerknes feedback is weak over the equatorial Atlantic. Because of the weak Bjerknes feedback and large atmospheric internal variability in equatorial zonal wind that drives thermocline depth variability, the predictability of the Atlantic Niño mode is much lower than that of ENSO.

Interannual SST variability is large off the Angola coast due to variability in the coastal current that advects a sharp front in mean SST. Anomalous coastal warm events are called Benguela Niños, named after the Angolan city at 12.5°S. Benguela Niño is tracked by SST anomalies averaged in the coastal Angola Benguela area (ABA) of 20°S−0°S, 8°E−15°E. ABA SST variance peaks in April. A coastal Benguela Niño often precedes the equatorial Atlantic Niño by a season, tied to the variability of the South Atlantic subtropical high (Lübbecke et al. 2018). A weakened South Atlantic subtropical high relaxes the easterly trades along the equator and the southerly alongshore winds off the Benguela coast. The resultant downwelling equatorial Kelvin wave and reduced coastal upwelling contribute to Benguela warming.

10.3 Meridional mode

In March–April, tropical Atlantic climate is nearly symmetric about the equator, with the marine ITCZ reaching the southernmost position of the year. With the mean SST uniformly warm across the equator between 5°S and 5°N, the Atlantic ITCZ is very sensitive to slight variations in cross-equatorial SST gradient. The leading EOF mode of FMA rainfall variability features a meridional dipole that represents anomalous displacements of the ITCZ from one year to another, much as in the tropical Pacific during the same season (see Section 9.4.4). This ITCZ dipole is accompanied by coherent SST-wind patterns that imply WES (wind-evaporation-SST) feedback (Fig. 10.8). An anomalous northward displacement of the ITCZ accelerates the southeast trades south and decelerates the northeast trades north of the equator. These wind changes reinforce the negative (positive) SST anomalies south (north) of the equator. The Atlantic meridional mode (AMM) refers to these coupled patterns of SST and wind associated with the meridional ITCZ dipole.

The SST anomalies of AMM occupy a broad subtropical region (25°S−25°N) beyond the meridional ITCZ dipole on either side of the equator. Over the southern lobe of AMM, surface wind anomalies are weak, but the coherent increase in low cloud cover (see Fig. 6.14) helps reinforce the negative SST anomalies by reducing incoming solar radiation. The same low cloud−SST feedback also operates over the northern subtropical lobe of increased SST but is weaker in magnitude, as measured by the ratio of cloudiness to SST anomalies.

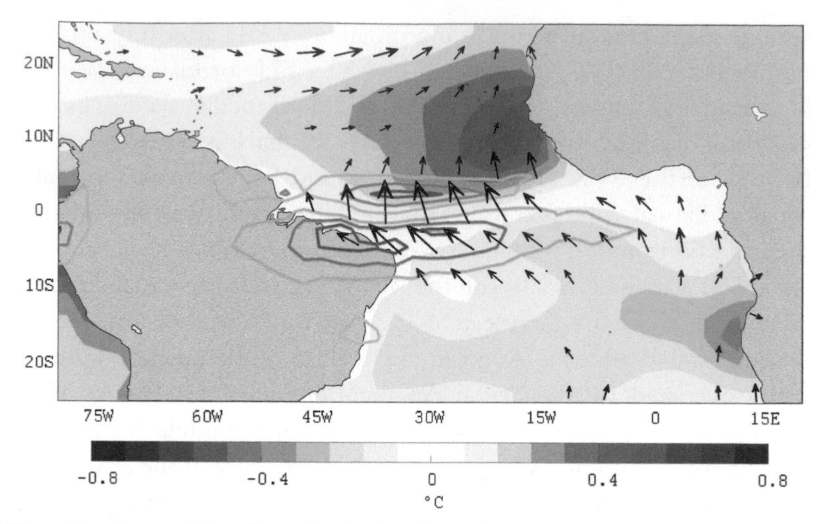

Fig. 10.8 The Atlantic meridional mode during boreal spring (FMA). Regressions against cross-equatorial sea surface temperature (SST) gradient (40°W–10°E, 10°S–10°N) for 1979-2017: SST *(color shading)*, surface wind velocity (maximum vector is 0.9 m/s), and precipitation (line contours at 0.5 mm/day interval, zero omitted, brown negative). *(Courtesy D.J. Amaya.)*

To track the AMM, we use the meridional wind at the equator (V_{eq}) and cross-equatorial SST difference (0–10°N minus 0–10°S), zonally averaged in 40°W–10°E. Both indices and their cross-correlation peak in March-April (Fig. 10.9) when the symmetric mean state allows antisymmetric disturbances to grow via WES feedback. The cross-correlation decreases from 0.9 in March to 0.3 in July as the northward-displaced mean ITCZ is not optimal for WES feedback.

To illustrate the importance of the mean ITCZ as the climatic axis of symmetry, we adopt a WES model of SST variations (the prime omitted) in Eq. 8.2,

$$\frac{\partial T}{\partial t} = aU - bT \tag{10.5}$$

Multiplying T on both sides yields an equation of squared variance of SST perturbations,

$$\frac{1}{2}\frac{\partial}{\partial t}\overline{T^2} = a\overline{UT} - b\overline{T^2} \tag{10.6}$$

where the overbar denotes the meridional integral. For a meridional SST dipole centered on the equator, the zonal wind response is largely positively correlated, $\overline{UT} > 0$ (Fig. 10.10).

Now consider the same SST dipole but displaced 10° northward as the mean ITCZ moves away from the equator. The wind response is no longer antisymmetric about either the mean ITCZ or equator. SST and zonal wind remain largely positively correlated north of the mean ITCZ but are poorly correlated to the south (dashed curves). As a

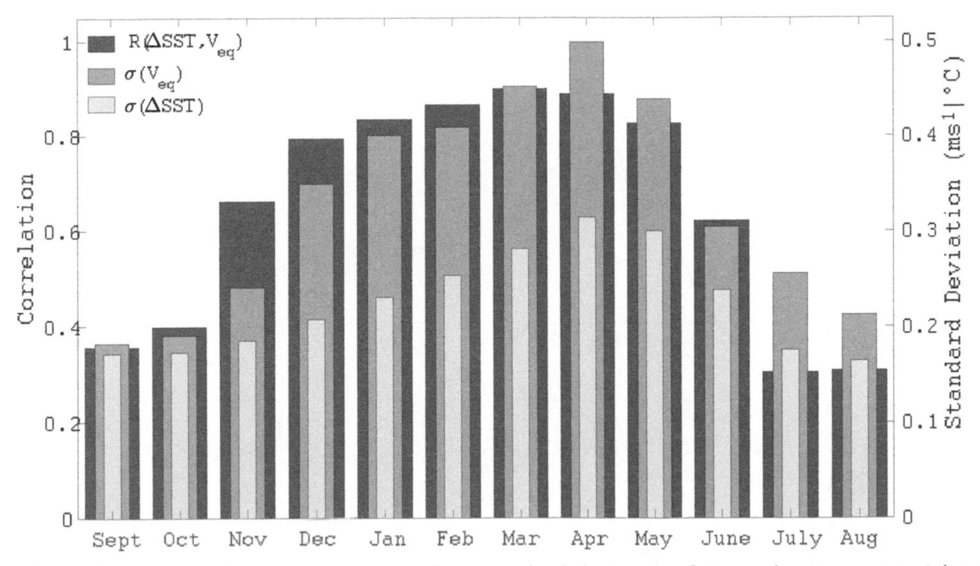

Fig. 10.9 Atlantic meridional mode seasonality. Standard deviation of V_{eq} and cross-equatorial sea surface temperature (SST) gradient (10°S–10°N) as a function of calendar month, along with their cross correlation. Atlantic zonal mean (40°W–10°E) and 3-month running mean data for 1950-2014 are used. *(Courtesy D.J. Amaya.)*

result, the meridionally integrated SST-wind covariance decreases, reducing the WES feedback onto the initial SST dipole compared to a symmetric mean state.

The previous calculation assumes that the SST forcing to the atmosphere is linearly proportional to the local SST anomaly. This assumption is invalid in seasons other than FMA. If convective anomalies develop only north of the equator, the WES feedback weakens even more.

10.4 Interactions with the Pacific

10.4.1 ENSO influence

The zonal propagation of tropospheric equatorial waves connects tropical ocean basins and allows them to interact. El Niño-induced tropospheric warming spreads in the east-west direction in the equatorial waveguide (the weak temperature approximation, see Section 3.6). Outside the equatorial Pacific, the associated subsidence causes rainfall to decrease and sea level pressure (SLP) to rise in the rest of the tropics. The rainfall decrease is especially pronounced over the maritime continent and South America. In the upper troposphere, the increased convective activity over the central Pacific excites an anticyclonic Rossby gyre on either side of the equator (aka the dumbbell), while a cyclonic dumbbell each develops over the maritime continent and South America with suppressed convection (Fig. 10.11). The Rossby dumbbell is broadly consistent

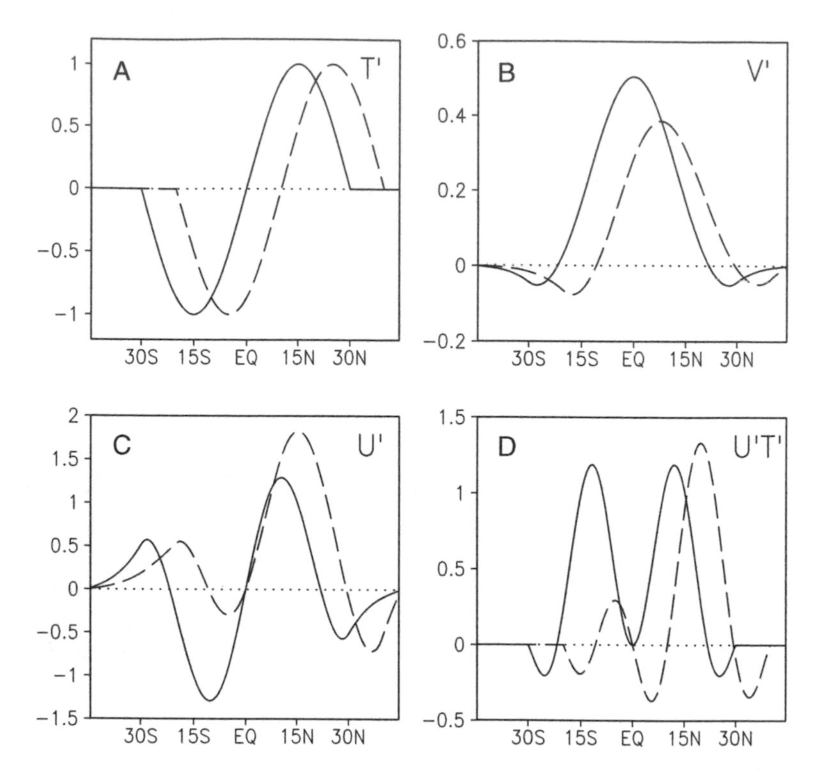

Fig. 10.10 Linear atmospheric response to (A) a meridional sea surface temperature (SST) dipole (°C); (B) meridional and (C) zonal wind velocity (m/s), and (D) destabilizing factor. The node of the SST dipole is either at the equator *(solid)* or 10°N *(long dashed)*. *(From Okajima et al. (2003.)*

with the Matsuno-Gill solution, but the observed eastward displacement relative to the convective heating is not.

In El Niño winter (DJF), the reorganization of tropical convection excites the barotropic Pacific-North American (PNA) pattern that lowers SLP in the subtropical North Atlantic around the latitudes of Florida. The reduced gradient between the PNA low in the subtropical North Atlantic and increased SLP in the equatorial Atlantic weakens the northeast trade winds, reducing evaporation and causing SST to rise over the tropical North Atlantic (see Fig. 10.11). In boreal spring (FMA), El Niño begins to dissipate, while a C-shaped wind pattern develops over the tropical Atlantic, suggestive of the WES feedback across the equator (see Fig. 10.11D).

The North Atlantic Oscillation (NAO), an atmospheric internal variability, is an additional forcing that modulates the winter northeast trade winds over the tropical North Atlantic (see Fig. 12.10A). The resultant SST anomalies over the tropical North Atlantic can trigger AMM development, displacing the Atlantic ITCZ. Both the ENSO and

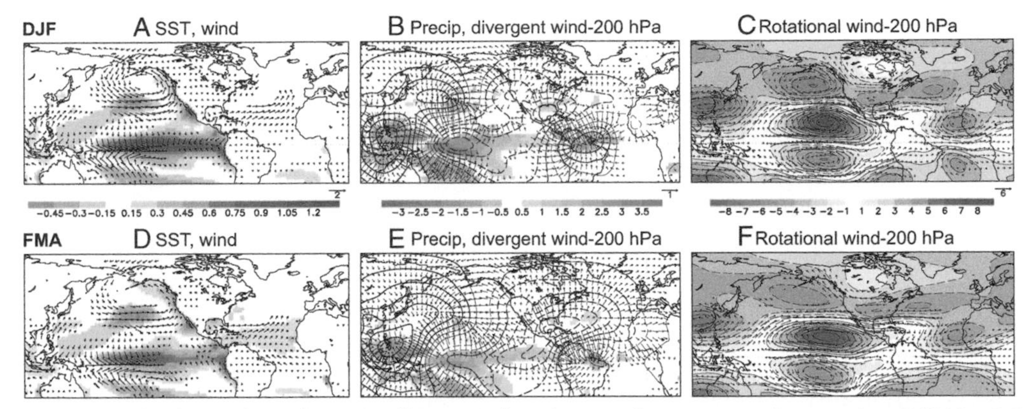

Fig. 10.11 El Niño and Southern Oscillation–induced anomalies as regressions against DJF Nino3.4 for 1982-2011 in *(top)* concurrent DJF and *(bottom)* subsequent FMA. (A) Sea surface temperature (*SST*; °C; shading) and surface wind (m/s; vectors); (B) precipitation (mm/day; shading), velocity potential [m^2/s; contours with contour interval (ci) 0.3 × 10^6 m^2/s], and divergent wind (m/s; vectors) at 200 hPa; (D) stream function (m^2/s; contours with ci = 10^6 m^2 /s) and rotational wind (m/s; vectors) at 200 hPa. (D–F) Same as (A–C) but for FMA. *(Adapted from García-Serrano et al. 2017; courtesy J. García-Serrano. © American Meteorological Society. Used with permission.)*

NAO forcing of the trades peak in boreal winter, which explains why AMM intensity peaks in FMA (see Fig. 10.9). The preferential excitation of AMM by subtropical trade wind variability is due to the positive WES feedback across the equator (Chang et al. 2001).

FMA is the rainy season for Brazil's Nordeste region (~40°W, 5°S) when the Atlantic ITCZ migrates south (see Fig. 10.1A). Nordeste rainfall experiences large interannual variability due to the influence of ENSO and the AMM, decreasing with El Niño conditions in the Pacific and/or a positive AMM (northward cross-equatorial SST gradient) (see Fig. 10.11E). The strong correlation with modes of tropical SST variability makes Nordeste rainfall predictable (Hastenrath and Greischar 1993).

10.4.2 Influence on the Pacific

Compared to the Pacific, the tropical Atlantic is one-third in basin size and one-half or less in interannual SST standard deviation. The ENSO influence on the tropical Atlantic is clear in the upper troposphere but is somewhat limited to the north of the equator near the surface possibly because of the blockage of low-level flows by the Andes. Recent modeling studies suggest pathways for tropical Atlantic SST variability to affect tropical Pacific climate.

Ding et al. (2012) conducted a pacemaker experiment with a global coupled general circulation model (GCM) by restoring SST toward observations over the tropical Atlantic (30°S–30°N) while leaving the rest of the world ocean fully coupled with the atmosphere. The ensemble mean Nino3 SST is most highly correlated with

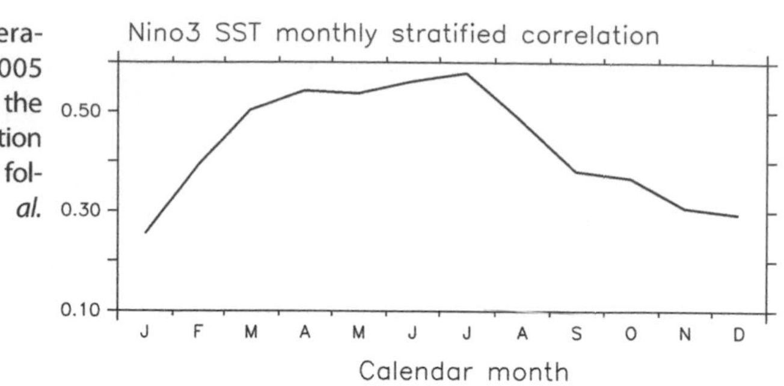

Fig. 10.12 Nino3 sea surface temperature (SST) correlation for 1970-2005 between observations and the ensemble-mean pacemaker simulation where tropical Atlantic SST variability follows observations. *(From Ding et al. (2012).)*

observations in early summer (April-July, r ~ 0.55 for 1970-2005) (Fig. 10.12). The correlation with Atlantic cold tongue SST in boreal summer shows that an Atlantic Niño tends to induce a La Niña state over the tropical Pacific (Fig. 10.13). The anomaly patterns over the tropical Pacific are remarkably similar between the tropical Atlantic pacemaker run and observations.

The atmospheric Kelvin wave is a conduit for the tropical Atlantic to affect the equatorial Pacific through the Indian Ocean. The warm tropospheric Kelvin wave induced by a tropical Atlantic warming is accompanied by anomalous easterlies in the lower troposphere

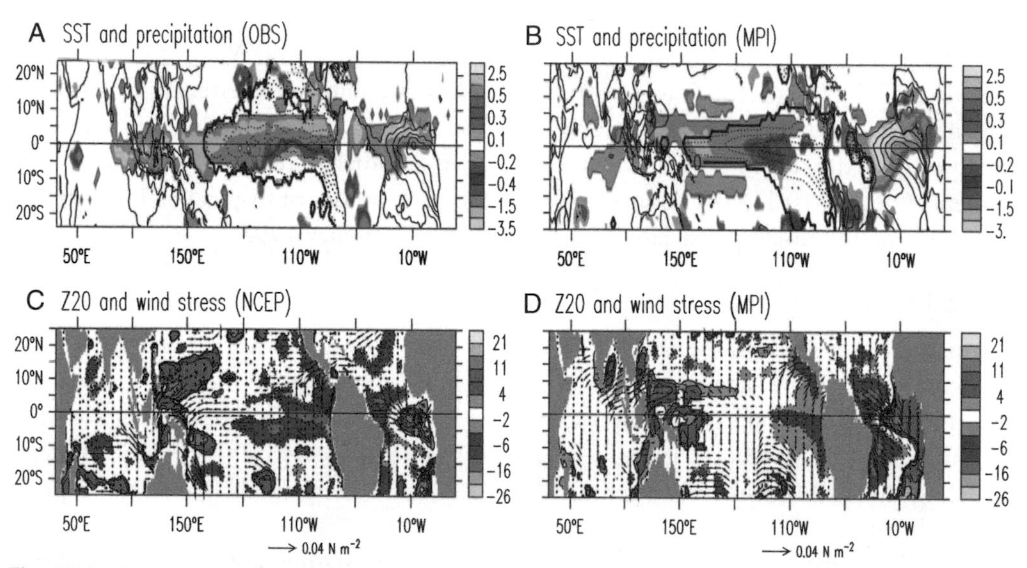

Fig. 10.13 (A, B) Sea surface temperature (SST; contours, °C) and precipitation (color shading, mm/day); (C, D) thermocline depth (shading, m) and wind stress (vector, Nm^{-2}) regressed onto Atlantic cold tongue SST in boreal summer; *(left)* from observations and *(right)* ensemble mean in the coupled model. The units for precipitation, thermocline depth, and wind stress are mm day^{-1}, m, and Nm^{-2}, respectively. *(From Ding et al. (2012).)*

(Fig. 10.14A), opposing the mean westerlies over the equatorial Indian Ocean. The resultant Indian Ocean warming strengthens the warm tropospheric Kelvin wave and the low-level anomalous easterlies over the western equatorial Pacific, which are favorable for La Niña conditions to develop (see Fig. 10.14B, C). The antisymmetric AMM, on the other hand, is not effective in exciting the equatorial Kelvin wave (Fig. 10.15). Instead, atmospheric Rossby waves through Central America seem important for the Atlantic SST anomalies to affect the subtropical Northeast Pacific (Ham et al. 2013) (Box 10.1). Despite the different interbasin pathways, a tropical North Atlantic warming leads to a La Niña-like cooling in the equatorial Pacific (see Figs. 10.14 and 10.15).

These recent studies suggest stronger cross-basin interactions in the tropics than previously thought (Mechoso 2020). While fully developed ENSO events are a major driver of climate anomalies around the world, variability in other smaller tropical basins could modulate tropical Pacific variability, including interannual ENSO and Pacific Decadal Oscillation.

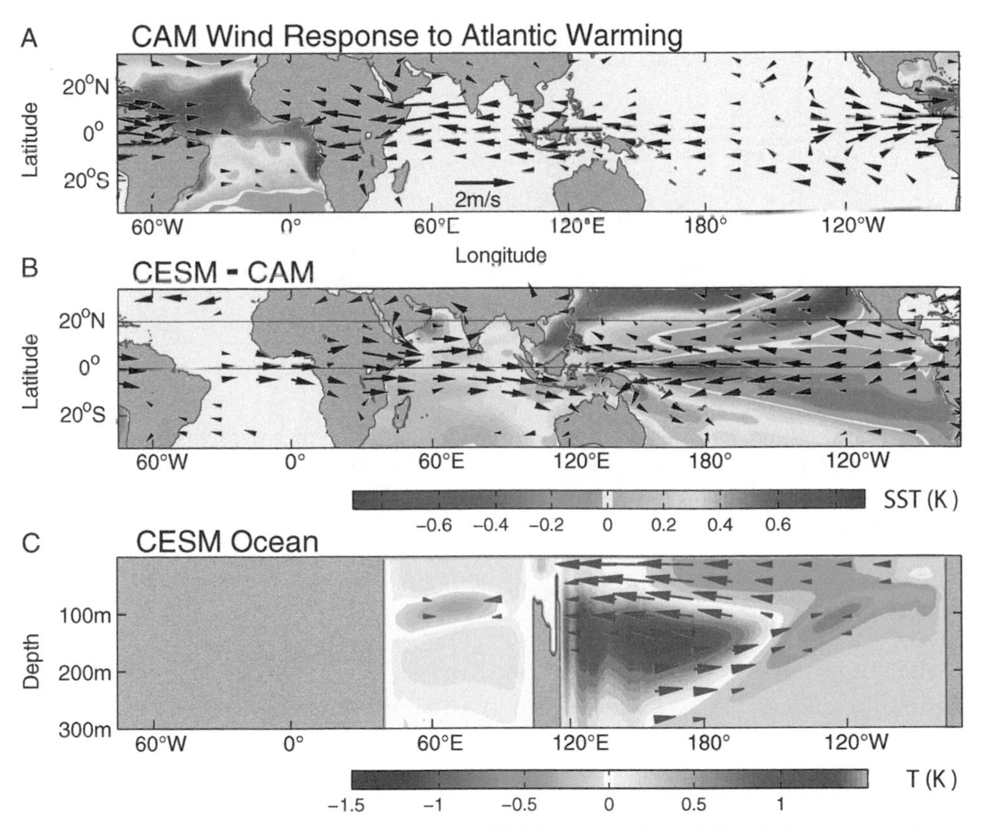

Fig. 10.14 (A) Atlantic sea surface temperature (SST) forcing and 850 hPa wind responses in atmospheric general circulation model (GCM). (B) Coupled minus atmospheric GCM differences in SST and 850 hPa wind. (C) Subsurface temperature and ocean current responses in the coupled model (vertical velocity magnified by 4000 times). *(Adapted from Li et al. (2016).)*

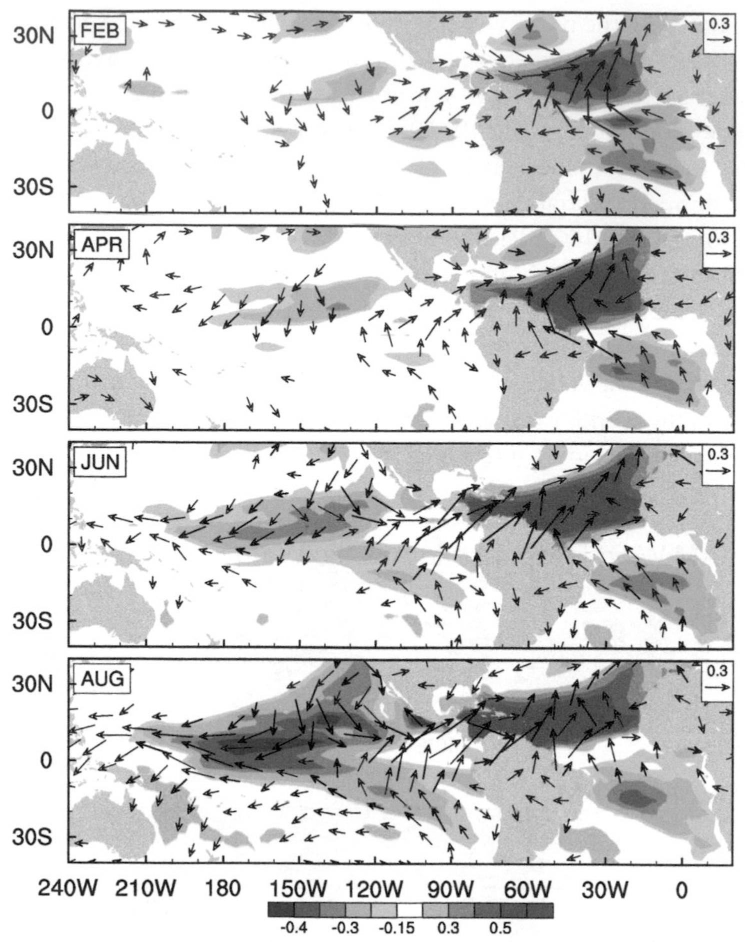

Fig. 10.15 First month-reliant singular value decomposition mode of ensemble spread in 850 hPa wind and sea surface temperature *(shading)* over the tropical Atlantic in perturbed initial condition ensemble forecast initialized in November, in correlation with the principal component. *(Adapted from Ma et al. (2021).)*

10.5 Climate modulation of tropical cyclones

Wei Mei contributed to Sections 10.5.1 and 10.5.4.

Tropical cyclones (TCs, wind speed >17 m/s) are severe storms that develop and grow over the ocean, with devastating impacts on coastal communities due to high winds, heavy precipitation, and storm surges. While TCs are a strongly nonlinear phenomenon individually, some statistics of the genesis, tracks, and intensity are subject to robust controls by the large-scale environment in the ocean and atmosphere. This section discusses these climatic controls. Emanuel (2003) is a concise review on the topic.

BOX 10.1 Coupled modes in forecast ensemble spread

For seasonal forecast, a coupled GCM is initialized each month with estimates of the ocean state and run typically for 12 months. Ocean initial conditions (e.g., SST and the thermocline depth) are the source of predictability. To sample the uncertainties, an ensemble of forecasts is made with slightly different initial conditions in the atmosphere. The size of such a perturbed initial condition ensemble (PICE) is typically 10. The PICE spread gives a measure of uncertainty in the prediction; a small spread indicates a high confidence in the prediction as represented by the ensemble mean, while a large spread indicates low predictability.

Insights into coupled dynamics can be gained by studying the PICE spread. This approach has the advantage of a very large sample size. If each member forecast is run for 12 months, each month during the study period has 120 forecasts (10 members $\times$ 12 months). The sample size increases by another factor of M for an ensemble of M models. Because ENSO dominates the tropics, the Atl3-Nino3 correlation is unstable since 1950, a period of reliable observations. With zero initial oceanic anomalies and a very large sample size, the PICE spread offers an unprecedented view of the behavior of weaker modes such as the AMM, including the effect on the tropical Pacific.

Fig. 10.15 shows the result from a singular value decomposition (SVD) analysis (see Section 1.5.2) of SST-850 mb wind velocity based on sequential monthly maps of PICE spread initialized in November. The analysis is performed in the tropical Atlantic domain, but the results are presented in the global tropics in terms of correlation with the SVD principal component. The northeast trade wind variability in the subtropical North Atlantic, as part of the NAO due to atmospheric internal dynamics (see Section 12.1.1), excites an AMM. The resultant tropical North Atlantic warming excites an atmospheric Rossby wave response that propagates across low-lying Central America into the tropical Northeast Pacific, inducing a cold North Pacific meridional mode (see Section 12.4.1). As the North Pacific meridional mode propagates southwestward due to WES feedback, the anomalous northeasterly winds on the south flank are favorable for La Niña development upon arriving at the equator.

Examples of similar PICE spread analysis over the North Indian and North Pacific oceans are presented in Sections 11.4.4 and 12.4.1.2, reaffirming the respective coupled mode suggested from limited observations.

TCs form over tropical oceans (in 5–25° latitude), except the South Atlantic and southeastern Pacific where SSTs are too low. The northwestern Pacific is the basin where TCs are most active, and the northeastern Pacific features the most TC genesis per unit area. TC genesis exhibits a strong seasonal dependence. It peaks in the late summer and early autumn in either hemisphere (i.e., July-October [JASO] in the Northern Hemisphere and December-March [DJFM] in the Southern Hemisphere).

10.5.1 Genesis potential

TCs are a highly organized convective system with rapidly rotating winds around a warm core. High SSTs (>26.5°C) are a necessary condition for TC genesis. The

SST affects TC genesis primarily by enhancing convective instability and promoting convection to moisten the midtroposphere. Axisymmetric TCs grow in intensity due to the following positive feedback: High wind speed increases surface enthalpy ($k = c_p T + Lq$) and the resultant warm core intensifies the storm. The surface energy flux that grows the storm is proportional to wind speed V while the energy dissipation due to the surface drag is proportional to V^3. When the latter eventually catches up with the former, the axisymmetric storm reaches the maximum intensity,

$$V_{pot}^2 = \frac{C_k}{C_d} \frac{T_s - T_o}{T_o} \left(k^* - k\right) \tag{10.7}$$

where C_k/C_d is the ratio of surface exchange coefficients of enthalpy and momentum, $(T_s - T_o)/T_o$ the thermal efficiency, T_s the SST, T_o the outflow temperature in the upper troposphere, and ($k^* - k$) represents the thermodynamic disequilibrium with k^* the saturated value of the specific enthalpy at the surface. Remarkably, the potential intensity (PI) of highly dynamic TCs is set by environmental thermodynamic variables: SST and the atmospheric temperature profile (Emanuel 1988). The spatial distribution of high potential intensity corresponds well to that of observed TC genesis.

While axisymmetric cyclones attain the potential intensity, real TCs are prevented from achieving their potential by environmental fields (e.g., vertical shear and steering flow) that drive them away from axisymmetry (Wang and Wu 2004). TC track prediction has improved steadily over time, but the intensity prediction has not kept the pace. If the potential intensity were the dominant factor, TC intensity prediction would be as straightforward as applying Eq. 10.7 on predicted TC tracks. The disparity in skill in predicting track and intensity implies strong effects of other environmental variables on TC genesis and intensification.

Four factors are important for TC genesis: thermodynamic potential intensity, vertical wind shear (V_{shear}), midlevel relative humidity (R_H in %), and low-level absolute vorticity ($\eta = \zeta + f$). Based on spatio-seasonal variations of TC genesis, Emanuel and Nolan (2004) proposed a genesis potential index (GPI):

$$GPI = \left|10^5 \eta\right|^{\frac{3}{2}} (1 + 0.1\, V_{shear})^{-2} \left(\frac{R_H}{50}\right)^3 \left(\frac{V_{pot}}{70}\right)^3 \tag{10.8}$$

The spatial distribution of the GPI agrees well with the climatologic frequency of TC genesis.

High relative humidity at midlevels reduces dry entrainment, favorable for TC development. High low-level vorticity helps trap the energy released in convection for TC development. TC genesis rarely occurs within 5° of the equator where the planetary vorticity f vanishes. Only sufficiently away from the equator (finite planetary vorticity *f*), the stretching of the lower tropospheric column due to the convective heating can spin up a cyclonic vortex.

10.5.2 Dynamics of wind shear

Vertical wind shear is often defined as the magnitude of the vector wind difference between 200 and 850 hPa, $V_{shear} = |\mathbf{V}_{200} - \mathbf{V}_{850}|$. Weak wind shear is necessary to sustain the vertical coherence of a vortex and reduce entrainment of dry air into the TC. A comparison of vertical wind shear and TC genesis in JASO shows that TCs develop where the wind shear is below 10 m/s (Fig. 10.16A).

The North Indian Ocean in summer features some of highest SSTs and deepest atmospheric convections on Earth. Yet TCs do not develop. The lack of TC activity during the high summer (July-August) is due to strong vertical wind shear between the low-level southwesterlies and upper-level easterlies (see Section 5.1.1), a shear pattern

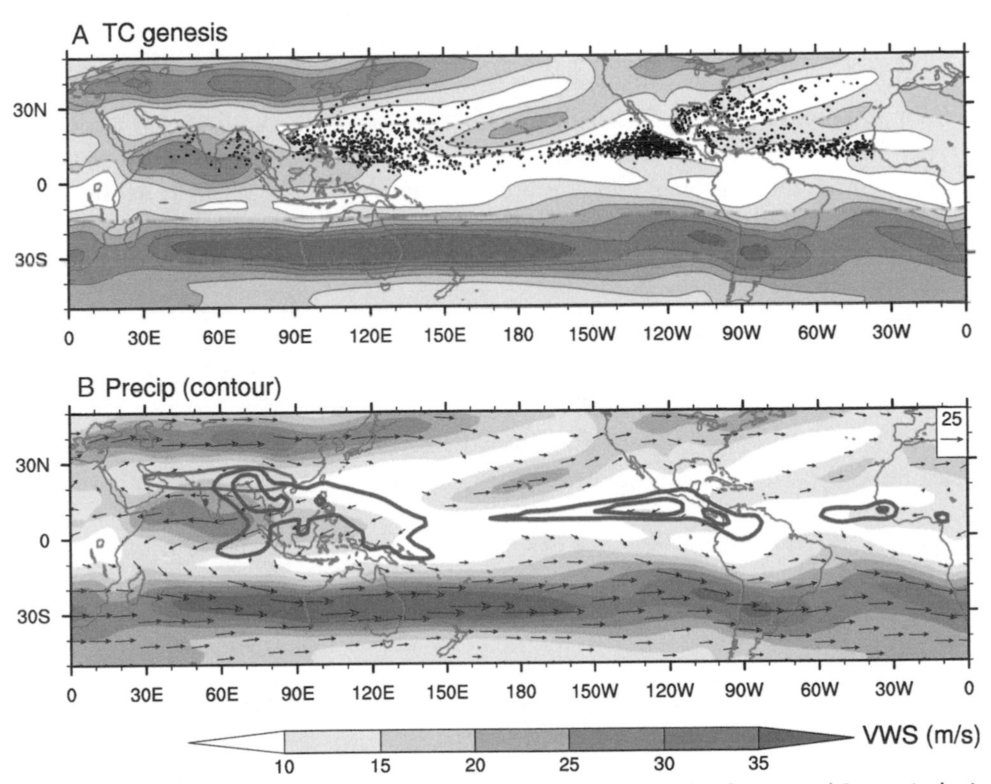

Fig. 10.16 Climatology for July-October (JASO) 1982-2020. Magnitude of 200-850 hPa vertical wind shear (VWS, line contours and shading, m/s), along with (A) tropical cyclone genesis (black dots). (B) Vertical shear vectors and precipitation (red line contours of 7 and 9 mm/day). The blue contour marks 12,500 m geopotential height at 200 hPa and the core of the Tibetan High. (A, Adapted based on Aiyyer and Thorncroft 2006; courtesy Y. Liang.)

explained by Gill's model with broad heating from the Bay of Bengal to northwestern tropical Pacific. Another region shielded by strong vertical wind shear is the central North Pacific near Hawaii, where the SSTs are high enough for TC genesis.

The Matsuno-Gill model is formulated for the baroclinic circulation or vertical shear in the tropics. Although it is the local scalar shear that matters for TCs, dynamically it is more straightforward to study the vector wind shear. The four factors for GPI (Eq. 10.8) are mutually related in the baroclinic atmospheric model of Matsuno (1966) and Gill (1980). An isolated ocean warming intensifies deep convection and upward motion, moistening the midtroposphere. It also induces low-level cyclonic vorticity and weakens the vertical shear in the convection. All these responses favor TC genesis. In reality the convective heating is distributed in complex ways, and the change in the scalar wind shear further depends on the climatologic distribution.

It is dynamically insightful to examine the vector wind shear field (see Fig. 10.16B). In the winter hemisphere, the wind shear is strong and westerly at the poleward terminus of a strong Hadley cell, in thermal wind balance with strong equator-to-pole temperature gradient. In the Northern Hemisphere, strong westerly shear extends from the subtropical North Atlantic all the way through East Asia, while strong easterly shear extends from the Philippines to the African Sahel. This distribution resembles the upper-level Tibetan High (see Fig. 5.4) excited by convective heating in the Asian summer monsoon. Likewise, convective heating over the Western Hemisphere warm pool (from the northeastern tropical Pacific to western tropical Atlantic) drives an upper-level anticyclone, resulting in a northeast-slanted band of high wind shear from Hawaii to California.

10.5.3 Interannual variability

ENSO affects North Atlantic hurricanes, with fewer storms—by a factor of 2—in El Niño than La Niña years, especially for major hurricanes (Fig. 10.17). During El Niño, the enhanced convective heating over the central-eastern tropical Pacific excites a Matsuno-Gill pattern in the upper troposphere with a Rossby dumbbell astride the

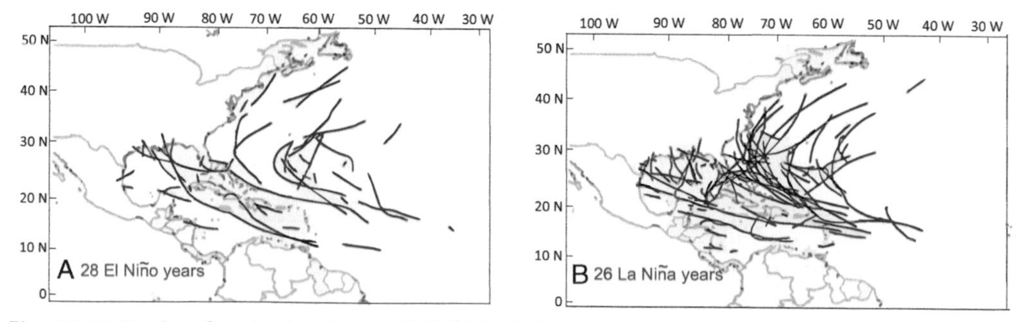

Fig. 10.17 Tracks of major hurricanes (1900-2009): (A) 43 storms over 28 El Niño years, and (B) 82 storms over 26 La Niña years. *(From Klotzbach (2011). © American Meteorological Society. Used with permission.)*

equator (Fig. 10.18C). The Kelvin wave tail intrudes eastward, resulting in anomalous easterly shear across the tropical Atlantic. Superimposed on the mean, the easterly shear anomalies strengthen the scalar shear over the Caribbean Sea and main development region (7.5°–20°N, 85°–15°W) (see Fig. 10.18A) where westward-moving African easterly waves often grow into TCs. The increased wind shear in the main development region reduces TC genesis during El Niño.

Thus SST variability alters the large-scale atmospheric circulation, which in turn modulates TC activity. High-resolution (~50 km in grid spacing) atmospheric GCMs forced with the observed SST variability successfully simulate the interannual-to-decadal variability in TC counts over the North Atlantic (Zhao et al. 2009). A long simulation reaches a correlation of 0.84 with observed North Atlantic hurricane counts during 1970-2010, including the extremely active seasons of 2005 and 2010 (Fig. 10.19). (Considerable uncertainties exist in observed counts in the presatellite era prior to 1966.)

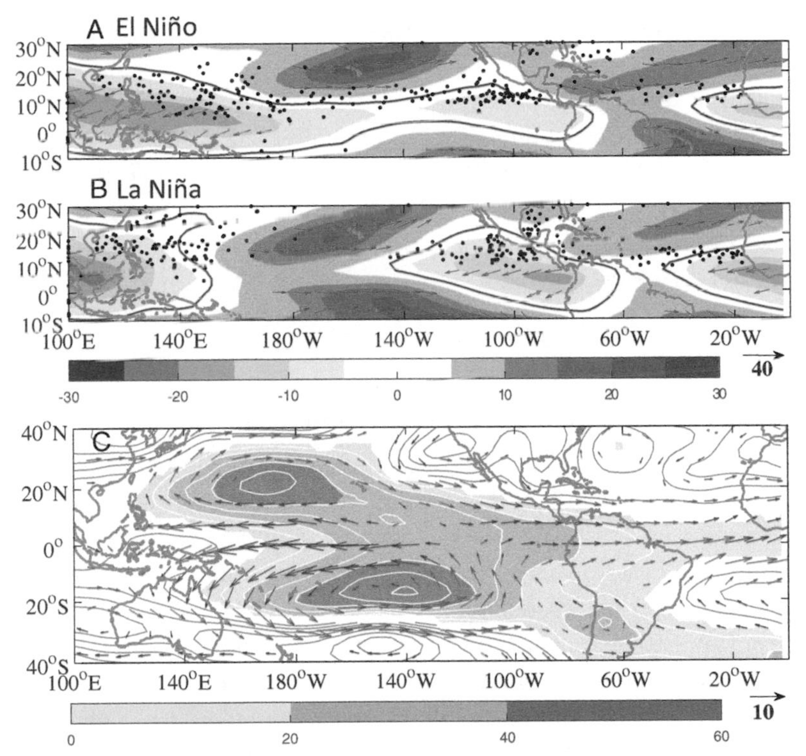

Fig. 10.18 (A) El Niño and (B) La Niña composites for JASO 1982-2020: shear vectors (m/s, zonal component in color shading, zero highlighted in red contours) and tropical cyclone genesis (*black dots*). (C) El Niño–La Niña differences in shear vectors and 850-200 hPa mean temperature (positive/negative in gray shading/contours). The anomalous shear vectors that strengthen/weaken the mean scalar shear are colored in red/blue. *(Courtesy Y. Shi and Y. Liang.)*

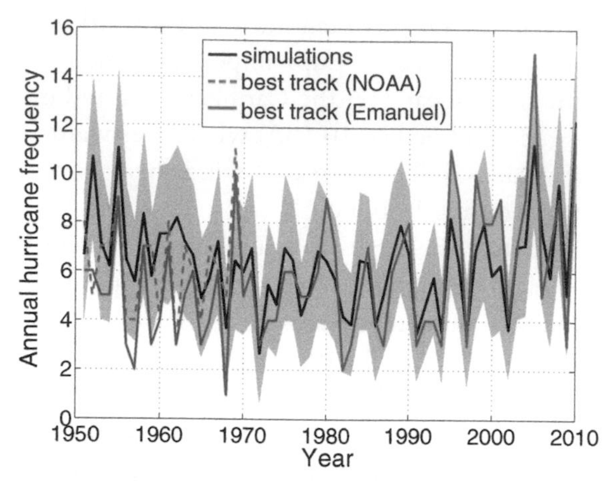

Fig. 10.19 North Atlantic tropical cyclone counts in observations and a high-resolution atmospheric general circulation model forced with observed sea surface temperatures. The gray shading represents one standard deviation among 100 member runs. The cross-correlation between observations and the ensemble-mean simulation is 0.84 for 1970-2010. *(Based on Mei et al. (2019))*

The simulation also captures an interdecadal oscillation in TC counts, a gradual decline from 1950 to around 1980 and a rebound afterwards. This coincides with the Atlantic Multidecadal Oscillation (AMO; see Section 12.3.2). When the SST over the tropical North Atlantic is anomalously high, a cyclonic wind shear forms with a slight northwestward displacement. These circulation anomalies tend to reduce the climatologic wind shear, producing a favorable environment for TC formation and intensification (Goldenberg et al. 2001).

10.5.4 Ocean feedback

The summer ocean mixed layer is shallow, only 20 to 40 m deep. TCs often entail marked sea surface cooling in the wake (Fig. 10.20A) due to intense surface heat loss, wind-induced Ekman upwelling, and vertical mixing. On the right side of the TC center, the winds are strong and change the direction in an anticyclonic sense that preferentially excites inertial oscillations above the seasonal thermocline. The strong vertical shear of the resultant inertial motions generates turbulence that entrains cold water and deepens the mixed layer (to $\sim$100 m deep). This preferential excitation of ocean inertial motions explains that cold wakes tend to form to the right of TC center (see Fig. 10.20A, B).

The sharp decrease in SST during and shortly after the TC passage reduces the surface energy flux from the ocean and is a negative feedback on TC intensity. Coupled regional ocean-atmosphere models forced with the large-scale environment consistently show that an interactive ocean reduces the storm intensity due to the formation of a cold wake (e.g., Bender et al. 1993).

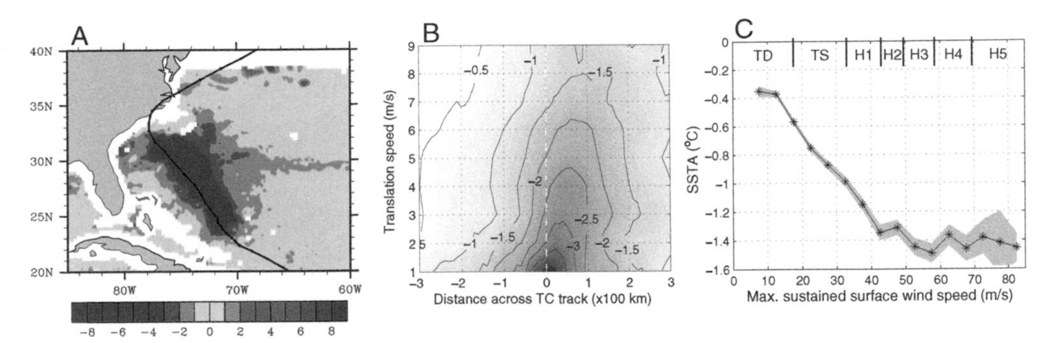

Fig. 10.20 (A) Sea surface temperature (SST) change (°C) for Hurricane Bonnie over a week of 19-26 August 1998. (B) Tropical cyclone (TC)–induced SST cooling (°C) as a function of cross-track distance and translation speed. (C) TC-induced SST cooling as a function of maximum wind speed. *(A, Adapted from Sriver and Huber (2007); B and C, Mei and Pasquero (2013). © American Meteorological Society. Used with permission.)*

The TC-induced SST cooling depends on several factors, including TC intensity, size, and translation speed as well as the prestorm ocean stratification. Generally, strong SST cooling occurs for stronger, larger, slower-moving TCs, and stronger stratification in the seasonal thermocline. A simple ocean mixed layer model forced by TC winds predicts a linear relationship between TC intensity and the magnitude of SST cooling. In observations, this relationship holds up to category 3 intensity, while the SST cooling levels off for stronger TCs (see Fig. 10.20C). The level-off has to do with the effect of TC translation speed: The SST cooling in the wake decreases with increasing translation speed (see Fig. 10.20B). In other words, only fast-moving TCs can grow to categories 4 and 5 by limiting the SST cooling. This is consistent with the observations that TC intensification rate increases as the SST cooling decreases and that TCs with a larger intensification rate on average move faster.

The upper ocean stratification relevant to TC intensification is often represented by the ocean heat content above the 26°C isotherm (H26). The isotherm is chosen because it is a typical environmental air temperature in the tropical region conducive to TC genesis (of SSTs >26.5°C). A cold wake with SSTs below 26°C would choke off the storm intensification. A modified potential intensity by replacing SST with vertical mean water temperature in the upper 80 m in Eq. 10.7 tracks the observed maximum intensity well, especially for slow-moving TCs that tend to generate a strong SST cooling (Lin et al. 2013).

Hurricane Katrina of late August 2005 in the Gulf of Mexico illustrates the upper ocean stratification effect. The Gulf Stream enters the Gulf from the south through Yucatán Channel between Mexico and Cuba, makes a semipermanent anticyclonic meander known as the Loop Current, and exits through the Florida Strait. The Loop Current is highly variable in time and often sheds anticyclonic, warm eddies that

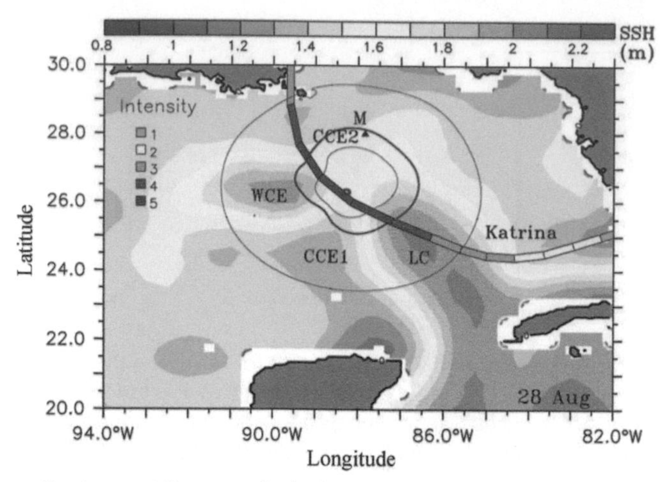

Fig. 10.21 Hurricane Katrina rapidly intensified when it passed over the warm Loop Current (LC) and a warm core eddy (WCE) in the Gulf of Mexico. The tropical cyclone track and intensity (color) on the background of sea surface height. The external, intermediate, and inner magenta circles stand for 10-m wind speed of 18, 28, and 33 m/s, respectively. *(From Jaimes and Shay (2009). © American Meteorological Society. Used with permission.)*

move northward. The Loop Currents and warm eddies are associated with the deepened thermocline and elevated sea level (Fig. 10.21). Katrina experienced rapid intensification over the Loop Current and a warm eddy to the northwest on its way to make landfall in New Orleans (see Fig. 10.21), making it the costliest Atlantic hurricane.

10.6 Summary

The equatorial Atlantic and Pacific are similar in several ways, both featuring a cold tongue that displays a pronounced annual cycle. The onset of the West African monsoon causes rapid equatorial cooling in the Atlantic in May and June, which is further amplified by the thermocline shoaling and Bjerknes (1969) feedback.

On interannual and longer timescales, no single mode seems to dominate. Instead, several mechanisms are responsible for tropical Atlantic variability. On the equator, Bjerknes feedback is weaker in the Atlantic than in the Pacific probably because of the smaller zonal width of the Atlantic basin and the rapid northward migration of the ITCZ after April. Thus the equatorial Atlantic SST variability is modest in amplitude and lasts only for a few months in boreal summer. The Atlantic Niño tends to cause rainfall to increase along the coasts of Guinea.

In addition to the equatorial mode, WES feedback gives rise to AMM variability, where cross-equatorial SST gradient and the marine ITCZ are coupled. The AMM affects rainfall over the surrounding continents, specifically over northeastern Brazil.

These tropical Atlantic modes are highly seasonal: the AMM develops in March–May when the equatorial Atlantic is uniformly warm; and the equatorial mode is most pronounced in the boreal summer coinciding with the season of the cold tongue and the shallow thermocline in the east.

In addition to well-documented ENSO effects on the tropical Atlantic through the PNA teleconnection and Walker circulation, there is increasing evidence that tropical Atlantic SST variability affects ENSO development in the Pacific through atmospheric Rossby waves across Central America and/or the equatorial Kelvin wave across the tropical Indian Ocean. Because it is a weaker signal than ENSO itself, much of this tropical Atlantic effect on ENSO is still under active research, and some of the details are probably sensitive to the spatial pattern and seasonal timing of the Atlantic anomalies.

Large-scale atmospheric-ocean environments affect the frequency, track, and intensity of TCs. Of particular importance is the vertical wind shear, which is subject to nonlocal equatorial wave dynamics (e.g., ENSO effect on Atlantic TCs). The ocean coupling is a negative feedback on TC intensity, which is a function of upper ocean thermal structure (e.g., the 26°C isotherm depth) and translation speed. North Atlantic TC counts tend to decrease as the vertical shear increases during El Niño summers. This relationship mediated by the atmospheric Kelvin wave is an important basis for seasonal prediction of North Atlantic hurricane activity.

Review questions

1. Compare the annual cycle of the equatorial cold tongue between the Pacific and Atlantic. How does the African continent affect the Atlantic annual cycle?
2. Cross-equatorial advection of easterly momentum is important for the seasonal cycle in zonal wind in equatorial oceans. Discuss why this mechanism causes an annual cycle in zonal wind in the equatorial Atlantic but a semiannual one in the equatorial Indian Ocean.
3. Estimate the annual and semiannual harmonics of the zonal wind based on Eq. 10.4, $\overline{v} = 4$ m/s, and $\widetilde{v} = 2$ m/s. Compare the estimates with observations in Fig. 10.3.
4. In which season is the Atlantic Niño mode most pronounced? What causes this phase locking?
5. The interannual SST variance in the equatorial cold tongue is much smaller in the Atlantic than Pacific. Why? Hint: Consider basin size and annual cycle in the mean thermocline depth.
6. Discuss coupled feedbacks involved in the tropical Atlantic meridional mode. Which season is this mode locked to? Explain briefly.
7. When is the rainy season in northeastern Brazil (Nordeste)? To which climate modes is its rainfall sensitive?

8. The first empirical orthogonal function of interannual rainfall variability over the tropical Atlantic during March-April represents an anomalous north-south shift of the ITCZ. What SST pattern is associated with this rainfall mode? Describe briefly ocean-atmospheric feedback for this mode.
9. What are favorable conditions for TC genesis? Why don't TCs form on the equator?
10. Why is it dynamically more insightful to study the vector wind shear even though it's the scalar shear that affects hurricanes?
11. Use the Matsuno-Gill model to discuss the far-reaching effect of the Asian summer monsoon on the vertical shear. What drives the vertical wind shear that shields Hawaii from hurricanes?
12. How does ENSO affect North Atlantic hurricanes? Discuss the response and effect of the vertical wind shear. Explain the equatorial wave dynamics involved.
13. How does the thermocline depth affect TC intensification?
14. Ocean dynamic theory predicts that the ocean response increases in magnitude with TC intensity. Why then in reality are Cate 4-5 TCs not accompanied by intense SST cooling in the wake?
15. The open ocean in summer features the following temperature profile: 28°C at the surface, with a 20-m mixed layer and a constant rate of decrease underneath to 20°C at 100 m. Suppose that TCs mix the upper 100-m water column. Calculate the SST after the TC mixing. Neglect the surface heat flux effect.
16. Now consider a shelf sea that is 40 m deep, connected to the open ocean and with the same vertical profile. How much surface cooling does the TC mixing cause? Compare with that over the open ocean.
17. Accra (5°33′N 0°12′W) is the capital city of Ghana on the south coast facing the equatorial Atlantic. Its monthly climatology is available on Wikipedia. Answer the following questions.
    - The daily mean temperature drops to 24.3°C in August. What causes the cooling in midsummer?
    - How is this cooling related to local rainfall? Monthly rainfall drops from the annual maximum of 221 mm in June to 28 mm in August.
    - How might rainfall be related between Accra and Niamey (13°31′N 2°8′E) of the Sahel during summer?

CHAPTER 11

Indian Ocean variability

Contents

The Indian Ocean (IO) differs from the Atlantic and Pacific in several important ways. The Asian continent drives a strong monsoon (see Chapter 5), and the monsoonal winds generate large seasonal variations in ocean currents, some with annual reversals in direction (e.g., the Somali Current). Equally important, the IO lacks steady easterly trade winds on the equator. Instead, the annual mean winds are weakly westerly on the equator as part of the western cell of the Walker circulation anchored by deep convection over the Maritime Continent (Fig. 11.1). As a result, there is no year-round equatorial upwelling in the IO. Instead, year-round upwelling takes place in the tropical South IO between the equatorial westerlies and southeast trades. Seasonal upwelling occurs in the Northern Hemisphere off East Africa and the Arabian Peninsula and east and west of the tip of India, and in the Southern Hemisphere off the west coast of Sumatra and Java Islands of Indonesia. As will become clear, these upwelling zones are important players in IO variability.

11.1 Seasonal cycle

Precipitation in the IO sector shows a pronounced seasonal migration in the meridional direction (Fig. 11.2), following the Sun and regions of warm sea surface temperature (SST) (see Fig. 5.1). In northern winter (DJF), tropical IO climate is nearly symmetric about the equator. The marine intertropical convergence zone (ITCZ) and meridional

Coupled Atmosphere-Ocean Dynamics: from El Niño to Climate Change
ISBN 978-0-323-95490-7, https://doi.org/10.1016/B978-0-323-95490-7.00011-4

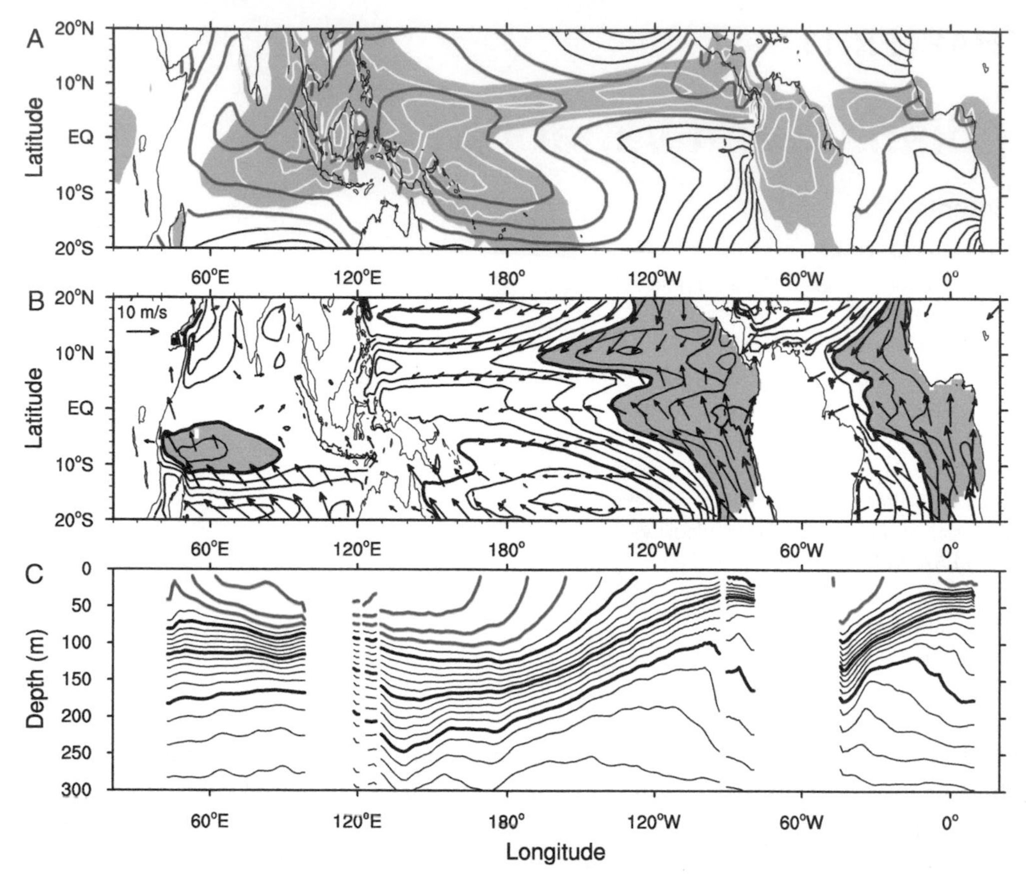

Fig. 11.1 The annual mean climatology: (A) sea surface temperature (SST; contours at 1°C intervals; red, thickened ≥27°C) and precipitation (shaded >4 mm/day; white contours at intervals of 2 mm/day); (B) surface wind velocity (vectors, m/s) and 20°C isotherm depth (20-m interval, shaded <100 m; 100- and 200-m contours thickened); and (C) ocean temperature at the equator (1°C interval; 15°C, 20°C, and 25°C thickened; ≥27°C in red). Based on ERA5, ERSST, GPCP, and ORAS4 climatology for 1979-2017. *(Courtesy C.Y. Wang.)*

SST maximum are slightly displaced south of the equator with the easterly trades on either side of the equator. Latent-heat loss caused by cool, dry air blowing off the Asian continent cools the North IO and South China Sea. While the DJF rainfall maximum is displaced south of the equator only by 5° latitude, the intense heating over the Asian continent draws monsoon convection much further poleward in JJA (see Fig. 11.2). Marked precipitation in the India-Bay of Bengal sector (70−100°E) penetrates as far north as 25°N. The strongly northward-displaced monsoon heating drives the southwest monsoon winds from the Arabian Sea through the South China Sea into the far western Pacific, much as depicted in Gill's baroclinic model (see Section 5.1.1). The intense southwest Findlater wind jet off East Africa causes strong summertime cooling in the western Arabian Sea as a result of upwelling off the Somali and Omani coasts. The upwelling cooling keeps atmospheric convection from the western Arabian Sea.

The seasonal migration of deep convection causes the meridional wind at the equator to reverse direction between DJF and JJA (see Fig. 11.2). This annual cycle in meridional wind drives a semiannual cycle in zonal wind on the equator. We ignore zonal variation for simplicity. Unlike the tropical Atlantic, the cross-equatorial asymmetry in annual-mean climate of the tropical IO is not very strong ($\overline{v} \approx 0$). The zonal wind perturbation at the equator due to the meridional advection in Eq. 10.3 becomes

$$u' = -\beta v'^2/\varepsilon \tag{11.1}$$

where the prime denotes the deviations from the annual mean. In other words, the cross-equatorial monsoon causes easterly acceleration on the equator, at both the southerly (JJA) and northerly (DJF) phases. Schematically (Fig. 11.3A), a southerly cross-equatorial flow in northern summer induces westerly lateral shear across the equator because of the beta effect. The southerly advection of the easterly momentum from the south of the equator causes an easterly acceleration on the equator. As a result, the equatorial westerlies decelerate during the summer and winter monsoon seasons and accelerate during the intermonsoon seasons (centered in May and November, respectively) (Fig. 11.4A). The advective correction (Eq. 11.1) is important only near the equator where the perturbation zonal wind vanishes in zonal-mean linear dynamics (Eq. 10.2). The semiannual zonal-wind harmonic is indeed trapped by the equator (see Fig. 11.3B), in support of the cross-equatorial advection mechanism.

The intensified equatorial westerlies drive strong eastward surface currents on the equator in May and November (see Fig. 11.4A), much as depicted by Yoshida's (1959) theory of an equatorial jet in an unbounded ocean in response to an abrupt onset of a spatially uniform zonal wind forcing (see Section 7.2.2). These equatorial jets during transitional periods between monsoons often exceed 1 m/s in speed and are commonly

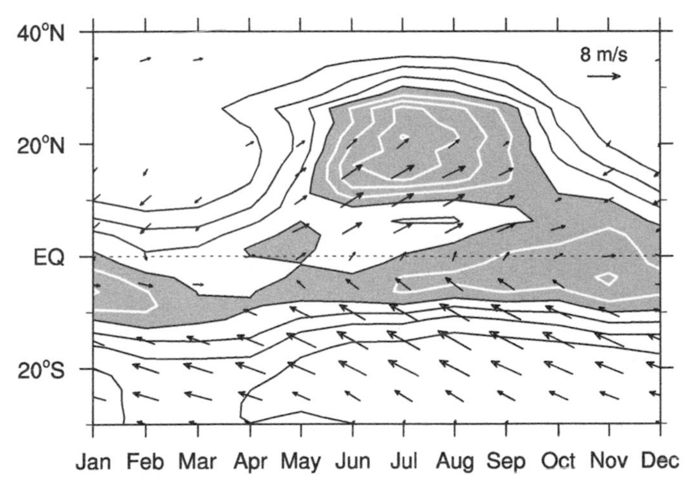

Fig. 11.2 Time-latitude section of rainfall and surface wind velocity (vectors in m/s) averaged in 70°–100°E. Rainfall contours at 1.5 mm/day intervals (gray shading >6.0 mm/day). *(Courtesy C.Y Wang.)*

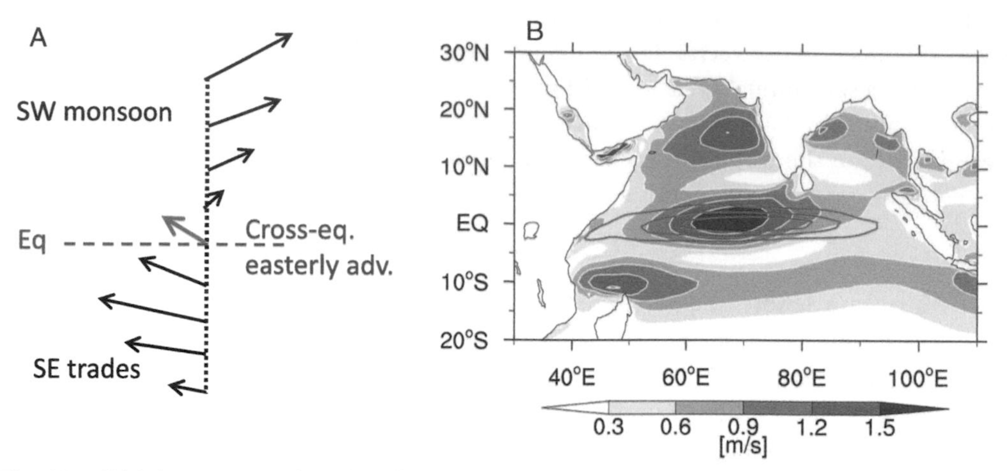

Fig. 11.3 (A) Schematic meridional profile of a southerly summer monsoon flow. (B) Semiannual harmonics for surface zonal wind (gray shading, m/s) and ocean current velocity averaged in top 50 m (red contours, every 0.1 m/s) based on ERA5 and ORAS4 (1979-2017). Updated based on Ogata and Xie (2011).

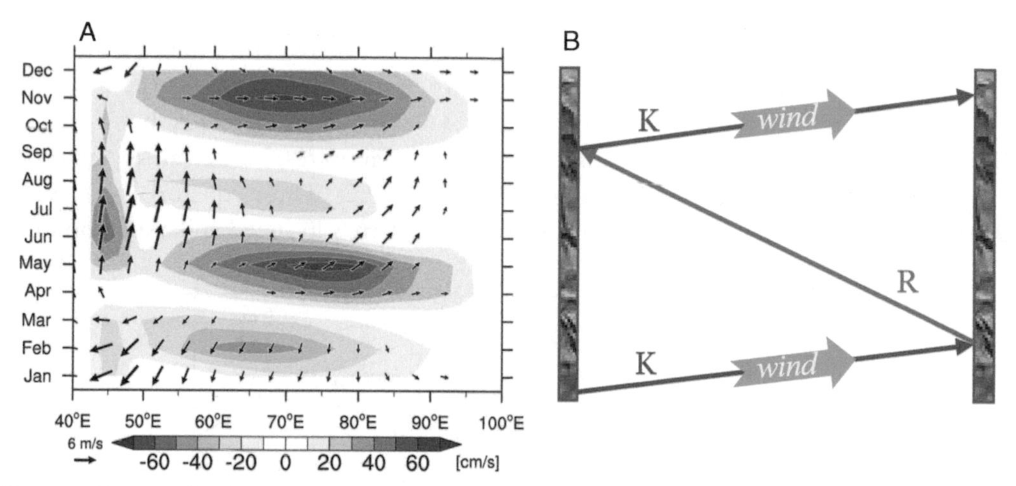

Fig. 11.4 Equatorial Indian Ocean seasonal cycle. (A) Longitude-time section of surface wind velocity (arrows, m/s) superimposed on surface zonal current (top 50 m averaged; color shading, cm/s). (B) Schematic for a resonance basin mode. *(A, Courtesy C.Y. Wang.)*

referred to as the Yoshida-Wyrtki jets or the Wyrtki jets, in honor of the discoverer (Wyrtki 1973) in the equatorial IO.

Wave reflections on the west and east boundaries of the equatorial IO cause resonance with the semiannual zonal wind forcing. The intensified westerly winds drive downwelling Kelvin waves, which reflect into downwelling Rossby waves on the eastern boundary. If the westerlies intensify just when the downwelling Kelvin waves return from the western boundary after the Rossby waves have criss-crossed the basin, a resonance occurs. Generally, the resonance occurs for a zonal wind forcing that oscillates at the periods

$$T = \frac{4L}{mc} \tag{11.2}$$

where L is the zonal width of the basin, c is the Kelvin wave phase speed, and m is an integer (Cane and Sarachik 1981). The time for the Kelvin wave and long Rossby wave of first meridional mode to cross the basin is L/c and $3\ L/c$, respectively (see Fig. 11.4B). The tropical IO is nearly in resonance with the second baroclinic mode ($c \sim 163$ cm/s) at the semiannual frequency (Han et al. 1999). This resonance is another reason why the semiannual Wyrtki jets dominate the equatorial IO.

11.2 Zonal mode: Indian Ocean dipole

Prior to 1997 it was widely assumed that the influence of the IO on climate variability, beyond its impact on the monsoons, was weak. There are several lines of evidence in support of this view. First, SST gradient is weak over the broad tropical IO except in the western Arabian Sea, making ocean horizontal advection ineffective in causing SST variability. Second, the thermocline is deep, with the 20°C isotherm generally deeper than 100 m. On the equator, the weak westerly winds keep the thermocline flat and deep, a condition that together with lack of ocean upwelling is not conducive to thermocline feedback on SST variability. Indeed, on the seasonal timescale, the semiannual Wyrtki jets have little effect on equatorial SST, which features a weak annual cycle in the eastern two-thirds of the basin.

This view of a climatically incept IO shifted in the fall of 1997 when dramatic changes took place in both the equatorial Pacific and IO (Fig. 11.5A). The characteristic cold tongue failed to develop and instead warm water with SST above 27°C spread across the equatorial Pacific from Indonesia to Ecuador. Over the equatorial IO where uniform high SSTs are expected, by contrast, a cold tongue emanated from Indonesia toward the west. The easterly trades ceased over much of the equatorial Pacific, while the equatorial westerlies were replaced by the easterlies across the equatorial IO, representing a major relaxation of the Walker circulation. The world was upside down! The tropical Pacific was IO-like, while the tropical IO was Pacific-like.

By several measures, 1997 was an extraordinary year for the IO and countries on its rim. Record rainfall was observed in East Africa during October and November. Severe flooding in Somalia, Ethiopia, Kenya, Sudan, and Uganda caused 2000 deaths and displaced hundreds of thousands of people. On the other side of the ocean, Indonesia suffered severe droughts at the same time, and wildfires broke out of control on several of its

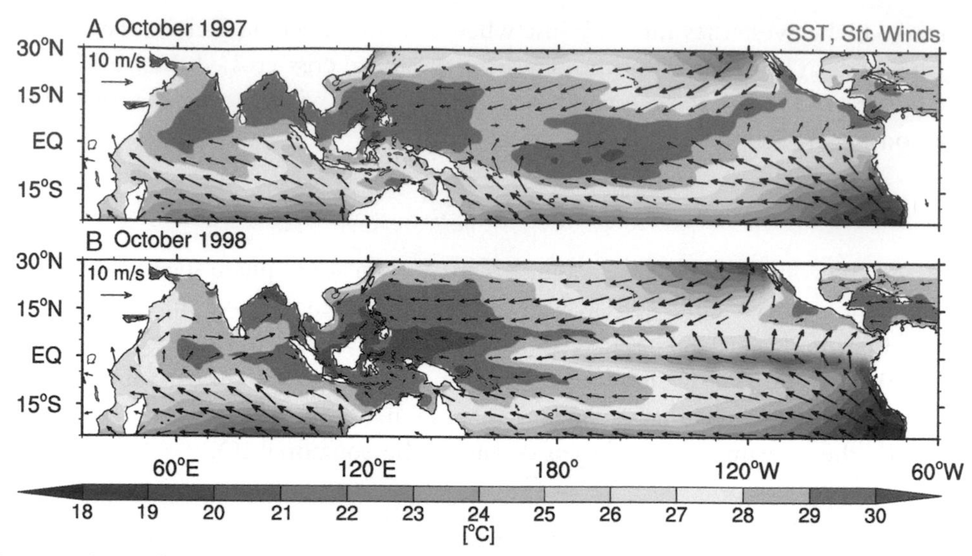

Fig. 11.5 Sea surface temperature (°C) and surface wind velocity (m/s) in (A) October 1997, and (B) October 1998. In the fall of 1997, the Pacific cold tongue failed to develop, but a cold tongue appeared in the equatorial Indian Ocean. Surface velocities with speed smaller than 2 m/s are not drawn. *(Courtesy C.Y. Wang.)*

islands. The smoke and haze they generated caused severe health problems in Indonesia and surrounding countries. Not all of these climatic anomalies were due directly to the El Niño of the century; unusual conditions in the tropical IO, specifically an anomalous cooling in the eastern equatorial IO (see Fig. 11.5A), contribute. The eastward Wyrtki jet failed to develop during July-November. Studies to uncover the physical cause of the extraordinary 1997 event led to rapid progress in understanding IO-atmosphere interaction and modes of climate variability.

While the annual-mean winds over the equatorial IO suggest that Bjerknes feedback is inactive, this does not rule out the possibility that seasonal upwelling opens a window of ocean-atmosphere coupling. Indeed, conditions are favorable for Bjerknes feedback during the boreal summer and fall in the southeast equatorial IO off Sumatra Island, sometimes leading to the development of La Niña—like ocean-atmospheric anomalies, namely, Indian Ocean dipole (IOD) events.

IOD events develop in June and peak in October, a seasonality due to ocean dynamics. The climatologic winds are southeasterly off the Sumatra coast from April-October (Fig. 11.6), reaching a peak during the summer at which time the winds have an easterly component all along the equator. The southeasterlies favor a shallow thermocline and coastal upwelling in the southeast equatorial IO, opening a seasonal window for thermocline feedback and the IOD. Indeed, major SST variance off the Indonesian coast is confined to this upwelling season. After October, the southeasterly winds weaken off

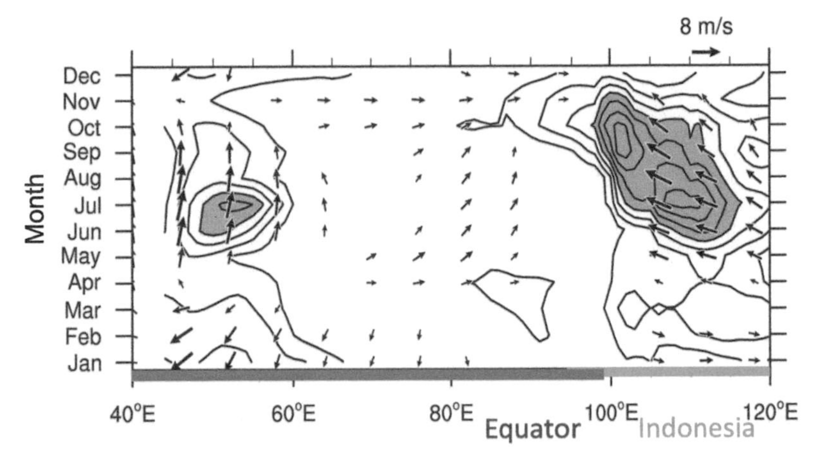

Fig. 11.6 Standard deviation of monthly sea surface temperature (contours at 0.1 K interval, shading ≥0.7 K), along with climatologic wind velocity (m/s) (speed <2.5 m/s are not drawn) on the equator *(blue x-axis)* and then along the west coast of Indonesia *(green)*, where a window for thermocline feedback opens in June-October. Updated based on Xie et al. (2002). *(Courtesy C.Y. Wang.)*

Sumatra, closing the window of upwelling and resulting in the rapid termination of IOD events.

A typical IOD event manifests itself through a zonal gradient of tropical SST, with marked cooling off Sumatra and moderate warming in the western ocean. As the IOD develops (September-November), an east-west dipole of anomalous rainfall is established over the tropical IO, with precipitation increasing in the west because of the low-level convergence associated with the anomalous equatorial easterlies, and vice versa in the east (Fig. 11.7). The rainfall dipole is an important element of Bjerknes feedback that sustains the IOD. Easterly wind anomalies blow from the cold/dry eastern equatorial IO to the warm/rainy western IO, lifting the thermocline in the eastern equatorial IO and amplifying the SST cooling there. To track IOD variability, a dipole mode index (DMI) is defined as the west (50−70°E and 10°S−10°N) minus east (90−110°E and 10°S−equator) difference in SST across the equatorial IO (Saji et al. 1999).

During El Niño, atmospheric convection shifts eastward and intensifies over the central-to-eastern equatorial Pacific, resulting in slow anomalous subsidence over the western Pacific and Maritime Continent. This large-scale shift in convection weakens the Walker circulation, with easterly (westerly) anomalies in the equatorial IO (Pacific) that drives IOD (El Niño) (Fig. 11.8B). Indeed, positive IOD events with the eastern cooling tend to occur with a developing El Niño in the Pacific (see Fig. 11.8A).

Most but not all IOD events are associated with El Niño and the Southern Oscillation (ENSO). Major positive IOD events (with anomalous cooling off Sumatra) of 1961 and

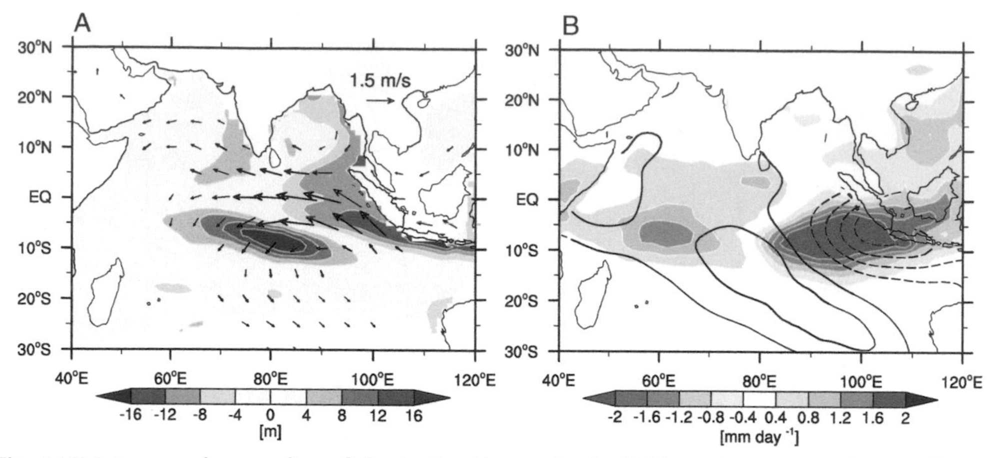

Fig. 11.7 Interannual anomalies of the Indian Ocean dipole (IOD) mode as regression coefficients against normalized PC1 of tropical IO *Z*20. (A) Surface wind (m/s) and 20°C isotherm depth (color shading, m). (B) Sea surface temperature (black contours, every 0.1 K, zero contour omitted) and precipitation (color shading, high values in white contours). Arrows are omitted when regression coefficients are <1.0 m/s. Updated based on Saji et al. (2006). *(Courtesy C.Y. Wang.)*

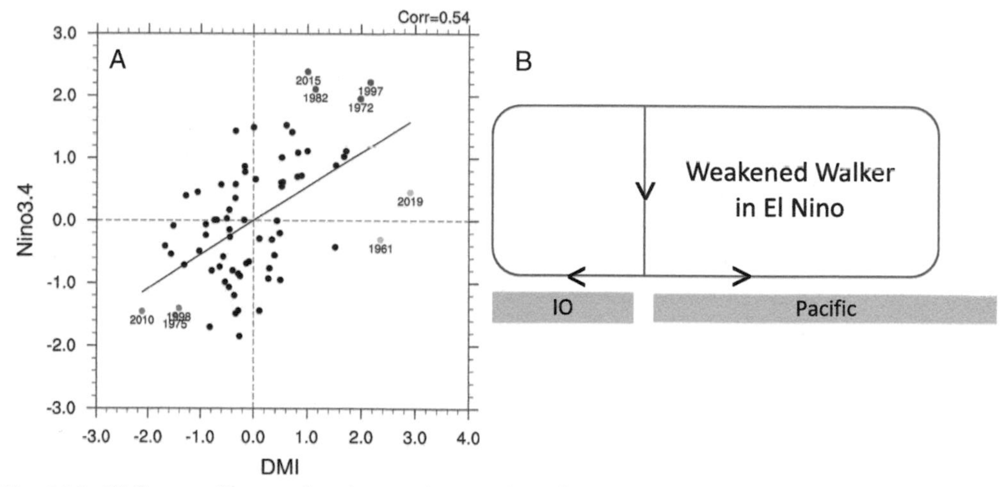

Fig. 11.8 (A) Scatter diagram for observed anomalies of SON DMI and NDJ Niño3.4 sea surface temperature for 1950-2020. (B) Schematic of the weakened Walker circulation during El Niño. *(A, Courtesy X.D. Wang.)*

2019 do not coincide with El Niño. In coupled climate models, IOD variability still exists even when tropical Pacific SST variability is artificially suppressed (Yang et al. 2015). This indicates that the IOD is an intrinsic mode of the IO forced, often but not always, by ENSO.

11.3 Basin mode

Empirical orthogonal function (EOF) decomposition of IO SST variability results in two modes (Fig. 11.9). EOF2 is the IOD mode, and principal component 2 (PC2) variance peaks in September. EOF1 explains 39% of the total variance and features a basinwide increase or decrease. PC1 is highly correlated with ENSO, with the variance peaking in February. While the so-called IO basin mode represents the basinwide warming following an El Niño event, it features distinct temporal evolution and mechanisms for SST variability in subbasins (e.g., between the North and South IO). This section highlights such subbasin variations in regional ocean mechanisms and coupling with the atmosphere.

11.3.1 Thermocline ridge

The southeast trades prevail year-round south of 10°S and are strongest at 20°S. Open-ocean Ekman upwelling takes place between the southeast trades and equator where the winds are weakly westerly in annual mean. From the Sverdrup relationship (Eq. 7.26),

$$\beta v = -\frac{\beta}{f} g' \frac{\partial h}{\partial x} = \frac{f}{H} w_E \qquad (11.3)$$

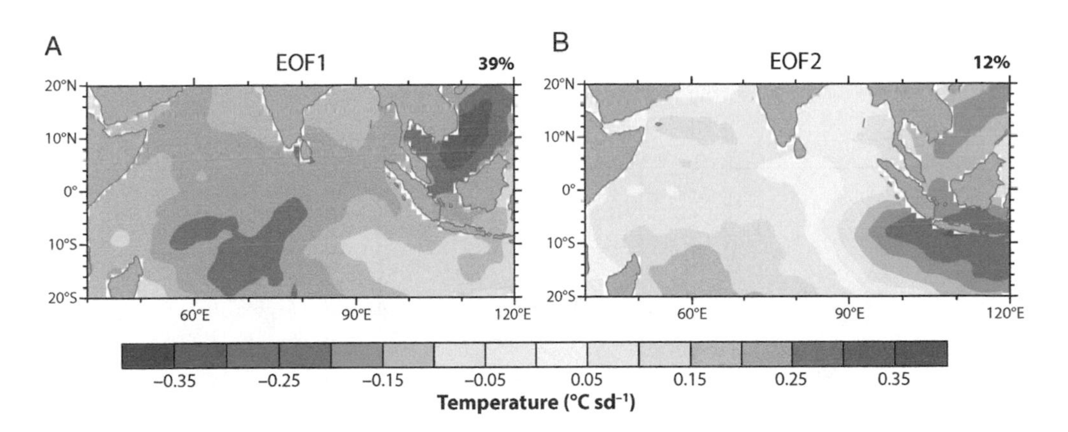

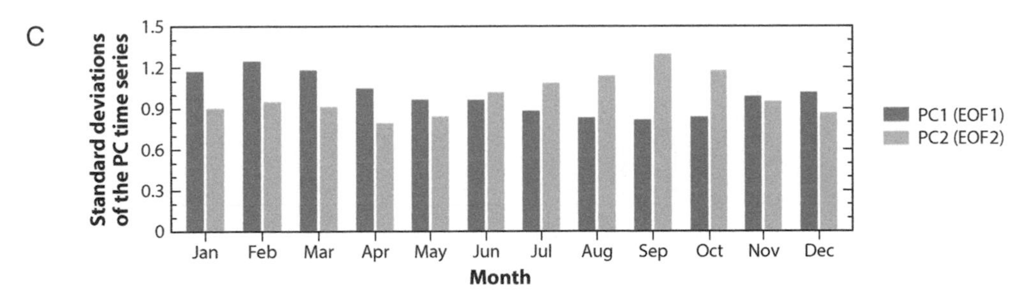

Fig. 11.9 (A, B) Leading empirical orthogonal function *(EOF)* modes of sea surface temperature variability over the tropical Indian Ocean, and (C) the standard deviations of their principal components as a function of calendar month. *(From Deser et al. (2010).)*

The Ekman upwelling [$w_E = \frac{1}{\rho} curl\left(\frac{\tau}{f}\right) > 0$] shoals the thermocline westward. The westward intensification due to the Rossby wave propagation forms a thermocline ridge in 5° to 10°S in the Southwest IO (see Fig. 11.1B). At the shallowest, $Z20$ is less than 80 m as compared to the background value of 120 m in the tropical IO. The thermocline ridge is often called the Seychelles Dome.

In the Seychelles Dome, the thermocline rises close to the surface. As a result, thermocline displacements strongly influence SST variability. Local correlation between $Z20$ and SST variability is high over the Dome, indicative of the thermocline feedback.

Toward the mature phase of El Niño (October-December), a region of anomalous, anticyclonic, wind-stress curl forms in the tropical Southeast IO (see Fig. 11.7A), caused by a Walker-type atmospheric bridge from the Pacific and local SST cooling associated with the IOD. The anticyclonic wind curls force downwelling Rossby waves in the Southeast IO, which, upon arrival in the Seychelles Dome several months later, deepen the thermocline and warm SST there. Thus the Southwest IO SST anomalies are largely determined by remotely forced changes in the depth of the thermocline. Embedded in a basinwide warming that peaks in February-April following El Niño, SST anomalies display a positive core that propagates westward with the downwelling Rossby waves (Fig. 11.10). The slow propagation of ocean Rossby waves prolongs the SST warming

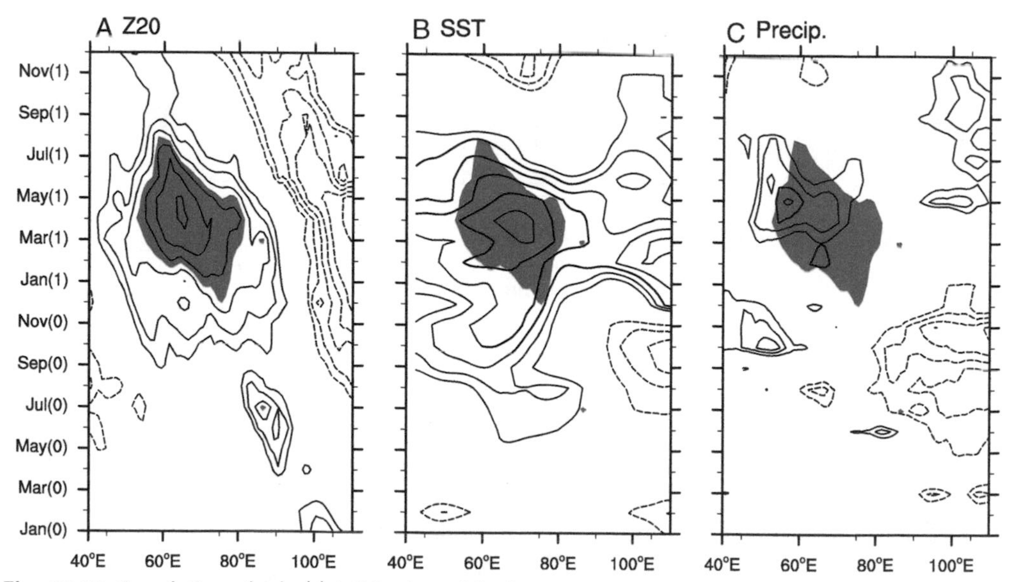

Fig. 11.10 Correlations (only |r| ≥0.3 plotted in line contours, 0.1 interval, negative dashed) with Nov(0)-Jan(1) Niño3.4 sea surface temperature (SST) as a function of longitude and calendar month: (A) $Z20$, (B) SST, and (C) precipitation, averaged in 8–12°S. The color shading indicates where $Z20$ correlation exceeds 0.55. Updated based on Xie et al. (2002). *(Courtesy C.Y. Wang.)*

over the Seychelles Dome through August(1) following El Niño. Here the numerals in the parentheses denote the developing (0) and decay (1) years of El Niño.

The ITCZ is displaced south of the equator during boreal winter and spring, covering the 5° to 10°S thermocline ridge. During post—El Niño March-August when the direct Pacific effect through atmospheric bridge weakens, the Southwest IO warming intensifies atmospheric convection in the region (see Fig. 11.10C), associated with a local cyclonic circulation and entailing anomalous northeasterlies across the equator in the broad North IO (Fig. 11.11).

December-April is the active tropical cyclone season for the Southwest IO. During and following an El Niño, the deepened thermocline and increased SST strengthen

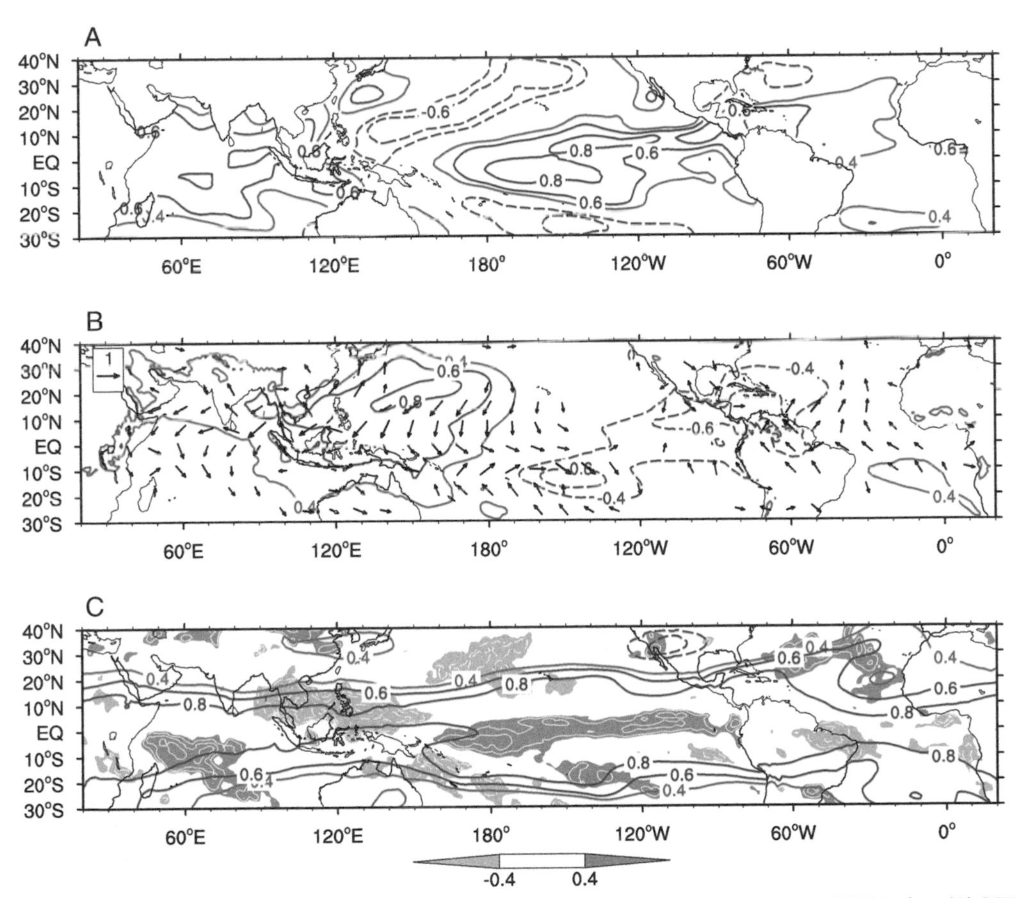

Fig. 11.11 MAM(1) correlation with the ND(0)J(1) Niño3.4 sea surface temperature (SST) index: (A) SST; (B) sea level pressure (contours) and surface wind velocity; and (C) precipitation (gray shade and white contours at intervals of 0.1) and tropospheric temperature (850—300 hPa average, line contours). *(Courtesy C.Y. Wang.)*

tropical cyclone activity in the region. The most intense tropical cyclone of the region, Fantala took place in April 2016, following the major El Niño event of 2015-16.

11.3.2 Wind-evaporation-sea surface temperature (WES) feedback in boreal spring

Except the Southwest IO where the thermocline feedback is important, surface heat flux changes, especially the wind-induced latent-heat flux, are the major cause of SST variability over the tropical IO (Klein et al. 1999). The North IO warms up concurrently with the growing El Niño. The North IO warming decays during the winter to spring and then reinvigorates in the boreal summer (Fig. 11.12A). Here we focus on spring anomalies while discussing the second summer peak of the North IO warming in the next section.

In MAM(1) following an El Niño, an antisymmetric atmospheric pattern appears over the tropical IO, with decreased (increased) precipitation and northeasterly (northwesterly) wind anomalies north (south) Of the equator. In the Northern Hemisphere, a large-scale anomalous anticyclonic circulation (AAC) stretches from the tropical Northwest (NW) Pacific through the North IO, accompanied by suppressed convection (see Fig. 11.11B, C). Mechanisms important for the atmospheric AAC in MAM(1) include:

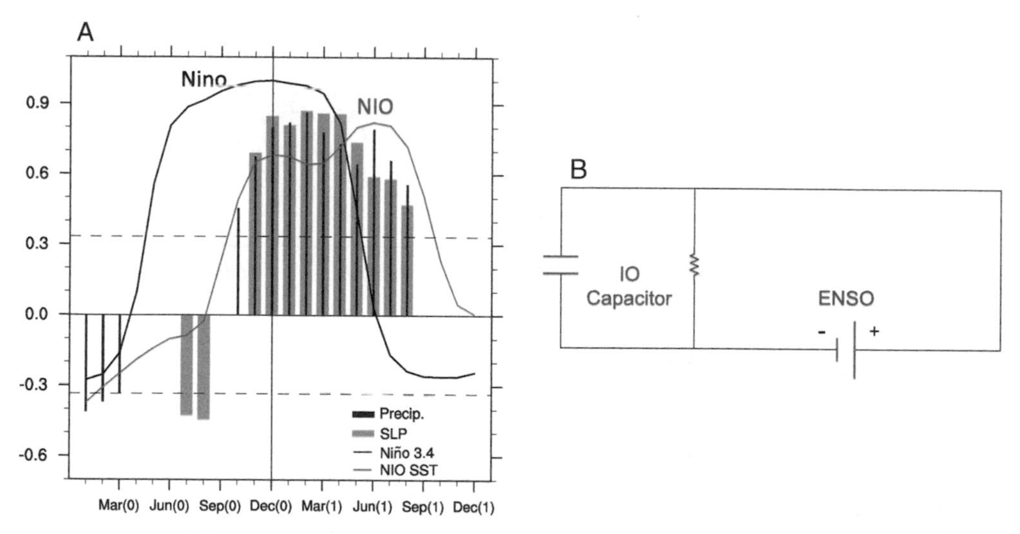

Fig. 11.12 (A) Correlations with NDJ Niño3.4 sea surface temperature (SST) as a function of calendar month from the El Niño and Southern Oscillation developing (0) to decay (1) year: Niño3.4 SST, North IO (5–25°N, 60–120°E) SST, sea level pressure (SLP) over the tropical NW Pacific (120–160°E, 10–30°N), and precipitation (sign reversed) near Guam (120–160°E, 10–20°N). OISST, CMAP precipitation, NCEP reanalysis SLP; 1982-2017, 3-month averaged with 9-year high-passed. Dashed lines indicate 95% confidence level. (B) Schematic of the capacitor effect. *(A, Courtesy C.Y. Wang.)*

- Seasonal modulations by the background state. As El Niño-induced convective and westerly wind anomalies in the central Pacific displace south of the equator (see Section 9.4.1), the equatorial Pacific warming causes sea level pressure (SLP) to rise on the winter side of the equator over the northern Indo-western Pacific (Stuecker et al. 2015).
- Local WES feedback over the NW Pacific. The anomalous northeasterlies on the southeast flank of the AAC strengthen the background trades and cool the NW Pacific (see Fig. 11.11) by enhancing surface evaporation and vertical mixing in the ocean. Negative SST anomalies over the tropical NW Pacific in turn strengthen the AAC by reducing convective activity (Wang et al. 2003).
- Southwest IO warming, anchored by slow-propagating ocean Rossby waves, intensifies atmospheric convection locally (see Fig. 11.11C), contributing to the antisymmetric wind pattern over the tropical IO. With the easterly trades in both hemispheres on the background, WES feedback amplifies the antisymmetric coupled pattern that includes the cooling tendency over the North IO (Wu et al. 2008; Du et al. 2009).

11.4 Post-ENSO summer capacitor effect

By JJA(1), SST anomalies have typically dissipated over the equatorial Pacific, but robust ENSO correlations remain in the atmosphere over the Indo-NW Pacific, including the subtropical AAC with suppressed convection inside (Fig. 11.13). On the NW Pacific island of Guam (145°E, 13.5°N), the summer rainy season starts with the onset of the NW Pacific monsoon (see Section 5.2.3). Summer rainfall on Guam is not correlated with concurrent ENSO, but with ENSO in the preceding winter (see Fig. 11.12A). This may appear peculiar but is consistent with the AAC in post—El Niño summers.

The AAC occupies the tropical NW Pacific from the peak phase of El Niño through the following summer (see Fig. 11.12A). As a result, typhoon activity is suppressed in the first half of a post—El Niño year. In the post—El Niño summer of 2016, the first typhoon of the year made landfall in Taiwan on July 8, while 10 typhoons have developed by July 9 in 2015. In another post—El Niño year of 1998, the first typhoon did not develop until August 2.

11.4.1 Indian Ocean effect on the atmosphere

In the post—El Niño summer [JJA(1)], the tropical IO warming stands out as the most robust oceanic anomalies in terms of correlation with the NDJ(0) El Niño index (see Fig. 11.13). This implies that the IO warming sustains atmospheric anomalies over the Indo-NW Pacific. General circulation model (GCM) experiments support this notion and generally simulate the AAC response to an IO warming during JJA. The El Niño effect on the JJA(1) AAC is not directly through atmospheric bridge but mediated by

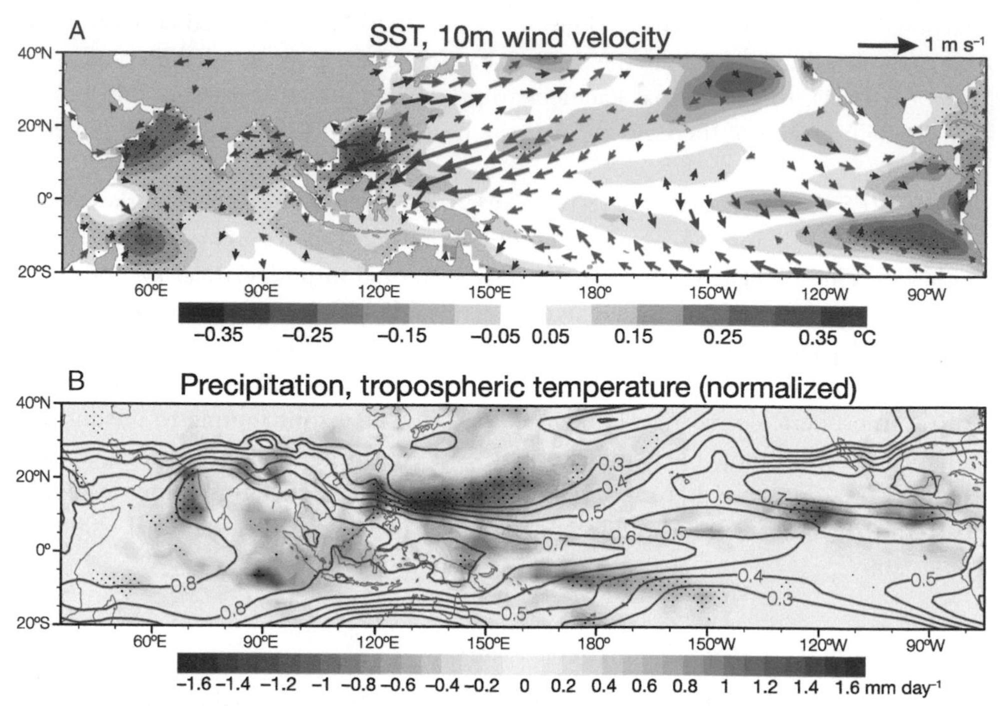

Fig. 11.13 Anomalies in the post—El Niño summer (JJA): (A) sea surface temperature (*SST;* color shading in K) and surface wind velocity (red/blue arrows strengthening/weakening the background wind, m/s); (B) precipitation (mm/day) and tropospheric temperature (850—250 hPa mean). *(From Xie et al. (2016))*

the IO like a capacitor: El Niño forces the IO to warm like a battery charging a capacitor, and then in post—El Niño summer the IO warming sustains the AAC like a discharging capacitor (see Fig. 11.12B).

How does the IO influence the subtropical NW Pacific? The IO warming excites a Matsuno-Gill pattern in tropospheric temperature, with a warm equatorial Kelvin wave propagating into the western Pacific (see Fig. 11.13B). Surface friction causes surface winds to converge onto low pressure in the equatorial Kelvin wave, a tendency most evident north of the equator in observations. The resultant surface divergence off the equator triggers convective feedback to amplify the coupled anomalies of circulation and precipitation.

A baroclinic model of the atmosphere, linearized around the JJA mean state, is used to illustrate the role of the Kelvin wave in mediating the AAC response. When deep convective heating is imposed over the tropical IO and symmetric about the equator, the low-level circulation anomalies are rather asymmetric, with a stronger response north than south of the equator (Fig. 11.14A). The Kelvin wave response is visible in the SLP field. On the northern flank of the Kelvin wave, surface friction turns the winds into

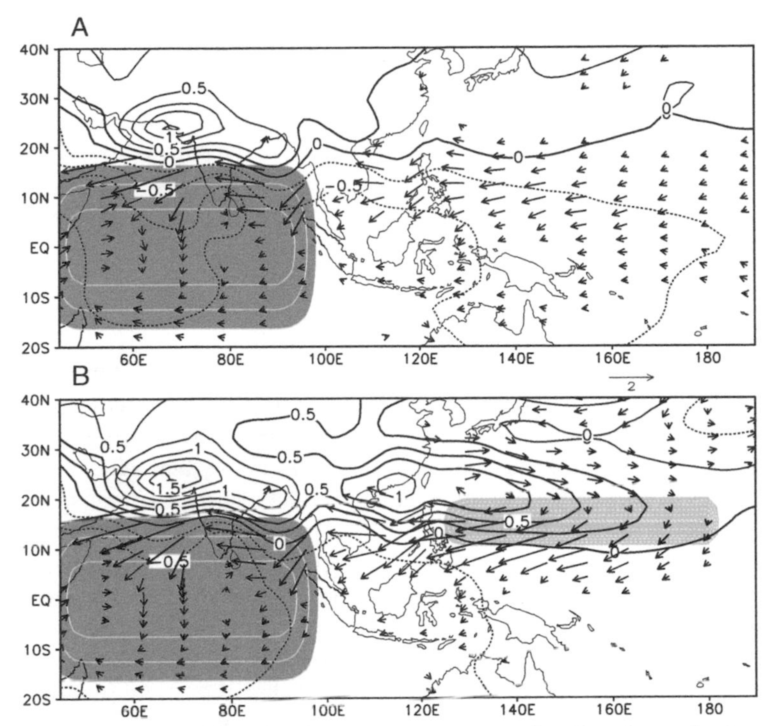

Fig. 11.14 (A) Linear atmospheric model response to a tropical Indian Ocean heat source *(orange shading and white contours)*: sea level pressure *(black contours)* and 1000 hPa wind. (B) Same as (A) but with convective feedback *(blue shading and white contours, negative dashed)* over the NW Pacific. *(From Xie et al. (2009).)*

northeasterly, causing surface divergence there. Box 11.1 describes what amplifies the circulation response north of the equator.

Next, we let deep convective heating over the NW Pacific be proportional to the regional average of surface convergence. The model response is now dominated by a strong anticyclone from the South China Sea to NW Pacific, where convection-circulation feedback amplifies the Kevin wave-induced Ekman divergence and maintains a strong diabatic cooling. The model response resembles observed anomalies during JJA(1) from the North IO to NW Pacific (see Fig. 11.14B), including anomalous easterlies in 10° to 20°N from the Arabian Sea to the International Date Line.

11.4.2 Regional ocean-atmosphere coupling

In post—El Niño summer, equatorial Pacific SST anomalies have largely dissipated, but peculiarly the North IO warming reinvigorates (see Fig. 11.12A). From an atmospheric perspective, the reinvigorated IO warming helps maintain the AAC and the anomalous northeasterlies over the North IO through the South China Sea. From an oceanic

BOX 11.1 An intrinsic mode to the summer monsoon

In summer (JJAS), the monsoon westerlies meet the easterly trades over the far western North Pacific (5–25°N). In a dry atmospheric model linearized around the observed mean state, this zonal wind confluence amplifies the wind response north of the equator even though the IO heating is symmetric about the equator (see Fig. 11.14A). The post–El Niño AAC features the easterly wind anomalies in the mean confluence zone ($-\frac{\partial \overline{u}}{\partial x}>0$) (Box Fig. 11.1D), a structure that allows the AAC to grow on the barotropic energy conversion (Eq. 9.15)

$$CK \approx -\frac{\partial \overline{u}}{\partial x}\overline{u'^2} \tag{B11.1}$$

where the prime denotes perturbations. In addition to the confluence of the mean flow, strong convective feedback in the Northern Hemisphere during JJA is another reason why the AAC develops north of the equator even though the tropospheric Kelvin wave from the tropical IO is symmetric (see Fig. 11.13). The convective feedback also explains why the AAC is more sensitive to the low-level confluence than the upper-level difluence on the southeast flank of the Tibetan High (see Fig. 5.4).

As a source of barotropic energy conversion, the zonal-wind confluence acts as an effective geographic anchor for stationary waves, including the Pacific-North American pattern in Section 9.5 and AAC here. In El Niño summers JJA(0), an anomalous cyclonic circulation develops in the tropical NW Pacific (see Box Fig. 11.1C), which is slightly displaced southward compared to the JJA(1) AAC. The JJA(0) anomalous circulation still taps into the barotropic energy conversion in the broad zonal-wind confluence. Thus the AAC shows a biennial tendency, in negative (positive) phase in El Niño developing (post–El Niño) summer.

In late July, deep convection jumps to fill the northern bulge of the western Pacific warm pool, centered at 20°N. The Ueda Jump (see Fig. 5.20) marks the final stage of the Asian summer monsoon development, connected to the withdrawal of Meiyu-Baiu over East Asia. While the high-SST (>28°C) preconditioning is necessary, the cyclonic circulation associated with the Ueda Jump is anchored on the south flank by the mean zonal-wind confluence (see Box Fig. 11.1B), energized by the barotropic energy conversion. Thus the Jump results from the joint effect of the seasonal SST increase and monsoon-trade wind confluence.

The Asian summer monsoon intraseasonal oscillation (MISO) is characterized by the eastward circumglobal and regional northward phase propagation (see Section 4.2.4). The circulation anomalies at the MISO phase 1-2 (see Box Fig. 11.1A) features an AAC that resembles the interannual one in post–El Niño summers (see Box Fig. 11.1D), with suppressed convection from Indo-China to the NW Pacific and active convection from India to the maritime continent. The MISO may be viewed as a moisture mode energized by barotropic energy conversion in the mean confluent flow (Wang et al. 2021).

Thus the AAC is an intrinsic mode to the Asian summer monsoon system, anchored on the south flank by the monsoon-trade wind confluence (Box Fig. 11.2) and energized by additional convective feedback. It affects the Meiyu-Baiu on the northwest flank by modulating the moisture transport at low levels and the Asian jet aloft. It is observed across a wide range of timescales from intraseasonal to interannual (see Box Fig. 11.1).

BOX 11.1 An intrinsic mode to the summer monsoon—cont'd

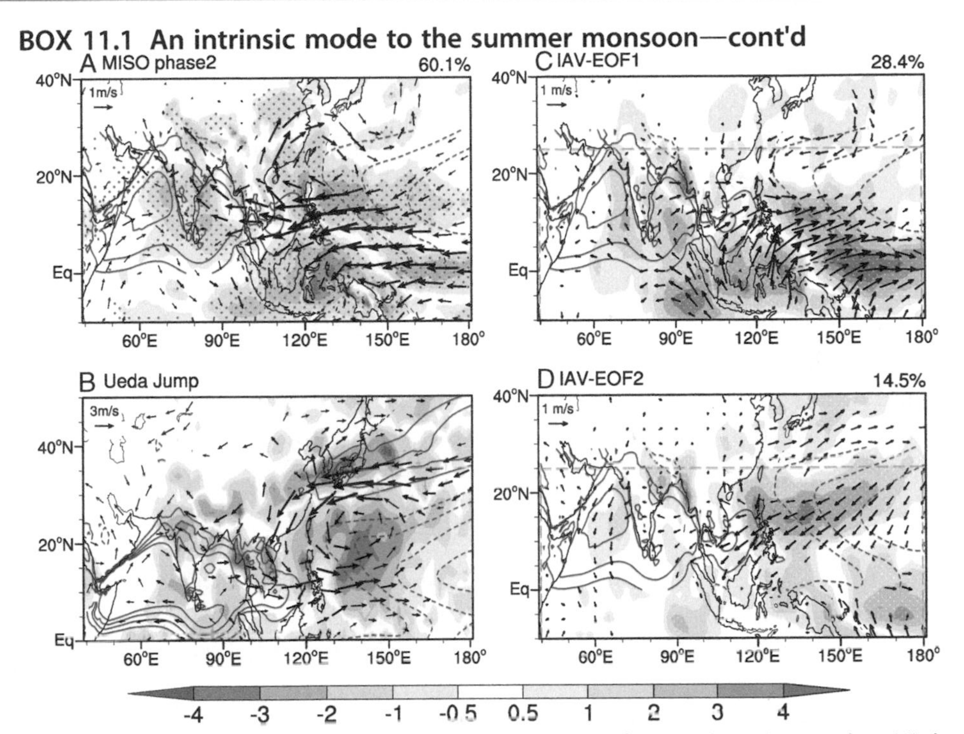

Box Fig. 11.1 Precipitation (color shading, mm/day) and 850 hPa wind (m/s) anomalies: (A) the summer monsoon intraseasonal oscillation at phase 2, and (B) the Ueda Jump (July 23—August 6 minus July, 3—17 difference). (C) First and (D) second empirical orthogonal function modes for JJA-mean precipitation variability over the domain of 40—180°E, 10°S—25°N during 1979-2018, as regressions against the normalized principal component (precipitation anomalies are doubled in magnitude for comparison). Blue line contours *(negative dashed)* in 1°S—30°N are the climatologic background 10-m zonal wind velocity for (A, C, D) JJA (at -5, -3, -1, 2, 5, 8 m/s), and (B) 3—17 July (at ±3, ±5, ±7, ±9 m/s). *(Courtesy Z.Q. Zhou, C.Y. Wang, and X.D. Wang.)*

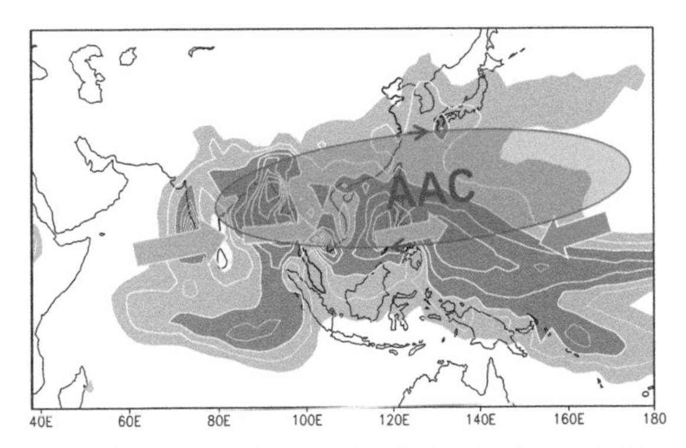

Box Fig. 11.2 Schematic of the anomalous anticyclonic circulation *(AAC)* spanning across the Asian summer monsoon, superimposed on the summer mean precipitation. The AAC is anchored on the south flank by the confluence of the monsoon westerlies *(orange block arrows)* and easterly trades *(blue block arrow)*.

perspective, on the other hand, the anomalous northeasterlies cause the North IO/South China Sea to warm for the second time by reducing surface evaporation as the background wind turns from northeasterly to southwesterly in May (see Fig. 11.13A).

The above circular argument suggests that the AAC and North IO warming are coupled and their interaction yields a positive feedback. This regional coupled mode is called the Indo-western Pacific ocean capacitor (IPOC). The WES feedback of the IPOC is conditional on the southwest monsoon over the summer North Indo-western Pacific Oceans. It is not a coincidence that the longest lasting ENSO-related anomalies are found in the Indo-western Pacific region. In fact, the recurrent coherent spatial structure of these ocean-atmospheric anomalies is a result of the IPOC.

Fig. 11.15 shows the local SST-precipitation correlation during JJA. Over the equatorial central Pacific, the correlation is high and positive, indicating a strong SST effect on atmospheric convection. Likewise, the correlation is significant and positive off Indonesia and over the Seychelles Dome (70°E–10°S). Over the Indo-NW Pacific warm pool, by contrast, the correlation is weak. This is often invoked to argue against regional ocean-atmosphere coupling.

EOF analysis supports that summer rainfall variability over the Indo-NW Pacific is coupled with SST variability, just not locally but from remote regions. Combined, the first two modes explain about 43% of interannual variance in summer rainfall within the analysis domain of 40° to 180°E, 10°S to 25°N. The leading EOF represents the response to developing ENSO and is most highly correlated with concurrent equatorial Pacific SST ($r = 0.81$ with Niño3.4 for 1979-2018) (Fig. 11.16, upper panels), while the second EOF corresponds to the post-ENSO IPOC and is most highly correlated with concurrent SST over the North IO and South China Sea (see Fig. 11.16, lower panels). Both EOFs are associated with robust SST patterns, just not collocated with major convective anomalies in

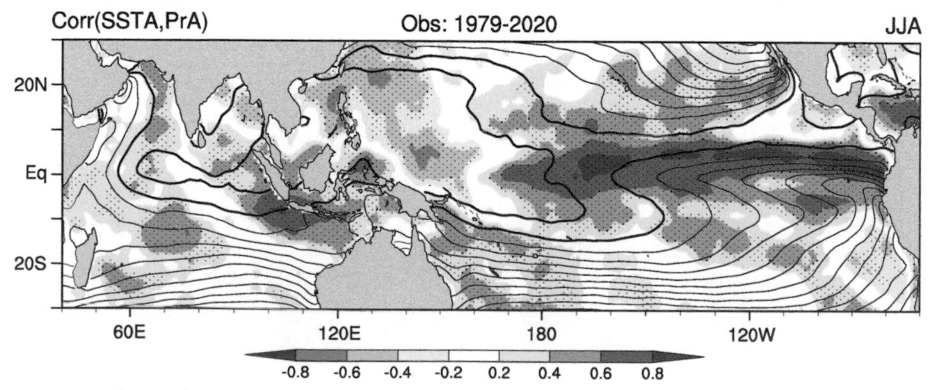

Fig. 11.15 Local correlation between JJA-mean sea surface temperature (SST) and precipitation *(color shading)*, superimposed on the mean SST (black solid contours, every 1°C; 28°C and 29°C contours thickened). *(Courtesy Z.Q. Zhou.)*

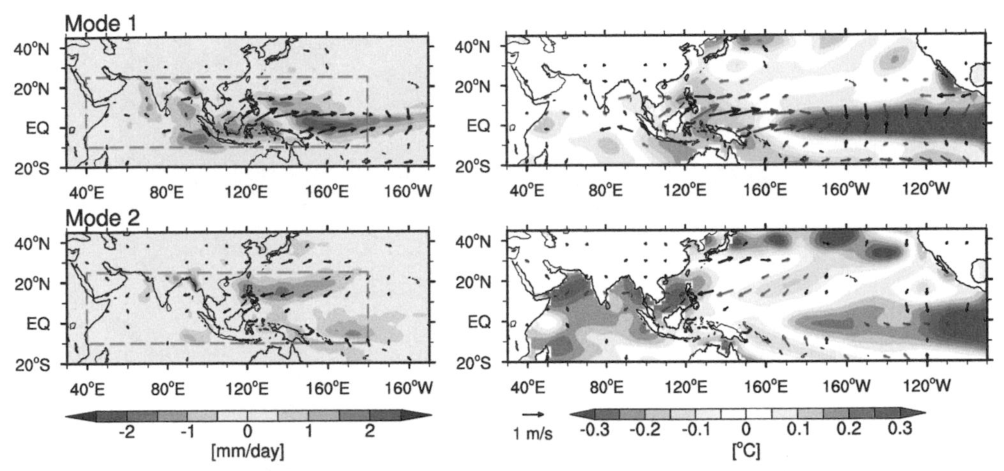

Fig. 11.16 First two empirical orthogonal function modes for JJA mean precipitation variability for 1979-2018 over the Indo-NW Pacific (brown dashed box: 40–180°E, 10°S–25°N), explaining 28.4% and 14.5% of the total interannual variance. Regressions against the principal component: *(left)* precipitation (m/day) and *(right)* sea surface temperature, along with surface wind velocity. *(right)* Red arrows indicate that the anomalous wind strengthens the background. *(Courtesy C.Y. Wang.)*

the Indo-NW Pacific domain. EOF1 features an eastward shift of the mean convective center, while the SST forcing is outside the EOF analysis domain in the equatorial Pacific. Likewise, EOF2 features the AAC and reduced rainfall over the tropical NW Pacific while the SST forcing is displaced to the west over the North IO and South China Sea. Around Guam (145°E–13.5°N), for example, the SST anomaly is nearly zero, while rainfall is significantly reduced (see Fig. 11.16, lower panels).

The top two precipitation EOFs and their PCs are well reproduced in atmospheric GCMs forced with observed SST variations. This reaffirms that the AAC is the atmospheric component of the coupled IPOC (Zhou et al. 2018).

11.4.3 Prediction

Coupled models initialized with observations show skills in predicting the IPOC at monthly to seasonal leads. The overall patterns of JJA precipitation EOF2 and SLP EOF1 are well produced (Fig. 11.17), with PC correlations between observations and the multiple-model ensemble (MME) mean forecast well above 95% significance level. The AAC is associated with decreased precipitation over the tropical NW Pacific and increased rainfall over the equatorial IO and maritime continent. In observations, rainfall increases from eastern China through Japan on the northwest flank of the AAC. The model prediction of East Asian anomalies is not as good as in the tropics because of internal atmospheric variability propagating along the East Asian westerly jet (Kosaka et al. 2012). In the meridional direction, AAC is associated

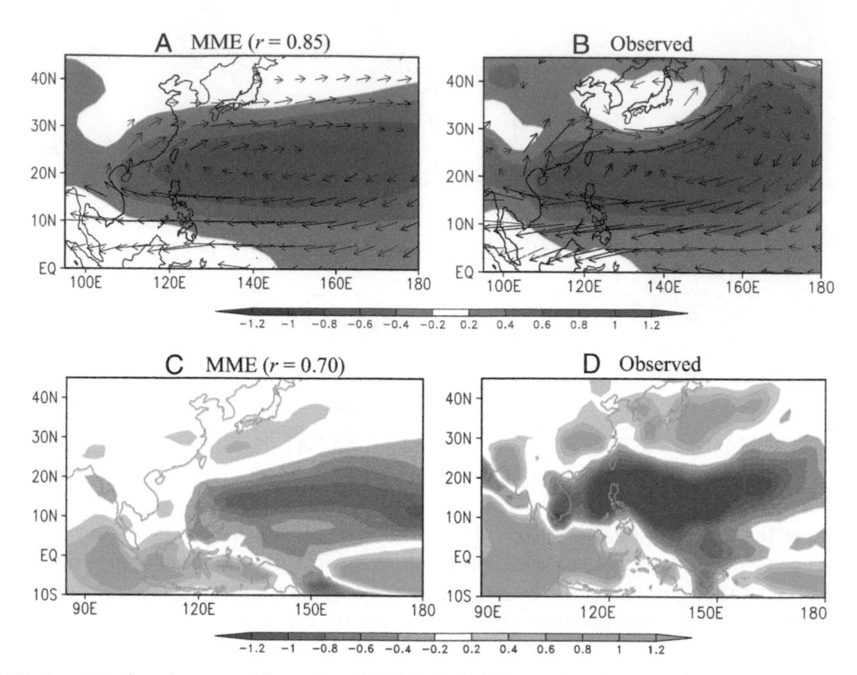

Fig. 11.17 Empirical orthogonal function 1 (EOF1) of JJA sea level pressure (SLP; *upper panel)* and EOF2 of JJA precipitation *(lower)* variability for 1980-2001, obtained from *(right)* observations and *(left)* 11-model mean prediction initialized in May. Temporal correlation in principal component (PC) between observations and multimodal ensemble *(MME)* prediction is noted in parentheses. Vectors represent the 850-hPa wind anomalies (m s^{-1}) regressed against the SLP PC. *(From Chowdary* et al. *(2010) https://doi.org/10.1029/2010JD014595.)*

with an anomalous cyclone over Japan. The meridional circulation/rainfall dipole is known as the Pacific-Japan pattern.

Insights into coupled dynamics can be gained by studying the perturbed initial condition ensemble (PICE) spread (see Box 10.1). Here, the singular value decomposition (SVD) is performed on sequential monthly maps of PICE spread initialized in February. The leading SVD mode of joint SST-850 mb wind velocity variability captures the IPOC mode. The resultant eigenvectors track the evolution of PICE spread (Fig. 11.18). In March-April, negative SST anomalies develop over the South China Sea and NW Pacific under northeasterly wind anomalies. As the background wind begins to turn southwesterly in May, the coupled positive SST-easterly wind anomalies develop first in the Arabian Sea-Bay of Bengal, grow and gradually move into the South China Sea. The apparent eastward propagation of the coupled anomalies is due to the eastward expansion of the mean southwest monsoon flow and the confluence with the easterly trades. The copropagation with the mean flow evolution supports that the barotropic conversion energizes the AAC (see Box 11.1). This IPOC mode in PICE spread is not

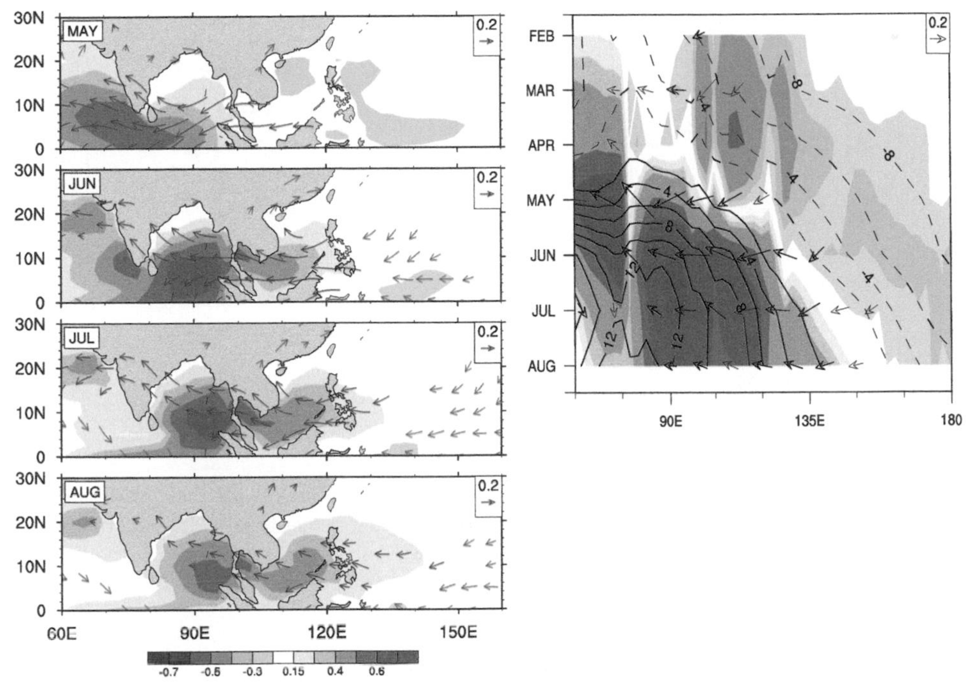

Fig. 11.18 *(left)* First month-reliant singular value decomposition (SVD) mode of ensemble spread in 850-hPa wind and sea surface temperature (SST) in the North Indian Ocean in perturbed initial condition ensemble spread initialized in February. The blue (red) vectors denote that the anomalous wind reduces (intensifies) the mean flow. *(right)* Hovmöller diagram of SST and wind anomalies in the SVD mode at 10°N, along with the mean wind velocity (contours at 2 m/s intervals, negative dashed, the zero contour omitted). Note that coupled SST-wind anomalies propagate eastward, together with the eastward expansion of the mean southwest monsoon wind. Based on Ma et al. (2017). *(Courtesy J. Ma.)*

associated with significant SST anomalies over the equatorial Pacific and results from positive feedback from the regional ocean-atmosphere interaction over the North IO-NW Pacific as described in this section.

11.5 Asian summer monsoon variability

The summer monsoon system brings much needed rainfall to populous South and East Asia. Here we discuss year-to-year variability in the monsoon, which are strongly influenced by ENSO and the regional IPOC mode.

11.5.1 India

All Indian Rainfall (AIR) is a widely used index based on rain gauges to track the overall rainfall over India. A year with the summer (JJAS) mean AIR above normal is called a

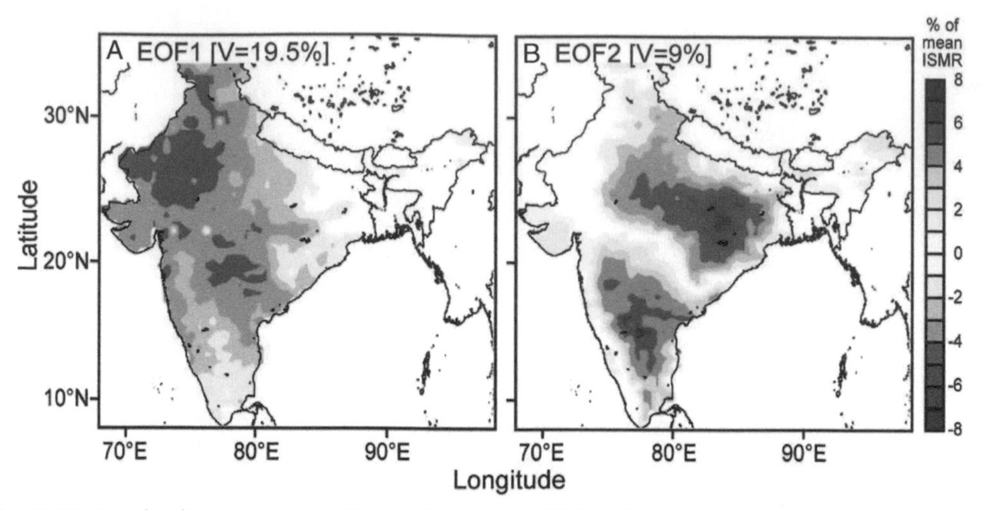

Fig. 11.19 Two leading empirical orthogonal functions *(EOFs)* of JJAS rainfall variability over India for 1900-2008. Shading indicates the rainfall anomalies, expressed as a percentage of the JJAS mean rainfall, associated with a principal component amplitude of +1 SD. *V* indicates the fraction of the domain-integrated variance explained by the modes. *(From Mishra et al. (2012)https://doi.org/10.1073/pnas.1119150109.)*

good monsoon, and vice versa. The JJAS AIR is highly correlated with ENSO, r(AIR, Niño3.4) = 0.63 over 1900-2018. In search for such a predictive relationship, Walker (1933) discovered the Southern Oscillation.

The leading EOF of JJAS-mean rainfall over India features a countrywide pattern with large loading in central and northwestern India (Fig. 11.19A). Not surprisingly, PC1 is highly correlated with the AIR index. It is also highly correlated with the concurrent ENSO, r[PC1, JJAS(0) Niño3.4] = 0.60. The second rainfall EOF shows a dipole between southwest and northeast India (see Fig. 11.19B) and tends to take place in post—El Niño summers. PC2 is correlated with concurrent North IO SST $r = 0.37$ as well as ENSO in the preceding winter. This signal in post-ENSO summers (an IPOC effect tracked by concurrent North IO SST) is not as widely known as the one concurrent with a developing ENSO event (a direct ENSO effect tracked by the concurrent Niño3.4) because EOF2 contributes little to AIR that tracks the country mean. This IPOC effect, together with the traditional ENSO effect on AIR, expands the window of Indian summer monsoon (ISM) predictability and holds the promise of improving regional rainfall prediction.

From June to September, the direct ENSO effect on the ISM grows, while the IPOC effect wanes. June is the sweet spot for both effects. A regression model, including the concurrent Niño3.4 and North IO SSTs, reproduces June variability over much of India (Fig. 11.20A). The IPOC effect on June surface temperature is especially large over central India (see Fig. 11.20B). In post—El Niño summer, the monsoon onset is delayed, prolonging the pre-monsoon heatwave and increasing June-average air temperature.

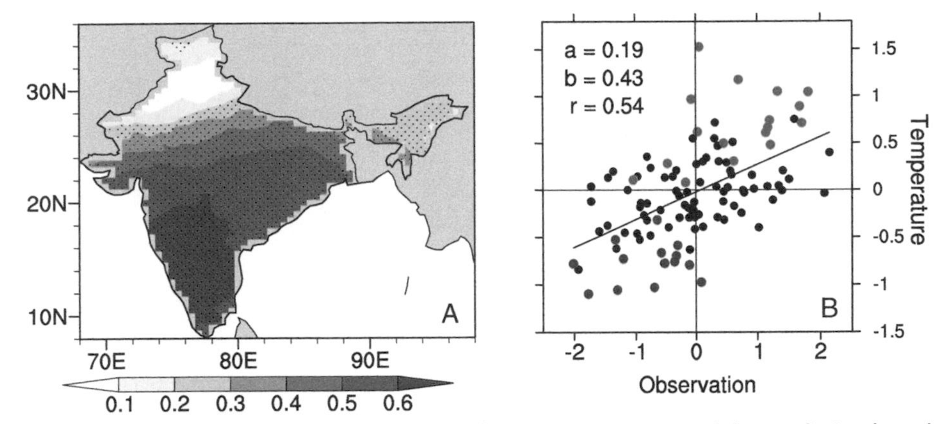

Fig. 11.20 Correlation between observed June surface air temperature and the prediction based on a multivariant regression, $T = a\text{ENSO} + b\text{NIO} + \varepsilon$, where ENSO and NIO denote concurrent JJA Niño3.4 and North Indian Ocean sea surface temperature (SST) indices, a and b denote the regression coefficients, and ε is the residual. (A) Distribution over India, and (B) for the area mean in central India. Red (blue) dots denote years with JJA NIO SST index above (below) one standard deviation. The correlation between observations and predictions is 0.54 for 1901-2018. *(From Zhou et al. (2019).)*

11.5.2 China

Over eastern China, the Meiyu rain band marches northward from June to July and then dissolves in late July (see Section 5.2). The leading EOF of July rainfall variability over eastern China features a meridional dipole between southern China and the Yangtze River basin (~30°N) (Fig. 11.21). The PC is correlated with antecedent ENSO. In post—El Niño summer, rainfall tends to increase along the Yangtze River, fed by the moisture transport by the anomalous southwesterlies on the northwestern flank of the AAC. Major Yangtze River floods of 1998 and 2016 each follow a major El Niño event.

The correlation between eastern China rainfall PC1 and antecedent winter Niño3.4 is marginal at 95% significant level ($r = 0.3$ for 1951-2020), due to large internal atmospheric variability in the subtropical monsoon of East Asia (see Section 12.1) compared to the tropical monsoon of India. The subtropical westerly jet, displaced north of Tibet in summer, serves as a waveguide for stationary waves called the Silk Road patterns, providing a conduit for intraseasonal variability over the North Atlantic and Europe to modulate atmospheric blocks that cause long-lasting heatwaves over Asia (Kosaka et al. 2012; Chowdary et al. 2019). The zonally wavy Silk Road patterns arise due to atmospheric internal variability with little predictability at monthly and longer leads.

The broad domain of 40° to 180°E, 10°S to 25°N encompasses major summer convective centers of the western Ghats, Bay of Bengal, South China Sea, and NW Pacific (see Box Fig. 11.2). The EOF modes over this broad Asian summer

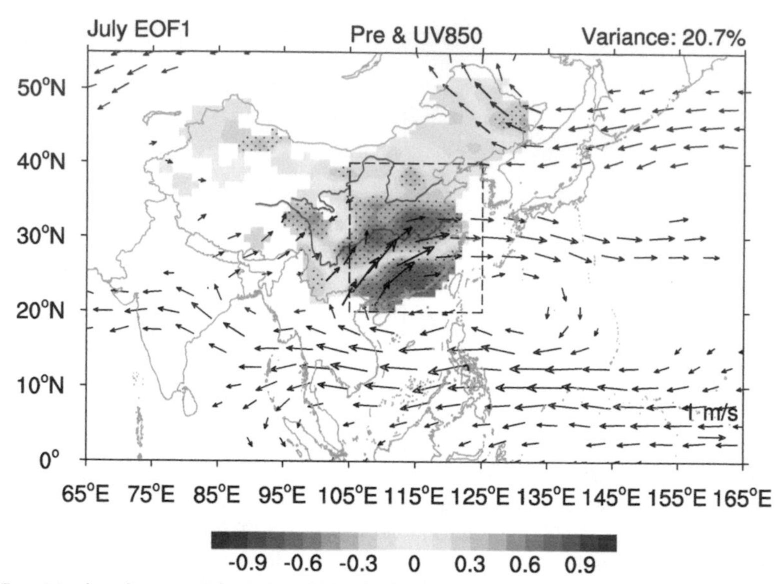

Fig. 11.21 Empirical orthogonal function *(EOF1)* of July rainfall variability *(color shading)* over eastern China *(gray dashed box)*, along with the 850-hPa wind velocity regression against principal component 1 (PC1). PC1 is correlated with the antecedent ND(0)J(1) Niño3.4 at $r = 0.3$ for 1951-2020. The blue curves denote the Yangtze and Yellow rivers. *(Courtesy X.D. Wang.)*

monsoon region (see Fig. 11.21) encompass the modes of smaller subsystems, say over India (see Fig. 11.16) or China (see Fig. 11.21). The broad monsoon modes are in turn forced by ENSO and modulated by regional ocean-atmospheric feedbacks over the Indo-western Pacific warm pool. Compared to those over the ocean, seasonal-mean rainfall anomalies are small on land, although the effects of land-atmospheric interactions could be important for shorter, subseasonal timescales.

11.6 Synthesis

Prior to the 1997 El Niño event, the IO was viewed as slave to ENSO with little feedback onto the atmosphere. The following observations seem to support this view. First, the leading EOF mode for SST explains 40% of interannual variance in the tropical IO basin and is highly correlated with ENSO at a season lag. Second, local SST-precipitation correlation is low. The 1997-98 El Niño event challenged this view by showcasing a major IOD event at the developing phase and a strong IPOC event in the following summer. In the equatorial IO, a cold tongue developed, and the Wyrtki jet failed to develop in boreal fall of 1997, while in 1998 summer SST averaged over the North IO-South China Sea (60—120°E, 5—25°N) rose by a record 0.4°C and China's Yangtze River suffered one of the worst floods.

The IO harbors two distinct regional coupled modes. The IOD mode arises from the Bjerknes feedback, enabled by the seasonal coastal upwelling off Indonesia in June-November despite a deep and flat equatorial thermocline in the annual mean. The IPOC mode involves the WES feedback between the AAC and SST from the North IO to the NW Pacific on the background westerly monsoon wind in summer (Box 11.1).

ENSO is the major driver for IO variability. Just as a drummer produces distinct sounds by striking different percussion instruments (Fig. 11.22), El Niño produces distinctive patterns of climate anomalies over the IO at various phases. As the sound of a drum or cymbal is determined by the physical structure and material, the spatiotemporal structure of an El Niño—induced mode is shaped by the positive feedback of the ocean-atmosphere interaction.

ACC is a recurrent pattern of atmospheric variability across a range of timescales, anchored in the Indo-NW Pacific by the monsoon mean flow (Box 11.1). The AAC and the IO warming are coupled and the resultant positive feedback makes the IPOC a preferred mode ENSO excites. The IPOC explains mysterious, recurrent atmospheric anomalies over the Indo-NW Pacific and South/East Asia that outlast El Niño itself and persist through the following summer.

A record-strong IOD took place in September 2019, followed by a major IPOC event that caused the historic Yangtze River flooding in 2020 summer (see Section 1.2). Unlike 1997-98, the 2019 IOD/2020 IPOC sequence took place in the absence of a major El Niño event (Fig. 11.23). The 2019 IOD contributed to the 2020 IPOC by exciting the

Fig. 11.22 Same forcing, different response. The coupled modes are cymbals and drums on the stage of climate variability. *(Courtesy W. Chapman.)*

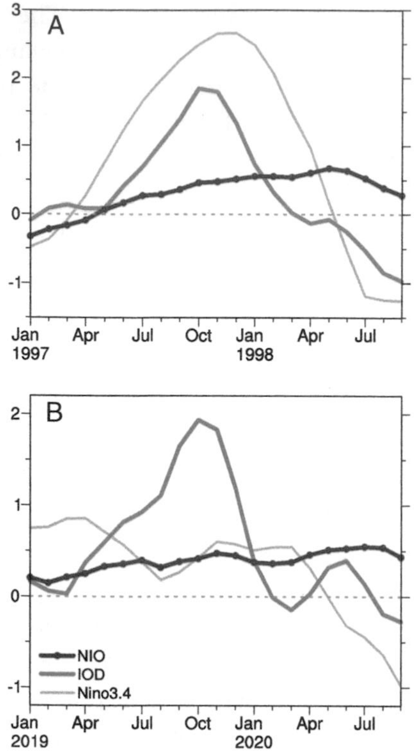

Fig. 11.23 Evolution of Nino3.4, Indian Ocean dipole *(IOD)*, and North Indian Ocean *(NIO)* sea surface temperature indices during (A) 1997-98 and (B) 2019-20. *(Courtesy Z.Q. Zhou.)*

slow-propagating downwelling Rossby waves with anticyclonic wind curls (see Fig. 11.7) in the South IO (Zhou et al. 2021). Nature showed us that the IOD and IPOC are intrinsic modes of regional climate variability that are often, but not always, forced by El Niño. Despite the lack of strong ENSO forcing, dynamic models initialized as early as April predicted the AAC and anomalously active Meiyu rain band of 2020 summer. This suggests regional predictability beyond ENSO. April ocean initial conditions—SST and thermocline depth—not only include the effects of the antecedant ENSO and IOD anomalies but are also what takes to predict an impeding ENSO event (see Section 9.7).

Review questions

1. Discuss why the zonal wind and ocean current are dominated by a semiannual cycle on the equator in the Indian Ocean. Consider the meridional advection of zonal momentum $v\frac{\partial u}{\partial y} - fv = -\varepsilon u$ by assuming $(\overline{u}, \overline{v}) = 0$, where the overbar denotes the annual mean.

2. The global zonal-mean wind is easterly on the equator, helping break the zonal symmetry over the tropical Pacific (see Section 8.2). Use Eq. 11.1 to explain that the annual-mean, global zonal-mean wind is easterly on the equator through the troposphere (see Fig. 2.10A). Hint: Consider the seasonal migration of the ITCZ.
3. The Atlantic ITCZ is displaced north of the equator. Describe the annual-mean meridional wind $\overline{v}$ and meridional profile $\overline{u}(y)$ across the equator over the central Atlantic. Discuss how the meridional advection causes an annual cycle in zonal wind over the equatorial Atlantic. Hint: Linearize around the annual mean state (refer to Section 10.1.3).
4. The westward-intensified thermocline ridge in the Southwest IO is unique among tropical oceans (see Fig. 11.1). Perennial Ekman upwelling exists under the ITCZ across the Pacific and Atlantic, but why is there no westward-intensified thermocline ridge? Refer to Section 7.2.6.
5. How do monsoons affect the seasonality of major Indian Ocean modes: Wyrtki jets, Indian Ocean dipole mode, and the interannual North Indian Ocean warming that outlives El Niño?
6. How does El Niño induce the IOD and IO basin modes?
7. Explain the difference in seasonality between the IO SST EOFs 1 and 2.
8. Based on the leading EOF modes of Indo-western Pacific rainfall variability and attendant SST and wind anomalies, make an argument that positive feedbacks from ocean-atmosphere interaction determine these recurrent patterns. How do these modes affect Asian monsoon regions over land?
9. Do these interannual modes resemble the spatial pattern of the climatologic summer monsoon, say in rainfall and low-level wind?
10. Compare the three tropical oceans regarding the seasonal cycle, degree of continental influence, and modes of interannual variability (e.g., Bjerknes feedback, zonal width effect, and seasonal phase locking).
11. In the Northern Hemisphere winter, the northeast winds prevail from the Arabian Sea through the tropical North Pacific. Can an atmospheric AAC provide positive feedback onto the anomalous warming of the North Indian Ocean?
12. Draw schematically the pressure and wind perturbations at 850 hPa of a warm tropospheric Kelvin wave. How does the friction affect the surface wind perturbations? How does this affect the wind convergence at the equator and 10° away on either side?
13. What benefits do you expect to see in seasonal forecast by including a fully coupled Indian Ocean? Consider both the summer during which a major ENSO event develops and the summer that follows.

CHAPTER 12

Extratropical variability and the influence on the tropics

Contents

Sea surface temperature (SST) over extratropical oceans often displays pronounced (multi)decadal variations that are correlated with those of surface wind and large-scale atmospheric circulation. The Pacific Decadal Oscillation (PDO) is defined as the leading empirical orthogonal function (EOF) of monthly SST* variability over the extratropical North Pacific (north of 20°N), where SST* $\equiv$ SST global mean. At the positive phase, a broad band of negative SST anomalies occupies much of the midlatitude North Pacific, surrounded by positive SST anomalies off the west coast of North America (Fig. 12.1A). The atmospheric Aleutian Low deepens at the surface, with the intensified westerly winds over the negative SST anomalies over the central North Pacific and anomalous south/southeasterlies over positive SST anomalies off the North American coast. The PDO index displays pronounced interdecadal variability, with marked phase transitions in the early 1920s, late 1940s, mid-1970s, and late 1990s. Covariations of the PDO and Aleutian atmospheric pressure (see Fig. 12.1B, C) bear remarkable similarity to those of Galapagos SST and Darwin atmospheric pressure associated with the coupled phenomenon of El Niño and the Southern Oscillation (ENSO; see Fig. 9.1).

One often hears the following argument: Because atmospheric memory is limited to less than a season, the ocean must set the timescale of long variability (Hypothesis 1). This

Coupled Atmosphere-Ocean Dynamics: from El Niño to Climate Change
ISBN 978-0-323-95490-7, https://doi.org/10.1016/B978-0-323-95490-7.00012-6

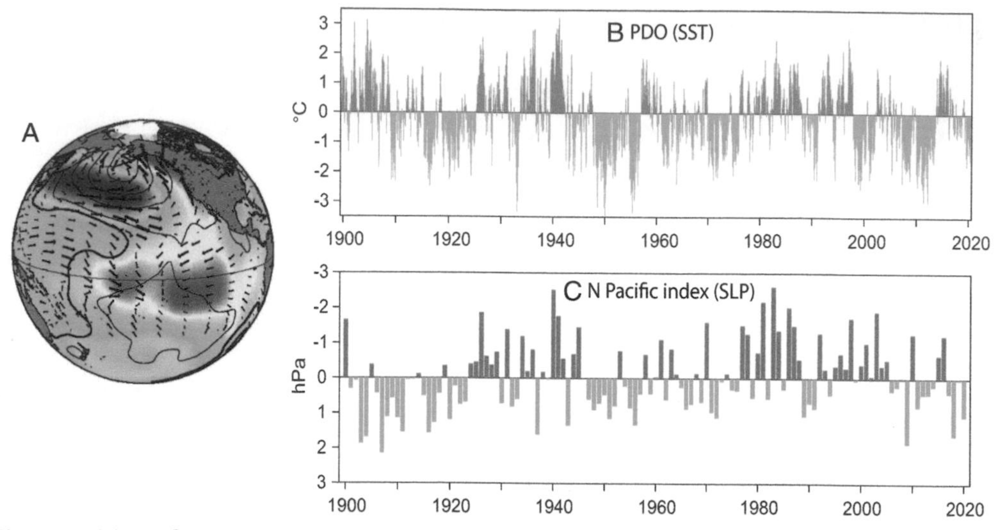

Fig. 12.1 (A) Pacific Decadal Oscillation *(PDO)* anomalies of sea surface temperature (*SST*; shading), surface wind, and sea level pressure (*SLP*; black contours). (B) Monthly PDO index and (C) winter (N-M) mean North Pacific index (SLP in 30°N–65°N, 160°E–140°W, normalized. Based on Trenberth and Hurrel 1994. *(A, From Mantua et al. (1997); B and C, courtesy Z.H. Song. © American Meteorological Society. Used with permission.)*

is true for tropical variability like ENSO for which large-scale ocean waves help set the preferred interannual timescale and are a source of predictability (see Chapter 9). In fact, there are reasons to believe that oceans embody longer timescales outside than in the tropics (e.g., the phase speed of long baroclinic Rossby waves decreases poleward as both the value of beta and density stratification decrease).

We will show that the abovementioned argument is not valid in the extratropics because of high levels of atmospheric internal variability that is largely independent of ocean. While atmospheric memory is indeed shorter than 1 month (as measured by lagged autocorrelation), its internal variability contains a wide range of timescales from subseasonal to multidecadal. The ocean, with a large thermal inertia (deep winter mixed layer) and slow Rossby wave dynamics, preferentially responds to low-frequency atmospheric forcing, resulting in apparent multidecadal variations in SST. The lack of strong oceanic feedback is consistent with low skills of coupled general circulation models (GCMs) in predicting atmospheric variability in the extratropics (see Section 9.7).

This chapter discusses fundamental distinctions in the origin of interannual variability in and out of the tropics. We start with atmospheric internal variability, and its role in forcing SST and thermocline anomalies.

12.1 Atmospheric internal variability

Eastward-moving storms—quasi-weekly transient eddies—dominate midlatitude weather variations. Averaged over 1 month, atmospheric variability is barotropic in the vertical as the standing waves in the westerly winds (see Section 9.6). Such month-to-month variability—stationary waves or eddies—arises spontaneously within the atmosphere, even in the absence of SST anomalies, growing on the instability of the mean westerly flow and/or the positive feedback with transient storm tracks.

EOF analysis in the Pacific and Atlantic sectors yields the Pacific-North American (PNA) and North Atlantic Oscillation (NAO) patterns as the leading modes of atmospheric variability, respectively. Each represents an accelerated/blocked jet at the positive/negative phase (Fig. 12.2). The blocking is a persistent (>1 week) weather pattern with an anomalous high in the north and anomalous low in the south, slowing down the westerlies and splitting the jet into two branches. Blockings are associated with long-lasting anomalous weather conditions, and their onset/decay is often hard to predict.

Atmospheric GCMs forced with observed evolution of SST/sea ice can be used to evaluate variability due to the surface boundary forcing and to internal dynamics of the atmosphere (see Section 1.4). To sample the chaotic internal variability, an atmospheric model is run multiple times to form an ensemble, with different initial conditions. The ensemble mean represents the SST forced response (hence potentially predictable), while the spread from the ensemble mean represents atmospheric internal variability independent of SST forcing.

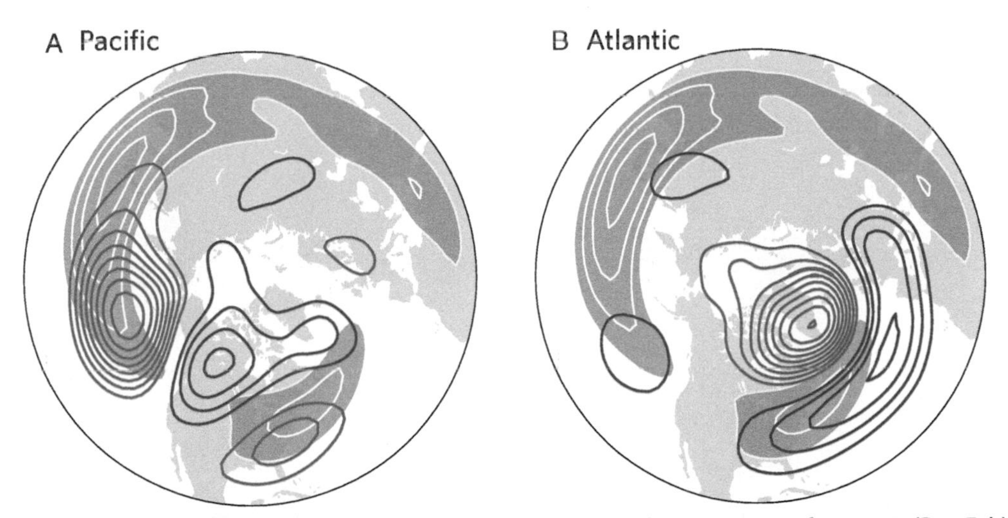

Fig. 12.2 (A) Pacific-North American and (B) North Atlantic Oscillation patterns for winter (Dec-Feb) 1948-2020. Regression maps (blue/red positive/negative) of the Northern Hemisphere (north of 20N) month-to-month variability in 500-hPa geopotential height against the principal component (PC1) in (A) the Pacific (120°E–80°W) and (B) Atlantic sectors (80°W–20°E), superimposed on the mean 300-hPa zonal wind (white contours at interval of 5 m/s, dark grey shading >20 m/s). PCs explain 28% and 34% of the total variance in the North Pacific and Atlantic domains, respectively. Adapted and updated based on Rennert and Wallace (2009). *(Courtesy Z.H. Song.)*

In the tropics, internal variability is small, and atmospheric variability is dominated by SST forcing (Fig. 12.3B). As a result, the SST-forced variability is highly correlated with observations. All this supports the view that the tropical atmosphere is tightly coupled with the ocean and the positive feedback from the coupling provides predictability at seasonal leads. This also explains a conundrum that climate is predictable, while deterministic weather predictability is limited to 2 weeks. We cannot predict whether and at what time thunderstorms will take place in Singapore 1 month from now, but the knowledge of a developing El Niño allows us to predict dry conditions for the month and possibly beyond. This is sometimes called the atmospheric predictability of the second kind that arises due to slowly evolving SST boundary condition, to distinguish from the predictability of weather for the next few days due to predictable atmospheric dynamics (e.g., the east advection by the westerly winds and westward-propagating Rossby waves). Predictability for weather and seasonal-mean variability arises from atmospheric and ocean initial conditions, respectively.

In the extratropics, atmospheric variability is dominated by the random internal component, except along the paths of the PNA and Pacific-South American (PSA) patterns where the SST-forced signal to noise (ensemble mean/spread) ratio approaches unit (see Fig. 12.3B). The PSA is the Southern Hemisphere counterpart of PNA and the ENSO-forced stationary wave trains that ride on the mean westerlies. For 500-hPa geopotential height during boreal winter, internal variability is especially high over the Arctic (including Greenland), the North Pacific near the Aleutians, and the subtropical North Atlantic around the Azores. The enhanced variance over these regions is suggestive of the PNA and NAO patterns. The leading EOF modes of internal variability indeed resemble NAO and PNA.

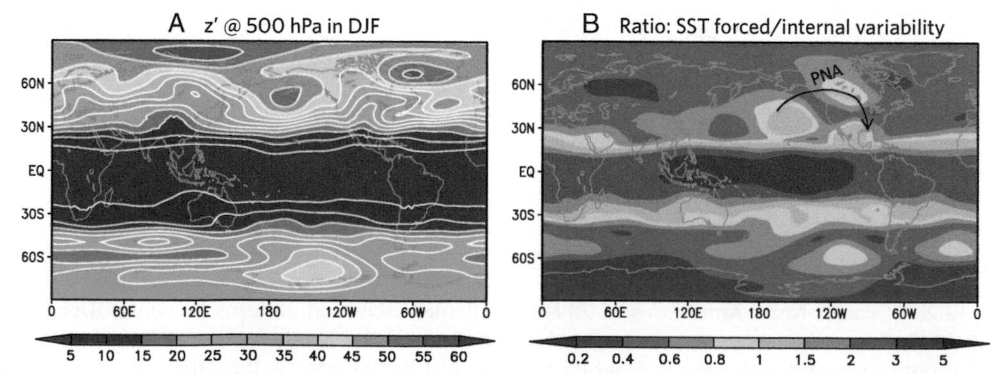

Fig. 12.3 (A) Standard deviation of the ensemble spread (m), and (B) the ratio of the ensemble mean to spread standard deviation, based on a 9-member AM2.1 atmospheric general circulation model simulation forced with observed sea surface temperature evolution for 1979-2011. *(Courtesy Y. Kosaka.)*

Thus atmospheric internal variability is random in time but of coherent large-scale spatial patterns. NAO and PNA are internal modes that arise from strong interactions between the westerlies and eddies. Internal variability is locally enhanced at so-called jet-exit regions where the upper-level westerlies decelerate ($\partial\overline{u}/\partial x < 0$) because barotropic energy conversion there (see Section 9.6) energizes stationary variability.

ENSO forces PNA variability by inducing upper-level divergent wind (as a Rossby wave source) over the subtropical North Pacific (see Fig. 9.24). The NAO, on the other hand, is predominantly an internal mode with little influence from SST forcing. This distinction has important implications for predictability: PNA anomalies are potentially predictable, while NAO is largely not predictable beyond the 2-week limit of weather prediction.

NAO is the sectoral expression of the so-called Northern Annular Mode (NAM), the latter defined as the leading EOF mode of sea level pressure (SLP) variability in the Northern Hemisphere poleward of 20°N. NAM is associated with strong zonal-mean variability, specifically a seesaw in zonal wind between middle and high latitudes, also known as the zonal index cycle. The Southern Annular Mode (SAM) features weaker zonal variations in the largely oceanic Southern Hemisphere.

In the tropics, deep convection couples the entire troposphere from the ocean surface to the tropopause. Horizontal temperature gradient and advection are both weak, and as a result, the response to convective heating follows the linear wave dynamics. Because convective heating peaks in the midtroposphere, tropical perturbations are of baroclinic vertical structure. Convection-circulation coupling is an important source of internal variability ranging from individual convective storms to the planetary Madden-Julian oscillation (MJO), but internal variability in the seasonal mean is generally much smaller than the SST forcing effect.

In the extratropics, the atmosphere is stably stratified, and convection in winter is limited to the lowest 1 to 2 km as in convective cloud streets following the passage of a cold front. This, together with the strong advection by the prevailing westerlies, renders the SST effect on the free troposphere weak and nonlocal. Because of strong temperature gradient, horizontal advection is a first-order effect. Internal variability is pronounced, ranging from extratropical storms, blocks to the stationary modes (e.g., NAO) of a wide range of timescales. Perhaps curiously, stationary perturbations are of a barotropic vertical structure despite a hydrostatically stable stratification. General, transient baroclinic eddies interact with low-frequency stationary perturbations in positive feedback.

12.2 Atmospheric forcing of SST: Lagged correlation diagnosis

SST variability tends to be negatively correlated with local wind speed: increased wind speed intensifies turbulent heat flux and vertical mixing, resulting in an SST cooling (see Fig. 12.1A). In the tropics, the negative correlation is indicative of WES (wind-evaporation-sea surface temperature) feedback, which amplifies coupled ocean-atmospheric anomalies antisymmetric about the equator as observed in the Atlantic Meridional Mode (AMM). Does this correlation generally imply positive feedback between the ocean and atmosphere (Hypothesis 2)? The answer is no in the extratropics in light of weak atmospheric response to local SST. We show that lagged cross-correlation can help infer ocean-atmospheric feedback.

12.2.1 Stochastic model 1 without positive feedback

Consider a simple mixed layer model for SST variability (Model 1) following Hasselmann (1976), forced by surface heat flux fluctuations due to atmospheric variability (say, in wind speed) F,

$$\partial T/\partial t = F - \lambda T \tag{12.1}$$

where λ is the damping rate ($\lambda^{-1} \sim$ a few months). Section 8.1.1 derives Eq. 12.1 by linearizing surface latent heat flux, but sensible heat flux becomes important outside the tropics (see Fig. 7.10). Here we solve Eq. 12.1 with the Fourier transformation,

$$(\mathrm{T},\ \mathrm{F}) = \int \left(\widetilde{T}, \widetilde{F}\right) \mathrm{e}^{\mathrm{i}\omega t}\, \mathrm{d}\omega \tag{12.2}$$

where the tilde denotes the Fourier transform and ω is the frequency. In the extratropics, the atmospheric forcing F is dominated by random internal variability with a decorrelation time less than 2 weeks. Consider a simple case where the atmospheric forcing is a white noise [$\widetilde{F}(\omega) = \widetilde{F}_0$=const.] without any feedback from SST. Substituting (2) in (1) yields

$$\widetilde{T}(\omega) = \widetilde{F}_0/(i\omega + \lambda) \tag{12.3}$$

The SST response lags the atmospheric forcing F by $\tau = \tan^{-1}\frac{\omega}{\lambda}$. The lag approaches zero at the low-frequency and 90° in phase at the high-frequency limit.

The SST power spectrum is

$$\left|\widetilde{T}\right|^2(\omega) = \left|\widetilde{F}_0\right|^2/(\omega^2 + \lambda^2) \tag{12.4}$$

SST preferentially responds to low-frequency atmospheric noise. The SST response $\left|\widetilde{T}\right|$ increases rapidly with decreasing frequency ($\propto 1/\omega$) for $\omega > \lambda$, and eventually reaches a high plateau of $\widetilde{F}_0/\lambda$ for $\omega << \lambda$. The timescale for the transition from the inverse linear relation to the low-frequency limit is λ^{-1}. Eq. 12.4 is called the red spectrum with enhanced power at low frequencies. While the extratropical atmosphere does not have memory beyond a month, its chaotic variability covers all timescales from hours to decades (white noise). The ocean thermal inertia integrates the noise with high SST variance at low frequencies.

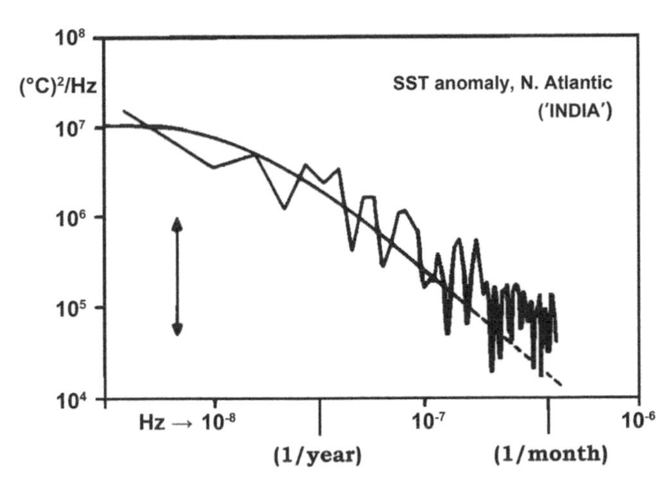

Fig. 12.4 Sea surface temperature *(SST)* spectrum at weathership (59°N, 19°W) for 1949-64, with 95% confidence interval. The smooth curve is from Model with $\lambda=$ (4.5 months)$^{-1}$. *(From Frankignoul and Hasselmann (1977).)*

Fig. 12.4 compares the SST power spectrum in the open North Atlantic (59°N) with the simple model (Eq. 12.3) forced by white-noise atmospheric variability. The agreement is remarkable considering the simplicity of the model. Note that the net surface flux variability (right-hand side [RHS] of Eq. 12.1) is red as it includes the SST damping. The atmospheric forcing (e.g., wind and relative humidity) component of the surface flux can be modeled as a white noise because of vanishing autocorrelation at lags beyond 1 month (Fig. 12.5A). The weak month-to-month correlation is the basis for the notion that the atmosphere does not have memory longer than 1 month. In the tropics, month-to-month correlation is not negligible for atmospheric variability (e.g., the Southern Oscillation), as it includes positive feedback from the ocean.

12.2.2 Lagged correlation

In Model 1 (see earlier), the SST response lags behind the atmospheric forcing. The lagged correlation between the atmospheric forcing and SST response (Eq. 12.7) proves illuminating with regard to causality.

Define lagged covariance $R_{xy}(\tau) = \langle x(t+\tau)y(t)\rangle$, where $\langle \cdot \rangle \equiv \frac{1}{N}\sum_{1}^{N}(\cdot)$, with N the number of data in the time series. Cross-correlation $r_{xy}(\tau) = R_{xy}(\tau)/\sqrt{R_{xx}(0)R_{yy}(0)}$, where $\sigma_x^2 \equiv R_{xx}(0)$ is the squared variance of time series $x(t)$. Multiplying Eq. 12.1 with $F(t-\tau)$, we obtain

$$\frac{\partial}{\partial \tau}R_{TF}(\tau) = R_{FF}(\tau) - \lambda R_{TF}(\tau)$$

Here we assume stationary statistics, $R_{xy}(\tau) = x(t+\tau)y(t) = x(t)y(t-\tau)$. Noting that

$$\frac{\partial}{\partial \tau}R_{TF}(\tau) = \frac{\partial}{\partial \tau}\langle \mathrm{T}(\mathrm{t}+\tau)\mathrm{F}(\mathrm{t})\rangle = \langle \mathrm{F}(\mathrm{t})\frac{\partial}{\partial t}\mathrm{T}(\mathrm{t}+\tau)\rangle = \langle \mathrm{F}(\mathrm{t}-\tau)\frac{\partial}{\partial t}\mathrm{T}(\mathrm{t})\rangle$$

we obtain

$$\frac{\partial}{\partial \tau}R_{TF}(\tau) = R_{FF}(\tau) - \lambda R_{TF}(\tau) \tag{12.5}$$

The lagged autocorrelation is

$$r_{FF}(\tau) = e^{-\nu|\tau|}$$

for atmospheric forcing and

$$r_{TT}(\tau) = e^{-\lambda|\tau|} \tag{12.6}$$

for the SST response. Here the e-folding decorrelation time is much shorter for the atmosphere than for the ocean—typically, ν^{-1} less than 1 month and λ^{-1} ~6 months (Fig. 12.5A).

Climate data are often averaged into monthly means. For $\lambda \ll \nu$, the lagged cross-correlation for monthly-mean data is

$$r_{TF}(m) \approx \begin{cases} 0, \quad m = -1, -2, \ldots \\ \sqrt{\tau_1^*/2}, \quad m = 0 \\ \sqrt{2\tau_1^*}exp(-m\tau_1^*), \quad m = 1, 2, \ldots \end{cases} \tag{12.7}$$

where $\tau_1^* = \lambda\tau_1$ is the nondimensional length of a month with $\tau_1 = 30$ days (Frankignoul and Hasselmann 1977). The lagged correlation r_{TF}(-1, 0, 1) = (0, 0.3, 0.5) for $\lambda =$ (6 months)$^{-1}$. The functional form of the lagged correlation can be predicted from the differential Eq. 12.5. Because the forcing term R_{FF} is nonzero only at m = 0, the solution R_{TF} grows from m = −1, peaks at m = 1, and then decays exponentially for m >1. Remarkably, the Model 1 result fits the observations over the midlatitude North Pacific quite well (see Fig. 12.5B, C).

Contemporaneous cross-correlation between atmospheric and SST variability is of a finite value, but this does not necessarily imply two-way coupling. One needs to examine the lagged cross-correlation to determine the causality. In Model 1, the lagged cross-correlation vanishes at m = −1. The causality is clear here: Atmospheric noise drives SST variability but without any feedback from the ocean. In the Hasselmann Model 1, SST is of some predictability because of the ocean thermal inertia and temporal

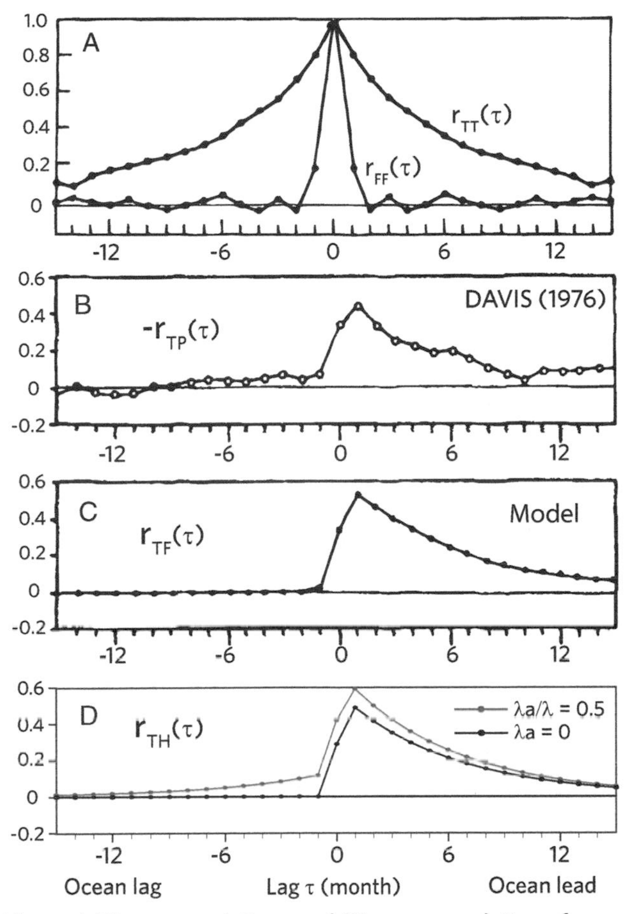

Fig. 12.5 Observed lagged (A) autocorrelations and (B) cross-correlation of sea surface temperature and sea level pressure empirical orthogonal functions over the North Pacific as a function of lag. (C) Theoretical cross-correlation for $\nu = (8.5\ \text{day})^{-1}$, $\tau = (6\ \text{months})^{-1}$. (D) Cross-correlation without *(black)* and with *(red)* ocean feedback $\lambda_a / \lambda = 0.5$. *(A and B, From Davis (1976) © American Meteorological Society. Used with permission; C, from Frankignoul and Hasselmann (1977); and D, courtesy Z.H. Song.)*

persistence, but atmospheric variability F is independent of SST and unpredictable by formulation.

12.2.3 Stochastic model 2 with ocean feedback

We can include ocean feedback ($\lambda_a T$) in the atmospheric forcing

$$H = F + \lambda_a T \tag{12.8}$$

This is appropriate for subtropical trade-wind regions, where the wind variability includes both atmospheric internal variability F (e.g., NAO) and the SST feedback (e.g., low-cloud and WES feedbacks). The SST equation (Model 2) is now

$$\frac{\partial T}{\partial t} = H - \lambda T = F - \widehat{\lambda} T \tag{12.9}$$

where the ocean feedback reduces the effective damping rate, $\widehat{\lambda} = \lambda - \lambda_a$.

All the analysis for Model 1 can be used for Model 2, with this reduced damping rate $\widehat{\lambda}$. From Eq. 12.8 we obtain the cross-covariance between atmospheric forcing and SST response

$$R_{TH}(\tau) = R_{TF}(\tau) + \lambda_a R_{TT}(\tau) \tag{12.10}$$

and the cross-correlation

$$\begin{aligned} r_{TH}(\tau) &= \frac{\sigma_F}{\sigma_H}\left[r_{TF}(\tau) + \frac{\sigma_T}{\sigma_F}\lambda_a r_{TT}(\tau)\right] \\ &= \sqrt{\frac{1}{1 + \left(2\widehat{\lambda}/\lambda_a + 1\right) r_a^2}}\left[r_{TF}(\tau) + r_a r_{TT}(\tau)\right] \end{aligned} \tag{12.11}$$

where $r_a = \lambda_a \sigma_T \big/ \sigma_F = \left(\lambda_a \big/ \widehat{\lambda}\right) r_{TF}(0)$, with σ_F and σ_T being the standard deviations of atmospheric forcing and SST response, respectively. On the RHS, the first term is asymmetric about τ much as in Model 1, while the second term is symmetric with a long decorrelation time $\widehat{\lambda}^{-1}$. At negative lags $\tau = m\tau_1$ ($\tau_1 =$ 1 month, and m $= -1, -2$ ...), the first term on the RHS vanishes, and the second term dominates, with the cross-correlation

$$r_{TH}(m) = \sqrt{\frac{r_a^2}{1 + \left(2\widehat{\lambda}/\lambda_a + 1\right) r_a^2}}\, r_{TT}(m), m \leq -1 \tag{12.12}$$

Thus positive feedback from the ocean has two important effects: (i) increasing the decorrelation timescale $(\lambda - \lambda_a)^{-1}$; and (ii) creating a positive correlation at lag -1 between the atmospheric and SST variability (Eq. 12.12). For $\lambda_a/\lambda \ll 1$,

$$r_{TH}(-1) \approx \left(\frac{\lambda_a}{\lambda}\right) r_{TF}(0) r_{TT}(-1)$$

In other words, the lag -1 cross-correlation is linearly proportional to and a measure of the ocean feedback coefficient λ_a.

12.2.4 Observed cross-correlation

With moderate feedback ($\lambda_a < \lambda$), the cross-correlation is asymmetric between negative and positive lags (the first term in Eq. 12.11), but significant positive correlations at negative lags are suggestive of positive oceanic feedback (see Fig. 12.5D). When the oceanic feedback approaches the thermal damping ($\lambda_a \lesssim \lambda$), the second term in Eq. 12.11 dominates, and the cross-correlation becomes symmetric between negative and positive lags. Such variations in the shape of the cross-correlation can be used to infer the strength of oceanic feedback.

Take the AMM as an example (Fig. 12.6). At 20°N, the concurrent cross-correlation between the zonal wind and SST is significant, but the correlation at lag -1 month (SST leading) is insignificant, suggesting a strong wind forcing of SST variability but with little ocean feedback at this subtropical latitude. The trade winds fluctuate on the southern flank of the subtropical Azores High as part of the atmospheric internal NAO (see Fig. 12.2B). At 10°N, the zonal wind-SST cross-correlation is still asymmetric but the correlation is positive and significant at negative leads, suggestive of positive feedback from SST. At the equator, the cross-correlation between cross-equatorial SST gradient

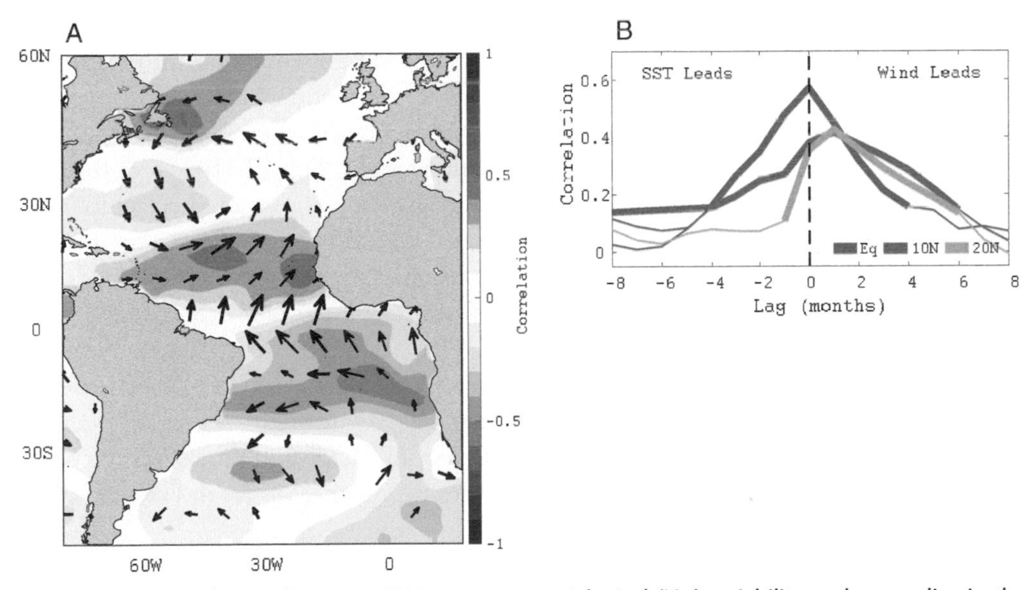

Fig. 12.6 (A) Correlations between FMA cross-equatorial wind (V_{eq}) variability and anomalies in the preceding winter (DJF): sea surface temperature (SST; color shading) and surface wind (vectors). (B) Lagged correlation between SST and zonal wind variability at 20°N (orange curve, >95% significance level thickened) and 10°N *(red)*; cross-equatorial SST difference (10°N–10°S) and meridional wind at the equator *(blue)*. Anomalies are zonally averaged across the Atlantic basin before the analysis (1979-2017). Correlation >95% significance thickened. Adapted based on Amaya et al. (2017). *(Courtesy D.J. Amaya.)*

and meridional wind becomes nearly symmetric and peaks at zero lag, indicative of a strong positive feedback between the ocean and the atmosphere. Chang et al. (2001) showed that subtropical trade wind variability preferentially excites the AMM because of the cross-eqiatorial WES feedback in the deep tropics.

Extratropical SST variability is coherent and of large scales in space because atmospheric variability is so (see Section 12.1). Over the North Pacific, the dominant atmospheric mode represents the rise and fall of SLP centered at the Aleutian Islands, while the dominant SST mode features a band of anomalous warming and cooling on the southern flank of the SLP mode (see Fig. 12.1). Furthermore, the SLP and SST principal components (PCs) are correlated contemporaneously, indicating that zonal wind variability drives SST through surface heat flux and Ekman advection. North Pacific SSTs decrease when a deepened Aleutian Low intensifies the surface westerly winds. The contemporaneous correlation does not necessarily mean a two-way coupling of SLP and SST variability (Box 12.1). A close look shows a mismatch in timescale between the ocean and atmospheric variability: While the SLP PC is basically uncorrelated from one month to another, significant autocorrelation persists for as long as 1 year in SST (see Fig. 12.5A). More importantly, the cross-correlation vanishes with SST leading but peaks with SLP leading by 1 month (see Fig. 12.5B). All this fits Hasselmann Model 1 of stochastic atmospheric forcing of SST variability without feedback from the ocean. The lack of oceanic feedback is broadly consistent with the atmospheric model result that atmospheric internal variability dominates the extratropics (see Section 12.1).

The Davis (1976) analysis covered a short 28-year period of 1947–74 when the PDO was in the cold phase and interannual variability dominated (Fig. 12.1B). Over the past century, the PNA (tracked by Aleutian atmospheric pressure) shows marked multidecadal variability (see Fig. 12.1C) that is highly correlated with PDO, $r \sim 0.7$ for winter (DJF) seasonal-mean data. This does not imply positive feedback from the SST PDO onto the atmospheric PNA. Instead, they both originate from tropical ENSO. Indeed, r(PDO, PNA) peaks with PNA leading (Newman et al. 2016). The PNA may be decomposed into ENSO-forced and atmospheric internal components. While the latter is close to being white noise with little month-to-month persistence, the former has a broad interannual peak with lower-frequency variability. The ocean's large heat capacity and slow Rossby wave dynamics further redden the tropically forced SST in the extratropical North Pacific.

While defined over the extratropical North Pacific, the PDO shows an interhemispheric pattern that resembles but is meridionally broader than ENSO in the tropics (see Fig. 12.1A). Remarkably the SST pattern is nearly a mirror image between the North and South Pacific. This interhemispheric symmetric pattern suggests that tropical Pacific SST variability plays a role in driving the PDO in both the extratropical North and South Pacific.

BOX 12.1 Evolving views on extratropical variability

Extratropical storms grow on baroclinic instability and are steered eastward by the prevailing westerlies, both set up by the meridional temperature gradient. Weather forecast is an initial value problem. Current weather over the North Pacific is the initial condition forecasters of North America pay close attention to. It was therefore natural to extend the method for climate forecast and hypothesize that North Pacific SST anomalies "provide a more or less geographically fixed area for cyclogenesis," thereby affecting large-scale atmospheric circulation and hence North American climate (Namias 1959). The North Pacific Experiment (NORPAX; 1967-85) was launched to investigate this North Pacific effect but ended up largely falsifying the hypothesis. The lagged correlation analysis of Davis (1976), conducted as part of NORPAX, showed that North Pacific SST anomalies are a result, rather than a cause, of Aleutian Low variability (see Fig. 12.5A, B). Stochastic models support this null hypothesis that the atmosphere forces SST variability but with little feedback from the ocean (Frankignoul and Hasselmann 1977).

Except hurricanes, weather forecasters of North America had paid little attention to tropical conditions until Bjerknes (1969) suggested that tropical variability (e.g., ENSO) arises from ocean-atmosphere interaction and drives changes in the Hadley circulation, the North Pacific westerlies, and North American climate further downstream. The Bjerknes hypothesis laid down the foundation for coupled ocean-atmosphere dynamics and seasonal forecast (see Box 9.1).

The dazzling success of coupled approach to ENSO dynamics in the 1980s, followed by equally spectacular success in the tropical Atlantic and Indian oceans (see Chapters 10—11), raised the hope to discover other coupled modes. Extrapolating for the extratropics, one might assume that ocean processes (e.g., baroclinic Rossby waves and thermocline ventilation) set the slow timescale for coupled variability and provide additional predictability. Standing in the way is, however, the null hypothesis developed in the 1970s on midlatitude variability (Davis 1976; Frankignoul and Hasselmann 1977): The atmosphere forces the ocean but with little feedback from the ocean.

Corroborating the null hypothesis, atmospheric model results show that the atmospheric response to extratropical SST perturbations is nonlocal and weak. The atmospheric response to extratropical basin-scale SST anomalies is weak for several reasons. First, SST effect is confined in the planetary boundary layer (1—2 km) because of the stable stratification (as opposed to the tropics where deep convection couples the SST effect through the entire troposphere). Second, the atmospheric response is nonlocal because of strong advection by the intense westerly jet and mixing by storms. Third, the noise (atmospheric internal variability) is high compared to the extratropical SST forcing.

The road in developing coupled dynamics has been winding with surprising turns. Just as NORPAX hit a roadblock in connecting North Pacific SST downstream to North American climate variability, ENSO dynamics blossomed and culminated in operational seasonal forecast of practical value (see Box 9.1). Even more surprisingly, North Pacific SST anomalies affect North American climate after all, just not directly but through the PMM effect on ENSO with a time delay of two to three seasons (see Fig. 12.21).

12.3 Ocean dynamic effects

This section considers additional mechanisms for SST variability. Extratropical oceans display a pronounced seasonal cycle. In winter, surface cooling deepens the mixed layer, while in summer the upper layer restratifies with a seasonal thermocline below a shallow surface mixed layer.

The winter mixing-summer restratification cycle introduces a mechanism for SST anomalies to persist from one winter to the next. Suppose that winds are anomalously strong in a winter, and the mixed layer is anomalously cool. The summer restratification shields the negative temperature anomalies below the seasonal thermocline from the direct atmospheric influence. In the following winter, the preserved temperature anomalies from the previous winter entrain into the mixed layer and affect its temperature. This reemergence mechanism contributes to winter-to-winter correlation in extratropical SST (Fig. 12.7).

Integrating the temperature equation from the surface to the base of the winter mixed layer ($z = H_m$) yields

$$\frac{\partial T_m}{\partial t} = -\boldsymbol{u}_g \cdot \nabla T_m + \frac{1}{\rho H_m}\left[\frac{Q}{c_p} + \frac{1}{f}(\boldsymbol{k} \times \boldsymbol{\tau}) \cdot \nabla T_m\right] - w_m \frac{\partial T}{\partial z}\Big|_{z=H_m} \tag{12.13}$$

where the first term on the RHS is the advection by geostrophic flow, and the second term in the brackets is advection by the Ekman flow. Here we assume that turbulent

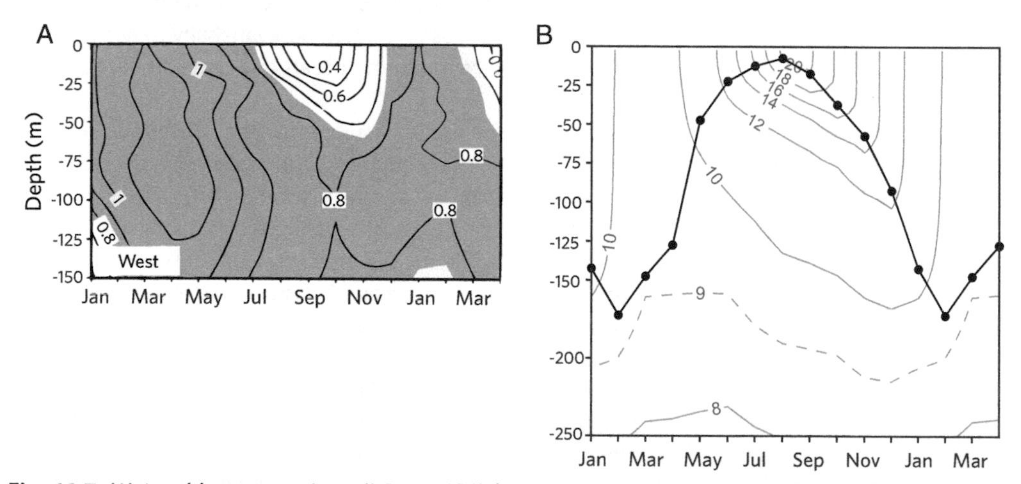

Fig. 12.7 (A) Lead-lag regressions (°C per 1°C) between temperature anomalies at the base point, located here at 5 m in April-May, and temperature anomalies from the previous January through the following April in the west Pacific region (38–42°N, 160–180°E). The contour interval is 0.1 and values >0.75 are shaded. (B) Same as (A) but for climatologic-mean temperature and mixed layer depth. *(A, From Alexander et al. (1999); B, courtesy Z.H. Song. © American Meteorological Society. Used with permission.)*

mixing is small below the winter mixed layer. Integrating Eq. 12.13 over 1 year forms the annual-mean heat budget.

Consider a simple case where the prevailing winds are westerly and the background SST varies only in the meridional direction. Eq. 12.13 can be linearized for perturbation temperature as

$$\frac{\partial T}{\partial t} = -v_g\frac{\partial \overline{T}}{\partial y} + upwelling - (a + a_E)U - bT \tag{12.14}$$

where the prime has been dropped for slow perturbations that vary on multiyear timescales (see Fig. 12.1), a is the WES coefficient (see Section 8.1), and $a_E = \frac{2}{\rho H_m}\frac{\overline{\tau}_x}{\overline{W}}\left(-\frac{\partial \overline{T}}{\partial y}\right)$ measures the Ekman advection, with $\overline{W}$ as the mean scalar wind speed and U the zonal wind perturbation. Thus in the mean westerly regime ($\overline{\tau}_x > 0$), the Ekman advection reinforces the WES effect. Outside the tropics, sensible heat flux is important, and its wind dependency can be absorbed in the WES term. A westerly wind anomaly causes SST to decrease by intensifying the background wind through turbulent heat flux and Ekman advection.

12.3.1 Ocean Rossby waves

SST variance over the North Pacific is large along 40°N. In the central North Pacific where geostrophic currents are weak, the SST variability is due mostly to surface processes (heat flux and Ekman advection). Along 40°N, the SST variance intensifies westward in the so-called Kuroshio-Oyashio Extension (KOE) region (Fig. 12.8), where the winter mixed layer is deep and geostrophic current variability is large. The combined effect makes subsurface processes (geostrophic advection and upwelling) important for

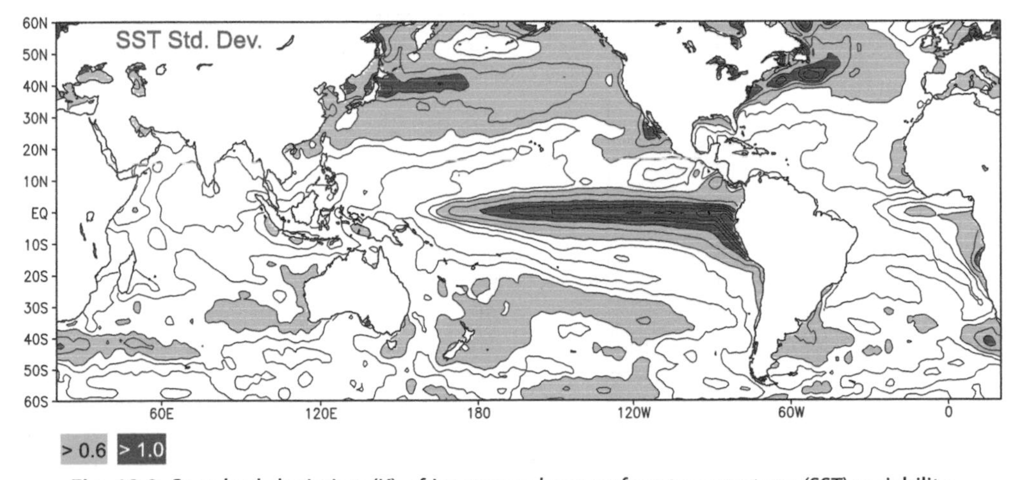

Fig. 12.8 Standard deviation (K) of interannual sea surface temperature *(SST)* variability.

SST. Temperature anomalies in this region have a deep vertical structure, and SST is highly correlated with the thermocline displacements and sea surface height.

Sea level ($\eta = g' h/g$) variability in the KOE follows baroclinic Rossby waves forced by wind fluctuations associated with the Aleutian Low (see Section 7.2.5),

$$\frac{\partial \eta}{\partial t} + c_R \frac{\partial \eta}{\partial x} = \frac{g'}{g} w_E - \gamma \eta \tag{12.15}$$

where the long Rossby wave phase speed c_R is typically 2.5 cm/s. As it takes ~5 years for the Rossby wave to cross the half-width of the North Pacific, the Rossby wave is another mechanism in favor of ocean response to low-frequency wind forcing. Forced with observed wind, the simple Rossby wave model captures thermocline displacements/dynamics height as observed by historical ship-based measurements of upper-ocean temperature and satellite altimeters (Fig. 12.9). This result also implies that sea surface height in the KOE is potentially predictable a few years in advance as baroclinic Rossby waves are forced in the central Pacific by fluctuations of the Aleutian low (Schneider and Miller 2001). Since the deep winter mixed layer opens a widow for subsurface variability (Rossby waves) to affect SST over the KOE, SST variability there features pronounced decadal timescales (see Fig. 12.1).

Cross-correlation between SST and surface heat flux (downward positive) is often a useful diagnostic regarding the cause of SST variability. In the simple case where the RHS of the Hasselmann model is due to surface heat flux, $Q = F - \lambda T$, the cross-

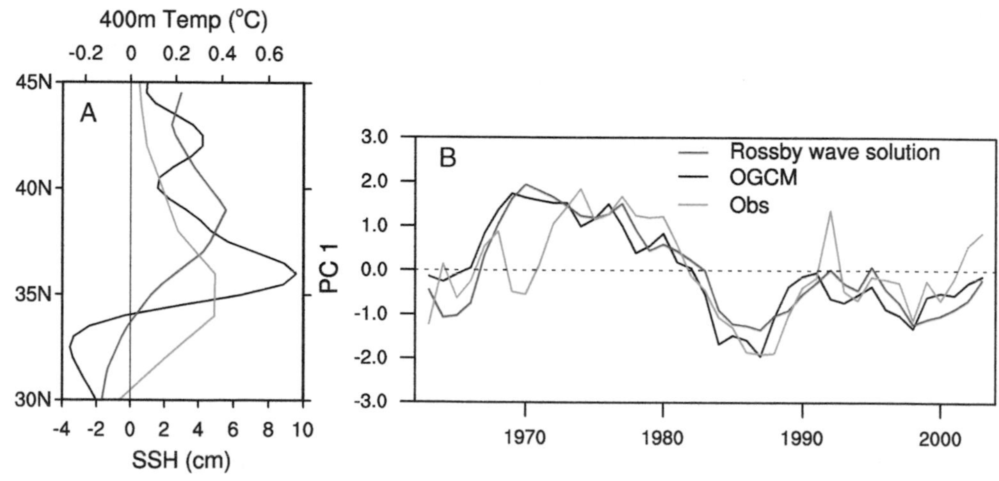

Fig. 12.9 (A) Leading empirical orthogonal function and (B) principal component of observed ocean temperature averaged in the top 400 m *(blue)*, sea surface height variability in an eddy-resolving ocean general circulation model *(black)*, and a zonally one-dimensional, linear Rossby wave model forced with observed wind stress variability *(red)*. All interannual anomalies averaged in 142-180°E Adapted based on Taguchi et al. (2007). *(Courtesy B. Taguchi.)*

correlation $r_{TQ} = 0$ since Q and T are 90° out of phase. In the other extreme (e.g., the KOE region), the advection by geostrophic currents dominates the RHS of Eq. 12.13, $R_{TQ} \approx -bR_{TT} < 0$ (i.e., the SST-surface heat flux correlation is negative). Thus r_{TQ} can help identify where ocean dynamics might be important for SST variability. High-resolution observations from space by satellite-borne cloud-penetrating microwave sensors reveal ubiquitous ocean forcing of atmospheric variability on the ocean mesoscale (100s km) near major ocean currents (Box 12.2), characterized by $r_{TQ}<0$.

BOX 12.2 Ocean front-atmosphere interaction

Over broad open ocean regions, subsurface dynamic effects (advection and upwelling) on SST are weak. Surface heat flux and Ekman advection are important for SST variability, with wind speed and SST variability negatively correlated in general (see Figs. 12.1 and 12.10A).

Narrow ocean jets often anchor sharp SST fronts (e.g., north of the equator over the eastern Pacific and Atlantic, along the Gulf Stream and Antarctic Circumpolar Current). In the atmospheric boundary layer (~1 km deep), air temperature and humidity adjust to the SST front through vertical mixing and lateral advection, while the response in the free troposphere may be complicated and weak. Because of the larger radius of deformation, atmospheric temperature cannot fully adjust to the sharp SST gradient, with a broader cross-frontal profile (Box Fig. 12.1C). As a result, the near-surface atmospheric static instability intensifies on the warmer flank, while decreasing on the colder flank of the ocean front. Because the surface drag maintains a vertical wind profile with higher wind speeds aloft, the enhanced vertical mixing accelerates the surface speed of the prevailing wind on the warmer flank, while surface wind decelerates on the colder flank. The hydrostatic SLP perturbations, displaced downwind due to the lateral advection, also contribute to this positive wind speed-SST correlation on the frontal scale, especially at moderate wind speeds.

Box Fig. 12.1B shows the surface static instability (SST-SAT) in the winter North Atlantic based on historical ship observations. The surface atmospheric instability and surface wind speed are locally enhanced on the warm flank of the Gulf Stream. Frequency of high-wind (>20 m/s) occurrence is high along the atmospheric storm track as the SST front is favorable for explosive cyclogenesis. The high-wind occurrence is locally enhanced as a result of increased atmospheric instability at shallow Flemish Cap (45°W–47°N) east of Newfoundland (Box Fig. 12.1A), where the warm North Atlantic Current is forced to make a cyclonic meander. This area was filmed in the 2000 blockbuster movie *The Perfect Storm* as the final fishing grounds of F/V *Andrea Gail* before sunken. The southern tip of Greenland is the windiest place on Earth as high mountains force tip jets.

The modulations of the atmospheric boundary layer apply not only to linear SST fronts but also to isolated ocean eddies: Surface wind intensifies locally over warm/anticyclonic eddies, while weakening over cold/cyclonic eddies. The temporal correlation between SST and wind speed is markedly positive on the short ocean mesoscale (see Box Fig. 12.2), say by applying high-pass (<1000 km) spatial filter. On the large/basin scale, by contrast, the causality and correlation are opposite with SST responding to, but not affecting, free-tropospheric wind variability (e.g., PNA). The positive wind-SST correlation characterizing the short-scale air-sea interaction is

(Continued)

BOX 12.2 Ocean front-atmosphere interaction—cont'd

especially high along major SST fronts as the underlying ocean jets energize ocean eddies, while the background SST gradients allow eddies to manifest in SST (Chelton and Xie 2010).

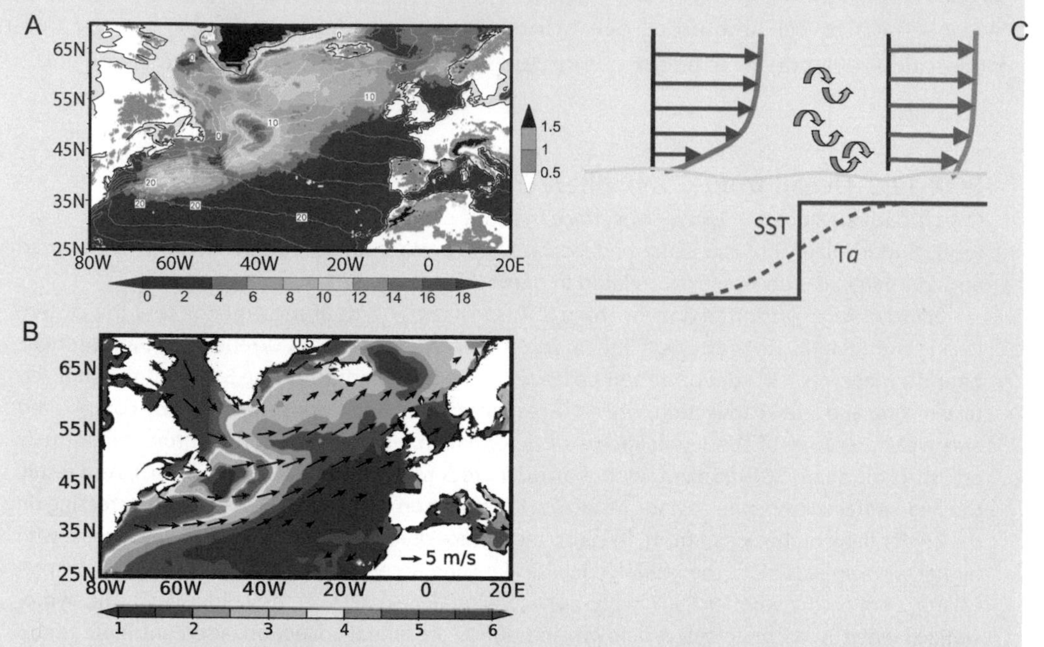

Box Fig. 12.1 Winter (DJF) climatology at sea surface. (A) High-wind frequency (bottom color bar in %) and sea surface temperature (SST; white contours at 2°C interval). Land topography is shaded (right color bar, m). (B) Wind velocity (arrows, m/s) and the sea–air temperature difference (color shading, °C). (C) Schematic of atmospheric adjustments across an SST front. *(A, From Sampe and (2007); B, courtesy L. Yang; C, courtesy M. Nonaka)*

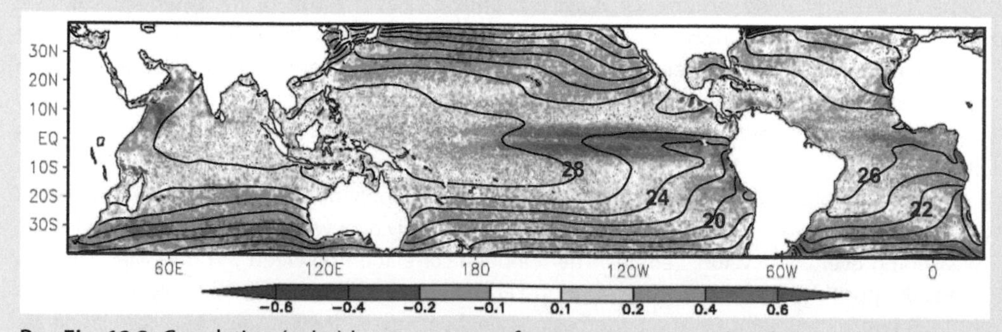

Box Fig. 12.2 Correlation (color) between sea surface temperature and surface wind speed, both zonally high-pass filtered to emphasize zonal scale shorter than 1000 km. Adapted based on Small et al. (2008). *(Courtesy Z.H. Song.)*

12.3.2 Atlantic Multidecadal Oscillation

The NAO is the dominant mode of atmospheric variability over the North Atlantic. At the positive phase, the meridional dipole in SLP intensifies the westerlies and steers winter storms toward Europe, causing anomalous conditions in the Mediterranean and North Africa. The intensified Icelandic Low advects cold polar air southward over northeastern Canada, while the southerly advection warms Scandinavia. By modulating the strength of the prevailing winds, the NAO induces a tripole pattern in SST. At the positive phase, SSTs decrease over the subpolar North Atlantic, increase in the midlatitudes east of the United States, and decrease west of West Africa (Fig. 12.10A). This tripole pattern dominates SST variability on the interannual timescale.

On multidecadal timescales, the SST pattern of the same sign over the entire North Atlantic basin becomes important (see Fig. 12.10B). At the positive phase, the Atlantic Multidecadal Oscillation (AMO) is associated with a northward displaced Atlantic intertropical convergence zone (ITCZ) with wet conditions over the African Sahel and enhanced Atlantic hurricane activity.

While atmospheric model results suggest that the NAO is due largely to atmospheric internal variability (Fig. 12.3), it has a surprising impact on the AMO through ocean dynamic adjustments, as illustrated by a pair of coupled model experiments forced with the NAO forcing of 50-year period. The same atmospheric GCM coupled to a motionless slab ocean model (SOM) in one experiment and to a fully dynamic ocean model (DOM) in the other. In SOM, the NAO forces the tripole pattern at a time delay of 1 to 2 years (Fig. 12.11). In DOM, by contrast, the NAO forces variability in Atlantic

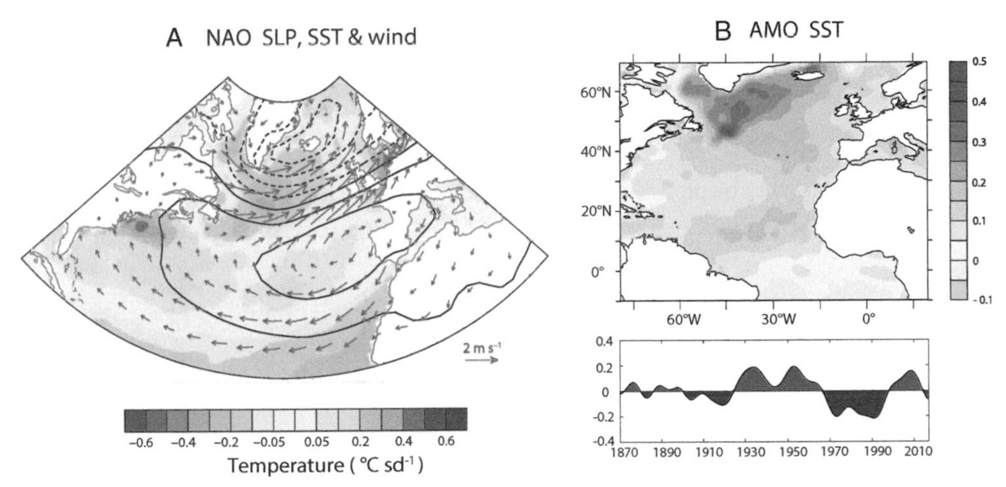

Fig. 12.10 (a) Sea level pressure *(SLP)*, wind, and sea surface temperature *(SST)* anomalies associated with the North Atlantic Oscillation. (B) SST anomalies of the Atlantic Multidecadal Oscillation (AMO), along with the AMO index time series *(bottom)*. *(A, From Deser et al. 2010; B, Zhang et al. (2019).)*

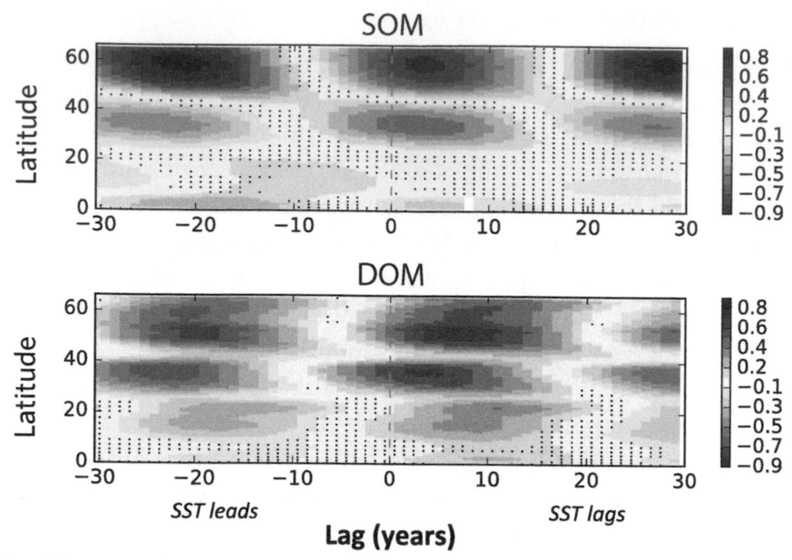

Fig. 12.11 Lead-lag correlations between the sea surface temperature *(SST)* response and the imposed 50-year periodic North Atlantic Oscillation forcing in slab ocean model *(SOM)* and dynamic ocean model *(DOM)* coupled experiments (CM2.1), zonally (60–20°W) and annually averaged. *(From Delworth et al. (2017). © American Meteorological Society. Used with permission.)*

Meridional Overturning Circulation (AMOC) and its heat transport, resulting in a basin-wide warming at a lag of about a decade.

An intensified AMOC is associated with positive SST and surface salinity anomalies in the subpolar North Atlantic (Fig. 12.12). Surface heat flux is negatively correlated with local SST over the extratropical North Atlantic. This surface heat flux damping of SST reaffirms the importance of ocean circulation variability in the AMO.

12.4 Extratropical influence on tropical climate

ENSO arises from the coupled Bjerknes feedback in the tropical Pacific and excites predictable climate anomalies outside the tropics (e.g., the PNA teleconnection pattern; see

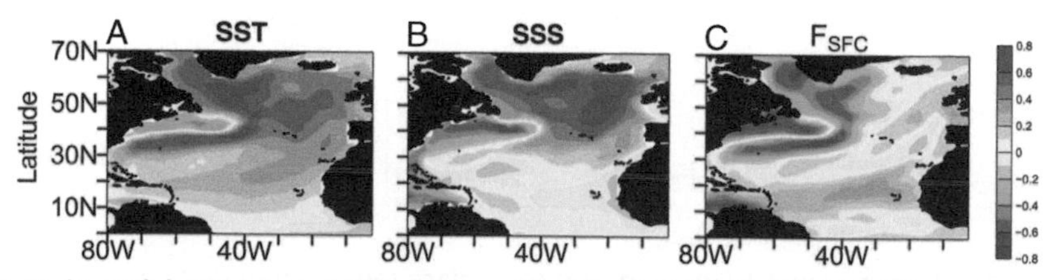

Fig. 12.12 Multimodel mean correlation with an Atlantic Meridional Overturning Circulation index at 26°N (10-year low-pass filtered): (A) sea surface temperature, (B) sea surface salinity, and (C) surface heat flux (F_{SFC}, downward positive) in CMIP5 unforced control simulation. *(From Yan et al. (2018).)*

Sections 9.5–9.6). This section discusses how atmospheric and ocean circulation variability outside the tropics can influence tropical climate, including ENSO and the ITCZ.

12.4.1 Pacific meridional mode

The subtropical trade winds are subject to both internal atmospheric variability and SST feedback. NAO, for example, modulates the northeast trades on the south flank of the Azores high and hence SST over the tropical North Atlantic (see Section 12.2.4). Likewise over the North Pacific, internal variability of the extratropical atmosphere causes the northeast trades to vary during the winter. The weakened northeast trades near Hawaii cause in situ SST to rise by reducing evaporative cooling. The positive SST anomalies induce a cyclonic circulation that is displaced westward because of atmospheric Rossby wave dynamics. The westerly wind anomalies on the south flank of the low-pressure circulation induce a warming tendency, causing the coupled SST-wind pattern to propagate southwestward (Fig. 12.13B). When approaching the equator in the central Pacific, the westerly wind anomalies of this Pacific Meridional Mode (PMM) can trigger El Niño by advecting warm water eastward and lowering the thermocline in the eastern equatorial Pacific. The Northeast Pacific stratus deck is at the bend of the PMM's hook-like SST pattern, and the stratus-SST feedback is important in energizing and persisting the PMM through JJA (sse Fig. 6.12).

The PMM is often defined as the leading mode for SST-surface wind covariability over the tropical and subtropical North Pacific, with concurrent ENSO removed (Chiang and Vimont 2004). (It is the second mode if ENSO is not removed, with ENSO being the first mode.) Thus PMM is not correlated with concurrent ENSO by definition, but PMM in boreal spring (MAM) is significantly correlated with ENSO in the subsequent seasons, suggesting that PMM triggers ENSO.

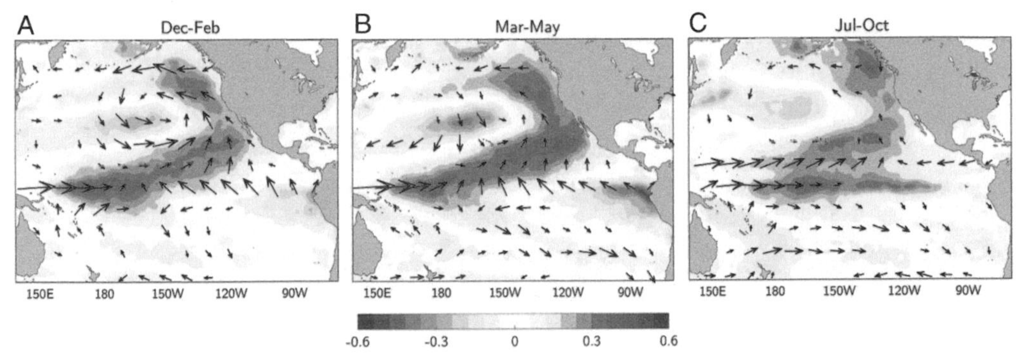

Fig. 12.13 Pacific Meridional Mode based on regressions against the MAM principal component of sea surface temperature (SST)—wind Maximum Covariance Analysis mode. Evolution of SST (color shading, °C) and surface wind anomalies: (A) December-February, (B) March-May, and (C) July to October. *(Adapted from Amaya et al. (2019).)*

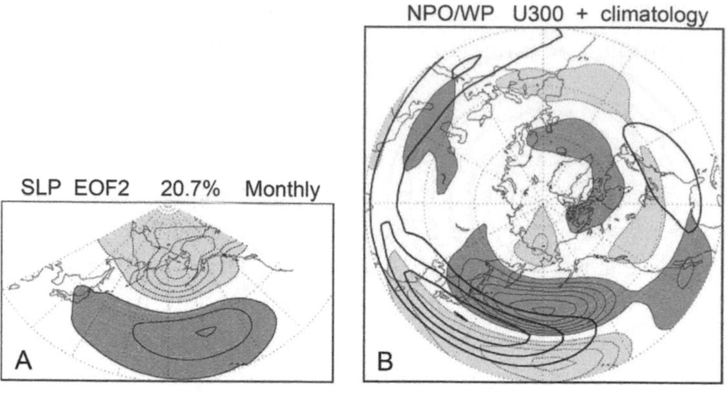

Fig. 12.14 The North Pacific Oscillation as (A) the second empirical orthogonal function *(EOF2)* of monthly sea level pressure *(SLP)* variability over the North Pacific (north of 20°N). (B) Regression of 300-hPa zonal wind *(gray shading and white contours)* superimposed on the climatology *(black contours). (From Linkin and Nigam et al. (2008). © American Meteorological Society. Used with permission.)*

The PNA and North Pacific Oscillation (NPO) are the first and second EOF modes of SLP and/or upper-tropospheric height variability in the North Pacific sector, respectively. The NPO is the Pacific counterpart of the NAO, featuring a SLP seesaw between the Bering Sea and midlatitude North Pacific (Fig. 12.14). Like NAO, NPO is associated with meridional displacements of the Asian-Pacific westerly jet in winter, whereas PNA varies the intensity and eastward extension of the jet. Like atmospheric internal variability in general, PNA and NPO are most pronounced in winter when the westerly jet is most intense and synoptic eddies are most active.

The NPO is associated with marked trade wind variability near Hawaii, which drives the PMM. While random and unpredictable itself, the NPO modulates ENSO as the PMM finds the way into the equatorial Pacific. The PMM imprints on SST and contributes to seasonal predictability of ENSO.

In perturbed initial-condition ensemble seasonal forecast (see Box 10.1), the uncertainty in Nino4 SST prediction can be traced in part back to the PMM excited by the random NPO in winter (Ma et al. 2017). In comparison, the South Pacific contribution to the ENSO forecast uncertainty is small. Atmospheric internal variability in the subtropical trade winds is stronger over the North than South Pacific because of stronger zonal variations in the mean westerly jet. Convective feedback in the ITCZ helps energizing atmospheric wind feedback (Amaya 2019), another reason in favor of the PMM from the North Pacific in affecting the equatorial ENSO.

12.4.2 Cross-equatorial energy transport

In steady state, the cross-equatorial energy transport needs to balance between the ocean and atmosphere. This gives rise to a simple theory for cross-equatorial asymmetry of zonal-mean climate. This zonal-mean energy theory complements the tropical theory for the northward-displaced ITCZ presented in Section 8.1.

Consider the zonal, column-integrated energy balance of the atmosphere (Eq. 2.9),

$$\frac{\partial}{\partial y}F_a = R_{TOA} - Q_{net} \tag{12.16}$$

where F_a is the meridional energy flux due to atmospheric motions, R_{TOA} is the net radiative flux at the TOA, and Q_{net} the net surface flux (both downward positive). Near the equator, atmospheric energy transport is accomplished mostly by the Hadley overturning circulation (Section 2.4),

$$F_a = \int_0^{ps} vmdp/g = V\Delta m \tag{12.17}$$

where V is the mass transport of the Hadley cell in the upper branch, and $\Delta m = m_u - m_l$ is the gross moist stability, which is the upper minus lower tropospheric difference in moist static energy (see Section 2.4). For a northward-displaced ITCZ, the lower branch of the cross-equatorial Hadley cell transports moisture onto the ITCZ, but the upper branch transports high dry energy back into the other hemisphere. With $\Delta m > 0$, the net moist energy transport (column integrated) follows the direction of the transport by the upper branch and is toward the hemisphere opposite to the ITCZ.

For an ocean that is in steady state,

$$\frac{\partial}{\partial y}F_o = Q_{net} \tag{12.18}$$

where $F_o = \rho c_p[vT]_o$ is the meridional heat transport by ocean currents, zonally and depth integrated (Eq. 2.11). With Eq. 12.18, Eq. 12.16 becomes

$$\frac{\partial}{\partial y}(F_a + F_o) = R_{TOA} \tag{12.19}$$

We are interested in cross-equatorial atmospheric energy transport as it determines the meridional asymmetry of the ITCZ through Eq. 12.17. Integrating Eq. 12.16 from the equator to poles yields

$$F_a + F_o = -\langle R_{TOA}\rangle_{SH}^{NH}/2 \tag{12.20}$$

at the equator, where $\langle R_{TOA}\rangle_{SH}^{NH} \equiv \langle R_{TOA}\rangle_{NH} - \langle R_{TOA}\rangle_{SH}$, with the angular brackets denoting the hemispheric integration.

Spanning a temperature range of 3°C to 10°C in the vertical, the AMOC transports energy northward across the equator. To compensate this AMOC forcing of $F_o|_{y=0} > 0$,

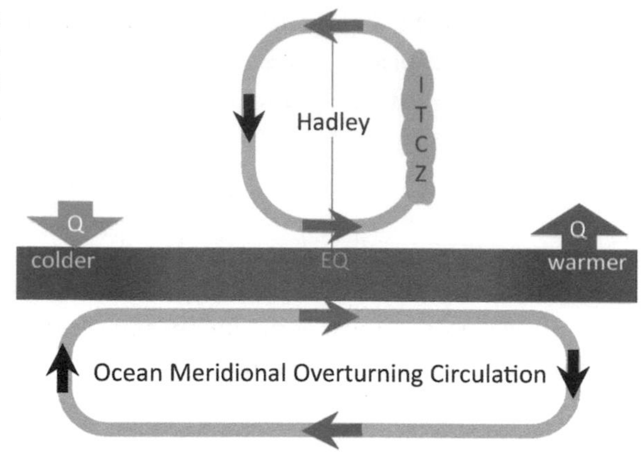

Fig. 12.15 Schematic of an anomalous Hadley circulation response to cross-equatorial heat transport by an ocean meridional overturning circulation. *(Based on Frierson et al. (2013).)*

the atmosphere needs to transport energy southward with the ITCZ displaced north of the equator (Fig. 12.15). Since the hemisphere with the ITCZ is warmer, we assume that the TOA radiative response (RHS of Eq. 12.20) damps the atmospheric transport, $\langle R_{TOA} \rangle_{SH}^{NH} / 2 = \alpha F_a$, with $\alpha > 0$. Eq. 12.20 becomes

$$F_a = -F_o/(1+\alpha) \tag{12.21}$$

at the equator.

Once the cross-equatorial atmospheric energy transport (F_a) is known, we can calculate the ITCZ displacement. In Eq. 12.16 we approximate the meridional derivative with a finite difference, $[0 - F_a(y=0)]/\delta y = R_{TOA} - Q_{net}$ at the equator. Here δy is the northward displacement of the ITCZ from the equator, and we assume that atmospheric energy transport vanishes at the ITCZ, $F_a|_{y=\delta y} = 0$. The ITCZ position is then given by

$$\delta y = -F_a/(R_{TOA} - Q_{net})|_{y=0} \tag{12.22}$$

Hence the ITCZ displacement is not only dependent on the cross-equatorial F_a but also on the net energy flux into the atmospheric column. The latter gives the meridional gradient of F_a. In observations, $R_{TOA} \sim 75$ W/m^2, and $Q_{net} \sim 50$ W/m^2 around the equator.

Model results suggest that interactive clouds, especially those capped by the trade wind inversion, act as positive feedback (with a reduced damping rate α), amplifying the ITCZ response compared to the case with fixed clouds (Kang et al. 2009). We note that α here represents radiative damping on the interhemispheric difference, not necessarily the same as that for global mean surface temperature ($-\lambda$) to be discussed in Section 13.1. The low-cloud feedback on interhemispheric temperature difference, for example, is probably strongly positive as the anomalous downdrafts in the cooler hemisphere are to increase low-cloud cover by increasing cloud-top radiative cooling (see Section 6.1) and with an enhanced capping inversion.

From Eq. 12.16, we have

$$-2F_a\big|_{y=0} = \langle R_{TOA}\rangle_{SH}^{NH} - \langle Q_{net}\rangle_{SH}^{NH} \qquad (12.23)$$

Asymmetric energy perturbations far away from the tropics can cause the ITCZ to shift because cross-equatorial atmospheric energy transport results from the interhemispheric difference in the net energy flux into the atmospheric column, regardless the latitude of the energy perturbations. The global energy theory here is broadly consistent with and complements the tropical WES feedback discussed in Section 8.1. The northwest-tilted coastal line of the Americas favors the upwelling on the Pacific coast south of the equator, and the resultant energy perturbation ($\langle R_{TOA}\rangle_{SH} > 0$) is consistent with a northward displacement of the ITCZ through Eq. 12.23.

Land surface is much more extensive in the Northern than in the Southern Hemisphere. The interhemispheric difference in surface albedo could potentially cause asymmetry in solar radiation. Satellite observations, however, show that the Northern—Southern Hemisphere asymmetry in surface albedo is compensated by larger could albedo in the Southern Hemisphere, mostly over the ocean. The hemispheric-integrated downward solar radiation at the TOA is nearly identical between the Northern and Southern Hemisphere (Voigt et al. 2013).

12.5 Deep meridional overturning circulation

The abovementioned energy theory shows that the heat transport by the interhemispheric ocean MOC is crucial in determining the global zonal-mean ITCZ position. Here we consider factors that place the sinking branch of the global MOC (Ferreira et al. 2018). Why is the Northern Hemisphere favored over the Southern, and why is the North Atlantic chosen over the North Pacific?

The Antarctic Circumpolar Current (ACC) flows eastward unimpeded in the top 1500 m in the Southern Ocean channel, where Scotia Ridge east of the Drake Passage is the tallest bathymetry. In climate models, the circumpolar Southern Ocean consistently locks the upwelling branch of the global MOC. Around 41 million years before present (mBP), South America detached from Antarctic Peninsular, and the opening of the Drake Passage caused the permanent glaciation of Antarctica.

In the Southern Ocean circumpolar channel, the equatorward surface Ekman flow under the prevailing westerly winds is compensated by a poleward geostrophic flow under the sill. Above the sill in the circumpolar channel, meridional geostrophic flow is not allowed in the zonal mean because of the vanishing zonal-mean zonal pressure gradient. A clockwise Deacon cell forms in the Eulerian mean (Fig. 12.16, blue). The cold

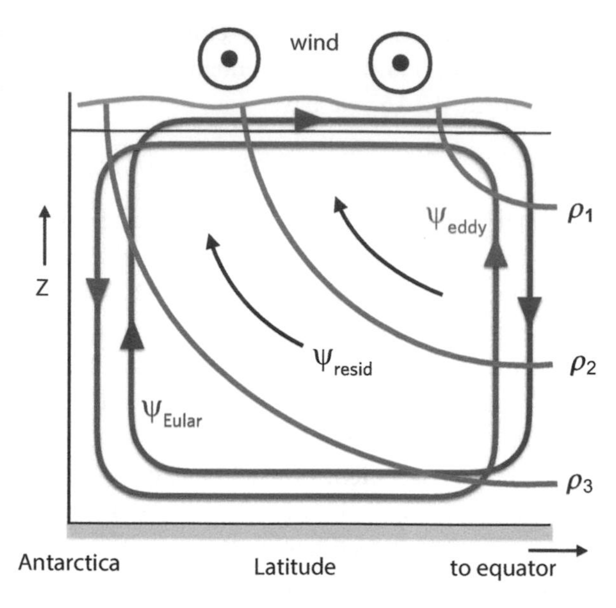

Fig. 12.16 Schematic of the Eulerian, eddy-induced ('bolus') and residual stream functions in a circumpolar channel. The clockwise Eulerian circulation *(blue)* is forced by the westerly winds, the eddy-induced circulation *(red)* opposes it, and the net, or residual, circulation *(slanted black arrows)* is nearly along isopycnals. *(From Vallis 2017.)*

advection by the equatorward surface Ekman flow and the warm advection by the poleward flow under the sill make the water column gravitationally unstable, and the convection would create vertical isopycnals in thermal wind balance with the ACC. In the real ocean, the vertical isopycnals generate baroclinic instability, and the resultant eddies flatten the isopycnals and create vertical density stratification. This is equivalent to a counterclockwise eddy-induced circulation (see Fig. 12.16, red).

The residual circulation is the sum of the Eulerian mean and eddy-induced bolus velocity, akin to the discussion of the eddy-induced Ferrel circulation of the midlatitude atmosphere (see Section 2.4.2). The residual overturning circulation that fluid particles follow differs from the Eulerian mean since eddies deform the isopycnals in the zonal direction. The residual circulation features the sloped upwelling in the circumpolar channel (see Fig. 12.16, black) much like the isopycnals that slope upward and outcrop in the winter mixed layer. Below the mixed layer, the sloping residual circulation connects to the deep MOC all the way to the subpolar North Atlantic, where the surface density in winter matches with that in the circumpolar channel of the Southern Ocean (Fig. 12.17). At the northern end of this interhemispheric MOC, deep convection takes place in the winter subpolar North Atlantic.

With the wind-induced sloped upwelling in the Southern Ocean, the global deep MOC needs to sink somewhere in the Northern Hemisphere. Subpolar surface salinity is much higher in the North Atlantic than in the North Pacific (Fig. 12.18) in favor of the

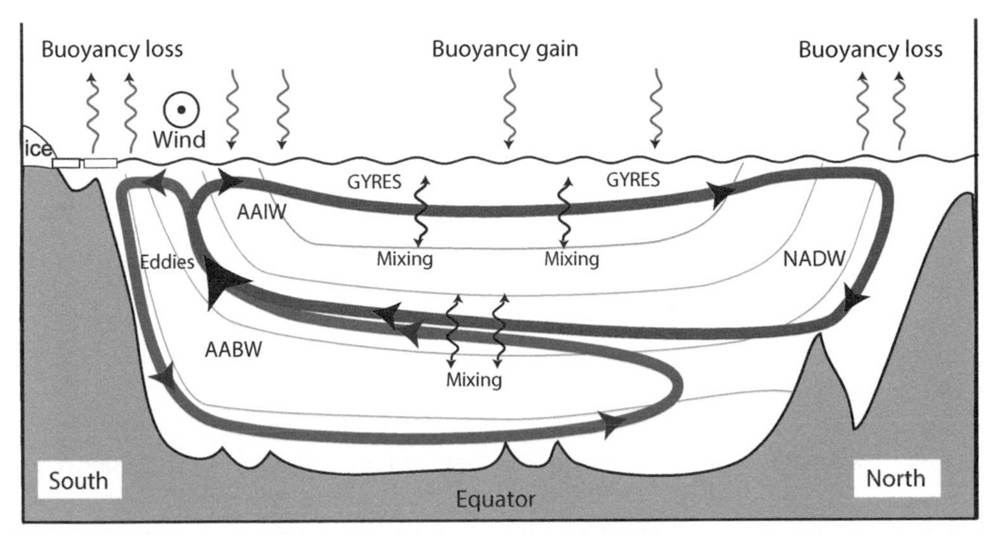

Fig. 12.17 The overturning circulation of the ocean and the main processes that produce it—winds, mixing, baroclinic eddies, and surface buoyancy fluxes. Thin blue curves are isopycnals. North Atlantic deep water *(NADW)* and Antarctic bottom water *(AABW)*. *(From Vallis 2017.)*

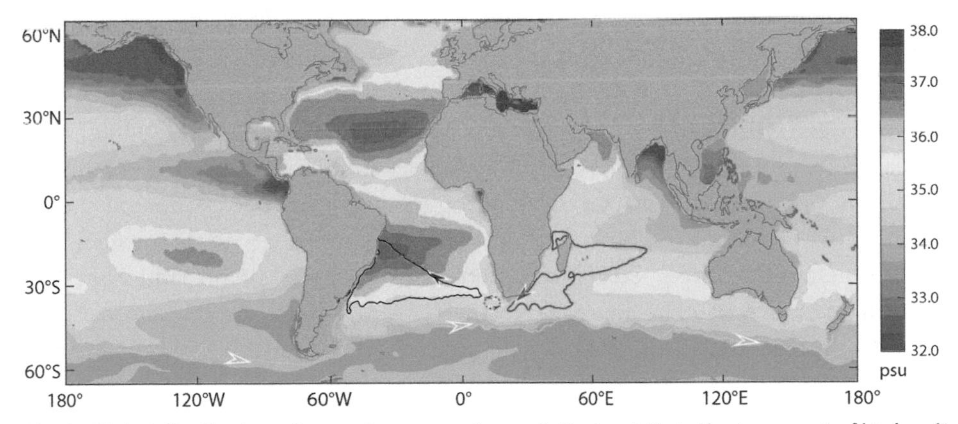

Fig. 12.18 Global distribution of annual-mean surface salinity *(psu)*. Note the transport of high-salinity water from the subtropical South Indian to Atlantic and the overall higher salinity of the Atlantic. Selected sea surface height contours of 1 m *(magenta)*, 0.58 m *(black)*, and 0 m *(cyan)* mark the Indian and Atlantic subtropical gyres, and the Antarctic Circumpolar Current, respectively. The dashed circles denote the eddy transport. *(Courtesy Z.H. Song.)*

North Atlantic for MOC sinking. Several factors contribute to the salinity difference between the Pacific and Atlantic basins.

1. Moisture transport between the Pacific and Atlantic basins is blocked by high mountains of the Rockies on North America and the Andes on South America. Low elevations on the Central American Isthmus, however, allow water vapor transported from

the Atlantic to Pacific by the northeast trade winds, resulting in higher salinity in the Atlantic.

2. The southern tip of Africa is in a subtropical region of high surface salinity. The ocean circulation is such that it transports the high-salinity subtropical Indian Ocean water into the Atlantic basin. Much of the transport is due to eddies that shed from the Agulhas Current retroflection and travel northwestward into the South Atlantic.

In support of the second factor, surface salinity is higher in the subtropical South Atlantic than Pacific (see Fig. 12.18). The Atlantic-to-Pacific salinity difference increases further in the northern subtropics, in support of the cross-Central American moisture transport (as in the first factor).

During the transition that started 20 kBP from the Last Glacier Maximum to the current warm Holocene, the global climate briefly returned to a cold state at 12 kBP, triggered by a massive discharge of glacier melt water into the subpolar North Atlantic. This Younger Dryas cold event causes an anomalous southward displacement of the ITCZ over the Atlantic and elsewhere (e.g., a weakened Asian summer monsoon) (Fig. 12.19). The freshwater discharge lowers surface density over the subpolar North Atlantic and temporarily shuts down the AMOC, reducing the cross-equatorial ocean heat transport F_0. The energy theory (Eq. 12.19) predicts the ITCZ's anomalous southward displacement. In comprehensive coupled GCMs, adding fresh water over the subpolar North Atlantic (water hosing) results in a massive cooling over the Northern Hemisphere and the ITCZ shifts southward across the tropical oceans (Fig. 12.20), in broad agreement with the energy theory. During the last glaciation, periodic discharges of massive icebergs into the North Atlantic took place in so-called Heinrich events due to the instability of continental ice sheets, resulting in abrupt climate change on the global scale similar to the Younger Dryas. Water-hosing experiments with coupled climate models simulate the strong correlation between Heinrich events and the Atlantic ITCZ (Liu et al. 2009). The massive Northern Hemispheric cooling associated with an AMOC shutdown was dramatically depicted in the 2004 Hollywood movie *The Day after Tomorrow*.

12.6 Summary remarks

In the extratropics, atmospheric variability is dominated by internal dynamics with little SST feedback. Significant contributions from tropical modes of coupled ocean-atmosphere interaction are identifiable, such as ENSO-induced PNA and PSA patterns. Atmospheric internal variability is random in phase but organized into coherent large-scale modes such as PNA and NAO. The resultant forced SST variability is of coherent spatial patterns and reddened in frequency by the large heat capacity of the winter mixed layer (e.g., in the central North Pacific) and slow Rossby wave dynamics (e.g., KOE).

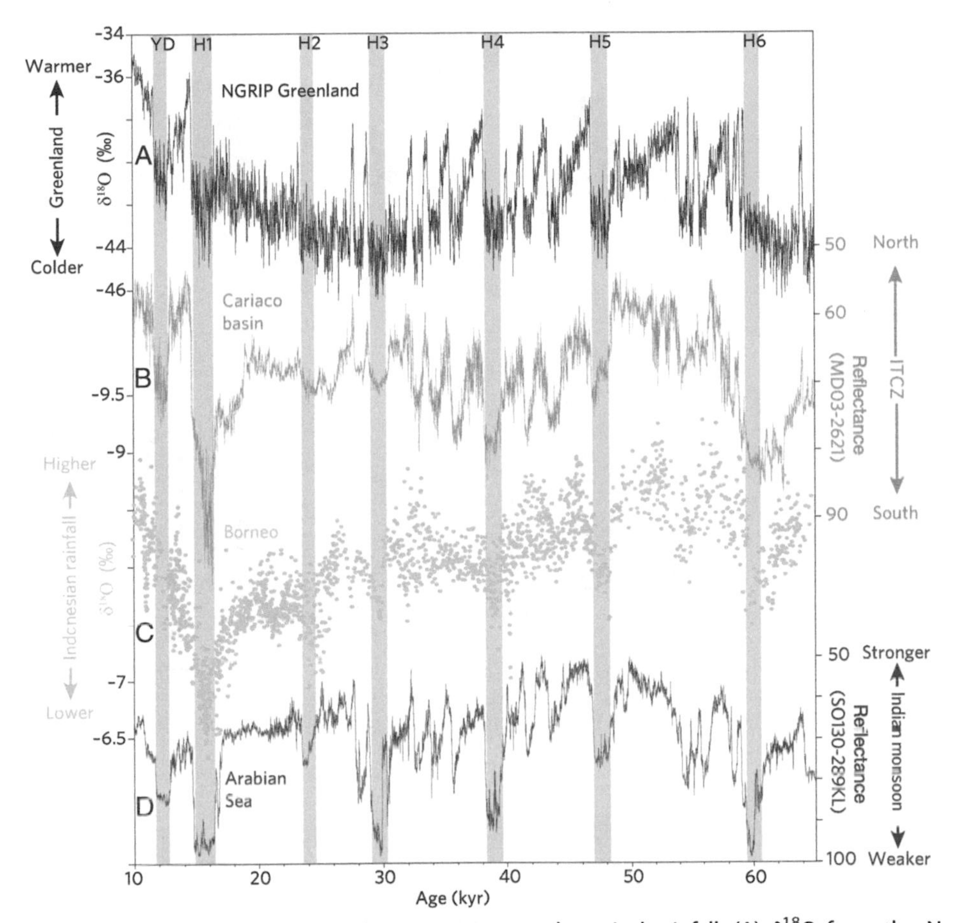

Fig. 12.19 Covariability between Arctic temperature and tropical rainfall. (A) $\delta^{18}O$ from the North Greenland Ice Core Project *(NGRIP)* is a proxy of Arctic temperatures. It indicates gradual warming after the Last Glacial Maximum (~20 kyr bp) and millennial-scale Dansgaard–Oeschger cycles. Cold intervals include the Younger Dryas (YD) and Heinrich stadials H1 to H6 *(grey shading)*. (B) Reflectance of Cariaco Basin sediments is interpreted as the intertropical convergence zone (ITCZ) excursion in boreal summer. Warm intervals indicated by the Greenland ice core are associated with a farther-northward boreal-summer ITCZ. (C) $\delta^{18}O$ in cave stalagmites from Borneo is a proxy of rainfall in the equatorial western Pacific. It is low during Heinrich stadials but less responsive to Dansgaard–Oeschger cycles. (D) Low reflectance of Arabian Sea sediments indicates high runoff from Indian monsoon rainfall during warm intervals indicated by the Greenland ice core. *(From Schneider et al. (2014).)*

Fig. 12.21 is a schematic of extratropical variability and the interactions with tropical climate. A simple stochastic model of ocean mixed layer temperature illustrates the reddening effect and simulates the observed spectrum of extratropical SST variability. It demonstrates that concurrent cross-correlation between SST and wind/SLP variability

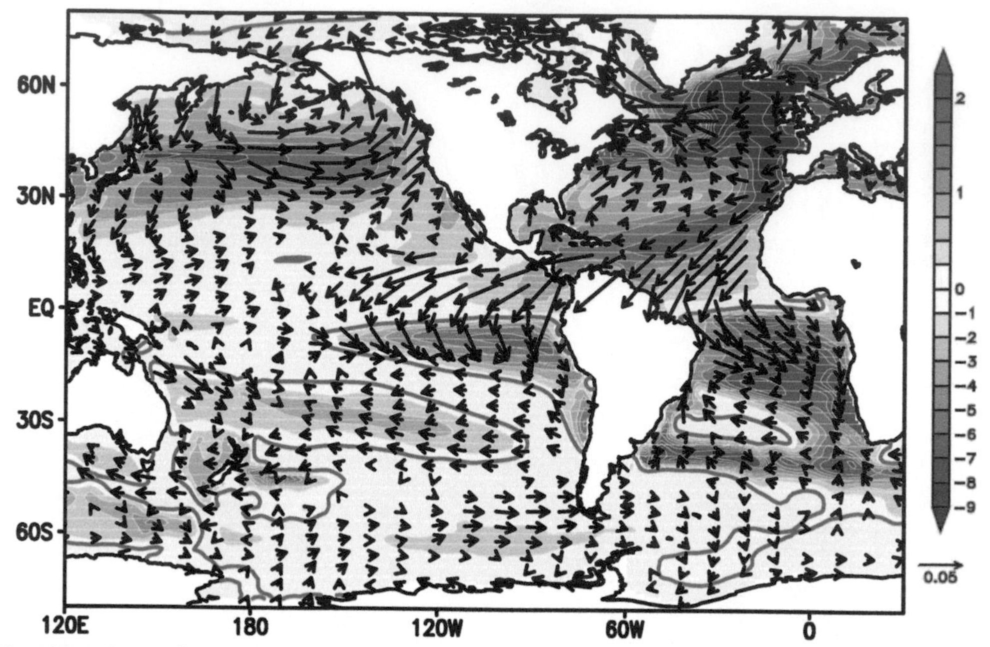

Fig. 12.20 Sea surface temperature (color shading, K) and wind stress (N/m^2) anomalies generated by the Atlantic Meridional Overturning Circulation shutdown due to the addition of fresh water (1 Sv) in 50–70°N of the North Atlantic. *(From Timmermann et al. (2007).)*

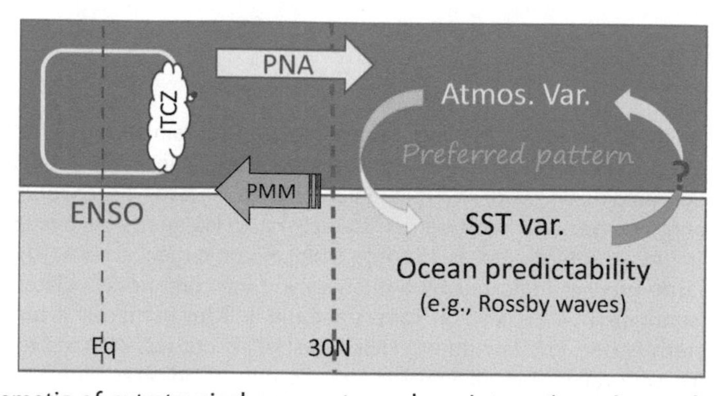

Fig. 12.21 Schematic of extratropical ocean-atmosphere interactions. Atmospheric variability is organized in large-scale spatial patterns and drives ocean variability *(yellow curved arrow)* with weak feedback from the ocean *(gray arrow)*. Predictability remains in slow ocean variables such sea surface temperature *(SST)* and subsurface temperature. Also shown are the Pacific-North American *(PNA)* teleconnection from the tropics as well as teleconnections into the tropics through the Pacific Meridional Mode *(PMM)* and cross-equatorial energy transport. Intertropical Convergence Zone (ITCZ).

is not necessarily indicative of SST feedback onto the atmosphere. Instead, one needs to examine the lagged cross-correlation, specifically the asymmetry about lag 0 and the coefficient at lag -1 month (SST leading). Although this model of stochastic atmospheric forcing precludes prediction of atmospheric variability, the slow ocean Rossby waves still permit predictability of important oceanic variability such as the intensity and position of the Kuroshio Extension.

The ITCZ and ENSO are traditionally considered as tropical phenomena and key drivers of the global climate system. Recent studies reveal marked extratropical influences on these tropical drivers. The PMM is an important conduit for random atmospheric internal variability in the extratropical North Pacific to affect ENSO behavior, although Bjerknes feedback is still the mechanism that grows ENSO over the equatorial Pacific. Likewise, the northward displaced ITCZ involves WES feedback across the equator but is not merely determined by tropical dynamics. Instead, the latitude of the zonal mean ITCZ is also sensitive to cross-equatorial ocean/atmospheric energy transport. Specifically, the zonal-mean ITCZ displaces north of the equator because the global MOC sinks in the subpolar North Atlantic and transports energy northward across the equator (see Fig. 12.15). Research is needed to explain pronounced zonal variations in the ITCZ.

Review questions

1. Use the geostrophic balance to figure out how PNA and NAO affect the intensity and latitude of the upper-level westerly jet.
2. Which of the following modes represents largely atmospheric internal variability and hence is not predictable at monthly/seasonal leads: ENSO, PNA, or NAO? Briefly discuss the evidence.
3. An atmospheric GCM (aGCM) can be run multiple times, with observed SST but perturbed intitial conditions. Why can the aGCM ensemble mean represent the SST effect?
4. What does the aGCM ensemble mean to spread ratio say about the origin of atmospheric variability and the role of SST variability? Compare the tropics and extratropics.
5. Why is the atmospheric internal variability random in time? Is it random in space, too?
6. In the midlatitude Northern Hemisphere, where is the internal atmospheric variability large, and what determines this geographical preference?
7. North Pacific SLP and the NAO both show pronounced multidecadal variability. Does this necessarily indicate that the ocean is important in setting the slow timescale? What additional analysis would you recommend to find this out?

8. What does the Hasselmann Model 1 say in terms of the relationship between atmospheric and ocean variability?
9. How is the atmospheric response to SST fundamentally different in and out of the tropics? What are the important factors that cause the differences?
10. Estimate the e-folding timescale (in days) for atmospheric and SST damping rates from the autocorrelations of Davis (1976).
11. Discuss factors that weaken the atmospheric response to extratropical SST anomalies.
12. The SST-wind speed cross-correlation in Region A is (-0.4, -0.6, -0.4) for lags (-1, 0, 1). (Lag -1 means atmosphere lagging by 1 month.) The correlation in Region B is (0, -0.4, -0.6). Which region has a stronger ocean feedback to the atmosphere?
13. What is the asymmetry in cross-correlation between lags -1 and +1 tell in terms of the relationship between atmospheric and ocean variability? What does that at lag -1 tell about the ocean-to-atmosphere feedback?

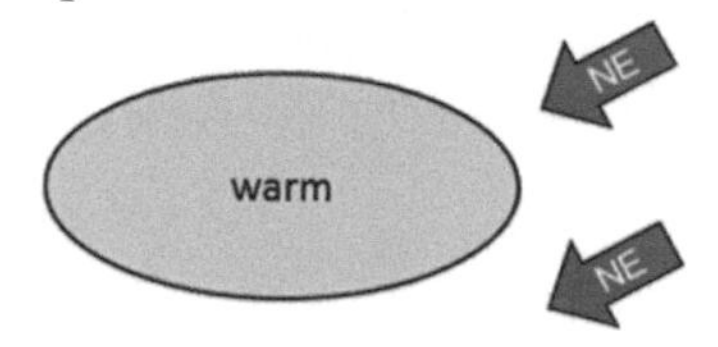

14. The figure above shows a positive SST anomaly in the background northeast trade winds (block arrows; 10°N–30°N). Draw a schematic to illustrate the Rossby response of the atmosphere in the surface wind velocity. Show that the atmospheric response produces a tendency of southwestward propagation in SST through (generalized) WES feedback.
15. Tropical SST can force atmospheric pressure variability in the Aleutian through the PNA teleconnection, which we take as the atmospheric forcing (F) in Hasselmann's Model 1. Tropical Pacific SST variability includes interannual and decadal components of typical periods of 4 and 40 years, respectively. Assume that the interannual to decadal ratio of tropical SST variability is 4:1. Use Model 1 ($\lambda = 1/6$ months) to calculate the ratio of midlatitude North Pacific SST variability.
16. Discuss how the North Atlantic Oscillation can affect the Atlantic ITCZ. The NAO is often considered an "extratropical" mode of atmospheric variability.
17. Discuss how the North Pacific Oscillation can affect ENSO.
18. How does the cross-equatorial energy transport by the ocean surface Ekman flow damp meridional displacements of the ITCZ?
19. The energy theory based on cross-equatorial transport applies to the zonal-mean perturbations. By itself, does it explain why the ITCZ is northward displaced in

the Western Hemisphere (eastern Pacific and Atlantic) but not in the Eastern Hemisphere?

20. Cross-equatorial ocean heat transport varies among the tropical ocean basins. Describe their contributions to the northward ocean transport.

21. Explain the global monsoon from the perspective of interhemispheric energy theory.

CHAPTER 13

Global Warming: Thermodynamic Effects

Contents

Atmospheric concentration of carbon dioxide (CO_2) has increased from the preindustrial levels of ~280 to above 420 parts per million (ppm, 10^{-6}) in volume, and global-mean surface temperature (GMST) has risen by more than 1°C since 1900 when reliable estimates based on instrumental measurements began (Fig. 13.1). Over the last millennium prior to the anthropogenic greenhouse warming, the so-called hockey stick curve (Mann et al. 1998) shows a relatively stable climate on the background, with a slow cooling trend if anything. Climate models predict that GMST is to rise by another 3°C to 5°C by the end of this century in business-as-usual scenarios without major efforts to curtail greenhouse gas (GHG) emissions. Earth climate experienced larger swings in the past (e.g., between ice ages and interglacial epochs), but the rapid pace of the current global warming is unprecedented and may be beyond the range to which natural ecosystems and human society can adapt.

Coupled Atmosphere-Ocean Dynamics: from El Niño to Climate Change
ISBN 978-0-323-95490-7, https://doi.org/10.1016/B978-0-323-95490-7.00013-8

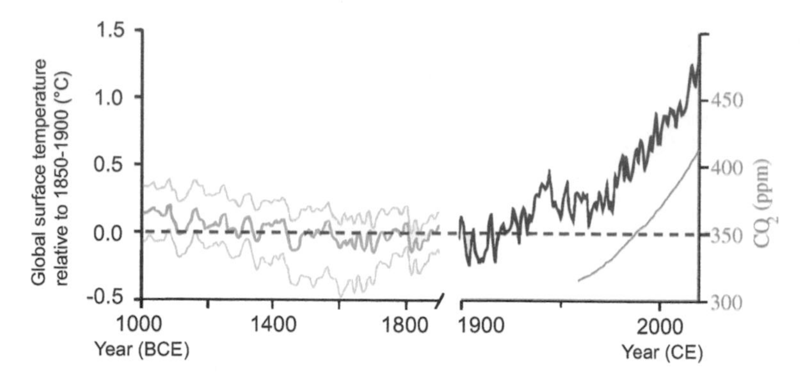

Fig. 13.1 Global-mean surface temperature over 1000–1900 CE *(green)*, and 1900–2020 CE *(purple curve)*, along with the annual-mean CO_2 concentration *(orange)*. Median of the multimethod reconstruction *(bold line)*, with 5th and 95th percentiles of the ensemble members *(thin lines)*. All temperatures relative to the 1850–1900 reference period. Adapted from Gulev et al 2022. *(Courtesy Y.F. Geng and S.M. Long.)*

Decades of research led to a strong scientific consensus as stated in the latest Sixth Assessment Report (AR6) of the Intergovernmental Panel on Climate Change (IPCC 2022a): "[H]uman influence has warmed the atmosphere, ocean and land." It continues: "Human-induced climate change is already affecting many weather and climate extremes in every region across the globe." The report (IPCC 2022b) further concludes, "Human-induced climate change has caused widespread adverse impacts and related losses and damages to nature and people, beyond natural climate variability."

For future climate projection, AR6 considers a range of emission scenarios representing various socioeconomic pathways (Box 13.1). To first order, the GMST increase at the end of the 21st century scales linearly with the radiative forcing, with the multimodel ensemble (MME) mean ranging from as little as 1.5°C to 5.0°C (see Box Fig. 13.1) depending on the strength of greenhouse gas mitigation the world community chooses to implement.

13.1 Climate feedback analysis

Consider an idealized radiative-convective equilibrium where the tropospheric lapse rate is a constant of $\Gamma_m = 6.5$ K/km. Effective radiative temperature is the blackbody temperature of the outgoing longwave radiation (OLR), and the corresponding height on the temperature profile is called the effective radiative height. With increased GHG concentrations, the atmosphere becomes more opaque to infrared radiation, and outgoing infrared radiation seen from space is from a layer of increased altitude. Calculations show that doubling the atmospheric CO_2 concentration is equivalent to lifting the effective radiative height by 150 m (Fig. 13.2A), or a reduction of 3.8 W/m^2 in OLR (the radiative forcing to be defined later) if atmospheric temperature is held unchanged.

BOX 13.1 Coupled Model Intercomparison Project and radiative forcing scenarios

The CMIP coordinates coupled ocean-atmosphere GCM experiments among modeling centers around the world and the distribution of model output in support of IPCC assessments. Core CMIP experiments include the following:

- Preindustrial control is a free run with the radiative forcing fixed at the preindustrial levels. Models simulate climate modes such as ENSO, PDO, and AMO.
- Historical simulation starts from the preindustrial control and is forced with the realistic evolution of GHGs, ozone (natural, volcanic, and anthropogenic), aerosols, and land use estimated from observations.
- Idealized greenhouse warming experiments with 2× or 4× CO_2, either abrupt or at the 1% per year rate.
- Future scenario projections start from the end of the historical simulation under several scenarios of future radiative forcing (Box Fig. 13.1).

Some CMIP models conduct ensemble simulations with different initial conditions to sample internal variability. Since the member runs share the same radiative forcing, the perturbed initial-condition ensemble (PICE) mean represents the radiatively forced response, while the spread the internal variability (e.g., ENSO and PDO). Not all models conduct the ensemble simulations. Alternatively, the MME mean, often with one run from each model, approximates the forced response, with the caveat that the radiative forcing and model physics differ among models.

CMIP phase 6 (CMIP6) uses five Shared Socioeconomic Pathways (SSPs): SSP1-1.9, SSP1-2.6, SSP2-4.5, SSP3-7, and SSP5-8.5. The numerals following the hyphen denote the radiative forcing (W/m^2) at year 2100 relative to the preindustrial. In all SSPs, anthropogenic aerosols are projected to decrease through the 21st century. SSP5-8.5 represents a business-as-usual scenario with little emission mitigation, while SSP1-2.6/1.9 assumes very aggressive reductions in GHG emissions to achieve the Paris Agreement target of limiting global warming to $<2.0/1.5°C$ relative to the preindustrial, with the radiative forcing that peaks around 2030/2050 and declines gradually thereafter. SSP2-4.5 represents a middle of the road future in emissions. SSP1-2.6, SSP2-4.5, and SSP5-8.5 roughly correspond to representative concentration pathways 2.6, 4.5, and 8.5 of CMIP5, respectively.

Expensive and sustained infrastructures are required to develop/improve climate models (scientific staff) and run simulations (big computers and support staff). The high costs limit the full-service modeling efforts to a small number of national centers. CMIP flattens the play field (Meehl et al. 2007), allowing anyone with modest computers to access simulations from latest comprehensive climate models. A boom of multimodel diagnosis ensures, especially from CMIP3 (~2006) onward. Important caveats include (1) robust results among models may suffer from common model errors and biases, and (2) intermodel correlations may not be physical.

(Continued)

BOX 13.1 Coupled Model Intercomparison Project and radiative forcing scenarios—cont'd

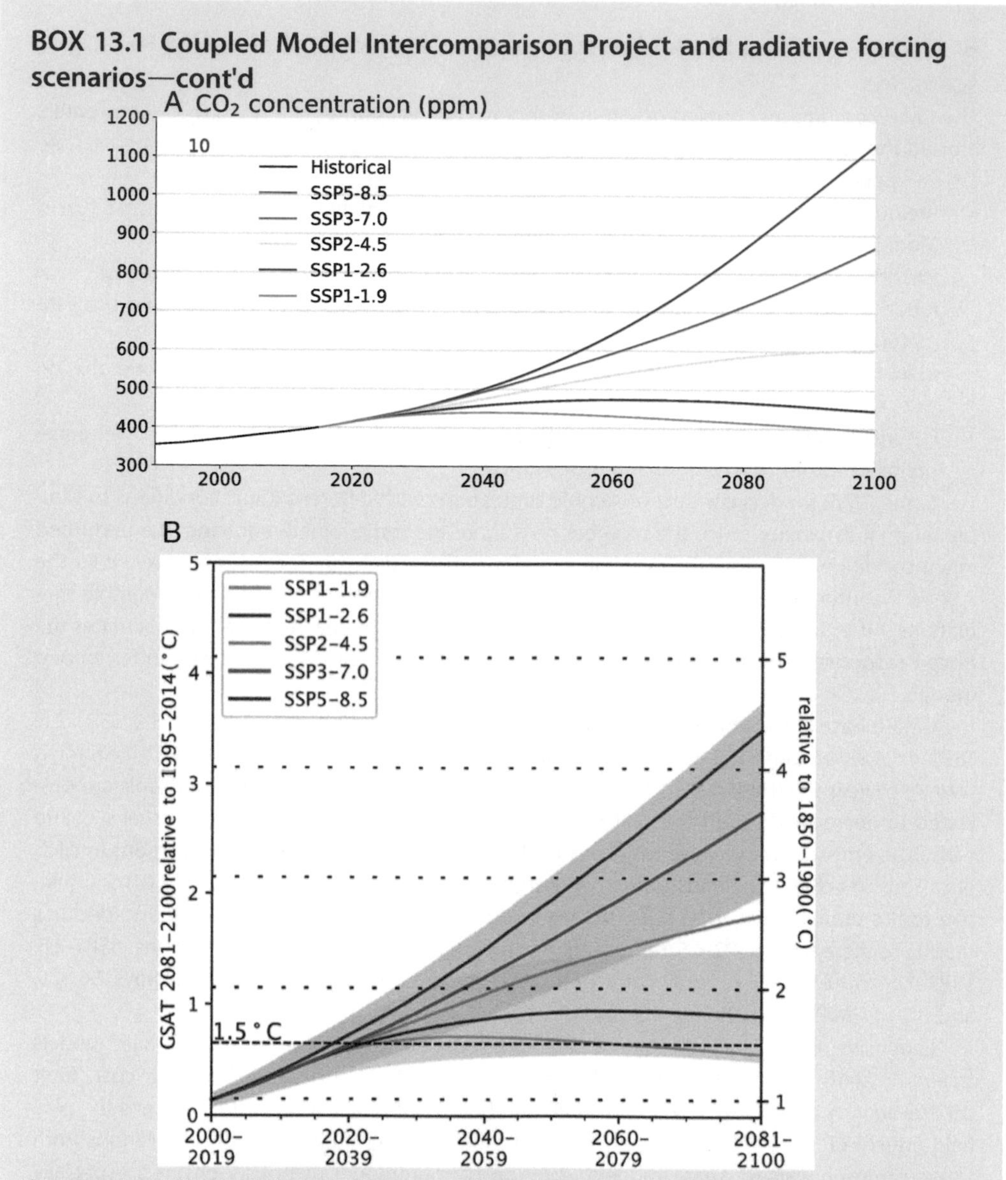

Box Fig. 13.1 (A) Projected CO_2 concentrations in the Shared Socioeconomic Pathways scenarios. (B) Projected global surface air temperature changes relative to the average over 1995–2014. *(A, Adapted from Arias et al. (2022); B, from Lee et al. (2022).)*

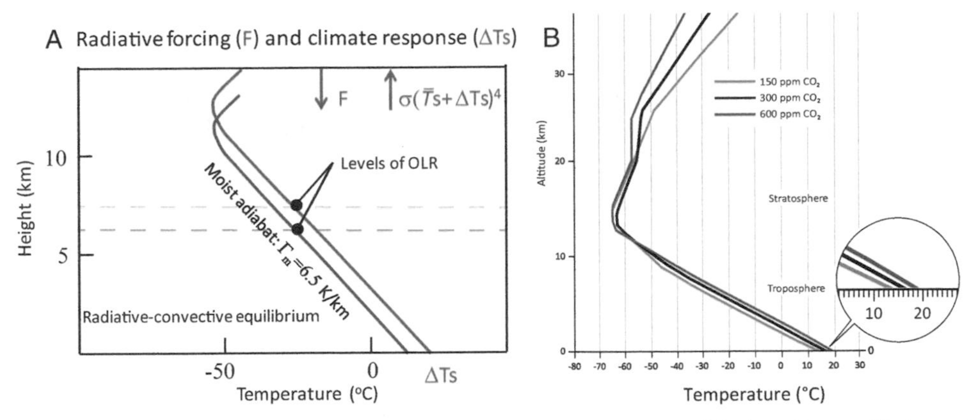

Fig. 13.2 (A) Atmospheric temperature profiles in the present climate *(blue)* and in equilibrium with increased CO_2 concentration *(red)*. The dot indicates the level where outgoing longwave radiation originates. Adapted from Houghton (2015). (B) Atmospheric temperature profiles with different CO_2 concentrations. Adapted based on Manabe and Wetherald (1967). (A, Courtesy S.M. Long; B, Johan Jarnestad/The Royal Swedish Academy of Sciences.)

To restore the top of atmosphere (TOA) radiative balance, the tropospheric temperature profile needs to shift to the right so that the temperature at the new effective radiative height has the same temperature as that before the CO_2 increase (see Fig. 13.2A). Without changes in the lapse rate ($\Gamma_m = 6.5$ K/km), this represents a temperature increase of 1 K for a CO_2 doubling that raises the effective radiative height by 150 m.

13.1.1 Equilibrium response

Let's start with a simple case where downward radiative flux at the TOA is a function of atmospheric GHG concentrations (G) and surface temperature (T), $R = R(G, T)$. Our convention is that downward flux to warm the planet is positive. If we perturb GHG concentrations by G', the temperature response that restores the TOA radiative balance is given by

$$R' = \frac{\partial R}{\partial G} G' + \frac{\partial R}{\partial T} T' = F' + \lambda_P T' = 0 \tag{13.1}$$

where $F' \equiv \frac{\partial R}{\partial G} G'$ is called the *radiative forcing* and represents the reduced upward infrared radiation as an increased GHG concentration makes the atmosphere more opaque to infrared radiation. The radiative forcing due to a CO_2 doubling is about 3.8 W/m^2. Note that the radiative forcing is proportional to the logarithm of GHG concentration, $F \approx 5.35 ln(G/G_0)$ W/m^2, where G_0 is the reference concentration. Here, $\lambda_P \equiv \frac{\partial R}{\partial T} = -4\sigma \overline{T}^3 < 0$ is called the *Plank feedback*, measuring the increased upward infrared radiation due to 1 K surface warming following the Stefan-Boltzmann law.

Atmospheric water vapor is a strong GHG and increases with temperature. Increased water vapor acts to amplify surface warming, constituting positive feedback. Generally, TOA radiative flux is a function of not only well-mixed GHG concentration and surface temperature through the Plank law but also of water vapor (W), cloud (C), and snow/ice albedo (A),

$$R = R(G, T; W, C, A) \tag{13.2}$$

An increase in GHG concentration causes surface temperature to increase, and the change in TOA radiative flux is given by

$$R' = F' + \lambda T' \tag{13.3}$$

where $\lambda = \lambda_P + \lambda_W + \lambda_C + \lambda_A$. Here $\lambda_W = \frac{\partial R}{\partial W}\frac{\partial W}{\partial T}$ is the water vapor feedback, $\lambda_C = \frac{\partial R}{\partial C}\frac{\partial C}{\partial T}$ the cloud feedback, and $\lambda_A = \frac{\partial R}{\partial A}\frac{\partial A}{\partial T}$ the ice/snow feedback. In a moist radiative-convective atmosphere, the tropospheric warming intensifies upward following a moist adiabatic profile. The upward intensified warming emits more infrared radiation into space than a vertically uniform warming. The increased damping effect is called the lapse rate feedback. The negative lapse-rate feedback is commonly combined, as is here, with the greenhouse effect of water vapor that amplifies surface warming. The combined feedback coefficient λ_W has the nice property of reduced uncertainty among climate models.

The cloud response to surface warming is complex and depends on cloud type and surface warming pattern (e.g., whether surface warming is larger over subtropical oceans with low clouds capped by an inversion or over deep convective regions). Cloud feedback suffers large uncertainties among models, but recent studies indicate that it is most likely positive (IPCC 2021). The uncertainty is associated with low-cloud cover over eastern subtropical oceans (see Section 6.3.2). Ice/snow albedo feedback is positive; surface warming melts ice and snow, and reduced ice and snow amplifies surface warming by reflecting less solar radiation. Over the Arctic, low clouds often form where sea ice melts, resulting a smaller change in effective albedo.

Table 13.1 shows the estimates of various feedback coefficients from an ensemble of models. The Plank feedback $\lambda_P \sim -3.2$ Wm^{-2}K^{-1}, the combined other feedback amounts to 2.0 Wm^{-2}K^{-1}, and the net *climate feedback* $\lambda \sim -1.2$ Wm^{-2}K^{-1}.

TABLE 13.1 Climate feedback parameters from CMIP6 MME: Planck (P), combination of water vapor and lapse rate (WV+LR), albedo (A), clouds (C), and sum of all feedbacks (Total).

Feedback	Total	Plank	Water vapor and lapse rate	Surface albedo	Clouds
W/m^2 K^{-1}	-1.2	-3.2	1.3	0.35	0.42

Adapted from Arias et al. (2022): fig TS.17. In Press.

When the TOA radiative balance is restored, the surface temperature change is given by

$$T_E = \frac{F}{-\lambda} \tag{13.4}$$

The response to a doubling CO_2 (F_{2x} ~3.8 Wm^{-2}) is called the *equilibrium climate sensitivity*, about 3 K for the MME mean. Owing to positive climate feedback, the temperature response is three times larger than with the Plank feedback alone.

Fig. 13.2B shows the results from a radiative-convective model that assumes atmospheric constant relative humidity. It gives an equilibrium climate sensitivity of 2.36 K. The simple model shows that increased CO_2 causes a tropospheric warming but a stratospheric cooling, a prediction verified by observations.

13.1.2 Transient response

The ocean's heat capacity is much larger than that of the rest of the climate system combined. Observations show that in response to anthropogenic radiative forcing F, more than 90% of the TOA energy imbalance ($N \equiv F + \lambda T$) is stored in the ocean—that is, approximately

$$\frac{dH}{dt} = F + \lambda T \tag{13.5}$$

where H is the rate of change in global ocean heat content (the globally integrated ocean temperature change) due to the ocean uptake of anthropogenic heat. (The rest of this section adopts the notation N, instead of R, for net TOA radiation as in the literature of climate change.) The ocean is stably stratified, and the greenhouse warming is initially mostly confined in the top mixed layer. The anthropogenic heat penetrates slowly into the deeper ocean through vertical mixing and advection by ocean meridional overturning circulations (see Chapter 2). It takes hundreds of years for the whole ocean-atmosphere system to approach equilibrium. In a scenario where atmospheric CO_2 concentration increases slowly at 1% per year—a rate close to the observed in recent decades—the surface temperature increase at the time of CO_2 doubling (~year 70) is called the transient climate sensitivity

$$T_T = T_E - \left(\frac{1}{-\lambda}\right)\frac{dH}{dt} \tag{13.6}$$

Obviously the transient climate sensitivity is smaller than the equilibrium sensitivity because of the ocean heat uptake. The ocean heat content change causes the global sea level (η) to rise,

$$\Delta\eta = \frac{\alpha}{\rho c_p}\Delta H \tag{13.7}$$

where α is the thermal expansion coefficient of water.

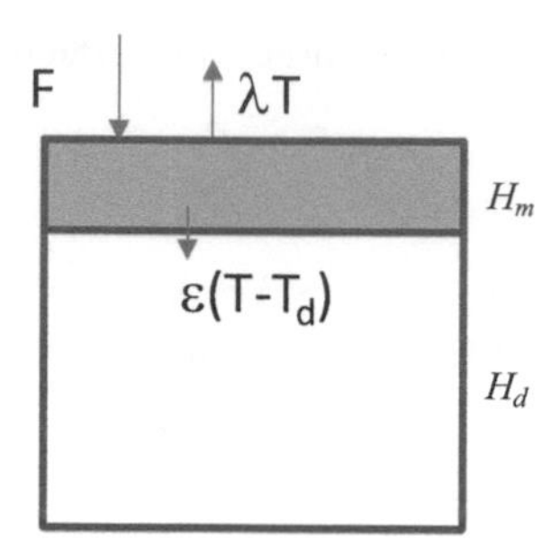

Fig. 13.3 Schematic of the two-layer ocean model.

To illustrate the ocean adjustment to a radiative forcing perturbation F, we consider a two-layer ocean model (Fig. 13.3): the mixed layer of depth H_m ~100 m and the deep layer of depth H_d ~4000 m (e.g., Held et al. 2010). The heat budgets for the top and bottom layers are

$$C_m \frac{dT}{dt} = F + \lambda T - \varepsilon(T - T_d) \tag{13.8}$$

$$C_d \frac{dT_d}{dt} = -\varepsilon(T_d - T) \tag{13.9}$$

where ε is the interfacial mixing coefficient, and $(C_m, C_d) = \rho c_p(H_m, H_d)$ is the layer heat capacity.

The deep ocean warming lags way behind and is negligibly small compared to the surface because of the large heat capacity. ($T_d/T \sim H_m/H_d \ll 1$). Eq. 13.8 is approximately

$$C_m \frac{dT}{dt} = F + (\lambda - \varepsilon)T \tag{13.10}$$

The mixed layer warming is damped by radiative feedback (λ) and mixing with the deep ocean (ε). The e-folding timescale of the radiative damping $C_m/|\lambda|$ ~11 years for $H_m =$ 100 m (see Review question 9[c]).

For slowly increasing radiative forcing (e.g., 1% per year increase in CO_2), the mixed-layer heat storage term (the left-hand side [LHS] of Eq. 13.8) is negligible. As a result, the mixed-layer temperature response to the radiative forcing is in quasi-equilibrium

$$T \approx F/(-\lambda + \varepsilon) = T_E/(1 - \varepsilon/\lambda) \tag{13.11}$$

The mixing coefficient ε is somewhat ambiguous but considered to be of the same order of magnitude as λ. The TOA radiative imbalance $N \equiv F + \lambda T \approx \varepsilon T$ is estimated at N ~0.87 Wm^{-2} from the change of global ocean heat content observed from Argo floats since 2005. With an estimated radiative forcing of $F = 2.72$ Wm^{-2} (IPCC

2022), we can estimate the ratio of the planetary energy imbalance to the radiative forcing,

$$N/F = (1 - \lambda/\varepsilon)^{-1}$$

With $N/F \sim 1/3$, we have $-\lambda/\varepsilon \sim 2$. The radiative damping on GMST is twice as large as the damping due to the deep ocean heat uptake.

13.1.3 Abrupt CO_2 increase experiment

The idealized experiment with an abrupt increase in CO_2 is useful to diagnose important parameters and shed light on the evolution of climate response. GMST increases rapidly following the CO_2 increase with an e-folding scale of $\tau_m = C_m/(-\lambda + \varepsilon)$ (see Review question 9). This fast response is followed by a slow response of the deep ocean warming, with an e-folding scale that is longer by a factor of H_d/H_m.

The real ocean is more complicated than the two-layer model, but important diagnostics can be obtained by focusing on the TOA radiative flux change from a preindustrial equilibrium

$$N = F + \lambda T \qquad (13.12)$$

Here we have omitted the prime for the change. Fig. 13.4 shows N as a function of GMST increase in an abrupt $2 \times CO_2$ experiment. The first-year mean already shows a warming of $\sim$1oC, but we can infer the intercept to estimate the radiative forcing F (blue dot). While it takes centuries for the model to reach a new TOA radiative

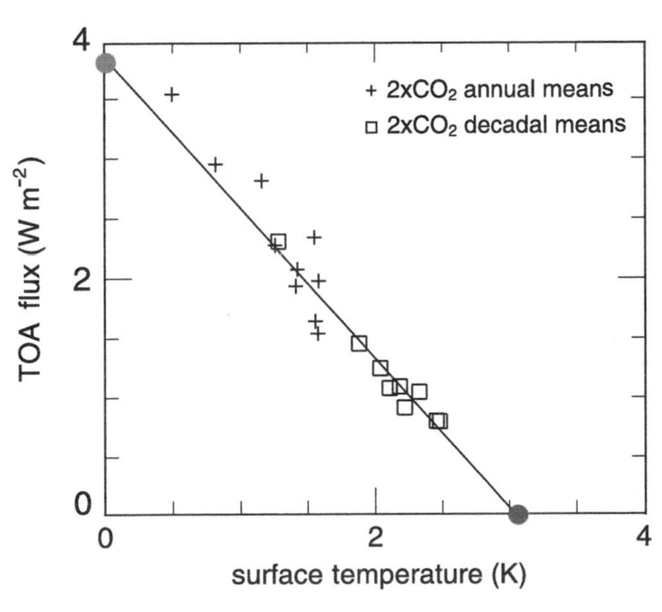

Fig. 13.4 Evolution of global average top of the atmosphere *(TOA)* net downward radiative flux (W/m^2) with global average surface air temperature change (K) in HadCM3 experiment of abrupt $2 \times CO_2$ increase. The blue dot denotes the estimated effective radiative forcing and the red dot the equilibrium climate sensitivity. Plus (square) symbols denote the annual (decadal) means for (after) the first decade. Adapted from Gregory et al. (2004). *(Courtesy Z.H. Song.)*

equilibrium, we can extrapolate from a simulation of a limited length (say, 70 years) to infer the intersection with the x-axis (red dot) as an estimate of the equilibrium climate sensitivity. The slope of the regression line gives the climate feedback parameter.

13.2 Global warming hiatus

During the 15-year period of 1998 to 2013, the atmospheric concentration of carbon dioxide rose by 30 ppm, but the rate of GMST increase slowed to 0.05°C/decade. This is much smaller than the rate of 0.2°C/decade during earlier periods from the 1970s to 1990s (Fig. 13.5). This slowdown in the rate of surface warming over an extended period came as a surprise to those who had expected a continual, if not intensified, global warming in the face of a steady increase in atmospheric GHG concentrations. Indeed, the MME mean simulation shows a steady increase in GMST in lockstep with GHGs. The apparent deviation of the observed GMST evolution from the multimodel mean results led some to question the validity of climate models and a key IPCC conclusion that increased GHG concentrations in the atmosphere cause the Earth to warm. Here we refer to the temporary slowdown in global surface warming as a hiatus.

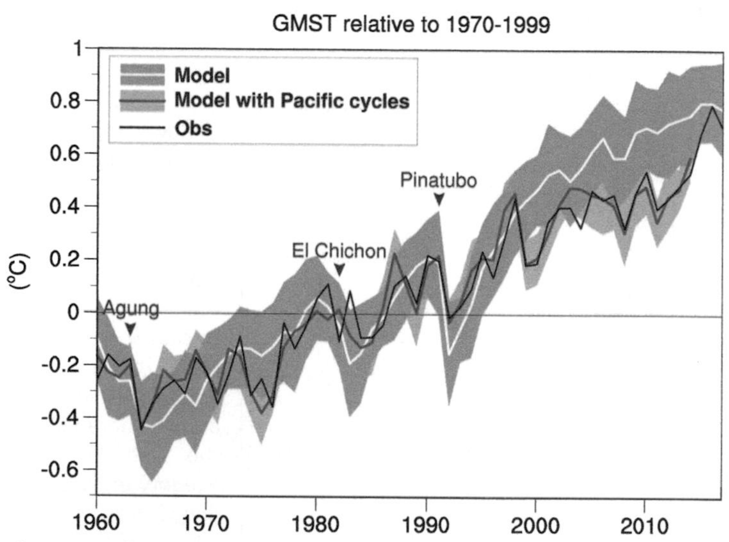

Fig. 13.5 Global mean surface temperature anomalies in observations *(black)*, historical (20-member mean in white line and spread in blue shading), and tropical Pacific pacemaker (10-member mean in red line and spread in orange shading) runs. Major volcanic eruptions are indicated. Updated from Kosaka and Xie (2013). *(Courtesy Y. Kosaka.)*

13.2.1 Tropical Pacific pacemaker effect

Internal variations of the climate system can cause GMST to increase and decrease, independent of the anthropogenic warming. This is obvious in the ups and downs from one year to another in the GMST record, caused chiefly by El Niño and the Southern Oscillation (ENSO). But an extended hiatus over 15 years is rare, especially in the face of the rapid increase in radiative forcing. The last time when the rate of 15-year GMST change was that low was during the so-called big hiatus period from the 1940s to 1970s, a time when the rate of increase in anthropogenic radiative forcing was much lower.

The Pacific SST change during the hiatus shows a Pacific Decadal Oscillation (PDO) —like pattern (see Fig. 12.1) with negative anomalies over the tropical Pacific (Fig. 13.6A). Indeed, a pacemaker experiment that forces SSTs over the tropical Pacific to follow the observed evolution (termed Pacific Ocean-global atmosphere [POGA] run) reproduces the recent hiatus remarkably well. The decadal trend in GMST is substantially reduced compared to the historical simulation without the tropical Pacific SST restoring (see Fig. 13.5). This demonstrates that climate models can reproduce the hiatus as long as they have the right phasing of PDO. The PDO is an internal mode of the coupled climate system. Because the PDO phase is random in individual member runs, the ensemble mean removes the PDO effect on GMST and retains only the radiative forced increase in GMST that is unabated during 1998-2013.

GMST is a convenient index tracking global climate change, but much information on the mechanism for GMST variability is lost in global averaging. By unpacking GMST in spatial and seasonal dimensions, we can identify the fingerprints of internal modes that pace GMST. Patterns of surface temperature change during the hiatus are in much closer agreement with observations in the pacemaker experiment than in the historical run with a fully coupled tropical Pacific Ocean (see Fig. 13.6). The comparison supports that the tropical Pacific cooling explains many regional changes in observations during the hiatus: cooling in the Northeast and Southeast Pacific, the V-shaped warming that extends from the equatorial western Pacific, warm and dry

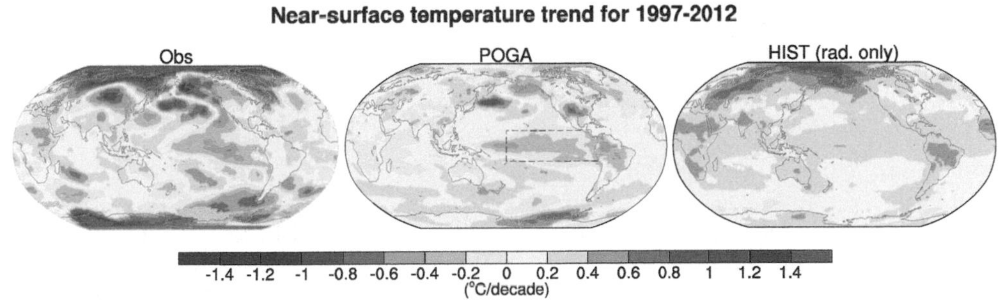

Fig. 13.6 Trends in surface temperature from 1997-2012 in observations *(left)*, tropical Pacific pacemaker *(middle)*, and historical *(right)* runs. (Courtesy Y. Kosaka.)

anomalies in the southern and southwest United States, wetter conditions over the Maritime Continent, and the drier central Pacific. These regional anomalies are distinct from radiatively forced response, confirming the spatial fingerprints of the PDO. The warm water piled up in the western Pacific by the intensified trade winds sets a favorable condition for tropical cyclone growth by limiting the cold wake. The anomalously deep thermocline contributed to the growth of the super typhoon Haiyan that made landfall on the Philippines in November 2013 (Lin et al. 2014). The high sea level further worsened the storm surge by the typhoon.

The tropical Pacific pacemaker effect is distinct from that by the tropical Atlantic or Indian Ocean. While the former causes SST in the entire tropics to respond in the same sign, the latter drives the tropical Pacific into an opposite phase to the rest of the tropical oceans (Fig. 13.7) (see Section 10.4.2). In other words, the warming in the tropical Atlantic or Indian Ocean does not guarantee an increase in GMST. The tropical Pacific is unique in inducing a large response in GMST as illustrated by the recent hiatus.

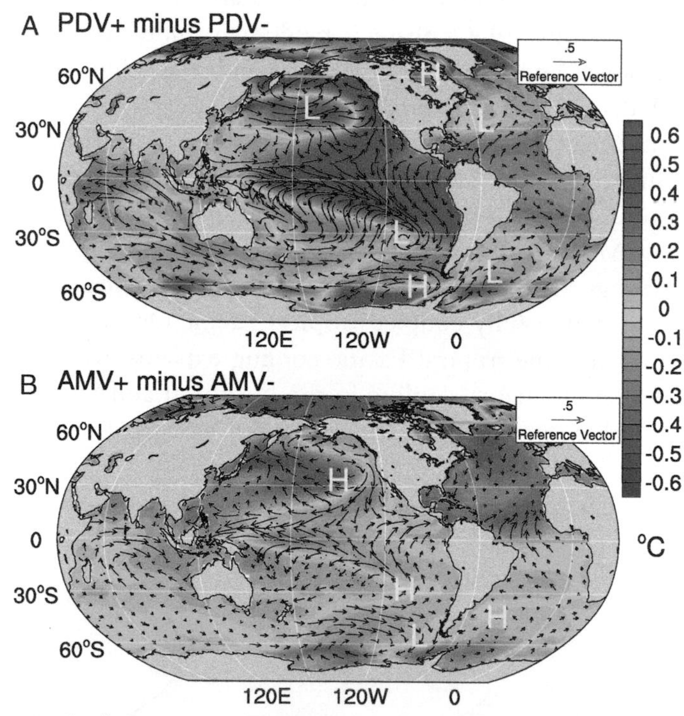

Fig. 13.7 Community Earth System Model response to (A) Pacific Decadal Oscillation and (B) Atlantic Multidecadal Oscillation sea surface temperature (SST) anomalies: SSTs (°C) and surface winds (m s^{-1}, scaling vector in upper right). Horizontal dark green lines demarcate area of specified pacemaker SSTs. Anomalous cyclonic circulations *(L)* and anticyclonic *(H)* correspond to centers of sea level pressure anomalies. Adapted from Meehl et al. (2021). *(Courtesy A. Hu.)*

13.2.2 Planetary energetics

From Eq. 13.5, the planetary energy budget during the hiatus of 1998–2013 (Δt) is

$$\frac{\Delta H}{\Delta t} = \Delta N = \Delta F + \lambda_F \Delta T_F + \lambda_I \Delta T_I \tag{13.13}$$

where the subscripts F and I denote the radiatively forced and internal variability. Here we did not assume a priori that climate feedback parameters are the same for the radiative forced (F) warming and internal (I) variability. In abrupt or 1% CO_2 increase experiments, ΔT_F and TOA radiative imbalance N are highly correlated (see Fig. 13.4), and λ_F is well defined. For unforced internal variability, by contrast, ΔT_I and TOA radiation N are poorly correlated at decadal and longer timescales (Fig. 13.8). We set $\lambda_I = 0$ for simplicity. This contrast in the global feedback parameter has to do with that in surface warming pattern related to cloud feedback: While greenhouse warming is to first order spatially uniform, internal variability is dominated by spatial variations (e.g., PDO in Fig. 12.1) and accompanied by little net TOA radiative perturbations. The SST pattern effect on global climate feedback is an area of active research (Armour et al. 2013; Zhou et al. 2017; Xie 2020).

The AMIP (Atmospheric Model Intercomparison Project) protocol refers to atmospheric GCM experiments forced by historical observations of SST and sea ice. By holding radiative forcing unchanged, one can estimate the TOA radiation change due to the observed SST change ΔN_{AMIP} and an apparent climate feedback

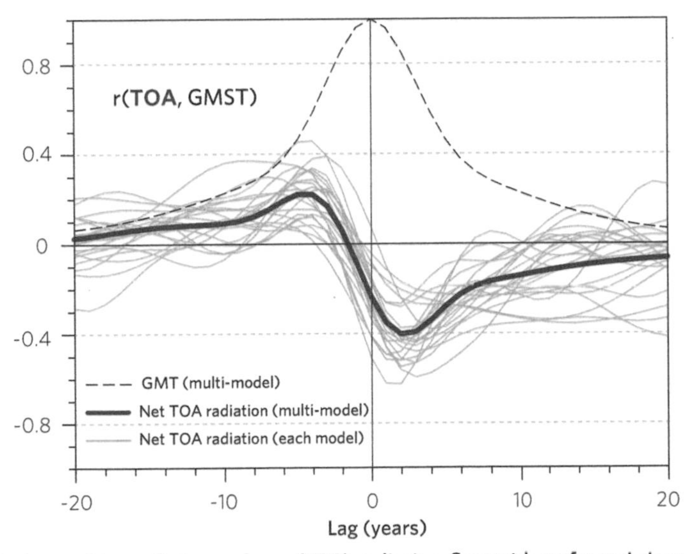

Fig. 13.8 Correlations of top of atmosphere *(TOA)* radiative flux with unforced decadal global-mean surface temperature (GMST) variability (the multimodel ensemble mean is thickened). GMST autocorrelation is dashed as reference. *(From Xie et al. (2016).)*

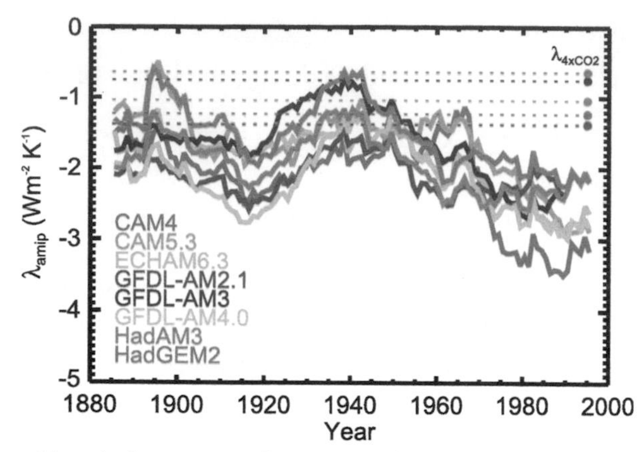

Fig. 13.9 Time series of λamip from atmospheric general circulation models in a sliding 30-year window, the year represents the center of the window. Colored circles with dashed horizontal lines show the feedback parameter values from abrupt 4 × CO_2 run. *(From Andrews et al. (2018).)*

$$\lambda_{AMIP} \equiv \frac{\Delta N_{AMIP}}{\Delta T} \approx \lambda_F \frac{\Delta T_F}{\Delta T} \tag{13.14}$$

During the hiatus period, $\Delta T < \Delta T_F$ so the AMIP method overestimates the climate feedback $|\lambda_{AMIP}| > |\lambda_F|$. Indeed, atmospheric model results consistently show an apparently increase in estimated climate feedback $|\lambda_{AMIP}|$ in recent decades (Fig. 13.9), even though the climate feedback parameter λ_F—estimated from the 4 × CO_2 experiment—is constant in each model. The large departures of λ_{AMIP} from λ_F are of concern, indicating that the observed SST warming patterns differ significantly from models because of biased physics or large internal variability.

13.2.3 Estimating anthropogenic warming

In the instrumental era, the rate of global surface warming displays marked interannual to interdecadal variations (see Fig. 13.1). To the extent that the tropical Pacific-pacemaker simulation captures the modulations of the warming rate, one can estimate internal variability of GMST from such pacemaker experiments and derive anthropogenic warming by subtracting the model-derived internal variability from observations (Fig. 13.10). Unlike the conventional model-based method, this new method of deriving forced GMST change is largely free of the uncertainties in radiative forcing and climate sensitivity. The new method yields an anthropogenic warming that is consistently above the raw observations. In 2013, the method puts the anthropogenic warming at 1.2°C referenced to the late 19th century much higher than the visual estimate of 0.9°C from raw data in 2013. The higher estimate from the POGA pacemaker run is now in line with the visual result

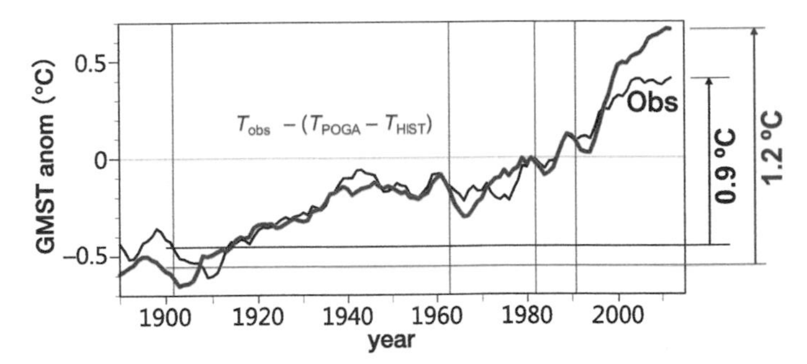

Fig. 13.10 Observed global-mean surface temperature (GMST; black) and the radiatively induced component estimated with the Pacific pacemaker simulation *(purple). (From Xie and Kosaka (2017) https://doi.org/10.1007/s40641-017-0063-0.)*

as GMST has since increased by 0.3°C, aided in part by the major El Niño event of 2015-16. The 1.2°C achieved anthropogenic warming heightens the challenges to meet the Paris Agreement's target of keeping GMST increase below 1.5°C relative to the preindustrial level.

13.3 Robust atmospheric changes due to thermodynamic effects

Greenhouse warming is accompanied by characteristic horizontal and vertical structures that are broadly consistent and robust across climate models. Water vapor change, for example, is largely a function of temperature. We call them the thermodynamic effects of global warming because they are not dependent on detailed structure of surface warming. The next chapter discusses regional changes that are sensitive to the spatial variations in surface warming and atmospheric circulation change (the dynamic effects).

13.3.1 Enhanced warming over land

Increased GHGs cause surface temperature to increase everywhere, but the magnitude of the warming varies in the horizontal. Over land, water availability limits surface evapotranspiration, and surface sensible heat flux intensifies the warming compared to ocean (Fig. 13.11) (Joshi et al. 2008). This is consistent with our experience that in the summer afternoon, a wet surface is cooler than a dry surface. Small heat capacity of land surface is often invoked to explain the land-sea contrast in the warming rate, but this does not seem to be a major reason. Consider a thought experiment by replacing land surface with a shallow ocean mixed layer (say, 0.1 m deep), a swamp of small heat capacity but unlimited in water supply for evaporation. Because both the swamp and ocean mixed layer (~100 m deep) is in quasi-equilibrium with the slowly increasing GHG forcing, the swamp to ocean warming ratio is near 1.0, rather than the MME mean land to ocean ratio of 1.5.

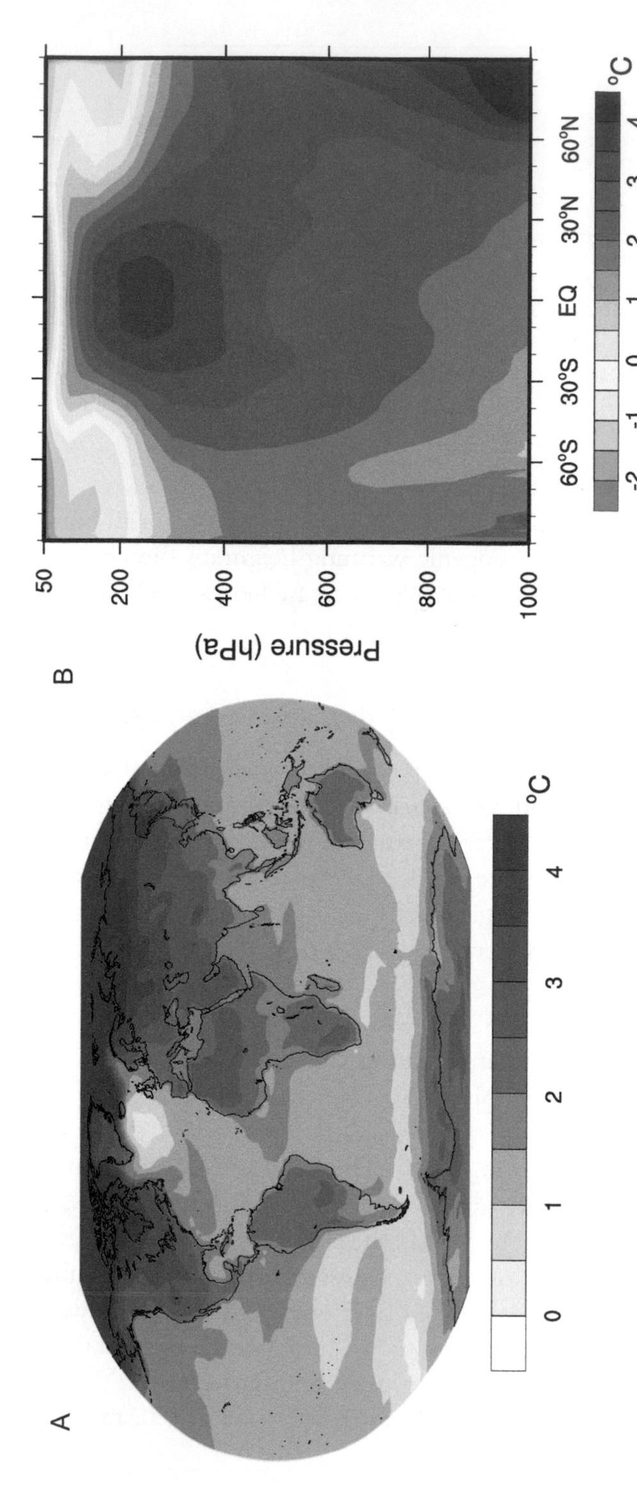

Fig. 13.11 Annual-mean temperature change in the 1% CO_2 run at year 70 (the time of CO_2 doubling): (A) at the surface, and (B) zonally averaged. CMIP6 multimodel mean is used. *(Courtesy Y.F. Geng.)*

13.3.2 Heatwaves

A moderate increase in mean temperature can lead to a large increase in the odds of extreme heat (Hansen et al. 1988). For simplicity, assume that temperature variability follows a Gaussian normal distribution,

$$f(T) = \frac{1}{\sigma\sqrt{2\pi}} exp\left[-\frac{1}{2}\left(\frac{T-\mu}{\sigma}\right)^2\right] \tag{13.15}$$

where μ is the mean and σ the standard deviation. By defining $x \equiv \frac{T-\mu}{\sigma}$, we obtain the standard normal distribution,

$$\varphi(x) = \frac{1}{\sqrt{2\pi}} exp\left[-\frac{1}{2}x^2\right] \tag{13.16}$$

The probability of a 1σ event ($x > 1$) to occur is 16%. For a mean temperature increase of $\Delta x = 0.5$ (corresponding to a rightward shift of the probability density function [PDF], brown dashed curve in Fig. 13.12), the probability for the same 1σ event in current climate increases by a factor of 2. For the same mean warming, a 3σ event increases its odds of occurrence by a factor of 4.6, from 1.35 in 1000 to 6.21 in 1000. Generally, the factor of increase is larger for a more extreme threshold x_E. In fact, it can be shown that the factor of increases is an exponential function of the extreme threshold

$$p(x > x_E - \Delta x)/p(x > x_E) \approx exp(x_E \Delta x), \text{ for } \Delta x/x_E \ll 1 \text{ and } x_E \gg 1 \tag{13.17}$$

Over land, the daily temperature PDF for summer is not Gaussian but skewed with a long tail at the warm end. Surface warming is accompanied with decreased relative humidity (R_H), both acting to dry the topsoil. This increasingly puts evapotranspiration in the water-limited regime (Box 5.1), resulting in a longer tail of extreme high temperature occurrences (a change in the shape of PDF, red curve in Fig. 13.12). This, along with the

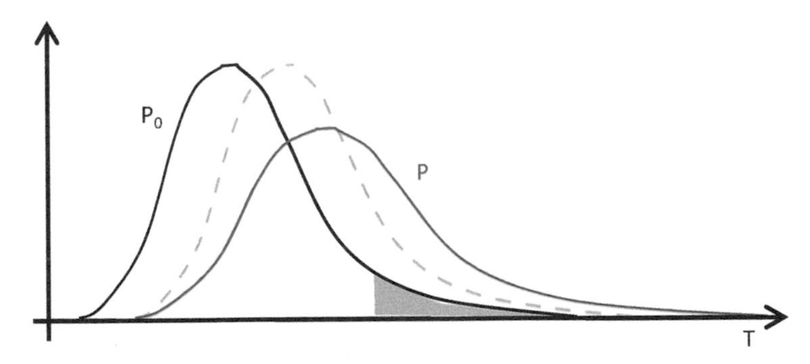

Fig. 13.12 Probability densityfunction (PDF) of daily temperature in summer: $f_0(T)$ in present and f(T) in warmer climate. Background warming shifts the PDF *(orange dashed)*, while the drying of topsoil increases the warm tail through positive feedback between temperature and soil moisture. The combined change increases the occurrences and exacerbates the severity of extreme heat.

mean warming that shifts the PDF, exacerbates the risks of extreme heat, as reported at increasing frequency around the world (2018 European heatwave, 2020 California wildfire, and 2021 US Pacific Northwest heatwave).

Vapor pressure deficit, VPD $= (1 - R_H)e_s$, is a good predictor of fire risk (see Section 6.4.2). It is an exponential function of temperature even with R_H unchanged. Extreme high temperatures and dry soil are the perfect recipe for wildfire. The rapid increase in fire burned area for the recent decades in California (Section 6.4.2) and many other regions is among the first symptoms that have emerged of background greenhouse warming.

13.3.3 Arctic amplification

Surface warming in the Arctic is enhanced due to ice/snow albedo feedback. In a warmer climate, the poleward moisture transport by atmospheric eddies intensifies, and the resultant energy convergence contributes to Arctic warming. The latter effect can be modeled as an eddy diffusion that flattens the meridional gradient of near-surface moist static energy $m = c_pT + Lq + gz$. In the extreme case of a meridionally uniform change Δm, the corresponding profile of temperature change follows,

$$\Delta T = \frac{\Delta m}{c_p(1 + b_e)} \tag{13.18}$$

where b_e is the inverse Bowen ratio,

$$b_e = \alpha \frac{Lq_0}{c_p} \tag{13.19}$$

In polar regions, the temperature change is amplified because mean moisture q_0 is low (Roe et a1. 2015).

The transient response in surface temperature is suppressed over the Southern Ocean (see Fig. 13.11), where the upwelling of deep water takes up anthropogenic heat very effectively (see Section 14.5). One can use an atmospheric GCM coupled with a motionless slab ocean model (SOM, typically 50 m deep) to diagnose the effect of ocean heat uptake. Here the surface heat flux (Q_{net}, known as the Q-flux) pattern is diagnosed from the coupled simulation of greenhouse warming and imposed as an external heat sink in SOM. Without Q-flux, the greenhouse warming is strongly amplified at both poles (red curve, Fig. 13.13), by a factor of 3 compared to the tropics. The Q-flux due to ocean heat uptake causes amplified cooling at both poles, although the magnitude of cooling is much larger around Antarctica because the subpolar ocean uptake is much larger when zonally integrated over the circumpolar Southern Ocean than over the North Atlantic. As a result, in a fully dynamical ocean model, the surface warming is amplified in the Arctic but quite flat in the Southern Hemisphere (black, Fig. 13.13).

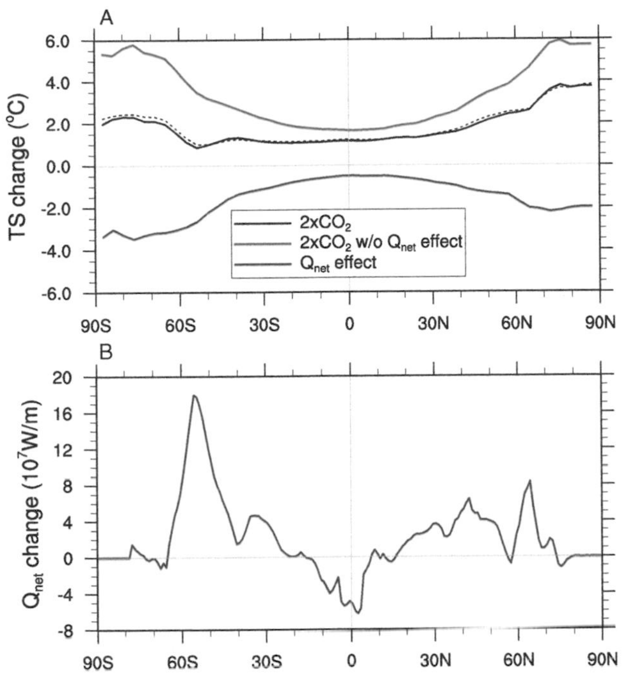

Fig. 13.13 (A) Zonal-mean surface temperature *(TS)* change in years 41-50 after an abrupt CO_2 doubling in CESM1 with a dynamic ocean model (DOM; black dashed). The slab ocean-atmosphere coupled model (SOM) response to the CO_2 doubling, when forced by Q_{net} diagnosed from the DOM, reproduces the DOM results *(black solid)*. Also shown is the SOM response to either the CO_2 *(red)* or Q_{net} *(blue)* forcing. (B) Zonally integrated Q_{net} change in the DOM. *(Adapted from Hu et al. (2022).)*

13.3.4 Hydrologic cycle

Radiative-convective equilibrium models are often used to predict the global mean response to GHG increase. An important assumption made in such calculations is that relative humidity in the atmosphere does not change. Three-dimensional coupled ocean-atmosphere model simulations generally support this assumption. With a constant relative humidity, atmospheric specific humidity (q) increases following the Clausius-Clapeyron (CC) equation

$$\frac{1}{q}\frac{dq}{dT} = \frac{L}{RT^2} = \alpha(T) \tag{13.20}$$

where α is nearly a constant of 0.06 to 0.07 K^{-1} for typical tropospheric temperatures. Model calculations show that the global atmospheric water vapor content increases with surface temperature at approximately this CC rate, $\Delta q/q = 0.07\Delta T$.

In climate models, global mean precipitation increases with surface temperature at a much slower rate because it is constrained by atmospheric radiative cooling. Like the TOA radiation $R_{TOA} = R_{TOA}(G, T)$, surface radiative flux (downward positive) is a function of GHG concentration and surface temperature $R_{sfc} = R_{sfc}(G, T)$. Small atmospheric heat capacity requires the global TOA radiative imbalance to equal the net heat flux at the surface,

$$R_{TOA}(G, T) = R_{sfc}(G, T) - (Q_E + Q_H) \tag{13.21}$$

where $Q_E = \rho_o LP$ and Q_H are the latent and sensible heat fluxes at the surface (upward positive), and P is precipitation. In warming climate, the change in surface evaporation and hence global precipitation is given as

$$\Delta Q_E = \Delta R_{sfc} - \Delta R_{TOA} - \Delta Q_H \tag{13.22}$$

Thus global precipitation increase is constrained by planetary energy budget, not by the CC rate of water vapor increase. Precipitation increases to compensate the increased radiative cooling of a warmer atmospheric column, while the sensible heat flux change is small (Fig. 13.14B). Radiative transfer calculations show that surface latent heat flux increases by $\Delta Q_E = 1.2\ \mathrm{W/m^2}$ for 1 K surface warming (see Fig. 13.14B). With the mean latent heat flux $\overline{Q}_E = 80\ \mathrm{W/m^2}$, this is equivalent to a percentage increase in global precipitation $\Delta Q_E/\overline{Q}_E = 0.015\Delta T$, in broad agreement with the rate of $\sim 2\%\ \mathrm{K^{-1}}$ in model simulations.

Consider a different situation that the CO_2 concentration G is increased abruptly while the surface temperature has not yet changed. An increased GHG concentration increases surface downward infrared radiation, but the surface radiation increase is smaller

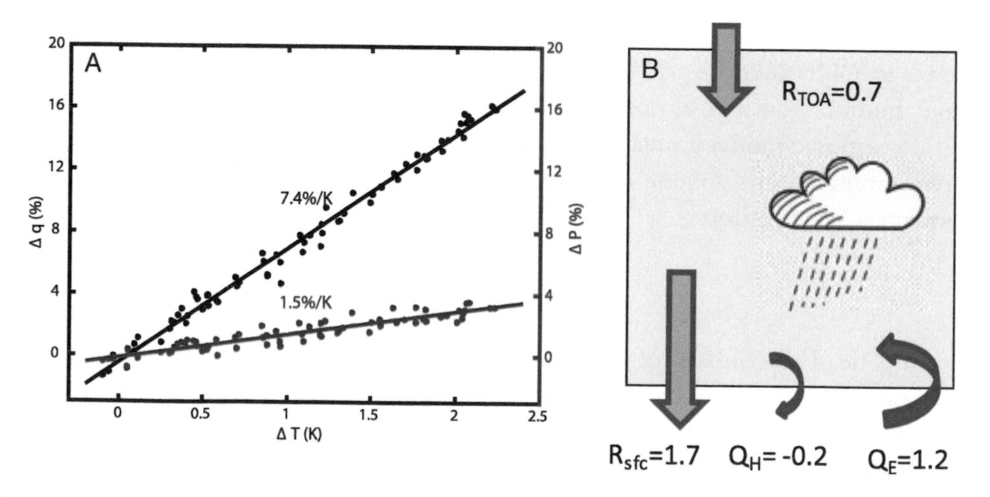

Fig. 13.14 Changes in global hydrologic cycle and energy budget in CESM2 1% CO_2 run. (A) Scatter-plot of the percentage change in global-mean column-integrated water vapor and precipitation vs the global-mean surface temperature change for the first 70 years. Lines of 7.4% K^{-1} and 1.5% K^{-1} are shown. (B) Energy flux change at top of atmosphere (TOA) and the surface for 1 K surface warming ($W/m^2\ K^{-1}$) at year 70. Adapted based on Pendergrass et al. (2014). *(Courtesy S.M. Long.)*

than the TOA change because the surface atmosphere is already optically thick with high moisture—that is, $\frac{\partial R_{sfc}}{\partial G} \ll \frac{\partial R_{TOA}}{\partial G}$.

$$\Delta Q_E \approx -\Delta R_{TOA}(G, T) = -\frac{\partial R_{TOA}}{\partial G}\Delta G = \Delta F \tag{13.23}$$

where we have neglected surface sensible heat flux change and the right-hand side (RHS) is the radiative forcing change. In the abrupt CO_2 increase experiment with global climate models, a decrease in global precipitation is indeed observed initially, with $\Delta T \sim 0$. As the climate warms, the TOA radiative imbalance decreases, while the downward radiative flux increases at the surface, resulting in increased surface evaporation and precipitation (Eq. 13.22).

13.3.5 Slowdown of Walker circulation

The slow rate of global precipitation (*P*) increase has important implications for the atmospheric circulation (Held and Soden 2006). The atmospheric water balance is approximate as $P = M \cdot q$, where M is the mass flux vented upward in deep convection and q specific humidity in the surface boundary layer. Perturbing both sides yields

$$\Delta M/M = \Delta P/P - \Delta q/q \tag{13.24}$$

With $\Delta P/P \sim 2\%/K$ and $\Delta q/q \sim 7\%/K$, we have $\Delta M/M \sim -5\%/K$. The mass flux in convection needs to decrease by 5% per degree surface warming. Most precipitation occurs in the tropics, where upward mass flux takes place in stationary convective regions such as the ITCZ and Indo-western Pacific warm pool. As the upward mass flux change is proportional to the upward vertical velocity change in these convective regions, the reduced upward mass flux implies a slowdown of the tropical overturning circulation.

With increasing GHG forcing, climate models indeed project a general weakening of divergent flow in the upper troposphere. In major convective regions of the Indo-western Pacific and tropical Africa, the upper-level flow shows a trend of convergence, while over the subsidence regions of the eastern Pacific it shows a trend of divergence (Fig. 13.15B). This mass-flux argument predicts a slowdown of tropical circulation but does not predict which overturning circulation would slow. Model simulations generally show a clear slowdown in the zonal Walker circulation but not so much in the meridional Hadley circulation, especially the southern Hadley cell (Fig. 13.15A) because of the change in SST gradient (see Section 14.3).

13.3.6 Extreme precipitation

If storms and other rain-making disturbances remain unchanged in circulation, the increase in moisture intensifies precipitation rate. Globally, the rate of increase in heavy rain rate is roughly 7%/K, consistent with the CC equation. As global-mean precipitation

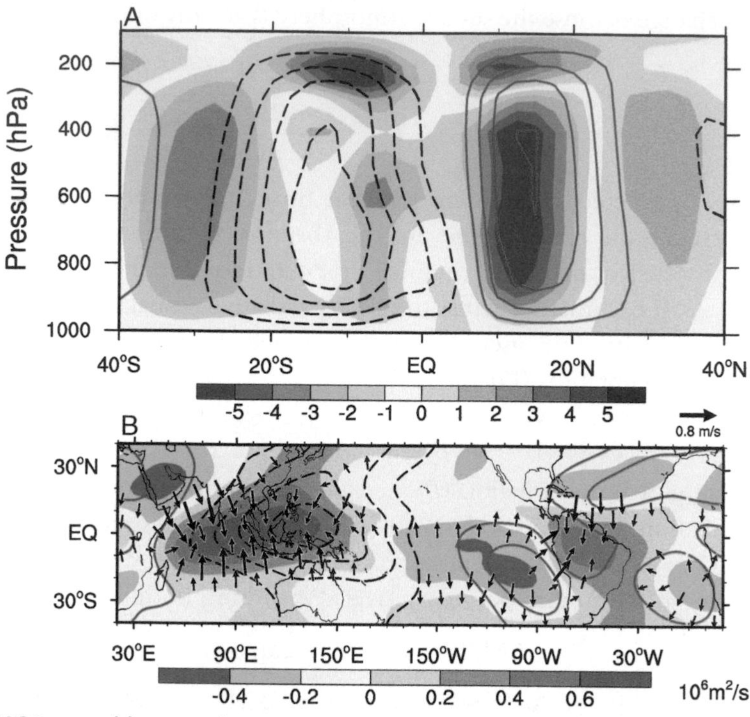

Fig. 13.15 CMIP6 ensemble, annual-mean atmospheric circulation change *(color shading)* at year 70 in the 1% CO_2 run, along with the preindustrial climatology (line contours, negative dashed). (A) Zonal-integrated stream function (109 kg/s) and (B) velocity potential at 250 hPa with the divergent wind change (vectors). *(Courtesy Y.F. Geng.)*

increases only at 2%/K, this implies a reduction in frequency of light/moderate rain. Generally, the PDF of precipitation is likely to change, with more heavy rain events in store (O'Gorman 2015).

13.3.7 Vertical structure of the tropospheric warming

In the tropics, tropospheric temperature adjusts toward a moist adiabatic profile set by moist static energy in the planetary boundary layer over convective regions. In a saturated, undiluted rising parcel, moist static energy m^* is constant in the vertical. Perturbed by a positive radiative forcing, the vertical profile of temperature increase is given by

$$\Delta T = \frac{\Delta m^*}{c_p\left(1 + b_e^*\right)} \tag{13.25}$$

where b_e^* is the inverse saturation Bowen ratio,

$$b_e^* = \alpha \frac{L q_{s0}}{c_p}$$

With Δm^* kep constant, ΔT must increase with altitude (see Fig. 13.11B) as the saturation humidity q_{s0} decreases rapidly in the free troposphere. Physically, temperature increase in the upper troposphere is greater than at the surface due to extra latent heat of condensation in the rising parcel (moisture content in the upper troposphere is negligible). It is interesting to note that the upward amplification in the tropics and polar amplification are based on the mixing of MSE in the vertical and horizontal by convection and synoptic eddies, respectively.

The upward amplification of tropospheric warming is a robust feature in global warming simulations (see Fig. 13.11B), but for a long time atmospheric soundings by satellites and weather balloons failed to verify this vertical structure. Recent results begin to support the amplified warming in the upper troposphere when various sampling and instrument errors are properly corrected (Santer et al. 2017).

13.3.8 Expansion of subtropical dry zones

The Hadley cells expand poleward in warming climate, as illustrated by the change in overturning stream function that peaks on the boundary between the Hadley and Farrell cells (see Fig. 13.15A). Eady's growth rate for baroclinic instability is $0.31 \frac{f}{N} \frac{\partial u}{\partial z}$, where f is the Coriolis parameter, u the zonal wind velocity, and N is static stability (Eq. 2.21). The increased (dry) static stability in the tropics and subtropics stabilizes baroclinic instability (with vertical wind shear unchanged) on which midlatitude storms grow. Lu et al. (2007) argue that storm tracks need to move poleward (for increased f) in response to an increased static stability. The poleward shift in storm tracks leads to an expanding Hadley circulation (Shaw 2019) for adiabatic warming in the downward branch to balance the divergence of poleward eddy heat flux (Eq. 2.21). The poleward expansion of the Hadley circulation is associated with a similar expansion of subtropical dry zones as indicated by the $P - E = 0$ contour.

Storm tracks arise from baroclinic instability in the midlatitude atmosphere. In the upper troposphere, the upward intensified warming in the tropics increases the meridional temperature gradient. At the surface, the transient temperature response is muted over the Southern Ocean (see Fig. 13.11A), and the increased surface baroclinicity causes the Southern Hemispheric storm tracks to intensify and displace poleward. In the Northern Hemisphere, the amplified surface warming in the Arctic opposes the increased temperature gradient increase in the upper troposphere, making the zonal-mean storm track response not as robust (Shaw et al. 2016), with large zonal variations due to changes in stationary eddies (Simpson et al. 2015).

13.4 Surface acceleration of the subtropical ocean gyre

From an oceanographic point of view, ocean changes are induced by three distinct surface forcings: temperature, salinity, and wind. One can isolate and compare these surface forcing effects on the ocean circulation by restoring an ocean GCM toward the SST and sea surface salinity (SSS) changes diagnosed from the Coupled Model Intercomparison Project (CMIP) MME mean. The thermodynamic effect of surface warming dominates the ocean surface circulation response. Equatorial oceans are the exception, where the wind change effect is large. Sea surface warming speeds up the ocean surface circulation globally, including the subtropical gyres and Antarctic Circumpolar Current (ACC) (Peng et al. 2022). This thermodynamic effect of the greenhouse warming is robust, insensitive to the SST warming pattern.

Here we use a 1.5-layer reduced gravity model to illustrate the response of the subtropical gyre. The meridional volume transport is caused by the local Ekman pumping, following the Sverdrup relation (see Section 7.2.5),

$$\beta h v = f w_E \tag{13.26}$$

where the flow is in geostrophic balance,

$$f v = \frac{\partial}{\partial x}(g' h) \tag{13.27}$$

Combining the above yields

$$\frac{\beta}{2f}\frac{\partial}{\partial x}\left(g' h^2\right) = f w_E \tag{13.28}$$

The subtropical gyre solution in the present climate $h_0(x,y)$ satisfies the Sverdrup transport (Eq. 13.26) and geostrophic balance (Eq. 13.27).

Surface warming causes an increase in the reduced gravity, $\Delta g' > 0$. The response of the subtropical gyre can be broken down into the following two-step adjustment.

1. Uniform surface warming that is vertically confined above the thermocline results in a sea level change pattern of $g\Delta\eta = h_0 \Delta g'$ and an acceleration of the subtropical gyre.
2. Without wind change, the thermocline must shoal for the Sverdrup transport (Eq. 13.26) to remain unchanged.

The dynamically consistent solution can be obtained by perturbing either side of Eq. 13.28 with $\Delta w_E = 0$,

$$\Delta h = -\frac{\Delta g'}{2 g'_0} h_0 \tag{13.29}$$

An enhanced stratification ($\Delta g' > 0$) shoals the dynamic thermocline and accelerates the surface flow. The corresponding sea level change due to the thermal expansion is given by

$$g\Delta\eta = \Delta(g' h) = h_0 \Delta g' + g'_0 \Delta h = h_0 \Delta g'/2 \tag{13.30}$$

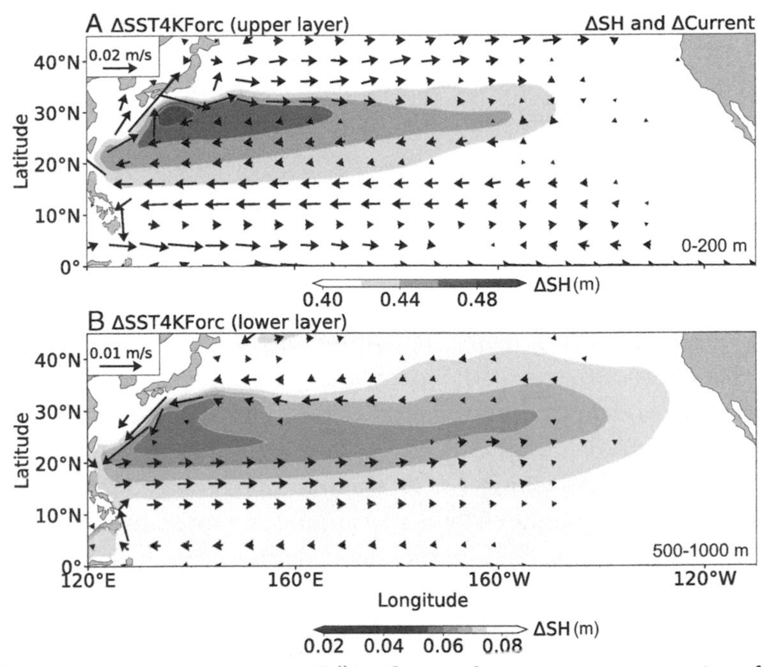

Fig. 13.16 Ocean current response to a spatially unform surface temperature warming of 4 K in (A) the top 200 m and (B) lower subtropical gyre (500–1000 m). Sea surface salinity and wind are kept unchanged. Changes in layer-mean current velocity (vectors, m/s) and steric height (SH, color shading, m). *(From Peng et al. (2022).)*

The shoaling thermocline corresponds to a deceleration of the lower subtropical gyre. The predicted acceleration (deceleration) of the upper (lower) subtropical gyre is reproduced in an ocean GCM simulation forced by a uniform surface warming (Fig. 13.16) and in comprehensive CMIP models (Wang et al. 2015). Note that the dynamic constraint imposed by the Sverdrup transport limits the ocean heat storage (Eq. 13.30) to only half of the amount allowed if temperature were a passive tracer with no effect on the ocean circulation (first term on the RHS or the previous step 1).

In the subtropical gyre, the surface warming of 4°C results in significant changes of $\Delta g'/g' \sim 0.4$, and $\Delta h/h_0 \sim 20\%$.

13.5 Discussion

Human-induced changes in atmospheric composition (GHGs and aerosols) have so far caused GMST to increase by about 1.2°C. The spatial pattern of surface temperature increase over the instrumental era (from the late 19th century) is to first order spatially uniform. The ocean has warmed (Cheng et al. 2020), and the large energy required for the ocean warming is consistent with independent estimates of radiative forcing and feedback at the TOA (Church et al. 2014). The thermal expansion due to the ocean warming and melting ice sheets/glaciers have caused the sea level to rise. All this led to the IPCC (2014) to conclude "the human effect on the climate system is clear."

The recent hiatus event illustrates that internal variability is large enough to cause considerable modulations of global warming rate over a decade and longer. The 2015-16 El Niño ended this hiatus event, and GMST in the ensuing 7 years (2015-21) was higher than any before 2014 (see Fig. 13.1). Hiatus research forged a closer integration of the climate variability and climate change communities, each with different foci and methods from the other. For example, the planetary energy budget is an important foundation of global warming research. The energy view of the hiatus highlighted the challenges in physical interpretation of the planetary energy budget on the decadal timescale. Specifically, the strong dependency of cloud-radiative effect on SST pattern raises the question of whether the cloud feedback on GMST change is well defined. The hiatus phenomenon provides a new impetus to understand, attribute, and predict decadal variability. On the regional scale, the superposition of forced warming and a negative PDO shift explains many observed changes during the hiatus: prolonged drought in the US Southwest (Delworth et al. 2015), accelerated Hadley cell expansion (Amaya et al. 2019), an intensified Walker circulation over the tropical Pacific, weak sea level change on the west coast of the Americas, and accelerated sea level rise in the tropical western Pacific (Church et al. 2014).

So far we have considered only cases of increasing GHG forcing. The Paris Agreement raises hopes for strong GHG mitigation policies. Unfortunately, it takes a long time to restore climate to preindustrial conditions even if we stop all the emissions immediately. Fig. 13.17 shows results from an earth system model with a ramp of CO_2 emissions to peak CO_2 values of 450, and 550 ppm, followed by zero emissions. Even with zero emissions, CO_2 concentration does not fall back to preindustrial level immediately but stays high for hundreds of years. Global mean temperature decreases even more slowly because of the large thermal inertia of the whole-depth ocean. Even as global surface temperature gradually stabilizes, the deep ocean continues to warm as long as $T - T_d > 0$ (Eq. 13.9). The thermal expansion causes sea level to rise for millennia. Thus the anthropogenic GHG emissions in the past have committed the world to significant amounts of sea level rise (on the order of 1 m) for centuries to come.

Review questions

1. What are the sources of predictability for weather forecast, seasonal climate prediction, and long-term climate projection, respectively?
2. Why can we predict climate beyond a month even though we cannot forecast weather beyond 2 weeks?

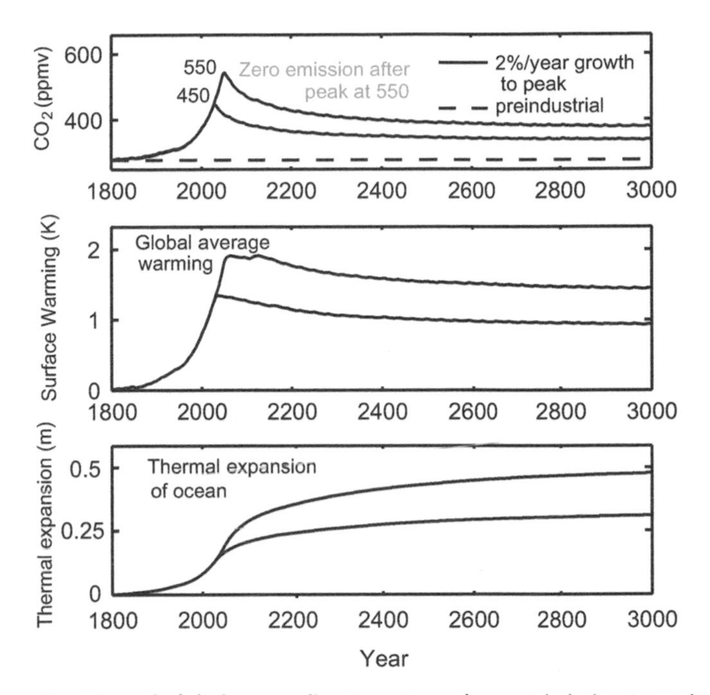

Fig. 13.17 Carbon dioxide and global mean climate system changes (relative to preindustrial conditions in 1765) from an earth system model. Climate system responses are shown for a ramp of CO_2 emissions at a rate of 2%/year to peak CO_2 values of 450 and 550 ppm, followed by zero emissions. *(top)* Falloff of CO_2 concentrations following zero emissions after the peak. *(middle)* Globally averaged surface warming (°C) for these cases. *(bottom)* Sea level rise (m) from thermal expansion only (not including loss of glaciers, ice caps, or ice sheets). *(Adapted from Solomon et al. (2009). (Courtesy Y.F. Geng.))*

3. GMST increases following El Niño. How does one tell greenhouse warming from naturally occurring internal variability in spatial pattern?
4. Does GMST always increase from one year to the next? Why?
5. What caused the rate of GMST increase to slow down in the early 2000s? How does the spatial distribution of surface temperature change during the hiatus help answer the question? Is global TOA energetics an essential driver for the hiatus?
6. Why does the tropical warming increase upward in the free troposphere? How does this help displace extratropical storm tracks?
7. Why does the rate of increase differ between global water vapor content and global-mean precipitation? What does this difference imply for the atmospheric overturning circulations?

8. Eq. 13.18 is the simplest model for Arctic amplification. Calculate the inverse Bowen ratio at 30°C and 0°C. What is the tropical to Arctic warming ratio the theory predicts? Is this ratio sensitive to the mean temperature for the Arctic? Why?
9. Consider the ocean mixed layer temperature response to an abrupt increase of radiative forcing F at $t = 0$.
   a. Show the solution to the two-layer model (Eq. 13.10) is $T = F/E\left[1 - exp\left(\frac{t}{\tau_m}\right)\right]$, where $E = -\lambda + \varepsilon$, and $\tau_m = C_m/E$. Draw a schematic of the solution as a function of time.
   b. Estimate the e-folding time τ of the mixed layer temperature response. Assume that $\lambda = -1.2\ W/m^2\ K^{-1}$, $\varepsilon = -\lambda/2$, and $H_m = 100$ m.
   c. Estimate the e-folding time and the equilibrium warming if the base of the ocean mixed layer were insolated.
10. Radiative forcing increases linearly in time, $F = F_{2x}t/\tau_{2x}$ where $\tau_{2x} = 70$ years is the time of doubling. This corresponds to the 1% CO_2 increase experiment. For $t > \tau_m$, the temperature response is approximately $T = \frac{F_{2x}}{E}\frac{t}{\tau_m}$. Show that the LHS of Eq. 13.10 is much smaller than either term on the RHS, thereby verifying the approximate solution.
11. The 1% CO_2 increase experiment approximates the historical evolution of anthropogenic radiative forcing. In light of the Paris Agreement, the 1% CO_2 decrease following a CO_2 doubling at year 70 is becoming relevant. Answer the following question based on the two-layer model (Eqs. 13.8 and 13.9).
   a. When does the GMST peak? Consider the fast-response approximation (Eq. 13.10).
   b. When does the deep ocean temperature peak? Use the solution (a) for GMST in Eq. 13.9.
   c. Is it possible to have thermosteric sea level rise while GMST decreases? When does the thermal expansion of the ocean stop?

CHAPTER 14

Regional climate change

Contents

14.1 Regional patterns of tropical rainfall change

To first order, surface warming caused by increasing greenhouse gas (GHG) concentrations is uniform (at least in sign), with some spatial variations in the magnitude (Fig. 14.1A). By contrast, the projected precipitation change is spatially highly variable, with regions of increase and decrease despite surface warming everywhere (see Fig. 14.1B). For example, rainfall is projected to increase over the equatorial Pacific and decrease over the subtropical Southeast Pacific. Precipitation is expected to increase in subpolar and polar regions, a result of enhanced poleward moisture transport by atmospheric eddies. Strong spatial variations pose greater challenge in predicting regional changes in rainfall than in temperature. The global mean is a good approximation for regional warming, but it is not representative of regional rainfall change. Predicting regional rainfall change needs to predict the spatial patterns of rainfall increase and decrease. Then, what determines the patterns of rainfall change?

Changes in precipitation have profound impacts on society and environment. Mammoth engineering projects are undertaken to ease mismatches between precipitation distribution and population centers, at great socioeconomic and environmental

Coupled Atmosphere-Ocean Dynamics: from El Niño to Climate Change
ISBN 978-0-323-95490-7, https://doi.org/10.1016/B978-0-323-95490-7.00014-X

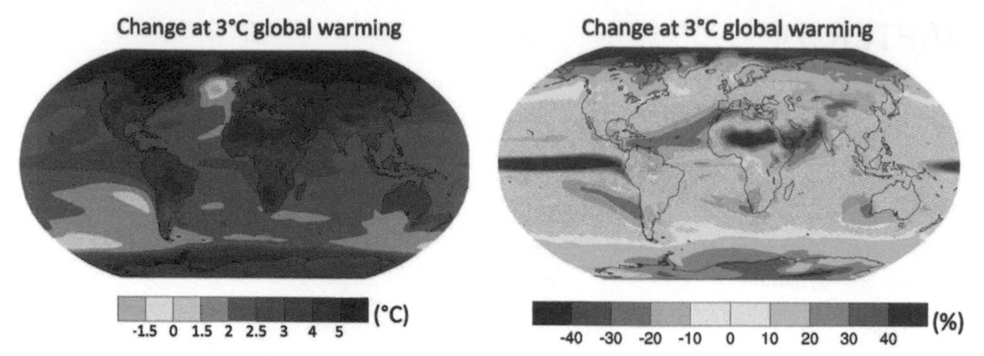

Fig. 14.1 Annual-mean surface temperature and precipitation (percentage, ΔP/P) change at 3°C global warming, based on the CMIP6 multimodel mean of SSP5-8.5. The inherent difficulty in predicting rainfall change is that it is spatially variable, increasing in some and decreasing in some other regions. Adapted from Lee et al. (2022). *(Courtesy Y.F. Geng.)*

costs. Examples include water diversion projects from water-rich northern to arid southern California, and from the Yangtze River to dry northern China.

The water vapor budget in an atmospheric column may be approximated as

$$\overline{\omega}\Delta q+\Delta\omega\overline{q} = \Delta(E-P) \tag{14.1}$$

where q is specific humidity at the cloud base (~1 km) and E the surface evaporation (Seager et al. 2010; Huang et al. 2013). Here we have neglected horizontal advection and moisture in the upper troposphere (see Section 3.1). Since ΔE is flat in space, the rainfall change dominates the spatial pattern of the right-hand side (RHS). On the left-hand side (LHS), the first term, called the thermodynamic component, represents the advection by mean vertical motion of increased water vapor due to atmospheric warming. The second term is called the dynamic component, representing rainfall change due to atmospheric circulation change. In the tropics, the circulation change includes a general slowdown proportional to the tropical mean warming (see Section 13.3.5) and the residual marked with an asterisk,

$$\Delta\omega = -\overline{\omega}\beta\Delta T+\Delta\omega^* \tag{14.2}$$

Analysis of CMIP5 projections suggests β ~0.04 K^{-1} (Chadwick et al. 2013). Eq. 14.1 becomes

$$(\alpha-\beta)\Delta T\overline{\omega}\,\overline{q}+\Delta\omega^*\overline{q} = \Delta(E-P) \tag{14.3}$$

If the residual circulation change ω^* is small, the earlier result suggests that rainfall increases where it currently rains, or a so-called "wet-get-wetter" pattern. The wet-get-wetter pattern emerges in atmospheric general circulation models (GCMs) if the ocean warming is horizontally uniform (Fig. 14.2A). In spatially uniform sea surface temperature (SST) increase (SUSI) experiments, rainfall increases along the intertropical convergence

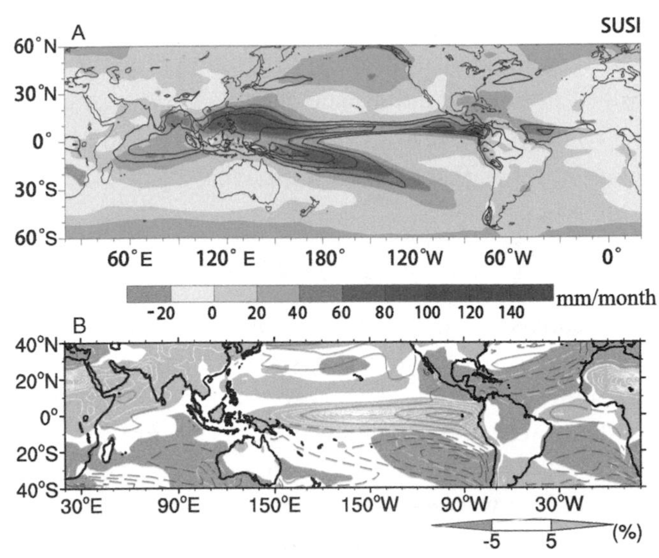

Fig. 14.2 Precipitation change (color shading, mm/month) for (A) spatial-uniform sea surface temperature (SST) increase *(SUSI)* of 4 K. All the results are scaled to a tropical (25°S—25°N) mean SST increase of 4 K, based on the ensemble average of 11 CMIP5 atmospheric general circulation models. Line contours are for climatologic precipitation (150, 200, 250, and 300 mm/month). (B) Annual-mean precipitation percentage change ($\Delta P/P$, shading and white contours at 20% intervals), and relative SST change (color contours at intervals of 0.3°C, the zero contour omitted) to the tropical (20°S—20°N) mean warming in 1% CO_2 run (year 101 ~ 150), shown as 37 CMIP6 model ensemble mean. The spatial correlation for the multimodel ensemble mean between the SST and precipitation fields is 0.49. Updated based on Christensen et al. (2014). *(A, From Xie et al. (2015); B, Courtesy Y.F. Geng.)*

zone (ITCZ) and South Pacific convergence zone (SPCZ), while decreasing slightly in the subtropics (20—40°) in either hemisphere associated with the poleward expansion of the Hadley circulation. When surface evaporation increase is factored in, the subtropical drying is quite severe. SUSI further captures the broad precipitation increase poleward 40°.

Annual rainfall change in coupled ocean-atmospheric GCMs is not well correlated with the mean rainfall. For example, mean rainfall is low over the equatorial cold tongue of the Pacific, but models project a pronounced increase in warmer climate. The strong deviations of projected rainfall change in coupled models from that in SUSI are due to the fact that ocean surface warming is not uniform but features marked spatial variations. Indeed, tropical rainfall change is correlated with the SST warming pattern (see Fig. 14.2B),

$$\Delta T^* = \Delta T - \langle \Delta T \rangle \tag{14.4}$$

where angular brackets denote the tropical mean and the asterisk the spatial deviations. ΔT^* is also called the relative SST change, referenced to the tropical mean.

In the tropics, the Coriolis force is weak, and the free troposphere cannot sustain much horizontal temperature gradient. The upper tropospheric warming is nearly uniform within 20°S–20°N, with the moist static energy (MSE) change set by the tropical mean, $\Delta m_u = c_p(1 + b_e)\langle \Delta T \rangle$, where $b_e = \alpha L\bar{q}/c_p \sim 2.4$ is the inverse Bowen ratio. The gross moist static instability change at a given location may be defined as the surface minus upper-level MSE difference,

$$\Delta(m_s - m_u) = c_p(1 + b_e)(\Delta T - \langle \Delta T \rangle) = c_p(1 + b_e)\Delta T^* \tag{14.5}$$

In other words, local convective instability change is proportional to the relative SST change and as a result, rainfall change follows a warmer-get-wetter pattern. Here a warmer region is referenced to the tropical mean warming at a given time.

Fig. 14.3 shows the results of precipitation decomposition (Eq. 14.1). By definition, the thermodynamic term follows the wet-get-wetter pattern, while the dynamic term is more complex and includes a wet-get-drier pattern because of the general slowdown of the atmospheric circulation. The sum of the thermodynamic term and circulation slowdown effect reduces the wet-get-wetter effect [by a factor of $(1 - \beta/\alpha) \sim 0.43$] (first term on LHS in Eq. 14.3), allowing the warmer-get-wetter or SST pattern effect (second term in Eq. 14.3) to emerge. Here we assume

$$-\Delta\omega^* \propto \Delta T^* \tag{14.6}$$

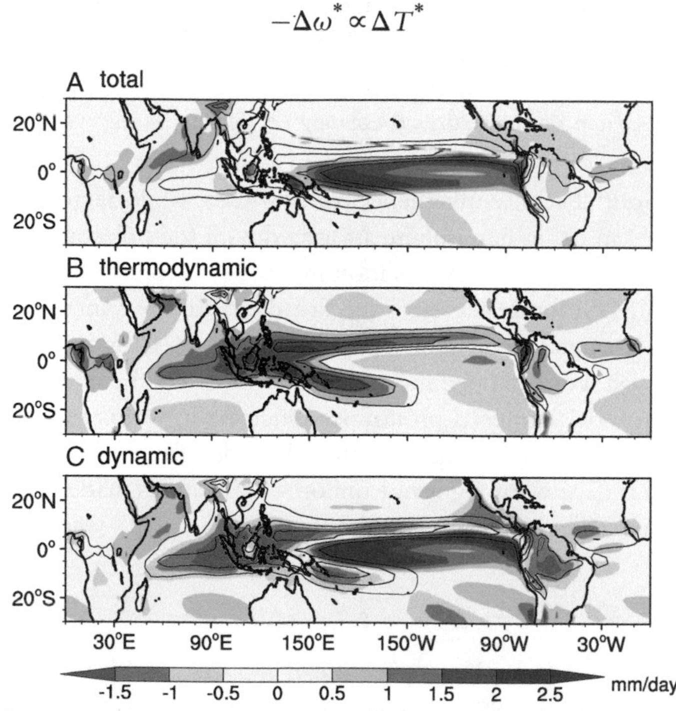

Fig. 14.3 Annual-mean precipitation change in 1% CO_2 run (year 101 ~ 150 mean; color shading; mm day^{-1}), superimposed on the mean *(thin gray line contours)*. (A) CMIP6 34-model ensemble mean, (B) the thermodynamic, and (C) dynamic components. *(Courtesy Y.F. Geng.)*

Namely, the ocean warming pattern drives atmospheric overturning circulation change in addition to a general thermodynamically induced slowdown of the mean circulation in the tropics.

14.2 SST pattern dynamics

The governing equation for local mixed layer temperature change is

$$C_m \frac{\partial T'}{\partial t} = D_o' + Q' \tag{14.7}$$

where $D_o = -\rho c_p \int_{H_m}^{0} \left(u\frac{\partial T}{\partial x} + v\frac{\partial T}{\partial y} \right) dz - Q_B$ represents the lateral advection and turbulent heat flux at the base of the ocean mixed layer (Q_B is also known as entrainment). $Q = Q_S + Q_L - Q_E - Q_H$ is the surface heat flux. Here the prime denotes perturbations. From an oceanic point of view, latent heat flux

$$Q_E = \rho_a L C_E W[q_s(T) - R_H q_s(T_a)] \tag{14.8}$$

includes both atmospheric forcing (e.g., wind speed W) and an SST response. Following Section 8.1, we linearize Eq. 14.8 with respect to an SST perturbation T',

$$Q_E' = Q_{Ea}' + \alpha \overline{Q}_E T' \tag{14.9}$$

where the second term on the RHS represents the SST response, the first term is the residual representing atmospheric forcing due to wind and relative humidity change, and α is the Clausius-Clapeyron coefficient. For simplicity, we have assumed $T_a' = T'$, which is valid over the open tropical ocean and implies a vanishing perturbation sensible heat flux. With this decomposition, Eq. 14.7 becomes

$$C_m \frac{\partial T}{\partial t} = D_o + Q_a - \alpha \overline{Q}_E T \tag{14.10}$$

where we have dropped the prime for clarity, and Q_a includes all the atmospheric "forcing" effect on surface radiation and evaporation, although the upward longwave radiation includes a weak additional damping on SST. The evaporative damping coefficient (last term in Eq. 14.10) is proportional to the mean evaporation. In the tropics, evaporative cooling is the dominant mechanism that offsets the downward radiative flux.

Local surface feedback analysis in Eq. 14.10—as opposed to the global top of atmosphere (TOA) feedback analysis in Section 13.1—is useful to understand ocean surface warming pattern (Xie et al. 2010). For slowly varying radiative forcing, the mixed layer heat storage term on the LHS is negligible, and the ocean dynamic effect can be inferred from the net surface heat flux change, $D_o' \approx Q'$. In other words, the net surface heat flux is determined by ocean dynamic effect, generally small except in the subpolar North Atlantic and Southern Ocean (see Section 14.4).

Over the tropical oceans, the mean latent heat flux varies from 100 Wm^{-2} on the equator (due to low winds and cold upwelling) to 150 Wm^{-2} in the subtropics, due to the strong trade winds and small relative humidity. In response to a uniform GHG radiative forcing, the reduced evaporative damping coefficient induces an enhanced SST response on the equator (Fig. 14.4). This equatorial peak in SST warming is especially pronounced over the Pacific, anchoring a band of marked rainfall increase through the warmer-get-wetter effect (see Fig. 14.2B).

Another robust SST pattern across models is that the ocean surface warms more slowly in the Southern than Northern Hemisphere. The reduced surface warming is especially pronounced in the subtropical Southeast Pacific. The southeast trades intensify over the weak warming region of the Southeast Pacific while the northeast trades weaken over the subtropical North Pacific where the SST increase is above the tropical mean (see Fig. 14.4). This suggests an interhemispheric WES (wind-evaporation-SST) feedback between the ocean warming pattern and trade wind intensity. Assuming that radiative perturbations are equal between the hemispheres, we have from Eq. 14.10

$$\overline{Q}_E(\delta W / \overline{W} - \alpha\delta T) = 0 \tag{14.11}$$

where δ denotes the cross-equatorial difference. The warming is reduced in the hemisphere with the intensified trade winds.

14.2.1 El Niño and Southern Oscillation (ENSO) changes

Models disagree on how SST variability of ENSO, either the magnitude or zonal pattern, responds to greenhouse warming, some showing an increase in variance, some a decrease, and still others no change. Despite the disagreement on SST variability, models show a

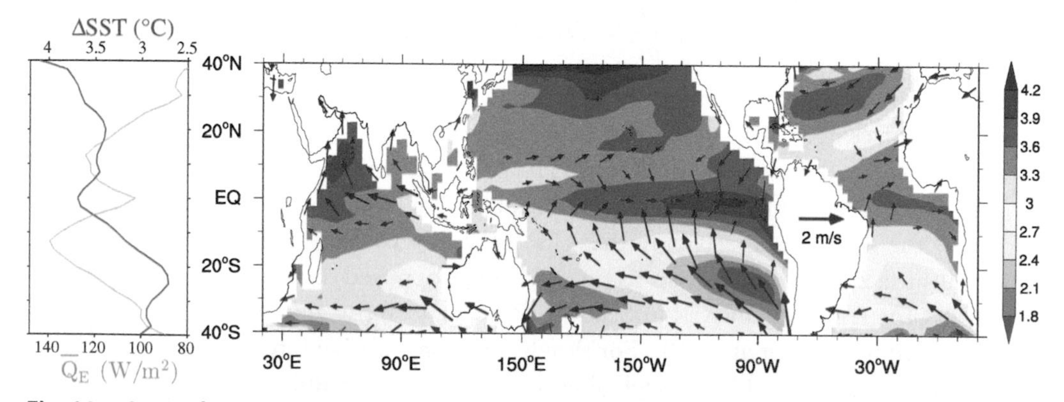

Fig. 14.4 Sea surface temperature (*SST*; °C) and surface wind (m/s) change in the 1% CO_2 run at year 140 (the time of 4 × CO_2), based on 31 CMIP6 model ensemble mean. *(left inset)* Zonal-mean SST change with the mean latent heat flux (W/m²). *(Courtesy S.M. Long and Y.F. Geng.)*

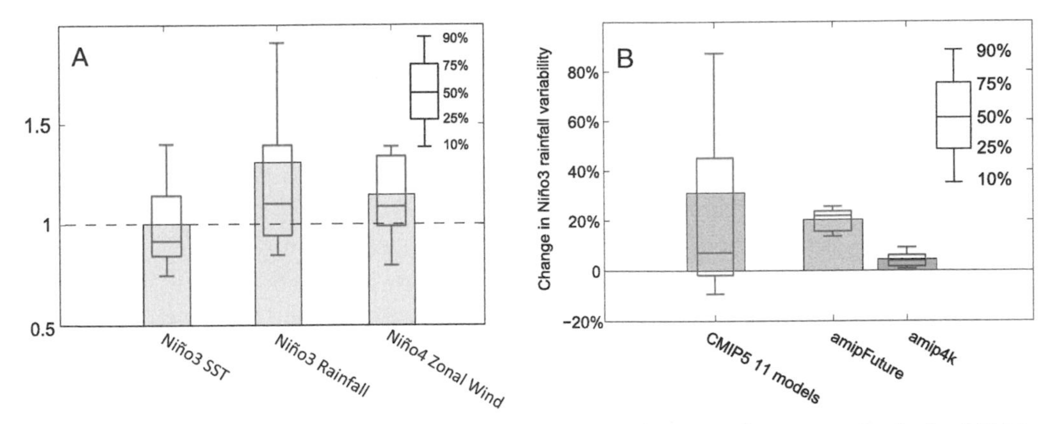

Fig. 14.5 (A) 21st- to 20th-century ratio of El Niño and Southern Oscillation amplitude for RCP8.5. CMIP5 multimodel ensemble mean *(bar)* and spread *(box-whisker)*. (B) Normalized increase in precipitation standard deviation over Nino3 in Amip4K and AmipFuture, and corresponding CMIP5 coupled models under RCP 8.5. AMIP 4K (Future) refers to the atmospheric GCM simulation with observed sea surface temperature (SST) variability, plus a uniform 4 K (spatially patterned) increase in background SST. Adapted from Zheng et al. (2016). See text for details. *(A, From Christensen et al. (2014); B, Courtesy Y.F. Geng.)*

robust increase in the occurrence of deep convective events over the eastern equatorial Pacific (Nino3 region) during El Niño (Fig. 14.5A).

Climatologically, Nino3 SST stays below the convective threshold ($\overline{T}^* < 0$). Deep atmospheric convection occurs only during extreme El Niño events (see Fig. 9.17) when Nino3 SST exceeds the convective threshold, $T' + \overline{T}^* > 0$, with the prime denoting El Niño anomalies and the overbar the mean. Most CMIP6 models project an eastward intensification of SST increase over the equatorial Pacific, with a positive relative warming over the Nino3 region ($\Delta\overline{T}^* > 0$) (see Fig. 14.2B). This mean warming pattern lowers the SST barrier to deep convection and increases the frequency of threshold exceedance even if the Nino3 SST variability remains unchanged.

Atmospheric GCMs capture the intensification of Nino3 precipitation variability in response to an SST warming pattern derived from the multimodel ensemble (MME) mean of 4 $\times$ CO_2 simulations (AmipFuture, global-mean SST change = 4 K), but the same atmospheric models fail to reproduce enhanced convective anomalies when forced with a uniform 4 K SST increase (see Fig. 14.5B). As interannual SST variability is identical between the AmipFuture and Amip4K runs, this result illustrates the role of the mean SST warming pattern in intensifying atmospheric variability of ENSO. The coupled model ensemble sharing the same atmospheric models shows a similar intensification of Nino3 rainfall variability, but the intermodel spread is much larger (see Fig. 14.5B) because of the differences among models in both the mean warming and interannual variability of equatorial Pacific SST.

The eastward shift of convective anomalies over the tropical Pacific induces a similar shift and intensification of the Pacific-North American (PNA) response to El Niño in boreal winter. The anomalous Aleutian low strengthens and shifts eastward to be closer to North America (Fig. 14.6). Among other changes, the tendency for California to become stormier and rainier during El Niño winters is likely to strengthen and become more robust in a warmer climate.

MME mean isolates the forced change common among climate models, while the ensemble spread represents uncertainties. Intermodel correlations could imply useful relationships and yield insights into physical processes and mechanisms upon careful investigations.

The uncertainty in projected change of ENSO SST variance among models is an example. In a warmer climate, El Niño–induced rainfall anomalies intensify over Nino3, implying a strengthened zonal wind feedback on El Niño (see Fig. 14.5A). Indeed, the intermodel spreads seem correlated between changes in the mean and variance of Nino3 SST. Namely, the uncertainty in projecting Nino3 SST variance is related to the uncertainty in the mean warming pattern, specifically in relative SST warming $\Delta\overline{T}^*$ among models (Fig. 14.7). The larger $\Delta\overline{T}^*$ is, the more intensified Nino3 convective variability and the atmospheric feedback on El Niño are.

Diversity in simulated ENSO among climate models reflects the fact that ENSO is sensitive to a wide variety of coupled feedbacks (Planton et al. 2021). Here we have focused on the nonlinear convective feedback in Nino3 region. In warming climate, the upper ocean becomes more stratified, strengthening the surface feedback (Ekman

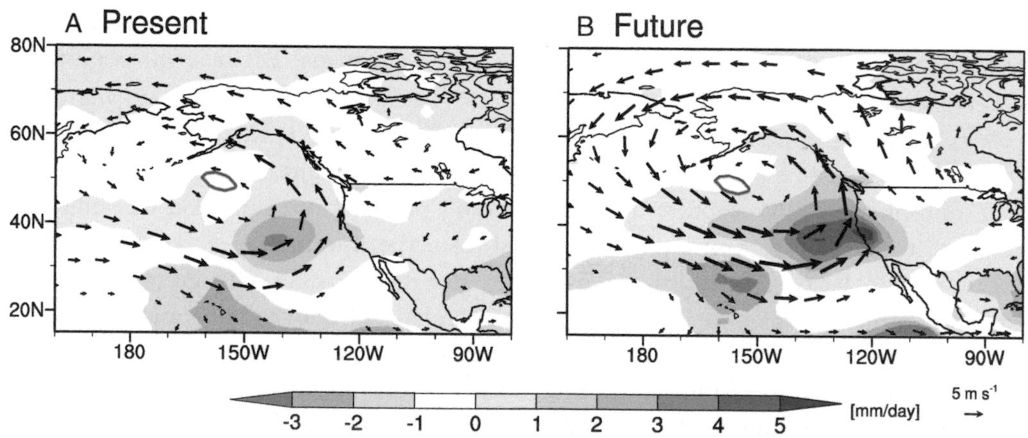

Fig. 14.6 El Niño–induced anomalies in DJF: precipitation (mm/day) and 850-mb wind velocity (m/s) in (A) present and (B) future climates as simulated by an atmospheric general circulation model (CAM4). The red ellipse marks the center of the El Niño–induced anomalous Aleutian low in present climate. The future climate is represented by an SST warming pattern derived from a RCP8.5 run. *(Based on Zhou et al. (2014).)*

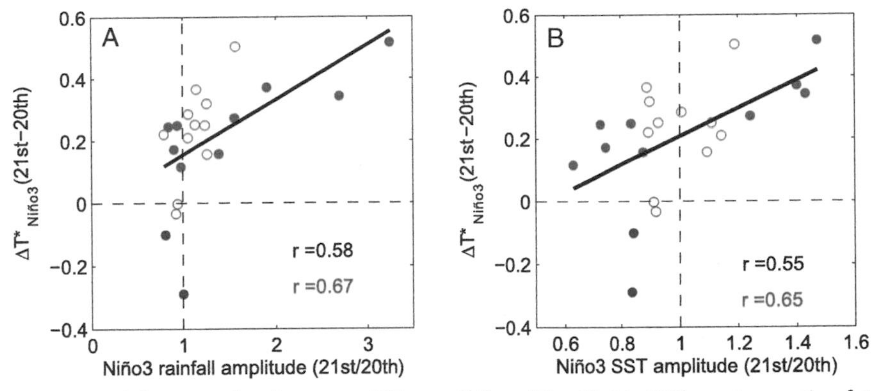

Fig. 14.7 Intermodel scatterplots between $DT^*_{\text{Niño3}}$ (°C) and the 21st to 20th century ratio of standard deviation for (A) precipitation and (B) sea surface temperature *(SST)* over Nino3 region. *(From Zheng et al. (2016).)*

dynamics terms in the perturbation SST Eq. 9.3) (Timmermann et al. 1999). See Cai et al. (2021) for a recent review.

14.2.2 El Niño–like warming

Most CMIP5 models favor an eastward intensification of SST increase under greenhouse warming (see Fig. 14.4), but they disagree on the magnitude of the zonal SST gradient change (see Fig. 14.7). From Eq. 9.3, SST change in the Nino3 region is governed by

$$\frac{\partial \Delta T}{\partial t} = -\left(\frac{\alpha \overline{Q}_E}{\rho C_p H_m} + \frac{\overline{w}_m}{2H_m}\right)\Delta T + \frac{\overline{w}_m}{2H_m}\frac{\partial \overline{T}_e}{\partial h}\Delta h + \frac{\Delta Q}{\rho C_p H_m} \tag{14.12}$$

where ΔQ is the surface heat flux perturbation due to the greenhouse effect. Here the zonal advection terms have been omitted. Three mechanisms are in play:

- *Evaporative damping* rate (first term in the parentheses) is weaker in the eastern than western equatorial Pacific, in favor of a weakened zonal SST gradient and hence positive ΔT^*. The cloud feedback on SST—positive (negative) for low (deep convective) clouds—also favors an eastward enhanced warming.
- *The slowdown of the Walker circulation* helps flatten the eastward shoaling thermocline, and the thermocline feedback (second from last term on RHS) reduces the zonal SST gradient (Vecchi and Soden 2007).
- *The ocean dynamic thermostat* refers to the damping on SST warming by the mean upwelling of the thermocline water subducted decades earlier and mixed with deeper water (second term in the parentheses). This upwelling damping strengthens the zonal SST gradient under greenhouse warming (Clement et al. 1996). This effect operates as the greenhouse forcing grows in time because the warming in the equatorial

thermocline lags behind. The effect weakens as the radiative forcing stabilizes and the thermocline warming catches up (Luo et al. 2017).

The relative importance of these mechanisms might differ among models and be influenced by model biases, say in the equatorial cold tongue and cross-equatorial winds.

The equatorial cold tongue extends too far westward in most CMIP5 models, reaching all the way to the Maritime Continent. This reduces the evaporative and convective cloud damping on SST over the western equatorial Pacific, biases that favor enhanced SST warming in the west. In fact, Li et al. (2016a) identified a correlation between the projected increase in east-west SST gradient and western Pacific mean precipitation among CMIP5 models (Fig. 14.8A).

Use of observed mean rainfall (the red line) to correct the biases in model projection yields a 0.4°C decrease in the east-west SST gradient, from an MME mean of 0.2°C. The use of observations to correct model projections based on an intermodel correlation between an observable quantity and model projections is called an *emergent constraint*. An intermodel correlation is only suggestive of a physical relationship. The physical mechanism needs to be rigorously tested were it used as an emergent constraint (Hall et al. 2019). Here, the cold biases in the western equatorial Pacific cause a dry bias in local precipitation and lowers the damping on the greenhouse warming.

14.2.3 Indian Ocean Dipole (IOD)—like warming

Under increasing greenhouse forcing, most CMIP5 models project a reduced surface warming (see Fig. 14.4) and decreased atmospheric convection (see Fig. 14.2C) over the southeast equatorial Indian Ocean, associated with an easterly wind change on the equator that lifts the thermocline off the west coast of Indonesia. This IOD-like pattern

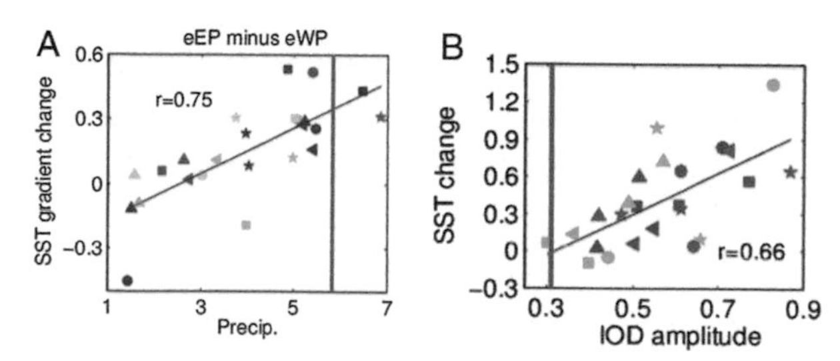

Fig. 14.8 Emergent constraints derived from 24 CMIP5 models under RCP8.5 scenario. (A) Relationships between the simulated mean precipitation (mm/day) in the equatorial western Pacific and projected changes in east-minus-west sea surface temperature *(SST)* difference across the equatorial Pacific (2°S–2°N). Eastern and western Pacific regions refer to 140°W–90°W and 140°E–170°W, respectively. (B) Scatterplot of the simulated present-day Indian Ocean Dipole *(IOD)* amplitudes vs projected changes for SON in the west-minus-east SST (°C) between the dipole regions. The red line denotes the observed value, and the intermodel correlation (*r*) is noted. *(A, From Li et al. (2016a); B, from Li et al. (2016b).)*

involves the Bjerknes feedback and is most pronounced during July-November when the coastal upwelling permits the thermocline feedback off Indonesia (see Section 11.2).

A robust pattern of projected future changes across models may not be realistic and could be an artifact of common model biases. The variance of the interannual IOD mode is generally too strong in CMIP5 models compared to observations, implying an excessive Bjerknes feedback in models. A correlation is identified between interannual IOD variance and the change in west minus east mean-SST difference among CMIP5 models (see Fig. 14.8B). If we use the observed interannual IOD variance (0.3 K) to adjust the model projections, the corrected projection of the west minus east difference in mean SST warming vanishes. Here the observed interannual IOD variance serves as an emergent constraint on the projected change in zonal SST gradient. The correction is important for projected rainfall change over the Maritime Continent and East Africa, much as in the interannual IOD (see Section 11.2).

14.3 Regional uncertainty due to atmospheric circulation change

While Section 14.2 identifies a few robust patterns of rainfall change in a warmer climate, intermodel uncertainties in projected rainfall change are quite large, especially in tropical and subtropical regions (Fig. 14.9A).

14.3.1 SST pattern

The water vapor budget of Eq. 14.1 can be applied to intermodel spread in projected rainfall change. Fig. 14.9 shows the results. The uncertainty in the thermodynamic component is small because models are built to simulate the observed mean overturning circulation ($\overline{\omega}$). The uncertainty in rainfall projections stems chiefly from that in projected atmospheric circulation change ($\Delta\omega$).

The SST warming pattern is an important driver of atmospheric circulation change. We investigate the relationship in the intermodel uncertainty between the two. For simplicity, we focus only on the zonal mean results from the 1% CO_2 run to sidestep uncertainties in aerosol forcing. The first singular value decomposition (SVD) mode of intermodel co-variability between $\Delta\omega$ and ΔT is antisymmetric about the equator (Fig. 14.10), with anomalous upward motion in the warmer hemisphere. While the ocean surface warming is larger in the Northern than Southern Hemisphere in the multi-model ensemble mean (see Fig. 14.4), the degree of this asymmetry in warming varies among models, inducing cross-equatorial Hadley circulation adjustments. The intermodel spread in cross-equatorial SST difference has been linked to that in the rates of Atlantic meridional overturning circulation (AMOC) slowdown and sea ice melt around Antarctica (Geng et al. 2022).

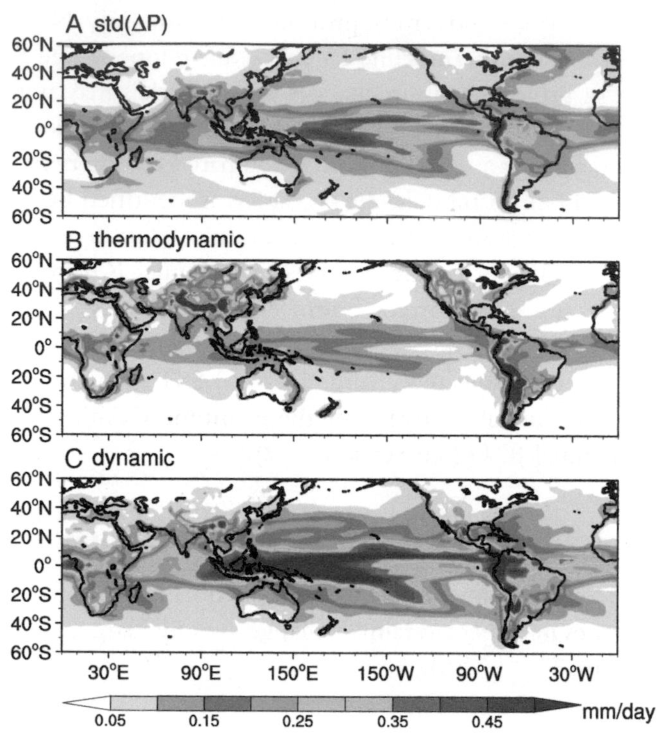

Fig. 14.9 Intermodel standard deviation of (A) precipitation change, (B) the thermodynamic, and (C) dynamic components in CMIP6 1% CO_2 run (year 101 ~ 150 mean). *(Adapted from Geng et al. (2020).)*

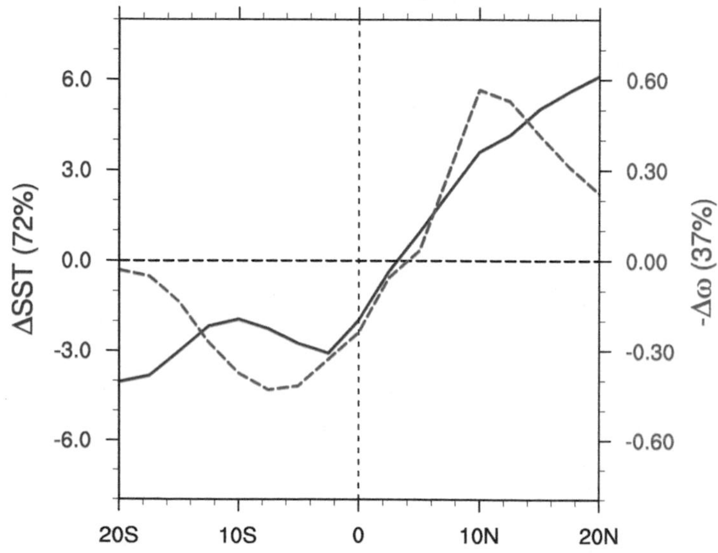

Fig. 14.10 First singular value decomposition mode of zonal-mean intermodel spread between sea surface temperature *(SST)* and 500-hPa vertical velocity changes in CMIP6 1% CO_2 run (year 101-150 mean). It explains 64% of the covariance. *(Adapted from Geng et al. (2022).)*

The second SVD mode of intermodel uncertainty features an equatorial peak in ΔT, driving a Hadley circulation change symmetric about the equator. While the ensemble-mean ocean warming peaks on the equator, the degree of this equatorial enhanced warming varies among models.

Tropical cyclones are severe storms of devastating impacts due to high wind, heavy rainfall, and storm surges. Globally, tropical cyclones are expected to become fewer in annual count, but they are to become more intense. The latter is consistent with the increase in potential intensity with SST but subject to the opposing effect of increasing ocean temperature stratification (see Section 10.5), which implies a stronger cold wake in warmer climate. The regional changes in tropical cyclone occurrence are uncertain, and this regional uncertainty is tied to the SST warming pattern (Zhao and Held 2012). Specifically, the relative SST change drives regional changes in convection and overturning circulation, affecting the genesis potential. Regions with positive $\Delta \overline{T}^*$ tend to see an increase in tropical cyclone activity.

14.3.2 Internal variability

Unforced internal variability contributes to multidecadal trends. We illustrate this effect using a large ensemble simulation with a coupled ocean-atmosphere GCM. All the member runs share the same radiative forcing, and each differs from the other only in initial conditions (Deser et al. 2020a). Any difference between member runs is due to internal variability of the ocean-atmospheric coupled system. Remarkably, multidecadal trend in winter surface temperature over the contiguous United States varies considerably among member runs in both the spatial pattern and regional mean (Fig. 14.11). The regional mean warming over the 56-year period ranges from 3°C in member 22 to 1°C in member 4. In member 4, temperature trend is even negative in parts of the Pacific Northwest and the northern Rockies. The variability in the US surface temperature trend is associated with that in the Aleutian low through advection by anomalous winds (see Fig. 14.11D). By assuming that the atmospheric circulation change induced by radiative forcing is small, a dynamical adjustment method is developed by removing the surface air temperature (SAT) trend associated with sea level pressure (SLP) trend based on a relationship derived from historical variability in observations or model simulations. Comparison with the ensemble mean, which represents the radiative forced change, confirms that this method successfully removes much of the internal variability effect on winter surface air temperature.

Large internal variability has hampered the detection of regional change patterns from observations. In eastern North America, winter SAT decreased over the 1950s to 1990s due to the phase changes in the North Atlantic Oscillation and Pacific Decadal Oscillation (PDO). Over the eastern equatorial Pacific, most CMIP6 models project a locally enhanced warming (see Fig. 14.4), but a cooling trend is observed during the satellite

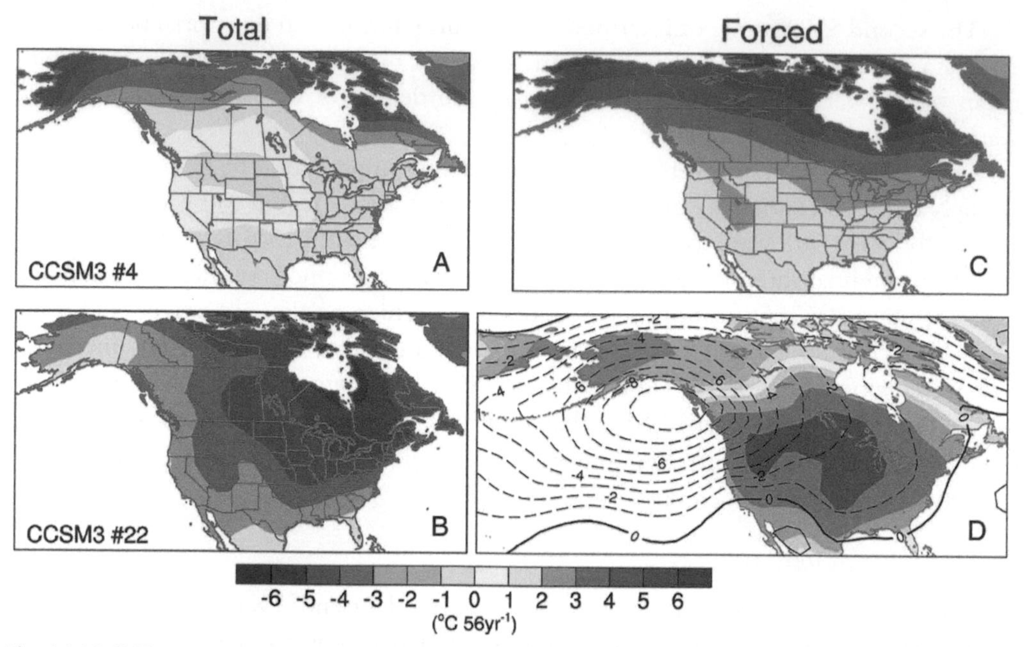

Fig. 14.11 DJF temperature trends during 2005-60 (°C per 56 years): (A, B) members 4 and 22, and (C) the 40-member ensemble mean from the A1B scenario, similar to RCP4.5. (D) Trends in the individual ensemble members (exclusive of 4 and 22) regressed on the 38 corresponding raw surface air temperature (SAT) trends averaged over the continental United States. SAT trends are indicated by colored shading and sea level pressure (SLP) trends by contours. Contour interval 1 hPa per 56 years. The zero contour is bold and dashed contours indicate negative SLP trends. *(From Wallace et al. (2016).)*

era (1980–), while the rest of the tropical oceans has warmed (see Fig. 13.9A). Is this inconsistency due to the internal PDO or does it signal that current models underestimate the upwelling damping effect on equatorial Pacific warming (see Section 14.2)? See Eyring et al. (2022, section 3.3.3.1) for a latest assessment.

14.4 Ocean heat uptake

For a slowly evolving radiative forcing (e.g., 1% per year increase in CO_2), the subsurface ocean is far from equilibrium and the slow ocean uptake of anthropogenic heat (Q_{net}) reduces the rate of global surface warming. From Eq. 13.5,

$$T = (F - Q_{net})/(-\lambda) \tag{14.13}$$

This section considers regional patterns of ocean heat uptake and discusses how they are shaped by the ocean circulation and drive the cross-equatorial atmospheric circulation.

14.4.1 Response to greenhouse forcing

In response to the increasing GHG forcing, strong ocean heat uptake is largely confined to the Southern Ocean and subpolar North Atlantic (Marshall et al. 2015) (Fig. 14.12), which are the upwelling and sinking regions of the global deep meridional overturning circulation (MOC), respectively. In the rest of the world ocean, the surface heat flux change is small as fast (~10 years) ventilation by the shallow MOC (see Fig. 12.17) keeps the upper ocean near thermal equilibrium with the warming atmosphere.

In the Southern Ocean, the prevailing westerly winds drive surface water equatorward, and the resultant upwelling pumps deep water (see Fig. 12.16) that is last in contact with the atmosphere 100s to 1000 years ago and has not been subject to anthropogenic warming. The mean upwelling damps the SST response and keeps the ocean mixed layer from being in a thermal equilibrium with the warming atmosphere, resulting in a large downward heat flux into the ocean. The surface water heated by this strong uptake in the Southern Ocean then flows equatorward and subducts into the thermocline as the Antarctic mode water is shielded again from the atmosphere. This equatorward advection causes a displacement between the meridional maxima of surface flux and ocean heat content changes (Armour et al. 2016) (Fig. 14.13, left).

The suppressed ocean warming in the upwelling zone and large ocean warming to the north imply an intensified Antarctic Circumpolar Current (ACC) by thermal wind. The Southern Ocean warming and surface acceleration of the ACC have been observed by Argo floats and satellite altimeters, respectively (see Fig. 14.13, right).

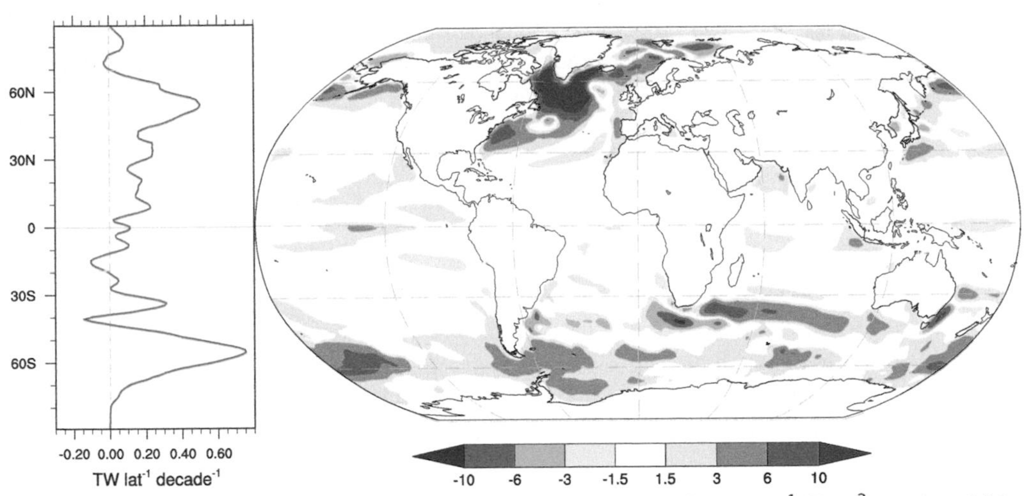

Fig. 14.12 The 1861-2005 trend in net surface heat flux (color shading in 10^{-1} W m^{-2} per decade) in CMIP5 historical greenhouse gas run. The inset is the zonally integrated heat uptake. Positive downward into the ocean. *(Adapted from Shi et al. (2018). (Courtesy J.R. Shi.))*

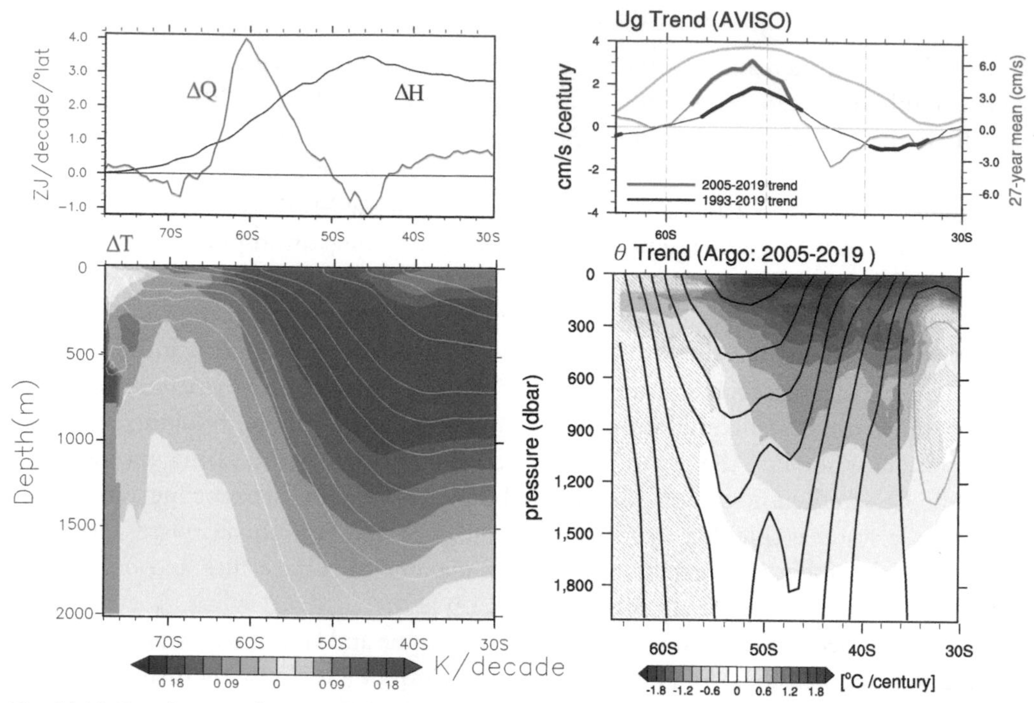

Fig. 14.13 Zonal-mean changes. *(left)* CESM abrupt 4 × CO_2 experiment. *(bottom)* Ocean temperature change superimposed on the mean potential density *(white contours)*. *(bottom)* Surface heat uptake and vertically integrated ocean heat content change. Adapted from Liu et al. (2018). *(right)* Observations. *(bottom)* Potential temperature trend (2005-19) and climatologic Ug (line contours starting from 0.5 cm/s at 1 cm/s interval, negative in gray) from Argo floats. Hatched are trends below the 95% confidence level from the two-tailed *t* test. *(upper)* Zonal geostrophic velocity (Ug) trend and the climatology (1993-2019) *(gray curve)* from satellite altimetry AVISO. Trends exceeding 95% significance level are thickened. *(A, Courtesy Y.F. Geng; B, From Shi et al. (2021).)*

In the subpolar North Atlantic, the northward heat transport by the Gulf Stream and North Atlantic Current sustains a strong heat release to the atmosphere in the climatology (see Fig. 2.9). The intense surface cooling during the winter causes deep convection, and the dense deep water forms and spreads around much of the global ocean. The greenhouse warming intensifies the poleward moisture transport in the atmosphere. The increased precipitation, along with the melting of sea ice and Greenland ice sheets, lightens the surface salinity (Fig. 14.14A) and density, reducing the deep-water formation and decelerating the AMOC. The decreased poleward heat transport reduces the radiatively forced ocean warming, creating a warming hole south of Greenland (see Fig. 14.1A). The slow ocean warming there results in a downward surface heat flux change (positive ocean heat uptake) (see Fig. 14.12).

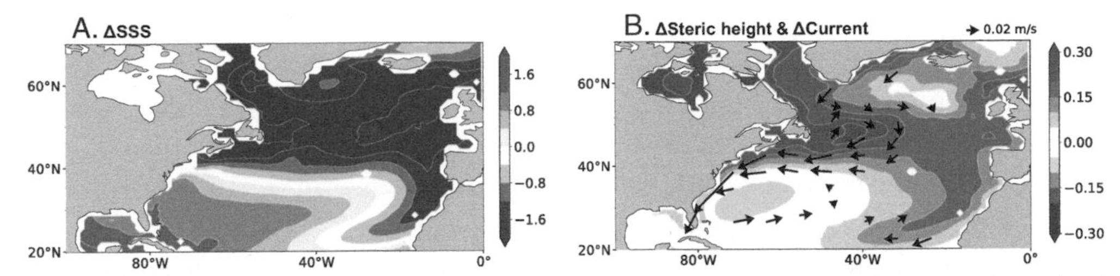

Fig. 14.14 Changes in abrupt 4 × CO_2 run (year 101-140 average): (A) sea surface salinity (psu); (B) surface steric height (m) and current velocity (m/s). CMIP6 30-model mean. *(Courtesy Q.H. Peng.)*

The AMOC slowdown causes the Gulf Stream to decelerate. The relaxed pressure gradient across the Gulf Stream exacerbates sea level rise on the northeast coast of North America (see Fig. 14.14B). The dynamic sea level pattern is very similar to the response in a water-hosing experiment where fresh water is added to the subpolar North Atlantic (Yin et al. 2009). This is in contrast with a surface acceleration of the North Pacific subtropical gyre in response to the increased density stratification—the enhanced sea level rise is kept offshore as baroclinic Rossby wave response (see Fig. 13.16).

14.4.2 Cross-equatorial energy transport

The spatial variations in ocean uptake of anthropogenic heat drives changes in the atmospheric circulation. The zonally and vertically integrated heat balance for the ocean over the whole depth (see Section 2.3) is expressed as

$$\frac{\partial H}{\partial t} + \frac{\partial F_o}{\partial y} = Q_{net} \tag{14.14}$$

Here the prime for perturbations has been dropped for clarity. The first term on the LHS is called the ocean heat storage and the RHS term the ocean heat uptake. The whole-depth heat storage term is important in transient adjustment to slow-varying radiative forcing (see Section 13.1). The ocean heat uptake is the sum of the heat storage and depth-integrated energy transport (F_o) divergence (second term on LHS). The zonally integrated energy balance for the atmospheric column is

$$\frac{\partial F_a}{\partial y} = R_{TOA} - Q_{net} \tag{14.15}$$

where the sign convention for flux is downward positive, and F_a is the northward energy transport by the atmospheric motions. Integrating over the hemispheres yields the atmospheric energy transport at the equator,

$$-F_a|_{y=0} = \langle R_{TOA} - Q_{net} \rangle_{SH}^{NH}/2 \tag{14.16}$$

where the angular brackets denote the hemispheric integration, and $\langle\cdot\rangle_{SH}^{NH} \equiv \langle\cdot\rangle_{NH} - \langle\cdot\rangle_{SH}$.

The ocean circulation shapes the ocean uptake of greenhouse heat. The subpolar North Atlantic heat uptake is due to the weakened AMOC, while the Southern Ocean uptake is due to the mean upwelling of the global deep MOC. Eq. 14.16 shows that the ocean uptake patterns drive the cross-equatorial atmospheric circulation if $\langle R_{TOA}\rangle_{SH}^{NH}$ is small. Under the GHG forcing, the zonally integrated heat uptake is slightly larger in the Southern Ocean than in the subpolar North Atlantic, but the difference in the hemispheric integral is small (see Fig. 14.12 inset). As a result, the cross-equatorial Hadley response to GHG forcing is weak (as discussed later [see Figs. 14.16B, 14.19]).

14.5 Aerosol effects

Combustion of fossil fuels emits carbon dioxide (CO_2) as well as aerosols and the precursors such as sulfur dioxide (SO_2). Upon reaction with water vapor, SO_2 forms sulfate, which scatters sunlight (the direct effect on radiation) and increases the number of cloud condensation nuclei and cloud droplets. With a given cloud liquid water content, the increased number of cloud droplets brightens the cloud (the indirect effect on radiation; see Section 6.3.2), as made visible by brightened ship tracks on solid stratus cloud decks. Thus the same chimneys of a coal-fired power plant emit both planet-warming CO_2 and cooling aerosols. Anthropogenic aerosols caused a global surface cooling of $0\sim0.8°C$ (IPCC 2022a). Fig. 14.15A shows the global-mean surface air temperature (GMST) response to GHG and aerosol forcing in a climate model.

The residence time of GHGs is long (100s years for CO_2 and 12 years for methane), during which atmospheric transport and mixing homogenize the concentrations. Aerosols, by contrast, have a residence time of a week. As a result, GHG concentrations are well mixed and nearly uniform in the atmospheric while aerosol concentrations are localized and high near the emission sources (see Fig. 14.15B). Specifically, anthropogenic aerosol loading is presently high in the Northern Hemisphere, over industrialized regions of North America and Europe, and rapidly industrializing regions of East and South Asia. Being localized, the aerosol forcing is very effective in driving changes in atmospheric circulation and rainfall, hence important for regional climate change.

In the Asian monsoon regions (India and China), the aerosol cooling induces downward motion and a reduction in summer monsoon rainfall (Wang et al. 2021). The dynamical component dominates the water vapor budget analysis of Eq. 14.1 because localized aerosol forcing drives strong anomalous atmospheric circulation (Li et al. 2015). This aerosol effect is not local, however, mediated by SST adjustments and ocean coupling, especially over South Asia (Bollasina et al. 2011).

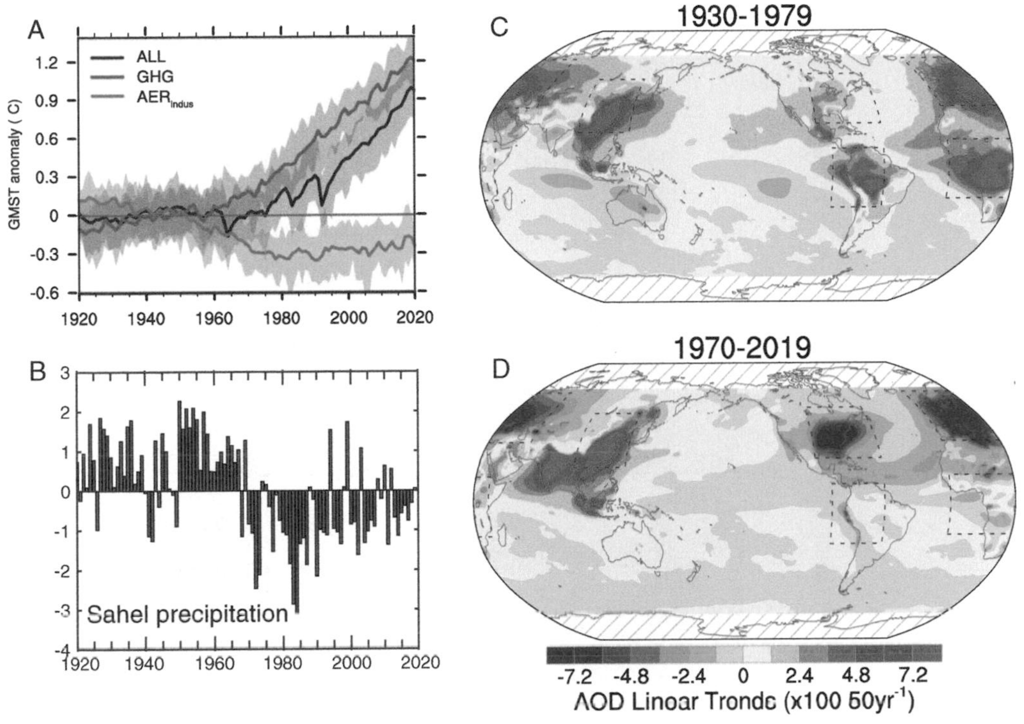

Fig. 14.15 (A) Annual global-mean surface air temperature *(GMST)* anomalies (°C) in response to all *(black curve)*, greenhouse gas (*GHG*; red), and industrial aerosol *(light blue)* forcing, from the Community Earth System Model v.1 (CESM1) large ensemble. Shading represents the ensemble spread. (B) Sahel (10–20°N, 20°W–10°E) rainfall anomalies. 50-year trends in industrial aerosol optical depth in CESM1 for (A) 1930-79 and (B) 1970-2019. *((A, C, D) From Deser et al. (2020b); B, Courtesy H. Wang. © American Meteorological Society. Used with permission.)*

14.5.1 Interhemispheric asymmetry

Energy theory introduced in Section 12.4.2 is useful in anticipating the cross-equatorial Hadley circulation response to negative TOA energy perturbations in the Northern Hemisphere. For simplicity, we first consider a 50-m deep, motionless slab ocean model (SOM) with $Q_{net} = 0$ in equilibrium. From Eq. 14.16, the atmospheric energy transport at the equator is

$$-2F_a|_{y=0} = \langle R_{TOA} \rangle_{SH}^{NH} \qquad (14.17)$$

The cross-equatorial energy transport balances the interhemispheric difference in TOA energy flux.

Consider purely reflective (nonabsorptive) aerosols in the Northern Hemisphere midlatitudes. To compensate for the reduced surface solar radiation in the Northern Hemisphere, Eq. 14.17 calls for a northward atmospheric energy transport across the equator

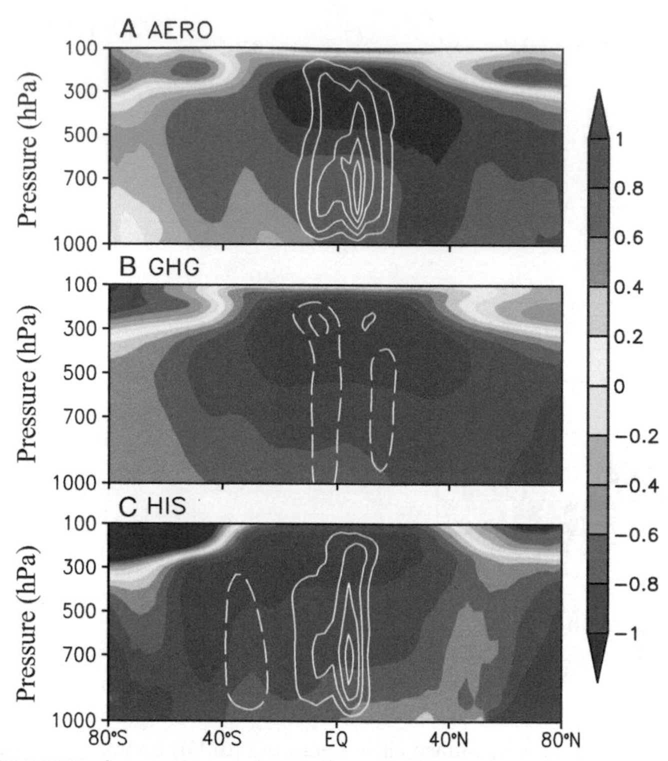

Fig. 14.16 The 2000-1950 changes in zonal-mean temperature and stream function (3×10^8 kg/s K^{-1}), normalized by tropical-mean tropospheric temperature change in CMIP5 historical simulation with (A) aerosol, (B) greenhouse gas (GHG), and (C) the full radiative forcing. *(Adapted based on Wang et al. (2016).)*

($F_a|_{y=0} > 0$). This is accomplished by a cross-equatorial Hadley cell with the rising branch and ITCZ displaced into the warmer Southern Hemisphere) (Fig. 14.16A).

The abovementioned energy theory is consistent with the SST view of coupled ocean-atmospheric dynamics in the first 12 chapters. In fact, the SST response to the aerosol forcing features a marked interhemispheric SST gradient, which drives the cross-equatorial Hadley circulation and ITCZ shift. The energy theory is convenient here as the change is driven by aerosol-induced energy perturbation. SST is still a key variable in coupled ocean-atmospheric adjustment.

If we neglect atmospheric absorption of solar radiation, reflective aerosols do not affect atmospheric temperature directly, but by cooling the surface first. The aerosol-induced cooling in the Northern Hemisphere midlatitudes (20−45°N) is not surface trapped but features a deep vertical structure (see Fig. 14.16A), suggestive of atmospheric eddy feedback. See Hwang et al. (2021) for a detailed discussion of SST and eddy adjustments, including changes in the subpolar westerly jet in the Southern Hemisphere where the local radiative forcing is small.

14.5.2 Ocean dynamic feedback

In the real ocean, the slow response of the deep ocean moderates the surface response to aerosol-induced radiative perturbations, compared to the SOM on which many early studies were based. Specifically, the ocean MOCs reduce the interhemispheric asymmetry in the atmospheric Hadley response. The aerosol cooling causes surface salinity to increase in the subpolar North Atlantic because of the reduced poleward transport of water vapor (opposite to the GHG-induced change). Both the temperature decrease and salinity increase intensify the deep water formation, and the intensified AMOC (Fig. 14.17B) increases the surface heat flux into the atmosphere. The surface flux in the subpolar North Atlantic acts against the interhemispheric asymmetry in TOA radiative perturbation due to Northern Hemisphere aerosols.

In the top 600-m ocean, a clockwise interhemispheric anomalous MOC develops (see Fig. 14.17A), transporting heat northward across the equator. The resultant upward

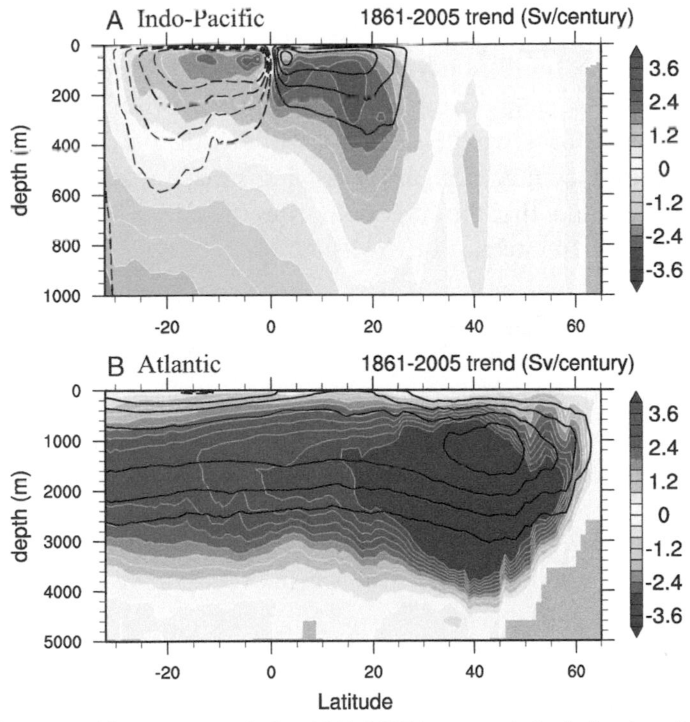

Fig. 14.17 CMIP5 ensemble mean trends for 1861-2005 in aerosol single forcing simulation. Zonal-integrated overturning stream function *(shading,* 10^6 m^3/s *per century)* in (A) the Indo-Pacific and (B) Atlantic basins (the present climatology in black contours). *(Courtesy J.R. Shi.)*

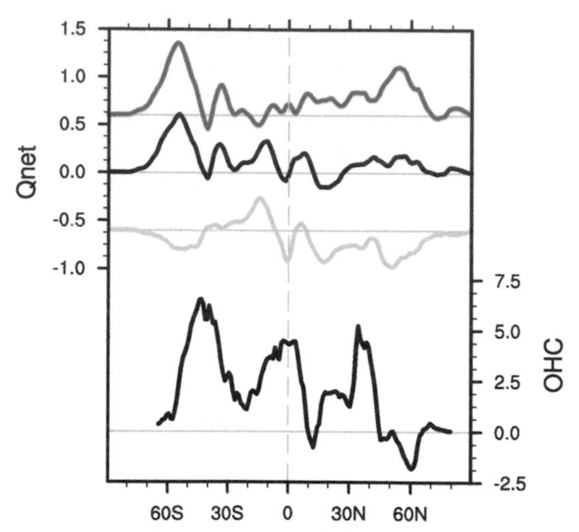

Fig. 14.18 Zonally integrated net heat flux trend at the sea surface (positive downward in 10^9 W/lat decade^{-1}) in the historical (black curve), greenhouse gas (red), and anthropogenic aerosol (cyan) experiments. Adapted based on Shi et al. (2018). *(Bottom)* Zonally integrated trend in ocean heat content (0-2000 m OHC, 10^7 W/m) from Argo for 2005-18. Updated from Roemmich et al. (2015). *(Courtesy J.R. Shi.)*

(downward) surface flux north (south) the equator (Fig. 14.18, blue curve) also acts to weaken the atmospheric energy transport. The anomalous MOC in the upper ocean is induced by both the surface wind change associated with the cross-equatorial Hadley cell and surface buoyancy flux change. By holding surface wind stress not to change, Luongo et al. (2022) show that the latter dominates the ocean heat transport.

Substituting Eq. 14.14 into Eq. 14.16 yields

$$-(F_a + F_o)|_{y=0} = \langle R_{TOA} - \frac{\partial H}{\partial t} \rangle_{SH}^{NH} \Big/ 2 \tag{14.18}$$

The northward heat transport by the anomalous shallow and deep MOCs in the ocean (see Fig. 14.17) reduces the need for the atmospheric energy transport northward across the equator. In other words, the ocean MOC change damps the ITCZ response to a hemispheric energy perturbation like aerosol cooling.

Generally, the surface heat flux represents the total ocean dynamic effect on the atmosphere, which is the sum of the heat storage and transport (Eq. 14.14),

$$-F_a|_{y=0} = \langle R_{TOA} - Q_{net} \rangle_{SH}^{NH} /2 \tag{14.19}$$

The ocean dynamic damping (the second term on RHS) reduces the cross-equatorial atmospheric energy transport, hence the ITCZ shift, by more than a factor of two (Fig. 14.19).

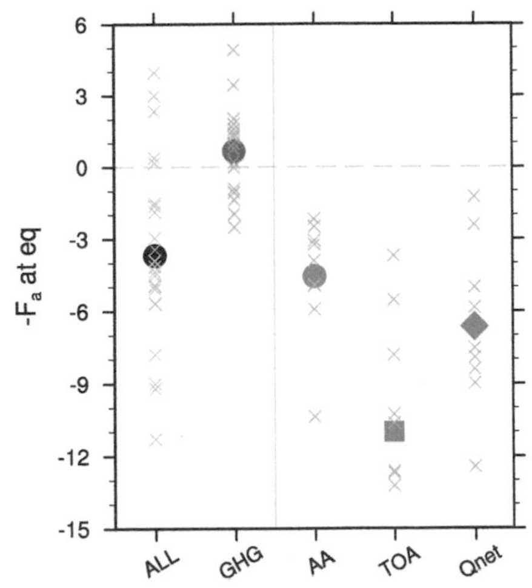

Fig. 14.19 Trend for 1861-2005 in cross-equatorial atmospheric energy transport (10^{12} W/decade, sign reversed) inferred as $-F_a|_{y=0} = \langle R_{TOA} - Q_{net}\rangle_{SH}^{NH} / 2$ in CMIP5 historical, greenhouse gas *(GHG)*—only, and anthropogenic aerosol *(AA)*—only runs. Negative values imply a northward atmospheric energy transport. For aerosol run *(blue symbols)*, the energy flux contributions at the top of atmosphere *(TOA)* and sea surface are also shown. CMIP5 multimodel ensemble mean in filled symbols. *(Courtesy J.R. Shi.)*

It is interesting to note that the ocean transport and mixing damp the global mean (Eq. 14.13) as well as interhemispheric (Eq. 14.18) responses to radiative energy perturbations. GMST and interhemispheric asymmetry—as measured by the cross-equatorial atmospheric energy transport—are lowest-order indices that track the state of global climate change.

14.5.3 Evolving distribution

The anthropogenic aerosol distribution displays complex spatiotemporal evolution that reflects society's struggle to balance economic development and public health. After the World War II up to the 1980s, the rapid economic growth in North America and Europe caused aerosols to increase rapidly, resulting in severe air pollution and acid rains. Society reacted by enacting clear air legislations to curb air pollution. The reduction of aerosol emissions from North America and Europe is offset by a rapid increase in emissions in Asia. As a result, the global/Northern Hemisphere aerosol loading begins to level off from the 1980s onward, and so does the global cooling effect.

Most studies so far focus on the first stage of the aerosol evolution (~the 1980s), which features a southward shift of the ITCZ (see Fig. 14.16A). The second stage, from the 1980s to present, is dominated by a geographic shift of aerosol loading from North America and Europe to Asia (see Fig. 14.15D), resulting in a North Atlantic warming and North Pacific cooling (Fig. 14.20). This may have contributed to the recent transition into positive AMO and negative PDO state (Watanabe and Tatebe, 2019), hence the global surface warming

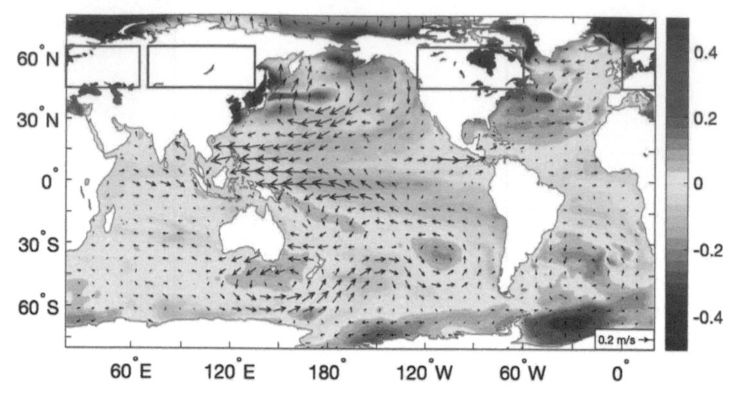

Fig. 14.20 Climate response to increased (decreased) insolation over North America and Europe (Asia), with zero zonal-mean radiative forcing: sea surface temperature (color shading, K) and wind velocity at 850 hPa (vectors, m/s) in a fully coupled general circulation model. *(From Kang et al. (2021).)*

hiatus (see Section 13.2). The zonal-shift mode of aerosol-induced climate change features an intensified Walker circulation across the equatorial Indo-Pacific oceans. In the tropical Atlantic, it drives multidecadal variations in cross-equatorial SST gradient and meridional displacements of the ITCZ, contributing to the multidecadal drought of the African Sahel during the 1950s to 1980s and the subsequent recovery (see Fig. 14.15B) (Hirasawa et al. 2020).

Dynamically, the Northern Hemisphere aerosol increase up to the 1980s excites a global meridional mode. Ocean dynamic damping weakens this zonal-mean energy mode by a factor of 2 to 3 (see Figs. 14.17 and 14.19), allowing weaker dependency on the aerosol distribution to manifest. The zonal-shift mode that dominates recent decades (the 1980s—), by contrast, is associated with little zonal-mean energy perturbation but involves instead basin-scale ocean-atmosphere interactions. Specifically, the basin-scale subtropical meridional mode of WES feedback, characterized by the equator-westward extension, is an important conduit for midlatitude aerosol forcing to influence the deep tropics, including the intensified Walker circulation (see Fig. 14.20).

Anthropogenic aerosols have long been identified as the largest source of uncertainty in radiative forcing. The uncertainty in aerosol-induced regional climate change may be decomposed into the magnitude and pattern uncertainties. While the magnitude uncertainty remains, the results here suggest that the spatial pattern of aerosol-induced climate change is quite robust, including the cross-equatorial Hadley cell and the acceleration of the AMOC.

14.6 Historical climate change

During the historical period of instrumental observations, anthropogenic climate change is largely the sum of the response to GHG and aerosol forcings. For 1750-2019, the global anthropogenic radiative forcing due to the anthropogenic increase in CO_2 is estimated at 2.2 Wm^{-2} for CO_2, -1.3 Wm^{-2} for aerosols, and 2.7 Wm^{-2} for other well-mixed greenhouse gases (e.g., methane) (IPCC 2022a).

Climate response to GHG forcing is symmetric across the equator compared to that of aerosols (see Fig. 14.16). For 1950-2000, GHGs cause a large tropical tropospheric warming of 1.03 K but only a weak cross-equatorial Hadley cell. Aerosols cause the tropospheric temperature to fall by -0.59 K and drive a strong clockwise Hadley cell across the equator (Table 14.1). As a result, the GHG forcing dominates the temperature response, but the aerosol forcing dominates the cross-equatorial Hadley adjustment in the full-forcing historical simulation.

Although the aerosol-induced reduction in surface solar radiation takes place mostly in the Northern Hemisphere, the zonal mean cooling extends as far southward as 60°S at the surface (see Fig. 14.16A). In the tropics, the tropospheric cooling shows an upward intensification that is nearly symmetric about the equator due to the moist adiabatic adjustment in the vertical and weak temperature gradient in the horizontal, respectively. This tropical tropospheric temperature structure resembles the response to GHG forcing.

The pattern similarity in GHG and aerosol-induced climate change extends to surface climate in the tropics, with a spatial correlation of -0.87 for SST and -0.76 for precipitation between the CMIP5 single-forcing simulations (Xie et al. 2013). The spatial similarity and opposite sign of the GHG and aerosol effects imply weak net precipitation change in the tropics during the period of 1950-2000 when aerosols increase in the Northern Hemisphere, given that the aerosol forcing is more effective in driving circulation and hence rainfall change.

Up to the early 2000s, the Southern Ocean dominates the ocean uptake of anthropogenic heat. Over the subpolar North Atlantic, the heat flux due to GHG and aerosol forcings happens to be of about equal magnitude so far, and the net change is close to zero (see Fig. 14.18). Over the Southern Ocean, by contrast, the aerosol effect is small with a large heat uptake in response to greenhouse warming. Argo observations capture the large ocean heat content increase (ΔH) in the Southern Ocean, with a peak at 45°S that is displaced equatorward of the surface uptake peak at 55°S much as simulated in models (see Fig. 14.18).

TABLE 14.1 Trends during 1950–2000 in tropical tropospheric (25°N ~ 25°S; 700 ~ 300 hPa) temperature (T, K) and interhemispheric Hadley circulation (ψ_{500}, stream function at 500 hPa, equator; 10^9 kg/s) in the historical (HIST) and greenhouse gas (GHG) and anthropogenic aerosol (AA) single-forcing runs. CMIP5 multimodel ensemble averages.

	HIST	GHG	AA
T	0.56	1.03	-0.59
ψ_{500}	2.16	-1.16	3.61

The dominance of the Southern Ocean in global anthropogenic heat uptake during the historical period is not representative of the greenhouse warming but results from the fortuitous proportion and evolution of the GHG and aerosol radiative forcing. This recognition is very important; if the dominance of the Southern Ocean heat uptake were due to the GHG forcing (which is nearly symmetric about the equator), the energy transport theory (Eq. 14.19) would predict a northward displaced ITCZ during the 20th century. In reality the aerosol cooling displaces the ITCZ southward, while the ocean uptake resists the southward displacement of the ITCZ.

14.7 Synthesis

Suki Manabe and his colleagues at the Geophysical Fluid Dynamics Laboratory carried out the first transient greenhouse warming experiment using a realistic coupled GCM in the late 1980s. Several key predictions from the early GCM simulation have materialized (Stouffer and Manabe 2017), including the larger warming on continents than ocean and enhanced warming in the Arctic. September Arctic sea ice extent reached a record minimum in 2007 and set a new record again in 2012. In contrast to the strong warming in the Arctic, SST warming in the Southern Ocean is damped by the upwelling of deep water just as predicted. Remarkably, Argo data detected the large heat uptake in the Southern Ocean associated with the muted surface warming. The slowdown of the AMOC has not materialized, however, not because their model was wrong, but because it did not include the anthropogenic aerosol effect.

Public health concerns over air quality prompted the reduction of anthropogenic aerosol emissions in North America and Europe. The same trend has started in Asia and elsewhere. The worldwide reduction in air pollution, together with the continued increase in GHGs, seems to be finally setting the AMOC on a declining trajectory. This will increase the contribution of the subpolar North Atlantic to the global ocean heat uptake (Fig. 14.21).

Because of the complex evolution of aerosol forcing in the spatial distribution and magnitude, unfolding regional changes will not always resemble those observed during the historical period—examples include the cross-equatorial Hadley circulation, Sahel rainfall (see Fig. 14.15B), AMOC, and ocean heat uptake. We can still learn from the past, not by simple extrapolation but at conceptual and process levels (Fig. 14.22). The ocean sets the pace of global warming and helps shape regional patterns of climate change through coupled feedbacks that give rise to ENSO and other recurrent patterns of spontaneous variability. As greenhouse warming grows in time, major climate events are being closely watched and analyzed. Expanding observations, careful analysis, and probing model experiments will continue to advance our understanding of the climate system and enable predictions at longer leads and higher accuracy.

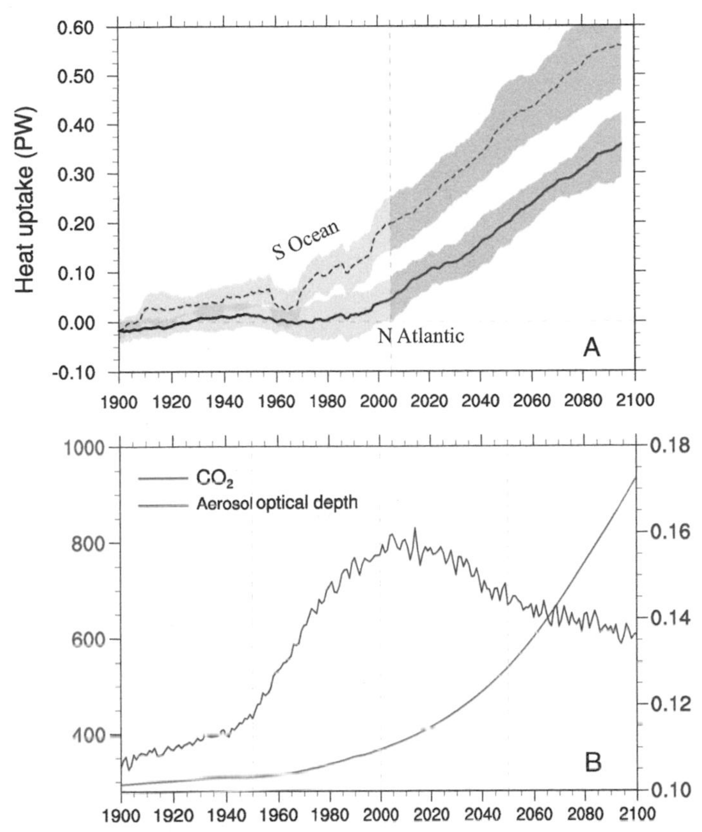

Fig. 14.21 (A) Area-integrated surface heat flux over the subpolar North Atlantic and Southern Ocean *(dashed)* in the historical *(black)* and RCP8.5 *(pink)* simulation. Shading denotes one standard deviation of CMIP5 ensemble spread. (B) Corresponding CO_2 concentration (red curve; ppm) and global mean aerosol optical depth at 550 nm *(blue)*. *(From Shi et al. (2018).)*

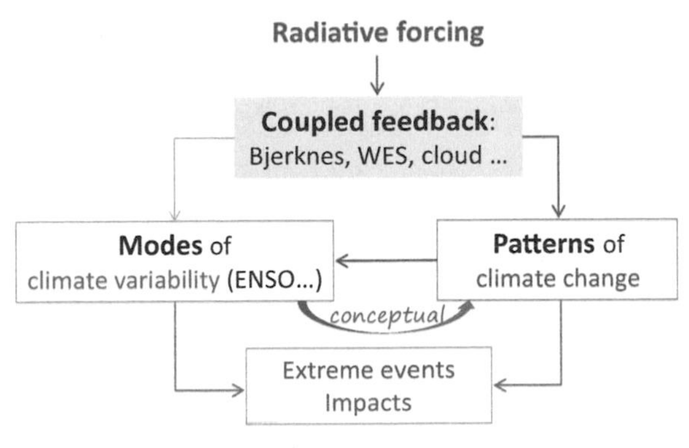

Fig. 14.22 Schematic of physical and conceptual connections between climate variability and change. They share common pattern dynamics due to coupled ocean-atmospheric feedbacks (highlighted in yellow). These feedbacks manifest in unforced climate variability, offering insights into dynamic processes important for regional climate change. Climate variability drives, and climate change exacerbates, extreme events of large societal impact.

Review questions

1. Is the global mean a good approximation of local precipitation change?
2. Ocean surface warming is to first order spatially uniform, but why do we care about the "small" spatial variations in SST change? In comparison, the land-sea difference in warming magnitude is much larger.
3. How does regional precipitation change under uniform ocean warming differ from that in fully coupled ocean-atmosphere models? What does this imply in terms of ocean's role in shaping regional climate change?
4. Identify SST patterns for the following atmospheric changes projected in climate models: (a) precipitation increase over the equatorial Pacific; and (b) intensified southeast trades and weakened northeast trades.
5. Why is the tropical mean SST an important reference for deep convection and tropical cyclogenesis?
6. Why is the convective threshold not static? How will it change with global warming?
7. How does studying climate variability internal to the coupled ocean-atmosphere system help understand and project regional climate change in the face of anthropogenic greenhouse warming? Give two examples.
8. What causes intermodel uncertainties in tropical rainfall? Is the SST warming pattern important?
9. To study El Niño's teleconnections, say on California rainfall, AGCM experiments prove useful by prescribing observed SST anomalies over the tropical Pacific, while setting SST at the climatology elsewhere (e.g., see Chapter 9). To study the effect of the projected tropical Pacific warming on the atmospheric circulation and rainfall, is it a good experimental design to simply prescribe the SST change over the tropical Pacific (and set it to zero elsewhere)? Why?
10. How would you modify the way the SST change is prescribed, to study the tropical Pacific warming pattern effect? Explain briefly.
11. Revisit the two-layer model of Fig. 13.3, where the heat exchange between the upper and deep ocean is represented as a constant mixing coefficient. Is this a good representation for the vast tropical/subtropical ocean? How would you revise this simplified view, and where does most of the exchange take place? Consider the greenhouse warming for simplicity (see Fig. 14.12).
12. What are the major differences in climate response to GHG and reflective aerosols?
13. In response to increased emissions of reflective aerosols, say over the mideastern United States, what local changes are likely to occur? What happens to the ITCZ? Use the energy theory to explain.
14. Explain how the ocean circulation change mediate the above ITCZ response to aerosols.

15. The ocean uptake/storage of anthropogenic heat is limited to the Southern Hemisphere extratropics in Argo data during 2006-2012. Is this heat uptake pattern characteristic of greenhouse warming?
16. Related, why has not the AMOC slowed down as predicted by models under GHG forcing? Use the energy theory to discuss the implications for ITCZ changes of correctly identifying the radiative forcing that causes the interhemispheric pattern of ocean heat uptake.
17. GMST and interhemispheric SST difference are two most important metrics of climate change. The full-depth dynamic ocean acts to damp both metrics in response to radiative forcing (say, Northern Hemisphere—confined aerosol forcing). Use energy theory to explain.

Epilogue

The ocean surface may seem flat and featureless, but sea surface temperature (SST) variations become apparent from a spaceborne infrared radiometer (see Fig 3.3), explaining why Panama is covered by dense forests while coastal Peru, just on the other side of the equator, is barren desert. This book showcases rich dynamics of our climate system arising from ocean-atmosphere coupling.

Spontaneous oscillations

The equatorial Pacific is vast and remote. Few of us have been to this part of the world, but most have felt its pulses through El Niño-Southern Oscillation (ENSO). The equator is where the Coriolis parameter vanishes, allowing the cold water to upwell and Kelvin waves to propagate. A ribbon of low SSTs gives the equator away and wipes out atmospheric convection across the eastern two-thirds of the equatorial Pacific (see Fig. 9.4A), depriving the Galapagos of "the beauty which generally accompanies such a position" (Keynes 2021). From the International Dateline to the Greenwich longitude, the intertropical convergence zone (ITCZ) is displaced north of the equator.

The circular argument led Bjerknes to suggest that ENSO is a coupled ocean-atmospheric phenomenon. The coupled feedback of his namesake—involving the east-west interaction of ocean upwelling and thermocline depth with atmospheric deep convection and winds—gives rise to ENSO as well as the Atlantic Niño and Indian Ocean dipole (IOD) (see Figs. 9.5, 10.6, and 11.7). ENSO is by far the most pronounced, due to the large zonal width and hence long wave transit time that keeps the Pacific off the balance and sets it to oscillate. The Bjerknes feedback and the delayed action of unbalanced ocean thermocline depth anomalies allow ENSO prediction at lead times of a few seasons.

Surface heat flux becomes important for SST variability off equatorial upwelling zones, over the broad tropics and subtropics. While the Bjerknes mode centers on the equator, wind-evaporation-SST (WES) feedback selectively amplifies antisymmetric perturbations across the equator. Such cross-equatorial WES feedback has been identified in meridional modes of interannual variability over all three tropical oceans (see Figs. 9.18C, 10.8, and 11.11), most pronounced during February-March-April when the mean climate is warm and nearly symmetric about the equator and moderate SST anomalies can cause large convective response in the atmosphere.

Extratropical atmospheric variability modulates the subtropical trade winds in the North Atlantic and North Pacific, with the Pacific meridional mode (PMM) as a conduit to affect ENSO. The subtropical WES feedback gives rise to the west-equatorward slanted pattern (see Fig. 12.13), energized by the low cloud-SST feedback under the Northeast Pacific stratus deck.

While SST modes are often discussed within ocean basins, the atmosphere is connected globally. IOD and ENSO are statistically correlated and physically connected through the Walker circulation. ENSO affects the tropical North Atlantic through the equatorial Kelvin wave as well as the Pacific—North American (PNA) pattern (see Fig. 10.18). Conversely, equatorial waves allow Atlantic SST variability to modulate ENSO. Such cross-basin interactions and coupling contribute to rich tropical variability, a topic of strong current interest.

Monsoons result from ocean-land-atmosphere interactions, but large-scale modes of interannual variability in the Asian summer monsoon are shaped mostly by coupled ocean-atmosphere dynamics. ENSO is the dominant driver, not only directly but also through the delayed effect of the Indo-western Pacific Ocean capacitor (IPOC). Encompassing the three submonsoon systems, The recurrent anomalous anticyclonic circulation is anchored and energized by the barotropic energy conversation in the mean confluence of the monsoon westerlies and easterly trades (see Box Fig. 11.2). Thus the IPOC is shaped by the summer monsoon and shapes the monsoon variability at the same time, although the role of land surface in large-scale variability remains fully explored.

In an audacious attempt to predict Indian summer rainfall, Walker discovered the Southern Oscillation, which by itself has little predictive value. Summer India rainfall variability is correlated with the antecedent May Southern Oscillation Index only marginally at $r=0.24$ for 1875-1930. The physical insights gained into coupled dynamics since Bjerknes (1969) have eventually realized Walker's dream of monsoon prediction. First, antecedent ENSO state is a skillful predictor for the timing of the Indian monsoon onset (the IPOC effect). Second, dynamic models show skills in predicting developing ENSO events by tapping into the ocean initial conditions. The predicted summer ENSO state then can be used to forecast monsoon rainfall (see Fig. 11.20). How far have we come!

Globally, ENSO is by far the most important predictor for climate variations at leads from a month to a season, although the teleconnection mechanism varies from baroclinic atmospheric waves in the tropics, the barotropic PNA pattern, to the coupled IPOC. Regional predictability can be further gained from other tropical modes as during the 12-month period from September 2019 to August 2020 over the Indo-western Pacific region (see Fig. 11.23).

Energy view

Unlike ENSO, which arises due to internal positive feedback, some climate anomalies are induced by radiative perturbations. Climate feedback analysis based on global energy conservation is useful in quantifying global mean surface temperature response to radiative perturbations. The radiative forcing due to anthropogenic increase of greenhouse gases is nearly spatially uniform, but the ocean heat uptake is confined in the subpolar North Atlantic and Southern Ocean upwelling zone. This, along with the subtropical low clouds, introduces the so-called spatial-pattern effect that complicates the global climate feedback analysis.

Cross-equatorial energy transport is a useful predictor of meridional asymmetry in climate response. In the deep tropics where the gross moist stability is small due to deep convection, the cross-equatorial Hadley circulation and ITCZ are highly sensitive to the interhemispheric difference in energy flux into the atmospheric column. The energy transport constraint offers a useful framework to study the meridional interactions across different dynamic regimes: the ITCZ in the deep tropics, WES and low-cloud feedback in the subtropics, midlatitude storm tracks, the Atlantic Meridional Overturning Circulation (AMOC) heat uptake in the subpolar North Atlantic, and polar sea ice. The atmospheric general circulation model (GCM) coupled with a slab ocean model shows some promise to diagnose climate response to sea surface heat flux distribution (Hu et al. 2022). This is akin to atmospheric GCMs as a tool to diagnose the atmospheric response to SST anomalies.

Outlook

Climate prediction faces different challenges than weather forecast, which benefits from daily validation where observations confront prediction. SST reconstructions based on instrumental observations extend back 150 years. The spatiotemporal sampling and coverage improved around 1950 and then again in the satellite era (1980∼). The SST reconstructions include ∼40 ENSO cycles, on par with the number of synoptic weather cycles one encounters in a single year at a station in the midlatitude storm track. Weather forecast based on the observations of 40 storms would be of some skill, but they would be far from what we enjoy today given blockings and other modes of low-frequency variability (MJO, ENSO, NAO, etc.). This illustrates how limited our observations of climate variability are, and how much more there is to learn about climate dynamics.

Earth climate experienced climate variability of much larger amplitudes on geologic timescales, including the recurrent glacial-interglacial cycles and precessional modulations of monsoons. For instance, global climate anomalies during Heinrich events (see Fig. 12.19) seem consistent with the cross-equatorial energy transport theory in response to an AMOC shutdown induced by massive freshwater discharges into the North Atlantic. Rapid expansion in the spatiotemporal coverage of paleoclimate proxy reconstructions has much to offer in terms of further insights into the climate system (Valdes et al. 2021), in regimes beyond observed during the instrumental era. Instrumental data, climate models, and paleoclimate proxies, each of their own limitations, together offer complementary and fuller views of the climate system. The resultant insights will help us better anticipate and prepare for unfolding climate change.

The societal need for monsoon forecast and curiosity at the equatorial Pacific oscillation between El Niño and La Niña led us to see the atmosphere and ocean as a coupled interactive system. The coupled view has since expanded to include other components of the Earth system: hydrology, glaciology, forestry, and biogeochemical cycle. Instrumental observations show that ocean-atmosphere coupling on which this book focuses is the core mechanism for climate variations on timescales from a season to a century, although the carbon cycle and glacier dynamics become important on longer timescales. As global warming has grown to be the most pressing problem of our time, climate dynamics as a field of study is evolving to embrace increasing societal need for actionable science and sustainable management of water, energy, forestry, and fisheries.

Now I see

Writing this book as a sequel to Gill (1982) has been in my plan for a long time. The book project kicked off in the summer of 2018 amid a record marine heatwave off La Jolla coast. Just as I was warming up for the spring quarter, California Governor Gavin "Newsom" ordered a statewide lockdown on March 19, 2020 in response to the growing COVID-19 pandemic. Like many teachers, I had to improvise teaching on Zoom, with students logging in from three continents.

"I once ... was blind, but now I see." On the empty steps of the grand Milan cathedral, Italian singer Andrew Bocelli sang to a world in despair. The song and the scene gave us courage to persevere. Spring came and went. During the ensuring 2 years, the manuscript underwent several revisions.

Tranquility afforded by working from home let me see new connections, at night in the home office and under dazzling sunshine on hiking trails: low-cloud feedback is nonlocal and joins forces with WES feedback (see Fig. 6.12) to make the PMM a key conduit for extratropical-to-tropical teleconnections (see Figs. 12.13 and 14.20). Zoom discussions across the Pacific led us to conclude "historic Yangtze flooding of 2020 tied to extreme Indian Ocean conditions" (Zhou et al. 2021) (see Fig. 11.23).

Highly effective vaccines developed at record speed gave hopes at the darkest moments of the pandemic. As the book project comes to an end, I'm warming up for spring quarter classes, in person with masks.

Shang-Ping Xie
19 March 2022, San Diego

References

Adames ÁF, Maloney ED. Moisture mode theory's contribution to advances in our understanding of the Madden-Julian oscillation and other tropical disturbances. *Curr Clim Change Rep*. 2021;7:72–85. https://doi.org/10.1007/s40641-021-00172-4.

Adames ÁF, Wallace JM. Three-dimensional structure and evolution of the MJO and its relation to the mean flow. *J Atmos Sci*. 2014;71:2007–2026.

Adames ÁF, Wallace JM, Monteiro JM. Seasonality of the structure and propagation characteristics of the MJO. *J Atmos Sci*. 2016;73:3511–3526.

Aiyyer AR, Thorncroft C. Climatology of vertical wind shear over the tropical Atlantic. *J Climate*. 2006;19: 2969–2983. https://doi.org/10.1175/JCLI3685.1.

Albrecht BA, Bretherton CS, Johnson D, et al. The Atlantic stratocumulus transition experiment—ASTEX. *Bull Amer Meteor Soc*. 1995b;76:889–904.

Alexander MA, Deser C, Timlin MS. The reemergence of SST anomalies in the North Pacific Ocean. *J Climate*. 1999;12:2419–2433.

Alford MH, Peacock T, MacKinnon JA, et al. The formation and fate of internal waves in the South China Sea. *Nature*. 2015;521(7550):65–69.

Amaya DJ. The Pacific Meridional Mode and ENSO: a review. *Curr Clim Change Reps*. 2019;5(4):296–307.

Amaya DJ, Bond NE, Miller AJ, DeFlorio MJ. The evolution and known atmospheric forcing mechanisms behind the 2013-2015 North Pacific warm anomalies. In: *A Tale of Two Blobs. Variations*. 14. 2016:1–6 [US CLIVAR].

Amaya DJ, DeFlorio MJ, Miller AJ, Xie S-P. WES feedback and the Atlantic Meridional Mode: observations and CMIP5 comparisons. *Clim Dyn*. 2017;49:1665–1679.

Amaya DJ, Kosaka Y, Zhou W, et al. The North Pacific pacemaker effect on historical ENSO and its mechanisms. *J Climate*. 2019:7643–7661.

Andrews T, Gregory JM, Paynter D, et al. Accounting for changing temperature patterns increases historical estimates of climate sensitivity. *Geophys Res Lett*. 2018;45:8490–8499.

Arias PA, Bellouin N, Coppola E, et al. Technical summary. In: Masson-Delmotte V, Zhai P, Pirani A, et al., eds. *Climate Change 2021: The Physical Science Basis. Contribution of Working Group I to the Sixth Assessment Report of the Intergovernmental Panel on Climate Change*. Cambridge University Press; 2022:33–144. https://doi.org/10.1017/9781009157896.002.

Armour KC, Bitz CM, Roe GR. Time-varying climate sensitivity from regional feedbacks. *J Climate*. 2013; 26:4518–4534.

Armour KC, Marshall J, Scott JR, et al. Southern Ocean warming delayed by circumpolar upwelling and equatorward transport. *Nature Geosci*. 2016;9:549–554.

Back LE, Bretherton CS. On the relationship between between SST gradients, boundary layer winds and convergence over the tropical oceans. *J Climate*. 2009;22:4182–4196. https://doi.org/10.1175/2009JCLI2392.1.

Batchelor GK, Hide R. *Adrian Edmund Gill. 22 February 1937-19 April 1986*. 34. Biog Mems Fellows Royal Society; 1988:221–258. https://doi.org/10.1098/rsbm.1988.0009.

Battisti DS, Hirst AC. Internal variability in a tropical atmosphere-ocean model: influence of basic state, ocean geometry and nonlinearity. *J Atmos Sci*. 1989;46:1687–1712.

Bauer P, Thorpe A, Brunet G. The quiet revolution of numerical weather prediction. *Nature*. 2015;525: 47–55. https://doi.org/10.1038/nature14956.

Behera SK, ed. *Tropical and Extratropical Air-Sea Interactions: Modes of Climate Variations*. Elsevier; 2021:300.

Bender MA, Ginis I, Kurihara Y. Numerical simulations of tropical cyclone-ocean interaction with a high-resolution coupled model. *J Geophys Res*. 1993;98(D12):23245–23263.

Berg A, Lintner BR, Findell KL, et al. Impact of soil moisture—atmosphere interactions on surface temperature distribution. *J Climate*. 2014;27:7976–7993.

Bollasina MA, Ming Y, Ramaswamy V. Anthropogenic aerosols and the weakening of the South Asian summer monsoon. *Science*. 2011;334(6055):502–505.

Boos W, Kuang Z. Dominant control of the South Asian monsoon by orographic insulation versus plateau heating. *Nature*. 2010;463:218–223. https://doi.org/10.1038/nature08707.

Bretherton CS. Insights into low-latitude cloud feedbacks from high-resolution models. *Phil Trans Royal Society A*. 2015;373:20140415.

Bretherton CS, Widmann M, Dymnikov VP, et al. The effective number of spatial degrees of freedom of a time-varying field. *J Climate*. 1999;12(7):1990–2009.

Busalacchi AJ, Takeuchi K, O'Brien JJ. Interannual variability of the equatorial Pacific—Revisited. *J Geophys Res*. 1983;88(C12):7551–7562. https://doi.org/10.1029/JC088iC12p07551.

Cai W, Santoso A, Collins M, et al. Changing El Niño–Southern Oscillation in a warming climate. *Nat Rev Earth Environ*. 2021;2:628–644. https://doi.org/10.1038/s43017-021-00199-z.

Cane MA, Sarachik ES. Forced baroclinic ocean motions. II-the linear equatorial bounded case. *J Marine Res*. 1977;35:395–432.

Cane MA, Sarachik ES. The response of a linear baroclinic equatorial ocean to periodic forcing. *J Mar Res*. 1981;39:651–693.

Cane MA, Zebiak SE, Dolan SC. 1986: Experimental forecasts of El Niño. *Nature*. 1986;321:827–832.

Chadwick R, Boutle I, Martin G. Spatial patterns of precipitation change in CMIP5: why the rich do not get richer in the tropics. *J Climate*. 2013;26:3803–3822.

Chang CP, Wang Z, Hendon H. The Asian winter monsoon. In: Wang B, ed. *The Asian Monsoon*. Springer; 2006:89–128.

Chang P, Ji L, Saravanan R. A hybrid coupled model study of tropical Atlantic variability. *J Climate*. 2001;14: 361–390.

Chapman W, Subramanian AC, Xie S-P, et al. Intraseasonal modulation of ENSO teleconnections: implications for predictability in North America. *J Climate*. 2021;34:5899–5921. https://doi.org/10.1175/JCLI-D-20-0391.1.

Chelton DB, Xie S-P. Coupled ocean-atmosphere interaction at oceanic mesoscales. *Oceanography*. 2010;23: 52–69.

Cheng L, Abraham J, Zhu J, et al. Record-setting ocean warmth continued in 2019. *Adv Atmos Sci*. 2020;37: 137–142. https://doi.org/10.1007/s00376-020-9283-7.

Chiang JCH, Vimont DJ. Analogous Pacific and Atlantic meridional modes of tropical atmosphere-ocean variability. *J Climate*. 2004;17:4143–4158.

Chowdary JS, Hu K, Srinivas G, et al. The Eurasian jet streams as conduits for east Asian monsoon variability. *Curr Climate Change Rep*. 2019;5(3):233–244.

Chowdary JS, Xie S-P, Lee J-Y, et al. Predictability of summer Northwest Pacific climate in eleven coupled model hindcasts: local and remote forcing. *J Geophys Res Atmos*. 2010;115, D22121. https://doi.org/10.1029/2010JD014595.

Christensen JH, Krishna Kumar K, Aldrian E, et al. Climate phenomena and their relevance for future regional climate change. In: Stocker TF, Qin D, Plattner G-K, et al., eds. *Climate Change 2013: The Physical Science Basis. Contribution of Working Group I to the Fifth Assessment Report of the Intergovernmental Panel on Climate Change*. New York: Cambridge University Press; 2014:1217–1308. https://doi.org/10.1017/CBO9781107415324.028.

Church JA, Clark PU, Cazenave A, et al. Sea level change. In: Stocker TF, et al., eds. *Climate Change 2013: The Physical Science Basis. Contribution of Working Group I to the Fifth Assessment Report of the Intergovernmental Panel on Climate Change*. Cambridge University Press; 2014:1137–1216.

Clarke A. *An Introduction to the Dynamics of El Nino and the Southern Oscillation*. Academic Press; 2008.

Clarke AJ. The reflection of equatorial waves from oceanic boundaries. *J Phys Oceanogr*. 1983;13: 1193–1207.

Clement AC, Seager R, Cane MA, Zebiak SE. An ocean dynamical thermostat. *J Climate*. 1996;9(9): 2190–2196.

Cobb KM, Westphal N, Sayani H, et al. Highly variable El Nino-Southern Oscillation throughout the Holocene. *Science*. 2013;339:67–70.

Cordeira JM, Stock J, Dettinger MD, et al. A 142-year climatology of Northern California landslides and atmospheric rivers. *Bull Am Meteor Soc*. 2019;100:1499–1509. https://doi.org/10.1175/BAMS-D-18-0158.1.

Cromwell T, Montgomery RB, Stroup ED. Equatorial undercurrent in the Pacific Ocean revealed by new methods. *Science*. 1954;119(3097):648–649.

Dai A. Drought under global warming: a review. *Wiley Interdisciplinary Reviews: Climate Change*. 2011;2: 45–65. https://doi.org/10.1002/wcc.81.

Davis RE. Predictability of sea-surface temperature and sea-level pressure anomalies over North Pacific Ocean. *J Phys Oceanogr*. 1976;6:249–266.

Delworth TL, Zeng F, Rosati A, et al. A link between the hiatus in global warming and North American drought. *J Clim*. 2015;28:3834–3845.

Delworth TL, Zeng F, Zhang L, et al. The central role of ocean dynamics in connecting the North Atlantic Oscillation to the extratropical component of the Atlantic Multidecadal Oscillation. *J Climate*. 2017;30: 3789–3805.

deMenocal PB, Tierney JE. Green Sahara: African humid periods paced by Earth's orbital changes. *Nat Ed Knowl*. 2012;3(10):12.

Deser C, Alexander MA, Xie S-P, Phillips AS. Sea surface temperature variability: patterns and mechanisms. *Ann Rev Marine Sci*. 2010;2:115–143. https://doi.org/10.1146/annurev-marine-120408-151453.

Deser C, Lehner F, Rodgers KB, et al. Insights from earth system model initial-condition large ensembles and future prospects. *Nat Clim Change*. 2020a;10:277–286.

Deser C, Phillips AS, Simpson IR, et al. Isolating the evolving contributions of anthropogenic aerosols and greenhouse gases: a new CESM1 large ensemble community resource. *J Climate*. 2020b;33:7835–7858.

Ding H, Keenlyside NS, Latif M. Impact of the equatorial Atlantic on the El Niño southern oscillation. *Climate Dyn*. 2012;38(9-10):1965–1972.

Ding Y, Chan J. The East Asian summer monsoon: an overview. *Meteorol Atmos Phys*. 2005;89:117–142. https://doi.org/10.1007/s00703-005-0125-z.

Du Y, Xie S-P, Huang G, Hu K. Role of air–sea interaction in the long persistence of El Niño–induced North Indian Ocean warming. *J Climate*. 2009;22:2023–2038.

Emanuel KA. An air-sea interaction model of intraseasonal oscillations in the tropics. *J Atmos Sci*. 1987;44: 2324–2340.

Emanuel KA. The maximum intensity of hurricanes. *J Atmos Sci*. 1988;45:1143–1155.

Emanuel KA. Tropical cyclones. *Ann Rev Earth Planet Sci*. 2003;31(1):75–104.

Emanuel KA, Nolan DS. Tropical cyclones and the global climate system. *Preprints*. 2004.

Eyring V, Gillett NP, Achuta Rao KM, et al. Human influence on the climate system. In: Masson-Delmotte V, Zhai P, Pirani SL, et al., eds. *Climate Change 2021: The Physical Science Basis. Contribution of Working Group I to the Sixth Assessment Report of the Intergovernmental Panel on Climate Change*. Cambridge University Press; 2022:423–552. https://doi.org/10.1017/9781009157896.005.

Ferreira D, Cessi P, Coxall HK, et al. Atlantic-Pacific asymmetry in deep water formation. *Ann Rev Earth Planet Sci*. 2018;46:327–352.

Flohn H. Large-scale aspects of the "summer monsoon" in South and East Asia. *J Meteor Soc Japan*. 1957;35A: 180–186.

Frankignoul C, Hasselmann K. Stochastic climate models, part II application to sea-surface temperature anomalies and thermocline variability. *Tellus*. 1977;29:289–305. https://doi.org/10.1111/j.2153-3490.1977.tb00740.x.

Frierson DMW, Hwang Y-T, Fuckar NS, et al. Contribution of ocean overturning circulation to tropical rainfall peak in the Northern Hemisphere. *Nat Geosci*. 2013;6:940–944.

García-Serrano J, Cassou C, Douville H, et al. Revisiting the ENSO teleconnection to the tropical North Atlantic. *J Climate*. 2017;30:6945–6957.

Geng YF, Xie S-P, Zheng XT, et al. CMIP6 intermodel uncertainty in interhemispheric asymmetry of tropical climate response to greenhouse warming: extratropical ocean effects. *J Climate*. 2022;35:4869–4882. https://doi.org/10.1175/JCLI-D-21-0541.1.

Gershunov A, Guzman Morales J, Hatchett B, et al. Hot and cold flavors of southern California's Santa Ana winds: their causes, trends, and links with wildfire. *Climate Dyn*. 2021;57:2233–2248. https://doi.org/10.1007/s00382-021-05802-z.

Gill AE. *Atmosphere-Ocean Dynamics*. Academic Press; 1982.

Gill AE. Some simple solutions for heat-induced tropical circulation. *QJR Meteorol Soc*. 1980;106:447–462. https://doi.org/10.1002/qj.49710644905.

Godfrey JS. On ocean spindown I: a linear experiment. *J Phys Oceanogr.* 1975;5:399–409.

Goldenberg SB, Landsea CW, Mestas-Nunez AM, Gray WM. The recent increase in Atlantic hurricane activity: causes and implications. *Science.* 2001;293:474–479.

Gregory JM, Ingram, WJ, Palmer MA, Jones GS, Stott, PA, Thorpe RB, Lowe JA, Johns TC, and Williams KD. A new method for diagnosing radiative forcing and climate sensitivity. *Geophys Res Lett.* 2004;31, L03205. https://doi.org/10.1029/2003GL018747.

Gulev SK, Thorne PW, Ahn J, et al. Changing state of the climate system. In: Masson-Delmotte V, Zhai P, Pirani A, et al., eds. *Climate Change 2021: The Physical Science Basis. Contribution of Working Group I to the Sixth Assessment Report of the Intergovernmental Panel on Climate Change.* Cambridge University Press; 2022: 287–422. https://doi.org/10.1017/9781009157896.004.

Hall A, Cox P, Huntingford C, et al. Progressing emergent constraints on future climate change. *Nat Clim Chang.* 2019;9:269–278. https://doi.org/10.1038/s41558-019-0436-6.

Halley E. A historical account of the trade winds, and monsoons, observable in the seas between and near the Tropicks, with an attempt to assign the phisical cause of the said winds. *Philos Trans R Soc London.* 1686; 16:153–168.

Halpern D. Observations of annual and El Niño thermal and flow variations at 0°, 110°W and 0°, 95°W during 1980–1985. *J Geophys Res.* 1987;92(C8):8197–8212. https://doi.org/10.1029/JC092iC08p08197.

Ham Y-G, Kug J-S, Park J-Y, Jin F-F. Sea surface temperature in the north tropical Atlantic as a trigger for El Niño/Southern Oscillation events. *Nat Geosci.* 2013;6:112. https://doi.org/10.1038/ngeo1686.

Han W, McCreary JP, Anderson DLT, Mariano AJ. Dynamics of the eastern surface jets in the equatorial Indian Ocean. *J Phys Oceanogr.* 1999;29:2191–2209.

Hansen J, Fung I, Lacis A, et al. Global climate changes as forecast by Goddard Institute for Space Studies three-dimensional model. *J Geophys Res.* 1988;93:9341–9364. https://doi.org/10.1029/JD093iD08p09341.

Harrison DE, Vecchi GA. On the termination of El Niño. *Geophys Res Lett.* 1999;26:1593–1596.

Hartmann DL. *Global Physical Climatology.* San Diego, CA: Academic Press; 1994.

Hartmann DL. *Global Physical Climatology.* 2nd ed. Elsevier Science; 2016.

Hasselmann K. Stochastic climate models part I. Theory. *Tellus.* 1976;28:473–485. https://doi.org/10.1111/j.2153-3490.1976.tb00696.x.

Hastenrath S, Greischar L. Further work on the prediction of northeast Brazil rainfall anomalies. *J Climate.* 1993;6:743–758.

Held IM, et al. Probing the fast and slow components of global warming by returning abruptly to preindustrial forcing. *J Clim.* 2010;23:2418–2427.

Held IM, Hou AY. Nonlinear axially symmetric circulations in a nearly inviscid atmosphere. *J. Atmos. Sci.* 1980;37:515–533.

Held IM, Soden BJ. Robust responses of the hydrological cycle to global warming. *J Climate.* 2006;19: 5686–5699.

Hirasawa H, Kushner PJ, Sigmond M, et al. Anthropogenic aerosols dominate forced multidecadal Sahel precipitation change through distinct atmospheric and oceanic drivers. *J Climate.* 2020;33(23): 10187–10204.

Hirst AC. Unstable and damped equatorial modes in simple coupled ocean-atmosphere models. *J Atmos Sci.* 1986;43:606–632.

Holton J. *An Introduction to Dynamic Meteorology.* 4th ed. Academic Press; 2004.

Horel JD, Wallace JM. Planetary-scale atmospheric phenomena associated with the Southern Oscillation. *Mon Wea Rev.* 1981;109:813–829.

Hoskins BJ, Karoly DJ. The steady-state linear response of a spherical atmosphere to thermal and orographic forcing. *J Atmos Sci.* 1981;38:1175–1196.

Hosoda S, Nonaka M, Tomita T, et al. Impact of downward heat penetration below the shallow seasonal thermocline on the sea surface temperature. *J Oceanogr.* 2015;71:541–556. https://doi.org/10.1007/s10872-015-0275-7.

Houghton J. *Global Warming: The Complete Briefing.* 5th ed. Cambridge University Press; 2015. https://doi.org/10.1017/CBO9781316134245.

Hu S, Xie S-P, Kang SM. Global warming pattern formation: the role of ocean heat uptake. *J Climate.* 2022; 35:1885–1899. https://doi.org/10.1175/JCLI-D-21-0317.1.

Hwang YT, Tseng H-Y, et al. Relative roles of energy and momentum fluxes in the tropical response to extratropical thermal forcing. *J. Climate.* 2021;34:3771–3786.

Huang RX. Surface/wind driven circulation. In: North GR, Pyle J, Zhang F, eds. *Encyclopedia of Atmospheric Sciences.* 2nd ed. Academic Press; 2015:301–314.

Inoue K, Back LE. Gross moist stability assessment during TOGA COARE: various interpretations of gross moist stability. *J Atmos Sci.* 2015;72(11):4148–4166.

IPCC. Summary for policymakers. In: Stocker TF, Qin D, Plattner G-K, et al., eds. *Climate Change 2013: The Physical Science Basis. Contribution of Working Group I to the Fifth Assessment Report of the Intergovernmental Panel on Climate Change.* New York: Cambridge University Press; 2014:1–30. https://doi.org/10.1017/CBO9781107415324.004.

IPCC. Summary for policymakers. In: Masson-Delmotte V, Zhai P, Pirani A, et al., eds. *Climate Change 2021: The Physical Science Basis. Contribution of Working Group I to the Sixth Assessment Report of the Intergovernmental Panel on Climate Change.* New York: Cambridge University Press; 2022a:3–32. https://doi.org/10.1017/9781009157896.001.

IPCC. Summary for policymakers. In: Pörtner H-O, Roberts DC, Tignor M, et al., eds. *Climate Change 2022: Impacts, Adaptation, and Vulnerability. Contribution of Working Group II to the Sixth Assessment Report of the Intergovernmental Panel on Climate Change.* Cambridge University Press; 2022b.

Jaimes B, Shay LK. Mixed layer cooling in mesoscale oceanic eddies during hurricanes Katrina and Rita. *Month Weather Rev.* 2009;137(12):4188–4207.

Jiang X, Adames AF, Kim D, et al. Fifty years of research on the Madden-Julian oscillation: recent progress, challenges, and perspectives. *J Geophys Res Atmosph.* 2020;125(17):e2019JD030911.

Jin FF. An equatorial ocean recharge paradigm for ENSO, part I: conceptual model. *J Atmos Sci.* 1997;54:811–829.

Johnson NC, Collins DC, Feldstein SB, et al. Skillful wintertime North American temperature forecasts out to 4 weeks based on the state of ENSO and the MJO. *Weather Forecast.* 2014;29:23–38. https://doi.org/10.1175/WAF-D-13-00102.1.

Joshi MM, Gregory JM, Webb M, et al. Mechanisms for the land/sea warming contrast exhibited by simulations of climate change. *Climate Dyn.* 2008;30:455–465.

Kang SM, Frierson DMW, Held IM. The tropical response to extratropical thermal forcing in an idealized GCM: the importance of radiative feedbacks and convective parameterization. *J Atmospheric Sci.* 2009;66(9):2812–2827.

Kang SM, Held IM, Frierson DMW, Zhao M. The response of the ITCZ to extratropical thermal forcing: idealized slab-ocean experiments with a GCM. *J Climate.* 2008;21:3521–3532.

Kang SM, Xie S-P, Deser C, Xiang B. Zonal mean and shift modes of historical climate response to evolving aerosol distribution. *Sci Bull.* 2021;66:2405–2411. https://doi.org/10.1016/j.scib.2021.07.013.

Kessler WS, Kleeman R. Rectification of the Madden–Julian oscillation into the ENSO cycle. *J Climate.* 2000;13:3560–3575.

Keynes RD, ed. *Charles Darwin's Beagle diary.* Cambridge University Press; 2021.

Kiladis GN, Thorncroft CD, Hall NMJ. Three-dimensional structure and dynamics of African easterly waves. Part I: observations. *J Atmos Sci.* 2006;63:2212–2230. https://doi.org/10.1175/JAS3741.1.

Kiladis GN, Wheeler MC, Haertel PT, et al. Convectively coupled equatorial waves. *Rev Geophys.* 2009;47:RG2003. https://doi.org/10.1029/2008RG000266.

Kilpatrick T, Xie S-P, Miller A, Schneider N. Satellite observations of enhanced chlorophyll variability in the Southern California Bight. *J Geophys Res Oceans.* 2018;123:7550–7563.

Kim K-Y, Hamlington B, Na H. Theoretical foundation of cyclostationary EOF analysis for geophysical and climatic variables: concepts and examples. *Earth-Science Rev.* 2015;150:201–218. https://doi.org/10.1016/j.earscirev.2015.06.003.

Klein SA, Hartmann DL. The seasonal cycle of low stratiform clouds. *J Climate.* 1993;6:1587–1606.

Klein SA, Soden BJ, Lau N-C. Remote sea surface temperature variations during ENSO: evidence for a tropical atmospheric bridge. *J Climate.* 1999;12:917–932.

Klotzbach PJ. El Niño - Southern Oscillation's impact on Atlantic Basin hurricanes and US landfalls. *J Climate.* 2011;24:1252–1263.

Köppen W. Volken E, Brönnimann S, trans. Die Wärmezonen der Erde, nach der Dauer der heissen, gemässigten und kalten Zeit und nach der Wirkung der Wärme auf die organische Welt betrachtet [The thermal zones of the earth according to the duration of hot, moderate and cold periods and to the impact of heat on the organic world]. *Meteorol Zeitschrift*. 1884/2011;20(3):351-360. https://doi.org/10.1127/0941-2948/2011/105.

Kosaka Y, Chowdary JS, Xie S-P, et al. Limitations of seasonal predictability for summer climate over East Asia and the Northwestern Pacific. *J Climate*. 2012;25:7574–7589.

Kosaka Y, Xie S-P. Recent global-warming hiatus tied to equatorial Pacific surface cooling. *Nature*. 2013; 501:403–407.

Koster RD, et al. Contribution of land surface initialization to subseasonal forecast skill: first results from a multi-model experiment. *Geophys Res Lett*. 2010;37:L02402. https://doi.org/10.1029/2009GL041677.

Lamjiri MA, Dettinger MD, Ralph FM, Guan B. Hourly storm characteristics along the US west coast: role of atmospheric rivers in extreme precipitation. *Geophys Res Lett*. 2017;44. https://doi.org/10.1002/2017GL074193.

Lau K, Chan PH. Aspects of the 40–50 day oscillation during the northern winter as inferred from outgoing longwave radiation. *Month Weather Rev*. 1985;113:1889–1909. https://doi.org/10.1175/1520-0493(1985)113<1889:AOTDOD>2.0.CO, 2.

Lee J-Y, Marotzke J, Bala G, et al. Future global climate: scenario-based projections and near-term information. In: Masson-Delmotte V, Zhai P, Pirani SL, et al., eds. *Climate Change 2021: The Physical Science Basis. Contribution of Working Group I to the Sixth Assessment Report of the Intergovernmental Panel on Climate Change*. Cambridge University Press; 2022:553–672. https://doi.org/10.1017/9781009157896.006.

Lengaigne M, Guilyardi E, Boulanger JP, et al. Triggering of El Niño by westerly wind events in a coupled general circulation model. *Climate Dyn*. 2004;23:601–620.

Lewis JM. Meteorologists from the University of Tokyo: their exodus to the United States following World War II. *Bull Am Meteor Soc*. 1993;74:1351–1360.

Li C, Yanai M. The onset and interannual variability of the Asian summer monsoon in relation to land–sea thermal contrast. *J Climate*. 1996;9:358–375.

Li G, Xie S-P, Du Y. A robust but spurious pattern of climate change in model projections over the tropical Indian Ocean. *J Climate*. 2016b;29:5589–5608.

Li G, Xie S-P, Du Y, Luo Y. Effect of excessive cold tongue bias on the projections of tropical Pacific climate change. Part I: the warming pattern in CMIP5 multi-model ensemble. *Climate Dyn*. 2016a;47: 3817–3831.

Li J, Xie S-P, Cook E, et al. El Nino modulations over the past seven centuries. *Nat Climate Change*. 2013;3: 822–826.

Li Q, Ting MF, Li C, Henderson N. Mechanisms of Asian Summer Monsoon changes in response to anthropogenic forcing in CMIP5 models. *J Climate*. 2015;28:4107–4125.

Li S, Banerjee T. Spatial and temporal pattern of wildfires in California from 2000 to 2019. *Sci Rep*. 2021;11: 8779. https://doi.org/10.1038/s41598-021-88131-9.

Li X, Xie S-P, Gille ST, Yoo C. Atlantic-induced pan-tropical climate change over the past three decades. *Nat Climate Change*. 2016;6(3):275–280. https://doi.org/10.1038/NCLIMATE2840.

Lin I-I, Black P, Price JF, et al. An ocean coupling potential intensity index for tropical cyclones. *Geophys Res Lett*. 2013;40:1878–1882.

Lin I-I, Pun I-F, Lien C-C. Category-6" supertyphoon Haiyan in global warming hiatus: contribution from subsurface ocean warming. *Geophys Res Lett*. 2014;41:8547–8553.

Lin X, Johnson RH. Kinematic and thermodynamic characteristics of the flow over the western Pacific warm pool during TOGA COARE. *J Atmos Sci*. 1996;53:695–715.

Lindzen RS, Holton JR. A theory of the quasi-biennial oscillation. *J Atmos Sci*. 1968;25:1095.

Linkin ME, Nigam S. The North Pacific Oscillation–West Pacific teleconnection pattern: mature-phase structure and winter impacts. *J Climate*. 2008;21:1979–1997. https://doi.org/10.1175/2007JCLI2048.1.

Liu JW, Yang L, Xie S-P. CALIPSO observed cloud-regime transition over the summertime subtropical northeast Pacific. *J Geophys Res Atmos*. 2022. https://doi.org/10.1002/2022JD036542.

Liu ZY, et al. Chinese cave records and the East Asia summer monsoon. *Quat Sci Rev*. 2014;83:115–128. https://doi.org/10.1016/j.quascirev.2013.10.021.

Liu ZY, Otto-Bliesner BL, He F, et al. Transient simulation of last deglaciation with a new mechanism for Bølling-Allerød warming. *Science*. 2009;325:310–314.

Liu W, Lu J, Xie S-P, Fedorov A. Southern Ocean heat uptake, redistribution and storage in a warming climate: The role of meridional overturning circulation. *J. Climate*. 2018;31:4727–4743. https://doi.org/10.1175/JCLI-D-17-0761.1.

Lu J, Vecchi GA, Reichler T. Expansion of the Hadley cell under global warming. *Geophys Res Lett*. 2007;34: L06805. https://doi.org/10.1029/2006GL028443.

Lübbecke JF, Rodríguez-Fonseca B, Richter I, et al. Equatorial Atlantic variability—modes, mechanisms, and global teleconnections. *WIREs Clim Change*. 2018;9:e527.

Luo Y, Lu J, Liu F, Garuba O. The role of ocean dynamical thermostat in delaying the El Niño-like response over the equatorial Pacific to climate warming. *J Climate*. 2017;30:2811–2827.

Luongo MT, Xie S-P, Eisenman I. Buoyancy forcing dominates cross-equatorial ocean heat transport response to hemispheric cooling. *J Climate*. 2022;35. https://doi.org/10.1175/JCLI-D-21-0950.1.

Ma J, Xie S-P, Xu H. Contributions of the North Pacific Meridional Mode to ensemble spread of ENSO prediction. *J Climate*. 2017;30:9167–9181. https://doi.org/10.1175/JCLI-D-JCLI-D-17-0182.1.

Ma J, Xie S-P, Xu H. Inter-member variability of the summer Northwest Pacific subtropical anticyclone in the ensemble forecast. *J Climate*. 2017;30:3927–3941. https://doi.org/10.1175/JCLI-D-16-0638.1.

Ma J, Xie S-P, Xu H, et al. Cross-basin interactions between the tropical Atlantic and Pacific in the ENSEMBLES hindcasts. *J Climate*. 2021;34:2459–2472. https://doi.org/10.1175/JCLI-D-20-0140.1.

Madden RA, Julian PR. Description of Global-Scale Circulation Cells in the Tropics with a 40–50 Day Period. *J. Atmos. Sci*. 1972;29:1109–1123.

Manabe S, Hahn DG, Holloway JL. The seasonal variation of the tropical circulation as simulated by a global model of the atmosphere. *J Atmos Sci*. 1974;31:43–83.

Manabe S, Wetherald RT. Thermal equilibrium of the atmosphere with a given distribution of relative humidity. *J Atmos Sci*. 1967;24(3):241–259.

Mann ME, Bradley RS, Hughes MK. Global-scale temperature patterns and climate forcing over the past six centuries. *Nature*. 1998;392(6678):779–787.

Mantua NJ, Hare SR, Zhang Y, et al. A Pacific interdecadal climate oscillation with impacts on salmon production. *Bull Am Meteorol Soc*. 1997;78(6):1069–1080.

Mariotti A, Baggett C, Barnes EA, et al. Windows of opportunity for skillful forecasts subseasonal to seasonal and beyond. *Bull Am Meteorol Soc*. 2020;101(5):e608–e625.

Marshall J, Scott JR, Armour KC, et al. The ocean's role in the transient response of climate to abrupt greenhouse gas forcing. *Clim Dyn*. 2015;44:2287–2299.

Maruyama T. Large-scale disturbances in the equatorial lower stratosphere. *J Meteor Soc Japan*. 1967;45: 391–408.

Matsuno T. Quasi-geostrophic motions in the equatorial area. *J Meteor Soc Japan*. 1966;44:25–43.

McCreary JP, Anderson DL. A simple model of El Niño and the Southern Oscillation. *Month Weather Rev*. 1984;112:934–946.

McPhaden MJ. Genesis and evolution of the 1997–1998 El Niño. *Science*. 1999;283:950–954.

McPhaden MJ, et al. The tropical ocean-global atmosphere observing system: a decade of progress. *J Geophys Res*. 1998;103(C7):14169–14240. https://doi.org/10.1029/97JC02906.

McPhaden MJ, Santoso A, Cai W, eds. *El Niño Southern Oscillation in a Changing Climate*. AGU/Wiley; 2020.

Mechoso CR, ed. *Interacting Climates of Ocean Basins Observations, Mechanisms, Predictability, and Impacts*. Cambridge University Press; 2020.

Meehl GA, Covey C, Delworth T, et al. The WCRP CMIP3 multimodel dataset: a new era in climate change research. *Bull Am Meteorol Soc*. 2007;88(9):1383–1394.

Meehl GA, Hu A, Castruccio F, et al. Atlantic and Pacific tropics connected by mutually interactive decadal-timescale processes. *Nat Geosci*. 2021;14:36–42. https://doi.org/10.1038/s41561-020-00669-x.

Mei W, Kamae Y, Xie S, Yoshida K. Variability and predictability of North Atlantic hurricane frequency in a large ensemble of high-resolution atmospheric simulations. *J Climate*. 2019;32:3153–3167. https://doi.org/10.1175/JCLI-D-18-0554.1.

Mei W, Pasquero C. Spatial and temporal characterization of sea surface temperature response to tropical cyclones. *J Climate*. 2013;26:3745–3765.

Meinen CS, McPhaden MJ. Observations of warm water volume changes in the equatorial Pacific and their relationship to El Niño and La Niña. *J Climate*. 2000;13:3551–3559.

Merrifield AL, Simpson IR, McKinnon KA, et al. Local and non-local land surface influence in European heatwave initial condition ensembles. *Geophys Res Lett*. 2019;46:14082–14092.

Mishra V, Smoliak BV, Lettenmaier DP, Wallace JM. A prominent pattern of year-to-year variability in Indian summer monsoon rainfall. *Proc Natl Acad Sci*. 2012;109:7213–7217. https://doi.org/10.1073/pnas.1119150109.

Mitchell TP, Wallace JM. The annual cycle in equatorial convection and sea surface temperature. *J Climate*. 1992;5:1140–1156.

Miyamoto A, Nakamura H, Miyasaka T, Kosaka Y. Radiative impacts of low-level clouds on the summertime subtropical high in the south Indian Ocean simulated in a coupled general circulation model. *J Climate*. 2021;34:3991–4007. https://doi.org/10.1175/JCLI-D-20-0709.1.

Molnar P, Boos WR, Battisti DS. Orographic controls on climate and paleoclimate of Asia: thermal and mechanical roles for the Tibetan Plateau. *Ann Rev Earth Planet Sci*. 2010;38:77–102.

Murakami T. *The Monsoon: Seasonal Winds and Rains (in Japanese)*. Tokyodo Publishing; 1986.

Nakazawa T. Tropical super clusters within intraseasonal variations over the Western Pacific. *J Met Soc Japan*. 1988;66:823–839. http://www.jstage.jst.go.jp/browse/jmsj/-char/en.

Namias J. Recent seasonal interactions between North Pacific waters and the overlying atmospheric circulation. *J Geophys Res*. 1959;64:631–646.

Neelin JD. *Climate Change and Climate Modeling*. Cambridge University Press; 2011.

Neelin JD, Held IM, Cook KH. Evaporation-wind feedback and low-frequency variability in the tropical atmosphere. *J Atmos Sci*. 1987;44(16):2341–2348.

Newman M, et al. The Pacific decadal oscillation, revisited. *J Climate*. 2016;29:4399–4427. https://doi.org/10.1175/JCLI-D-15-0508.1.

Norris JR, Leovy CB. Interannual variability in stratiform cloudiness and sea surface temperature. *J Climate*. 1994;7:1915–1925.

Ogata T, Xie S-P. Semiannual cycle in zonal wind over the equatorial Indian Ocean. *J Climate*. 2011;24:6471–6485.

O'Gorman PA. Precipitation extremes under climate change. *Curr Climate Change Rep*. 2015;1(2):49–59.

Okajima H. *Orographic Effects on Ttropical Climate in a Coupled Ocean-Atmosphere General Circulation Model*. PhD dissertation. Department of Meteorology, University of Hawaii; 2006.

Okajima H, Xie S-P, Numaguti A. Interhemispheric coherence of tropical climate variability: effect of climatological ITCZ. *J Meteorol Soc Japan*. 2003;81:1371–1386.

Okumura Y, Xie S-P. Interaction of the Atlantic equatorial cold tongue and African monsoon. *J Climate*. 2004;17:3588–3601.

Okumura Y, Xie S-P. Some overlooked features of tropical Atlantic climate leading to a new Nino-like phenomenon. *J Climate*. 2006;19:5859–5874.

Okumura YM. ENSO diversity from an atmospheric perspective. *Curr Climate Change Rep*. 2019;5(3):245–257.

Ooyama K. Numerical simulation of the life cycle of tropical cyclones. *J Atmos Sci*. 1969;26(1):3–40.

Pan LL, Munchak LA. Relationship of cloud top to the tropopause and jet structure from CALIPSO data. *J Geophys Res*. 2011;116, D12201. https://doi.org/10.1029/2010JD015462.

Pedlosky J. *Geophysical Fluid Dynamics*. 2nd ed. New York: Springer-Verlag; 1987.

Pendergrass AG, Hartmann DL. The atmospheric energy constraint on global-mean precipitation change. *J Climate*. 2014;27:757–768.

Peng Q, Xie S-P, Wang D, et al. Coupled ocean-atmosphere dynamics of the 2017 extreme coastal El Nino. *Nat Comm*. 2019;10:298. https://doi.org/10.1038/s41467-018-08258-8.

Peng Q, Xie S-P, Wang D, et al. Eastern Pacific wind effect on the evolution of El Nino: implications for ENSO diversity. *J Climate*. 2020;33:3197–3212.

Peng Q, Xie S-P, Wang D, et al. Surface warming-induced global acceleration of upper ocean currents. *Sci Adv*. 2022;8:eabj8394. https://doi.org/10.1126/sciadv.eabj8394.

Philander SG. *El Nino, La Nina, and the Southern Oscillation*. Academic Press; 1990.

Philander SGH, Pacanowski RC. The oceanic response to cross-equatorial winds (with application to coastal upwelling in low latitudes). *Tellus*. 1981;33:201–210.

Philander SGH, Seigel AD. Simulation of the El Niño of 1982–1983. In: Nihoul J, ed. *Coupled Ocean Atmosphere Models*. Elsevier; 1985:517–541.

Philander SGH, Yamagata T, Pacanowski RC. Unstable air-sea interaction in the tropics. *J Atmos Sci*. 1984; 41(4):604–613.

Planton YY, Guilyardi E, Wittenberg AT, et al. Evaluating climate models with the CLIVAR 2020 ENSO metrics package. *Bull Am Meteorol Soc*. 2021;102(2):e193–e217.

Ralph M, Dettinger M, Waliser D, Rutz J, eds. *Atmospheric Rivers*. Springer International Publishing; 2020.

Rasmusson EM. *Milestones on the Road to TOGA. The 27th Conference on Climate Variability and Change*. American Meteorological Society; 2015.

Rasmusson EM, Carpenter TH. Variations in tropical sea surface temperature and surface wind fields associated with the Southern Oscillation/El Niño. *Month Weather Rev*. 1982;110:354–384.

Rennert KJ, Wallace JM. Cross-frequency coupling, skewness, and blocking in the Northern Hemisphere winter circulation. *J Climate*. 2009;22:5650–5666.

Richter I, Behera SK, Doi T, et al. What controls equatorial Atlantic winds in boreal spring? *Climate Dyn*. 2014;43:3091–3104.

Richter I, Xie S-P, Morioka Y, et al. Phase locking of equatorial Atlantic variability through the seasonal migration of the ITCZ. *Climate Dyn*. 2017;48:3615–3629.

Rodwell MJ, Hoskins BJ. Monsoons and the dynamics of deserts. *QJR Meteorol Soc*. 1996;122:1385–1404. https://doi.org/10.1002/qj.49712253408.

Roe G, Feldl N, Armour K, et al. The remote impacts of climate feedbacks on regional climate predictability. *Nat Geosci*. 2015;8:135–139.

Roemmich D, Church J, Gilson J, et al. Unabated planetary warming and its ocean structure since 2006. *Nat Climate Change*. 2015;5:240–245.

Rosati A, Miyakoda K. A general circulation model for upper ocean simulation. *J Phys Oceanogr*. 1988;18: 1601–1626.

Rykaczewski RR, Checkley DM. Influence of ocean winds on the pelagic ecosystem in upwelling regions. *Proc Nat Acad Sci*. 2008;105(6):1965–1970.

Saji NH, Goswami BN, Vinayachandran PN, Yamagata T. A dipole mode in the tropical Indian Ocean. *Nature*. 1999;401:360–363.

Saji NH, Xie S-P, Yamagata T. Tropical Indian Ocean variability in the IPCC 20th-century climate simulations. *J Climate*. 2006;19:4397–4417.

Sampe T, Xie S-P. Large-scale dynamics of the Meiyu-Baiu rain band: environmental forcing by the westerly jet. *J Climate*. 2010;23:113–134.

Sampe T, Xie S-P. Mapping high sea winds from space: a global climatology. *Bull Am Meteorol Soc*. 2007;88: 1965–1978.

Santer BD, Fyfe JC, Pallotta G, et al. Causes of differences in model and satellite tropospheric warming rates. *Nat Geosci*. 2017;10:478–485.

Sarachik E, Cane MA. *The El Niño-Southern Oscillation Phenomenon*. Cambridge University Press; 2010.

Sardeshmukh PD, Hoskins BJ. The generation of global rotational flow by steady idealized tropical divergence. *J Atmos Sci*. 1988;45:1228–1251.

Schneider T, Bischoff T, Haug G. Migrations and dynamics of the intertropical convergence zone. *Nature*. 2014;513:45–53.

Schneider N, Miller AJ. Predicting western North Pacific ocean climate. 14 (20),. *J. Climate*. 2001;14: 3997–4002.

Seager R, Naik N, Vecchi VA. Thermodynamic and dynamic mechanisms for large-scale changes in the hydrological cycle in response to global warming. *J Climate*. 2010;23:4651–4668.

Shaw TA. Mechanisms of future predicted changes in the zonal mean mid-latitude circulation. *Curr Climate Change Rep*. 2019;5:345–357. https://doi.org/10.1007/s40641-019-00145-8.

Shaw TA. On the role of planetary-scale waves in the abrupt seasonal transition of the Northern Hemisphere general circulation. *J Atmos Sci*. 2014;71:1724–1746.

Shaw TA, Baldwin M, Barnes EA, et al. Storm track processes and the opposing influences of climate change. *Nat Geosci*. 2016;9:656–664. https://doi.org/10.1038/NGEO2783.

Shi J, Tally LD, Xie S-P, Peng Q, Liu W. Ocean warming and accelerating Southern Ocean zonal flow. *Nature Clim Change*. 2021;11:1090–1097.

Shi J, Xie S-P, Tally LD. Evolving relative importance of the Southern Ocean and North Atlantic in anthropogenic ocean heat uptake. *J Climate*. 2018;31:7459–7479, 2018.

Sikka DR, Gadgil S. On the maximum cloud zone and the ITCZ over India longitude during the southwest monsoon. *Month Weather Rev*. 1980;108:1840–1853.

Simmons AJ, Wallace JM, Branstator GW. Barotropic wave propagation and instability, and atmospheric teleconnection patterns. *J Atmos Sci*. 1983;40:1363–1392.

Simpson I, Shaw T, Seager R. A diagnosis of the seasonally and longitudinally varying midlatitude circulation response to global warming. *J Atmos Sci*. 2014;71:2489–2515.

Small RJ, deSzoeke S, Xie S-P, et al. Air-sea interaction over ocean fronts and eddies. *Dyn Atmos Oceans*. 2008;45:274–319.

Sobel A, Maloney E. Moisture modes and the eastward propagation of the MJO. *J Atmos Sci*. 2013;70: 187–192.

Solomon SG, Plattner K, Knutti R, Friedlingstein P. Irreversible climate change due to carbon dioxide emissions. *Proc Nat Acad Sci*. 2009;106(6):1704–1709. https://doi.org/10.1073/pnas.0812721106.

Song ZH, Xie S-P, Xu L, et al. Deep winter mixed layer in the Southern Ocean: role of the meandering Antarctic Circumpolar Current. *J Climate*. 2022.

Sperber KR, Annamalai H, Kang IS, et al. The Asian summer monsoon: an intercomparison of CMIP5 vs. CMIP3 simulations of the late 20th century. *Climate Dyn*. 2013;41:2711–2744. https://doi.org/10.1007/s00382-012-1607-6.

Sriver RL, Huber M. Observational evidence for an ocean heat pump induced by tropical cyclones. *Nature*. 2007;447(7144):577–580.

Stouffer RJ, Manabe S. Assessing temperature pattern projections made in 1989. *Nat Climate Change*. 2017; 7(3):163–165.

Stuecker MF, Jin F-F, Timmermann A, McGregor S. Combination mode dynamics of the anomalous Northwest Pacific anticyclone. *J Climate*. 2015;28:1093–1111.

Suarez MJ, Schopf PS. A delayed action oscillator for ENSO. *J Atmos Sci*. 1988;45(21):3283–3287.

Taguchi B, Xie S-P, Schneider N, et al. Decadal variability of the Kuroshio Extension: observations and an eddy-resolving model hindcast. *J Climate*. 2007;20:2357–2377.

Talley LD, Pickard GL, Emery WJ, Swift JH. *Descriptive Physical Oceanography: An Introduction*. 6th ed. Boston: Elsevier; 2011.

Tanimoto Y, Kajitani T, Okajima H, Xie S-P. A peculiar feature of the seasonal migration of the South American rain band. *J Meteorol Soc Japan*. 2010;88:79–90.

Timmermann A, et al. Increased El Nino frequency in a climate model forced by future greenhouse warming. *Nature*. 1999;398:694–697.

Timmermann A, Okumura Y, An S-I, et al. The influence of a weakening of the Atlantic meridional overturning circulation on ENSO. *J Climate*. 2007;20:4899–4919.

Trenberth KE, Hurrell JW. Decadal atmosphere-ocean variations in the Pacific. *Climate Dyn*. 1994;9: 303–319.

Ueda H, Yasunari T, Kawamura R. Abrupt seasonal change of large-scale convective activity over the western Pacific in the northern summer. *J Meteorol Soc Japan*. 1995;73:795–809.

Valdes PJ, Braconnot P, Meissner KJ, Eggleston S, eds. Paleoclimate modelling intercomparison project (PMIP): 30th anniversary. *Past Global Changes Mag*. 2021;29:61–108. https://doi.org/10.22498/pages.29.2.

Vallis GK. *Atmospheric and Oceanic Fluid Dynamics: Fundamentals and Large-Scale Circulation*. 2nd ed. Cambridge University Press; 2017.

van der Wiel K, Matthews AJ, Stevens DP, Joshi MM. A dynamical framework for the origin of the diagonal South Pacific and South Atlantic convergence zones. *QJR Meteorol Soc*. 2015;141:1997–2010. https://doi.org/10.1002/qj.2508.

Vecchi GA, Soden BJ. Global warming and the weakening of the tropical circulation. *J Climate*. 2007;20(17): 4316–4340.

Vera CS, Higgins W, Amador J, et al. Toward a unified view of the American monsoon systems. *J Climate.* 2006b;19:4977–5000.

Vinayachandran PN, Murty VSN, Ramesh Babu V. Observations of barrier layer formation in the Bay of Bengal during summer monsoon. *J Geophys Res.* 2002;107(C12):8018. https://doi.org/10.1029/2001JC000831.

Voigt A, Stevens B, Bader J, Mauritsen T. The observed hemispheric symmetry in reflected shortwave irradiance. *J Climate.* 2013;26(2):468–477. https://doi.org/10.1175/JCLI-D-12-00132.1.

Walker GT. Seasonal weather and its prediction. *Nature.* 1933;132(3343):805–808. https://doi.org/10.1038/132805a0.

Wallace JM, Deser C, Smoliak V, Phillips AS. Attribution of climate change in the presence of internal variability. In: Chang CP, Ghil M, Latif M, Wallace JM, eds. *Climate Change: Multidecadal and Beyond. World Scientific Series on Asia-Pacific Weather and Climate.* 2015;6:1–29. https://doi.org/10.1142/9789814579933_0001.

Wallace JM, Gutzler DS. Teleconnections in the geopotential height field during the Northern Hemisphere winter. *Month Weather Rev.* 1981;109:784–812.

Wallace JM, Hobbs PV. *Atmospheric Science: An Introductory Survey.* 2nd ed. Academic Press; 2006.

Wallace JM, Kousky VE. Observational evidence of Kelvin waves in the tropical stratosphere. *J Atmos Sci.* 1968;25:900–907.

Wallace JM, Smith C, Bretherton CS. Singular value decomposition of wintertime sea surface temperature and 500-mb height anomalies. *J Climate.* 1992;5(6):561–576.

Wang B, Biasutti M, Byrne MP, et al. Monsoons climate change assessment. *Bull Amer Meteorol Soc.* 2021;102. https://doi.org/10.1175/BAMS-D-19-0335.1.

Wang B, Ding Q. Global monsoon: dominant mode of annual variation in the tropics. *Dyn Atmos Oceans.* 2008;44:165–183.

Wang B, Lee J-Y, et al. Advance and prospect of seasonal prediction: assessment of the APCC/CliPAS 14-model ensemble retroperspective seasonal prediction (1980-2004). *Climate Dyn.* 2009;33:93–117. https://doi.org/10.1007/s00382-008-0460-0.

Wang B, Wu RG, Li T. Atmosphere-warm ocean interaction and its impacts on Asian-Australian monsoon variation. *J Climate.* 2003;16:1195–1211.

Wang C, Picaut J. Understanding ENSO physics—a review. In *Earth's Climate: The Ocean–Atmosphere Interaction. Geophys Monogr.* 2004;147:21–48.

Wang G, Xie S-P, et al. Robust warming pattern of global subtropical oceans and its mechanism. *J. Climate.* 2015;28:8574–8584. https://doi.org/10.1175/JCLI-D-14-00809.1.

Wang H, Xie S-P, Tokinaga H, et al. Detecting cross-equatorial wind change as a fingerprint of climate response to anthropogenic aerosol forcing. *Geophys Res Lett.* 2016;43:3444–3450. https://doi.org/10.1002/2016GL068521.

Wang M, Du Y, Qiu B, et al. Mechanism of seasonal eddy kinetic energy variability in the eastern equatorial Pacific Ocean. *J Geophys Res Oceans.* 2017;122:3240–3252. https://doi.org/10.1002/2017JC012711.

Wang YQ, Wu CC. Current understanding of tropical cyclone structure and intensity changes – a review. *Meteorol. Atmos. Phys.* 2004;87:257–278. https://doi.org/10.1007/s00703-003-0055-6.

Watanabe M, Tatebe H. Reconciling roles of sulphate aerosol forcing and internal variability in Atlantic multidecadal climate changes. *Climate Dyn.* 2019;53:4651–4665. https://doi.org/10.1007/s00382-019-04811-3.

Webster PJ. *Dynamics of The Tropical Atmosphere and Oceans.* Wiley-Blackwell; 2020.

Wettstein JJ, Wallace JM. Observed patterns of month-to-month storm-track variability and their relationship to the background flow. *J Atmos Sci.* 2010;67(5):1420–1437. https://doi.org/10.1175/2009JAS3194.1.

Wheeler M, Kiladis GN. Convectively coupled equatorial waves: analysis of clouds and temperature in the wavenumber–frequency domain. *J Atmos Sci.* 1999;56:374–399.

Wheeler MC, Hendon HH. An all-season real-time multivariate MJO index: development of an index for monitoring and prediction. *Month Weather Rev.* 2004;132:1917–1932.

Wheeler MC, Nguyen H. Equatorial waves. In: In North GR, Pyle J, Zhang F, eds. *Encyclopedia of Atmospheric Sciences.* 2nd ed. Academic Press; 2015:102–122.

Williams AP, Abatzoglou JT, Gershunov A, et al. Observed impacts of anthropogenic climate change on wildfire in California. *Earth's Future*. 2019;7:892–910.

Wood R. Stratocumulus clouds. *Month Weather Rev*. 2012;40:2373–2423.

Wu R, Kirtman BP, Krishnamurthy V. An asymmetric mode of tropical Indian Ocean rainfall variability in boreal spring. *J Geophys Res*. 2008;113(D5):79–88.

Wu X, Okumura YM, Deser C, DiNezio N. Two-year dynamical predictions of ENSO event duration during 1954-2015. *J Climate*. 2021;34:4069–4087.

Wyrtki K. An equatorial jet in the Indian Ocean. *Science*. 1973;181(4096):262–264.

Wyrtki K. El Niño—the dynamic response of the equatorial Pacific Ocean to atmospheric forcing. *J Phys Oceanogr*. 1975;5:572–584.

Xiang B, Zhao M, Jiang X, et al. The 3–4-week MJO prediction skill in a GFDL coupled model. *J Climate*. 2015;28:5351–5364. https://doi.org/10.1175/JCLI-D-15-0102.1.

Xie S-P. Ocean warming pattern effect on global and regional climate change. *AGU Adv*. 2020;1. e2019AV00013. https://doi.org/10.1029/2019AV000130.

Xie S-P. On the genesis of the equatorial annual cycle. *J Climate*. 1994;7:2008–2013.

Xie S-P. Satellite observations of cool ocean-atmosphere interaction. *Bull Am Meteorol Soc*. 2004a;85: 195–208.

Xie S-P. The shape of continents, air-sea interaction, and the rising branch of the Hadley circulation. In: Diaz HF, Bradley RS, eds. *The Hadley Circulation: Past, Present and Future*. Kluwer Academic Publishers; 2004b:121–152.

Xie S-P. Westward propagation of latitudinal asymmetry in a coupled ocean-atmosphere model. *J Atmos Sci*. 1996;53:3236–3250.

Xie S-P, Annamalai H, Schott FA, McCreary JP. Structure and mechanisms of South Indian Ocean climate variability. *J Climate*. 2002;15:864–878.

Xie S-P, Carton JA. Tropical Atlantic variability: patterns, mechanisms, and impacts. In: Wang C, Xie S-P, Carton JA, eds. *Earth Climate: The Ocean-Atmosphere Interaction. Geophys Monograph*. 2004;147:121–142.

Xie S-P, Deser C, Vecchi A, et al. Global warming pattern formation: sea surface temperature and rainfall. *J Climate*. 2010;23:966–986.

Xie S-P, Deser C, Vecchi GA, et al. Towards predictive understanding of regional climate change. *Nat Clim Change*. 2015;5:921–930.

Xie S-P, Hu K, Hafner J, et al. Indian Ocean capacitor effect on Indo-western Pacific climate during the summer following El Nino. *J Climate*. 2009;22:730–747.

Xie S-P, Kosaka Y. What caused the global surface warming hiatus of 1998-2013? *Curr Clim Change Rep*. 2017;3:128–140. https://doi.org/10.1007/s40641-017-0063-0.

Xie S-P, Kosaka Y, Du Y, et al. Indo-western Pacific Ocean capacitor and coherent climate anomalies in post-ENSO summer: a review. *Adv Atmos Sci*. 2016;33:411–432. https://doi.org/10.1007/s00376-015-5192-6.

Xie S-P, Kosaka Y, Okumura YM. Distinct energy budgets for anthropogenic and natural changes during global warming hiatus. *Nat Geosci*. 2016;9:29–33.

Xie S-P, Kubokawa A, Hanawa K. Evaporation-wind feedback and the organizing of tropical convection on the planetary scale. Part I: quasi-linear instability. *J Atmos Sci*. 1993;50:3873–3883.

Xie S-P, Lu B, Xiang B. Similar spatial patterns of climate responses to aerosol and greenhouse gas changes. *Nat Geosci*. 2013;6:828–832.

Xie S-P, Peng Q, Kamae Y, et al. Eastern Pacific ITCZ dipole and ENSO diversity. *J Climate*. 2018;31: 4449–4462. https://doi.org/10.1175/JCLI-D-17-0905.1.

Xie S-P, Philander SGH. A coupled ocean-atmosphere model of relevance to the ITCZ in the eastern Pacific. *Tellus*. 1994;46A:340–350.

Xie S-P, Saiki N. Abrupt onset and slow seasonal evolution of summer monsoon in an idealized GCM simulation. *J Meteor Soc Japan*. 1999;77:949–968.

Xie S-P, Saito K. Formation and variability of a northerly ITCZ in a hybrid coupled AGCM: continental forcing and ocean-atmospheric feedback. *J Climate*. 2001;14:1262–1276.

Xie S-P, Xu H, Saji NH, et al. Role of narrow mountains in large-scale organization of Asian monsoon convection. *J Climate*. 2006;19:3420–3429.

Yamagata T. Stability of a simple air-sea coupled model in the tropics. In: Nihoul JCJ, ed. *Coupled Ocean-Atmosphere Models*. Elsevier; 1985:637−657.

Yan X, Zhang R, Knutson TR. Underestimated AMOC variability and implications for AMV and predictability in CMIP models. *Geophys Res Lett*. 2018;45:4319−4328, 2018.

Yanai M, Esbensen S, Chu J-H. Determination of bulk properties of tropical cloud clusters from large-scale heat and moisture budgets. *J Atmos Sci*. 1973;30:611−627.

Yanai M, Maruyama T. Stratospheric wave disturbances propagating over the equatorial Pacific. *J Meteorol Soc Japan*. 1966;44:291−294.

Yanai M, Tomita T. Seasonal and interannual variability of atmospheric heat sources and moisture sinks as determined from NCEP−NCAR reanalysis. *J Climate*. 1998;11:463−482.

Yang L, Xie S-P, Shen P, et al. Low cloud-SST feedback over the subtropical northeast Pacific and the effect on ENSO variability. *J Climate*. 2022. https://doi.org/10.1175/JCLI-D-21-0902.1.

Yang Y, Xie S-P, Wu L, et al. Seasonality and predictability of the Indian Ocean dipole mode: ENSO forcing and internal variability. *J Climate*. 2015;28:8021−8036.

Yasunari T. Cloudiness fluctuations associated with the Northern Hemisphere summer monsoon. *J Meteorol Soc Japan*. 1979;57:227−242.

Yasunari T, Miwa T. Convective cloud systems over the Tibetan Plateau and their impact on meso-scale disturbances in the Meiyu/Baiu Frontal Zone. *J Meteorol Soc Japan*. 2006;84:783−803.

Yeh TC, Lo SW, Shu PC. The wind structure and heat balance in the lower troposphere over Tibetan Plateau and its surroundings. *Acta Meteor Sin*. 1957;28:108−121.

Yin J, Schlesinger M, Stouffer R. Model projections of rapid sea-level rise on the northeast coast of the United States. *Nat Geosci*. 2009;2:262−266. https://doi.org/10.1038/ngeo462.

Yoshida K. A theory of the Cromwell current and equatorial upwelling. *J Oceanogr Soc Japan*. 1959;15:154−170.

Young AP, Flick RE, Gallien TW, et al. Southern California coastal response to the 2015−2016 El Niño. *J Geophys Res: Earth Surface*. 2018;123:3069−3083. https://doi.org/10.1029/2018JF004771.

Zebiak SE. Air-sea interaction in the equatorial Atlantic region. *J Climate*. 1993;6:1567−1586

Zebiak SE, Cane MA. A model El Nino-Southern Oscillation. *Month Weather Rev*. 1987;115:2262−2278.

Zhang R, Sutton R, Danabasoglu G, et al. A review of the role of the Atlantic Meridional Overturning Circulation in Atlantic Multidecadal Variability and associated climate impacts. *Rev Geophys*. 2019;57:316−375.

Zhang S, Xie S-P, Liu Q, et al. Seasonal variations of Yellow Sea fog: observations and mechanisms. *J Climate*. 2009;22:6758−6772.

Zhao M, Held IM. TC-permitting GCM simulations of hurricane frequency response to sea surface temperature anomalies projected for the late-twenty-first century. *J Climate*. 2012;25:2995−3009.

Zhao M, Held IM, Lin SJ, Vecchi GA. Simulations of global hurricane climatology, interannual variability, and response to global warming using a 50-km resolution GCM. *J Climate*. 2009;22:6653−6678.

Zheng XT, Xie S-P, Lv LH, Zhou ZQ. Inter-model uncertainty in ENSO amplitude change tied to Pacific Ocean warming pattern. *J Climate*. 2016;29:7265−7279.

Zhou Z-Q, Xie S-P, Zhang GJ, Zhou W. Evaluating AMIP skill in simulating interannual variability of summer rainfall over the Indo-western Pacific. *J Climate*. 2018;31:2253−2265. https://doi.org/10.1175/JCLI-D-17-0123.1.

Zhou Z-Q, Xie S-P, Zhang R. Historic Yangtze flooding of 2020 tied to extreme Indian Ocean conditions. *PNAS*. 2021;118:e2022255118.

Zhou Z-Q, Xie S-P, Zheng X-T, et al. Global warming-induced changes in El Nino teleconnections over the North Pacific and North America. *J Climate*. 2014;27:9050−9064.

Zhou W, Xie S-P, Zhou Z-Q. Slow preconditioning for abrupt convective jump over the summer Northwest Pacific. *J Climate*. 2016;29:8103−8113. https://doi.org/10.1175/JCLI-D-16-0342.1.

Zhou C, Zelinka MD, Klein SA. Analyzing the dependence of global cloud feedback on the spatial pattern of sea surface temperature change with a Green's function approach. *J Adv Model Earth Sys*. 2017;9:2174−2189.

Zhou Z-Q, Zhang R, Xie S-P. Variability and predictability of Indian rainfall during the monsoon onset month of June. *Geophys Res Lett*. 2019;46:14782−14788.

Index

Note: Page numbers followed by "f" indicate figures, "t" indicate tables, and "b" indicate boxes.

A

B

C

D

E

F

G

N

O

P

T

U

V

W

Y

Z